I0789945

Silicon Materials Science and Technology X

Editors:

H. Huff, *retired*
SEMATECH
Austin, Texas, USA

H. Iwai
Tokyo Institute of Technology
Yokohama, Japan

H. Richter
IHP Microelectronics
Frankurt, Germany

Sponsoring Division:

 Electronics and Photonics

Published by
The Electrochemical Society
65 South Main Street, Building D
Pennington, NJ 08534-2839, USA

tel   609 737 1902
fax   609 737 2743

www.electrochem.org

ecstransactions ™

Vol. 2 No. 2

Copyright 2006 by The Electrochemical Society, Inc.
All rights reserved.

This book has been registered with Copyright Clearance Center, Inc.
For further information, please contact the Copyright Clearance Center,
Salem, Massachusetts.

Published by:

The Electrochemical Society, Inc.
65 South Main Street
Pennington, New Jersey 08534-2839, USA

Telephone 609.737.1902
Fax 609.737.2743
e-mail: ecs@electrochem.org
Web: www.electrochem.org

Printed in the United States of America

PREFACE

1. Overview

This Electrochemical Society Proceedings contains the papers presented at the *Tenth International Symposium on Silicon Materials Science and Technology*, held May 7-12, 2006, in Denver. Sponsored by the Electronics & Photonics Division of The Electrochemical Society, the *Silicon Symposia* series has provided a unique historical record of the progress in our understanding of silicon materials and related device/integrated circuit (IC) process fabrication issues for almost 40 years. The symposium continues to be truly international: Based on the lead author's affiliation, there in one paper from China, seven papers from Europe, 14 from Japan, one from Korea, one from Taiwan and 19 from the United States, surely not unexpected in view of the escalating globalization of the microelectronics industry. Concurrently, the necessity for an increased emphasis of the scientific understanding of the silicon materials properties, physico-chemical processes during silicon material fabrication and conversion into an IC as well as the selective application of design of experiment (DOE) methodologies in the fabrication of both silicon materials and ICs has become prevalent, at Integrated Device Manufacturers (IDM's) as well as universities and national research institutes. The continued coordination of industrial organizations and research institutes and the formation of consortia is indicative of the resulting awareness that cost-effective fabrication and enhanced manufacturing effectiveness throughout all facets of the microelectronic industry will require pre-competitive coordination.

This activity reflects the continuing recognition of the complexity and stimulating scientific and engineering challenges in the microelectronics industry. Indeed, 50% (22 of 44) of the 44 contributions are from universities, national research institutes or consortia, testifying to these opportunities. In that regard, most of the papers touched upon some aspect of modeling the phenomena under consideration. These modeling studies, from crystal growth through the unit IC process fabrication steps and their integrated effect, are contributing to the performance and yield metrics of both first silicon out and the continued pervasiveness of IDMs product lines. The common thread throughout the symposium, however, continues to be the characterization, annihilation and, in some cases, the selective utilization of defects to achieve superior IC performance, yield and reliability. These defects include, but are not restricted to, bulk silicon crystals, silicon wafers and epitaxial silicon films. Several examples are point-defects such as silicon vacancies and self-interstitials in the Czochralski (Cz) grown crystal, isotopic silicon enrichment in the epitaxial layer, chemical impurities such as oxygen, nitrogen and carbon (and their agglomerates with point defects) as well as structural imperfections introduced in the starting materials as well as during device/IC processing. Indeed, it is the interactive effects of these defect complexes that have led to an amazingly rich compendium of impurity-defect complexes, both detrimental and, in some cases, particularly favorable for enhanced device/IC performance, yield and reliability. Particular emphasis continues to be placed on the interactive effects of starting materials properties and the effect of multiple unit IC processing steps on the electronic and optical-electronic device configurations performance and reliability — point/extended

iii

defects, gettering, modeling and IC process integration issues — in the System Large-Scale-Integration (System LSI) era. Since many of the phenomena discussed are structure sensitive, the "process-structure-property" approach continues to be the unifying principle in describing IC electronic and opto-electronic characteristics. That is, the fabrication process determines the material structure, which then determines the material properties and device characteristics. Indeed, materials are the *sine qua non* of electronic and opto-electronic devices and circuits. To a significant extent, this approach has been the driver during the last 50 years [1,2].

Continuing the theme of materials and their fabrication into useful devices, circuits and systems, The Electrochemical Society and most of the learned technical societies have seen a virtual exponential growth in the number of symposia devoted to a host of specialized materials and topics, in which the terms "nano, nano-materials, nano-fabrication nano-devices," etc. have, perhaps, been excessively utilized. Nevertheless, the continued guidance of Moore's law as we move beyond planar silicon CMOS into 3D (vertical) device/IC configurations and into the nano-era and the drive to miniaturization, top-down and/or bottom-up approaches, will be of the utmost importance, in the coming nano-electronics era [3,4]. Nevertheless, and most interestingly, the contributions to this *Tenth Silicon Symposium* have increasingly reverted to silicon (Si), related silicon-germanium (Si-Ge) materials and silicon-on-insulator (SOI), per se, in contra-distinction to recent *Silicon Symposia* that had gravitated to a plethora of additional and alternative materials for the gate dielectric, gate electrode, low-resistance contact materials, etc. While this explosion of examining virtually the entire periodic table (somewhat of an exaggeration), especially for an alternative high-mobility channel material exhibiting the requisite transport properties (in conjunction with Si and Si-Ge) in the coming decade will become increasingly important, the realization that the host Si and Si-Ge materials are still especially relevant, and the entrance of SOI materials into the mainstream IDM products, is reflected in the sessions for this *Tenth Silicon Symposium: 300mm, Process Development and Modeling, Materials and Process Integration, Integrated Metrology and Diagnostics*, and *Strained Silicon*. The challenges associated with the above are but one facet of the evolving role the *International Technology Roadmap for Semiconductors* (ITRS) has achieved and continues to play in our industry. It should also be noted that the utilization of silicon materials for the solar industry has now exceeded its utilization for ICs. A major economic issue for both the solar energy industry's roadmap and the ITRS is the challenge of bringing up additional polysilicon facilities.

The Plenary session leads off with four invited reviews. G. Dan Hutcheson of VLSI Research presents *Forty Years of Moore's Law: Ever Smaller Transistors and Ever Larger Wafers*. D. Park of Samsung then follows with a review of the *Present and Future of Si-Based Transistor Technologies for Memory*. Y. Nara and colleagues from Selete review *Front-End Process Technology for High-Performance 65/45nm CMOS Devices*. Finally, L. Pfizner of the Fraunhofer Institute presents a vision of *Trends in European R&D — Advanced Process Control Down to Atomic Scale for Micro-and Nanotechnologies*.

The development of silicon material technologies and their fabrication into ICs are then reviewed. Twelve additional invited papers summarize the most significant developments and outstanding challenges to date in the sessions noted above. Twenty-eight contributed papers delve further into the details appropriate to the sessions. Because of the limited times available for the presentations, the papers in the symposium were selected to provide a complementary view of the subject matter within each session. In that regard, the *Silicon Symposium* proceedings complements other symposia from the Electronics & Photonics Division, often in conjunction with the Dielectric Science and Technology Division, on: *Defects in Silicon, High-Purity Silicon, Wafer Cleaning, Chemical Vapor Deposition, The Physics and Chemistry of SiO₂ and The Si-SiO₂ Interface, Silicon Nitride and Silicon Dioxide Thin Insulating Films, Physics and Technology of High-k Gate Dielectrics, Dielectric Material Integration for Microelectronics, Crystalline Defects and Contamination: Their Impact and Control in Device Manufacturing, Advanced Short-Time Thermal Processing for Si-Based CMOS Devices, ULSI Process Integration, Silicon-on-Insulator, Quantum Confinement and Nanostructures* and *Analytical and Diagnostic Techniques for Semiconductor Materials, Devices and Processes.*

II. Technical Agenda

Detailed chapter introductions have been developed by the session co-chairs, per se. Our purpose here is simply to present a macroscopic over-view of the Silicon Symposium.

300mm Wafers

Silicon crystal growth and wafer preparation are, perhaps, more important than ever with the 300mm diameter era in full swing and the possible introduction of the 450mm diameter wafer about 2012, per the 2005 edition of the ITRS. As noted in the 2002 *Ninth Silicon Symposium*, the maintenance of a two-year technology cycle, as tactically envisioned in the 2003 ITRS, partially mitigates the need for the 450mm wafer diameter size change, potentially delaying its introduction on a nearly year-by-year basis. Assessment of the relevant high-level strategies as regards implementation of the 450mm diameter wafer will require significant cooperation among the IDM's, wafer and equipment suppliers. The industry economic and productivity models may be one useful component facilitating that decision.

Generic issues as regards point defects are presented in one invited and two contributed papers followed by consideration of issues associated with the growth and defect formation in heavily doped crystals. The lessons learned in the 300mm transition as well as critical issues anticipated for the 450mm era are the presented. Now that SOI has entered high-volume device fabrication, it is appropriate for an invited paper to review 300mm SOI wafers. Finally, wafer strength and related issues are presented for 300mm wafers in an invited paper followed by a contributed paper on wafer warp introduced during wire-saw slicing.

Process Development and Modeling

Hisham Massoud of Duke University will be presented with the 2006 *Electronics & Photonics Division* award during the meeting and will present his invited paper on the *Growth Kinetics and Electrical Properties of Ultrathin Silicon Dioxide Layers.* This is followed by two additional invited papers and four contributed papers continuing the assessment of various point defects and chemical impurities in silicon, including their role during crystal growth, oxygen precipitation, and gettering of Cu and Fe during IC fabrication. The role of the ^{30}Si isotope in epitaxial layers and $Si_{1-x}Ge_x$ strain issues are also introduced in the session.

Materials and Process Integration

This session applies the knowledge of a variety of advanced materials for the successful fabrication of useful device/IC configurations, taking into consideration their impact and modification during IC fabrication. The stage is set via three invited papers discussing high-k gate dielectrics, multiple gate electrodes and enhanced channel mobility via strain integration methodologies. This is followed by five contributed papers further addressing defect issues associated with strain engineering, ultra-thin SOI and strained SOI (sSOI), surface diffusion in silicon trenches as well as the impact of defects on flash memory characteristics, and low-resistance contact technology.

Integrated Metrology and Diagnostics

More than ever, the characterization of materials and the unit IC process technologies in the fabrication of the IC, will be a leading issue for our industry. Two invited papers address in-line inspection technology and yield models for advanced process control. The development of standardization methodologies (important as ever) are then presented in three contributed papers for N, C, bulk micro-defects and denuded zones (DZ). A spectrum of additional characterization techniques are then presented via five contributed papers on x-ray reciprocal space mapping for strained $Si/Si_{1-x}Ge_x$ on bonded SOI, SOI low-frequency noise and traps, 1/f noise at the HfO_2/Si interface, copper chemical mechanical polishing (CMP) issues and the tuning of surface adsorbents on the silicon surface as a result of the gas velocity and chamber humidity.

Strained Silicon

The invited and three contributed papers on variously configured strained silicon materials (as fabricated in the substrate/film structure and/or process-induced during IC fabrication) present a current snapshot of one of the major initiatives being investigated by the Silicon Wafer Engineering and Defect Science (SiWEDS) consortium. This session is especially relevant inasmuch as strained silicon p-channel (pMOSFETs) and strained silicon n-channel (NMOSFET) methodologies are now available for selected high-performance products at the 65nm technology generation as well as the coming

45nm technology generation. (The technology generation has been defined as the DRAM half pitch or the logic physical gate length, wherein the latter is slightly smaller than 50% of the DRAM half-pitch, although these definitions are anticipated to change.)

III. Outlook

The opportunities appear boundless as the microelectronics industry approaches single digit nm physical channel lengths in research devices this decade (with anticipated production during the second decade of the 21st century). Improved control of the silicon wafer characteristics and IC fabrication processes, coupled with a deeper understanding of the interrelationships among silicon material, IC fabrication processes and device configuration/IC performance continues to be the key to superior IC product performance. These studies afford stimulating and challenging opportunities. Significant progress in the development of cost-effective silicon materials in conjunction with non-silicon channels for enhanced carrier transport in the future, with the requisite quality, however, is possible only if the IDMs, the wafer supplier and the equipment producers closely collaborate. We trust that these proceeding share some of the excitement in the world's largest industry — electronics.

IV. Acknowledgments

To preserve the highly current nature of the Symposium, the text has been reproduced digitally using the authors' own submissions, thus significantly shortening the publication cycle to three months. This would not have possible without the dedicated and timely efforts of the authors and the Silicon Symposium session co-chairs:

Dieter Gilles	Siltronic
Ulrich Gosele	Max Planck
Laszlo Fabry	Wacker
T. Hattori	Musashi Institute of Technology
Walter Huber	SUMCO-USA
Shuji Ikeda	ATDF, SEMATECH
Paul Packan	Intel
Mark Rodder	Texas Instruments
Eicke Weber	U. of California (Berkeley)
Rick Wise	Texas Instruments

The strained silicon session was especially developed through the initiative of George Rozgonyi of NCSU. We also appreciate the assistance of Sorin Cristoloveanu (IMEP), David Gilmer (Freescale Semiconductor), Chenming Hu (U. of California, Berkeley) and Jim Hutchby (SRC) during the early stages of developing the *Silicon Symposium*.

Most especially, we appreciate the extensive secretarial assistance of Mrs. Helen Huff throughout the course of developing this symposium. Without her superb support, furthermore, these proceedings would not have been issued in so timely a fashion.

The support of the Electronics & Photonics Division and The Electrochemical Society Office staff have contributed significantly, as usual, to the development of both the symposium and proceedings. We would be remiss, however, if we did not explicitly acknowledge the assistance of Valerie Yacko, Meetings and Program Coordinator, John Lewis, ECST Production Manager and Mary Yess, Deputy Executive Director for their extensive and cheerful support in ensuring the reality of both the meeting and proceedings volume for the *Tenth Silicon Symposium*. As usual, furthermore, we are indebted to the numerous industrial, research institutes and university colleagues, internationally, who have continued to contribute to the formulation and content of the *Silicon Symposia*. We dedicate these proceedings to the memory of Professor Al Tasch (1941-2004), our respected colleague.

Finally, it is also appropriate at this *Tenth Silicon Symposium* to thank the symposium co-chairs/co-editors and session co-chairs/assistant editors (see below) who have contributed, from its inception, to the warm reception afforded the *Silicon Symposium Series*.

Howard R. Huff   March, 2006
Hiroshi Iwai
Hans Richter

---

(1) H.R. Huff, An Electronics Division Retrospective (1952-2002) and Future Opportunities in the Twenty-First Century, J. Electrochem. Soc., **149**, S35-S58 (2002)
(2) R.L. Opila and D.W. Hess, A Century of Dielectric Science and Technology, J. Electrochem.Soc., **150**, S1-S10 (2003)
(3) H.R. Huff, (editor), Silicon: Into the Nano Era, Interface, **14**, No. 1, (2005)
(4) H. Wong and H. Iwai, The Road to Miniaturization, Physics World, **18**, No. 9, 40-44 (2005)

1969 Silicon Symposium—Co-Chairs/Co-Editors: Rolf R. Haberecht and Edward L. Kern; Session Co-Chairs/Assistant Editors: R.N. Hall, D.K. Hartman, H.F. Matare, Gunther H. Schwuttke, Elizabeth Tarrants and R.L. Weisberg

1973 Silicon Symposium—Co-Chairs/Co-Editors: Howard R. Huff and Ronald R. Burgess; Session Co-Chairs/Assistant Editors: Ken Bean, W. Murray Bullis, L. Clark, Somanath Dash, M.L. Joshi, Else Kooi, John E. Lawrence, Bernie T. Murphy, Francois Padovani, Gunther H. Schwuttke, Erhard Sirtl and Pei Wang

1977 Silicon Symposium—Co-Chairs/Co-Editors: Howard R. Huff and Erhard Sirtl; Session Co-Chairs/Assistant Editors: Ken Benson, Jan Bloem, W. Murray Bullis, H. Herzer, Carl M. Osburn, Ken Pickar, K.V. Ravi, George A. Rozgonyi, Yoshiyuki Takeishi, Fritz G. Vieweg-Gutberlet, Pei Wang and John Zoutendyk

1981 Silicon Symposium—Co-Chairs/Co-Editors: Howard R. Huff, Rudolph J. Kriegler and Yoshiyuki Takeishi; Session Co-Chairs/Assistant Editors: Takao Abe, Dimetri A. Antoniadis, Martin G. Buehler, Gilbert J. Declerck, Dan J. DiMaria, G. Foti, Milt Gosney, Ted I. Kamins, Lou E. Katz, Bernd O. Kolbesen, Carl M. Osburn, Dave S. Perloff, Al F. Tasch and R.G. Wilson

1986 Silicon Symposium—Co-Chairs/Co-Editors: Howard R. Huff, Takao Abe and Bernd Kolbesen; Session Co-Chairs/Assistant Editors: Keith G. Barraclough, George A. Brown, John R. Carruthers, Cor L. Claeys, Bob A. Craven, Michael I. Current, Bruce E. Deal, Richard B. Fair, L.C. Kimerling, Seigo Kishino, R.B. Marcus, Rafael Reif, G.R. Srinivasan and K. Wada

1990 Silicon Symposium—Co-Chairs/Co-Editors: Howard R. Huff, Keith G. Barraclough and Jun-ichi Chikawa; Session Co-Chairs/Assistant Editors: Werner Bergholz, G.F. Cerofolini, D. W. Greve, C.R. Helms, K. Izumi, H. Jacob, Junji Matsui,Bernie S. Meyerson, P. Gill, Dieter K. Schroder, Fumio Shimura, T. Suzuki, Masaharu Watanabe and Dim Wolters

1994 Silicon Symposium—Co-Chairs/Co-Editors: Howard R. Huff, Werner Bergholz and Koji Sumino; Session Co-Chairs/Assistant Editors: John G. Borland, L.T. Canham, Ulrich Gosele, Peter O. Hahn,S. Kawado, N. Koshida, Wolfgang Schroter, Tom O. Sedgwick, Tom J. Shaffner, R. Sivan, R. Takiguchi, Shin-Ichiro Takasu, Hideki Tsuya, Eicke R. Weber, K. Yamabe and W. Zulehner

1998 Silicon Symposium—Co-Chairs/Co-Editors: Howard R. Huff, Ulrich Gosele and Hideki Tsuya; Session Co-Chairs/Assistant Editors: George Celler, Laszlo Fabry, Dieter Gilles, Takeshi Hattori, Marc Heyns, Akihiko Ishitani, Seiichi Isomae, Shigeyuki Kimura, Sumio Kobayashi, Karl-Heinz Kusters, Hisham Massoud, Yosiaki Matsushita, James Meindl, Akira Ohsawa, John Poate, Hans Queisser, Wolfgang Schroter and Andrew Steckel

2002 Silicon Symposium—Co-Chairs/Co-Editors: Howard R. Huff, Laszlo Fabry and Seigo Kishino; Session Co-Chairs/Assistant Editors; Wilfried von Ammon, Simon Deleonibus, Alain Diebold, Steven Hillenius, Sunao Ishihara, Hiroshi Iwai, Hiroshi Koyama, Witek Maszara, Paul Mertens, Hans Richter, Tatsuhiko Shigematsu and Phil Tobin

ECS Transactions, Volume 2, Number 2
Silicon Materials Science and Technology X

Table of Contents

Preface   iii

Plenary

* Forty Years of Moore's Law: Ever Smaller Transistors and Ever Larger Wafers   3
   G. Hutcheson

* Present and Future of Si-Based Transistor Technology for Memories   11
   D. Park and B. Ryu

* Front-End Process Technology for hp65/45 CMOS Devices   27
   Y. Nara, F. Ootsuka and K. Nakamura

* Trends in European R&D - Advanced Process Control Down to Atomic Scale   33
for Micro- and Nanotechnologies
   L. Pfitzner, M. Schellenberger, R. Oechsner, G. Roeder and M. Pfeffer

300 mm

Introductory Remarks: 300 mm   57
   D. Gilles and W. Huber

* Parameters of Intrinsic Point Defects in Silicon Based on Crystal Growth, Wafer   61
Processing, Self- and Metal-Diffusion
   V. V. Voronkov and R. Falster

An Atomically Accurate Model for Point Defect Aggregation in Silicon   77
   T. Sinno, W. Haeckl and W. Von Ammon

Growth Technologies for 300 mm Arsenic Heavily Doped Silicon Single   89
Crystals
   H. Tu, Q. Zhou, G. Zhang, X. Dai, Z. Wu and T. Jia

Defect Formation Behaviors in Heavily Doped Czochralski Silicon   95
   W. Sugimura, T. Ono, S. Umeno, M. Hourai and K. Sueoka

* *invited paper*

xi

* Wafer Strength and Slip Generation Behavior in 300 mm Wafers   109
  T. Ono, W. Sugimura, T. Kihara and M. Hourai

Warp of Silicon Wafers Produced from Wire Saw Slicing: Modeling,   123
Simulation, and Experiments
  P. Gupta and M. Kulkarni

Lessons Learned from the 300 mm Transition   135
  S. Kramer, J. Draina, D. Fandel and J. Ferrell

Discussion on Issues toward 450 mm Wafer   155
  M. Watanabe, T. Fukuda, A. Ogura, Y. Kirino and M. Kohno

* 300 mm SOI for High Volume Manufacturing   167
  G. Pfeiffer, M. Haag, M. Schmidt, R. Krause, P. Tsai and J. Lee

Process Development and Modeling

Introductory Remarks: Process Development and Modeling   185
  U. Goesele and P. Packan

* Growth Kinetics and Electrical Properties of Ultrathin Silicon Dioxide Layers   189
  H. Z. Massoud

* Oxygen Vacancies at the $Si(001)/SiO_2$ Interface   205
  A. Korkin, J. Greer, T. M. Hendersen, G. Bersuker and R. Bartlett

Defect Dynamics in the Presence of Oxygen in Growing Czochralski Silicon   213
Crystals
  M. Kulkarni

Defect Dynamics in the Presence of Nitrogen and Oxygen in Growing   229
Czochralski Silicon Crystals
  M. Kulkarni

* Analytical Modeling of the Interaction of Vacancies and Oxygen for Oxide   247
Precipitation in RTA Treated Wafers
  G. Kissinger, J. Dabrowski, A. Sattler, C. Seuring, T. Mueller, H. Richter and
  W. Von Ammon

First Principles Calculation for Cu Gettering by Dopant or Dopant-Vacancy   261
Complex in Silicon Crystal
  K. Sueoka, S. Ohara, S. Shiba and S. Fukutani

* invited paper

Analysis of Internal Gettering of Iron Based on the Nucleation Model of Iron Precipitation 275
K. Nakamura and J. Tomioka

Silicon Self-Diffusion in Heavily B-Doped Si Using Highly Pure 30Si Epitaxial Layer 287
S. Matsumoto, S. R. Aid, S. Seto, K. Toyonaga, Y. Nakabayashi, M. Sakuraba, Y. Shimamune, Y. Hashiba, J. Murota, K. Wada and T. Abe

Modeling Growth Behavior for $Si_{1-x}Ge_x$ from SiH_4 and GeH_4 by CVD 299
X. Yang and M. Tao

Materials and Process Integration

Introductory Remarks: Materials and Process Integration 313
S. Ikeda and M. Rodder

* Scaled CMOS with SiON and High-k 317
K. Ishimaru, M. Takayanagi, T. Watanabe, S. Inaba, M. Fujiwara and D. Matsushita

* Multiple Gate MOSFETs 329
W. P. Maszara, Z. Krivokapic, Q. Xiang and M. Lin

* Mobility Enhancement and Strain Integration in Advanced CMOS 341
C. H. Diaz

Defect Engineering Considerations for Strained Silicon Substrates 349
C. Claeys, G. Eneman, M. Scholz, R. Loo, P. Verheyen, K. De Meyer and E. R. Simoen

Modeling of Morphological Changes by Surface Diffusion in Silicon Trenches 363
T. Mueller, D. Dantz, W. Von Ammon, J. Virbulis and U. Bethers

Thermal Agglomeration of Ultrathin SOI and SSOI Films: A Quantitative Stability Study and Physical Model to Guide Ultrathin SOI Process Design 375
D. T. Danielson, J. Michel and L. Kimerling

Impact of Defects in Silicon Substrate on Flash Memory Characteristics 391
Y. Hirano, K. Yamazaki, F. Inoue, K. Imaoka, K. Tanahashi and H. Yamada-Kaneta

Low-Resistance Ti/n-type Si(100) Contacts by Monolayer Se Passivation 401
J. Zhu, X. Yang and M. Tao

* invited paper

xiii

Integrated Metrology and Diagnostics

Introductory Remarks: Integrated Metrology and Diagnostics 413
  L. Fabry and T. Hattori

* Breakthrough of In-Line Inspection Technology in Volume Production for 65 nm 415
  Node and Beyond
  Y. Yamazaki

* Scenario for a Yield Model Based on Reliable Defect Density Data and Linked 433
  to Advanced Process Control
  A. Nutsch and R. Oechsner

Standardization of Measurement of Nitrogen Concentration in CZ Silicon 453
Crystals
  N. Inoue, A. Karen, H. Yagi, K. Masumoto, M. Shinomiya, K. Kashima, K.
  Eifuku, M. Koizumi, T. Takahashi, T. Takenawa and K. Shingu

Infrared Absorption Measurement of Carbon Concentration in Silicon Crystals 461
  N. Inoue, M. Nakatsu and V. Akhmetov

Standardization of Characterization of Bulk Microdefects and Denuded Zones in 471
Annealed CZ Si
  R. Takeda, N. Inoue, K. Moriya, K. Kashima, K. Nakashima, M. Kato, S.
  Kitagawa, T. Ono, H. Urushido, N. Nango and V. Akhmetov

X-Ray Reciprocal Space Mapping and Synchrotron Radiation Topography of 485
Strained $Si/Si_{1-x}Ge_x$ on Bonded SOI
  T. Ma, H. Tu, G. Hu, B. Shao and A. Liu

SOI Low Frequency Noise and Interface Trap Density Measurements with the 491
Pseudo MOSFET
  V. A. Kushner, J. Yang, J. Choi, T. Thornton and D. Schroder

$1/f$ Noise as a Tool to Assess Fermi Level Pinning (EF) at the HfO_2/poly-Si 503
Interface in High-k n-MOSFETs
  P. Srinivasan, E. Simoen, L. Pantisano, C. Claeys and D. Misra

Study of Inhibition Characteristics of Slurry Additives in Copper CMP Using 515
Force Spectroscopy
  M. K. Keswani, H. Lee, S. Babu, U. Patri, Y. Hong, L. Economikos, M.
  Goldstein, L. Borucki, A. Philipossian and Y. Zhuang

* invited paper

Influence of Gas Velocity and Humidity on Diethyl Phthalate Adsorption and   523
Desorption on Silicon Surface
   H. Habuka, M. Tawada, K. Suzuki, T. Takeuchi and M. Aihara

Strained Silicon

Introductory Remarks: Strained Silicon   539
   E. Weber and R. Wise

Probing Nanoscale Local Lattice Strains in Advanced Si CMOS Devices by   541
CBED: A Tutorial with Recent Results
   M. Kim, J. Huang, P. Chidambaram, R. Irwin, P. Jones, J. Weijtmans, E.
   Koontz, Y. Wang, S. Tang and R. Wise

Local Strain Measurement on Strained Si/SiGe Heterostructures Using   549
Convergent Beam Electron Diffraction Analysis
   W. Zhao, G. Duscher and G. Rozgonyi

Analysis of Nanoscale Stress in Strained Silicon Materials and Microelectronics   559
Devices by Energy-Filtered Convergent Beam Electron Diffraction
   P. Zhang, A. Istratov, H. He, J. Ager, C. Nelson, E. Stach, J. Mardinly, C.
   Kisielowski, E. Weber and J. Spence

* Threading vs Misfit Dislocations in Strained Si/SiGe Heterostructures: 569
Preferential Etching and Minority Carrier Transient Spectroscopy
   J. Lu, R. Zhang, G. Rozgonyi, E. Yakimov, N. Yarykin and M. Seacrist

Author Index   579

* *invited paper*

Facts about ECS

The Electrochemical Society (ECS) is an international, nonprofit, scientific, educational organization founded for the advancement of the theory and practice of electrochemistry, electrothermics, electronics, and allied subjects. The Society was founded in Philadelphia in 1902 and incorporated in 1930. There are currently over 7,000 scientists and engineers from more than 70 countries who hold individual membership; the Society is also supported by more than 100 corporations through Corporate Memberships.

The technical activities of the Society are carried on by Divisions. Sections of the Society have been organized in a number of cities and regions. Major international meetings of the Society are held in the spring and fall of each year. At these meetings, the Divisions and Groups hold general sessions and sponsor symposia on specialized subjects.

The Society has an active publications program that includes the following.

Journal of The Electrochemical Society — JES is the peer-reviewed leader in the field of electrochemical and solid-state science and technology. Articles are posted online as soon as they become available for publication. This archival journal is also available in a paper edition, published monthly following electronic publication.

Electrochemical and Solid-State Letters — ESL is the first and only rapid-publication electronic journal covering the same technical areas as JES. Articles are posted online as soon as they become available for publication. This peer-reviewed, archival journal is also available in a paper edition, published monthly following electronic publication. It is a joint publication of ECS and the IEEE Electron Devices Society.

Interface — *Interface* is ECS's quarterly news magazine. It provides a forum for the lively exchange of ideas and news among members of ECS and the international scientific community at large. Published online (with free access to all) and in paper, issues highlight special features on the state of electrochemical and solid-state science and technology. The paper edition is automatically sent to all ECS members.

Meeting Abstracts (formerly Extended Abstracts) — Abstracts of the technical papers presented at the spring and fall meetings of the Society are published on CD-ROM.

ECS Transactions — This online database provides access to full-text articles presented at ECS and ECS-sponsored meetings. Content is available through individual articles, or as collections of articles representing entire symposia.

Monograph Volumes — The Society sponsors the publication of hardbound monograph volumes, which provide authoritative accounts of specific topics in electrochemistry, solid-state science, and related disciplines.

For more information on these and other Society activities, visit the ECS website:

www.electrochem.org

SESSION 1
PLENARY

2

Forty Years Of Moore's Law:
ever smaller transistors and ever larger wafers

G. Dan Hutcheson

CEO, VLSI Research Inc, Santa Clara, CA 95054, USA

It has been forty years since Gordon Moore first posited what would one day come to be known as Moore's Law. Gordon's ideas were more than a forecast of an industry's ability to improve; they were a statement of the ability for semiconductor technology to contribute to economic growth and even the improvement of mankind in general. More importantly, Moore's Law set forth a vision of the future that harnessed the imaginations of scientists and engineers to make it all possible.

Today, we take many of the benefits of Moore's Law for granted. Yet if you look behind the curtains of the new breakthrough sciences, as well as many of the mundane, you will find semiconductors working. Much would not be possible without the relentless progress of the semiconductor industry doubling performance for the same price every two years or so and that is what Moore's Law is all about.

The miracles of nano-biology and genetic engineering would not be possible had Moore's Law not brought affordable computing power to the table. While our children play video games on black boxes filled with chips, professionals in the medical sciences use the same technology to visualize complex models of drug interaction and even to unlock genetic codes. The Lewis and Clarks of today's time don't use optics to map the landscape; they use computer visualization tools to map the human genome. Meanwhile, imagine traffic congestion without computer chips to turn the lights green when you drive up. It may seem mundane, but computer chips keep America moving efficiently. Without chips, cell phones would not be there to bring help to our loved ones in unexpected emergencies or simply to make that call to bring home a quart of milk. Communication is vital to the economy and chips have greatly expanded our abilities here. Computers are the engines of America's productivity surge that has held inflation down since the nineties and the engines of computers are semiconductors. Chips provide better automotive power-train control systems that make for fun cars that pollute less. Computer chips are replacing film in digital cameras, saving untold amounts of chemical pollution. Meanwhile chips are being attached to animals in the wild, so we gain an even deeper understanding of the world around us. Chips make smart bombs smart . . . and Moore's law makes them smarter. Chips are critical to our national defense, also making unmanned aircraft possible and saving untold lives on the battlefield. These breakthroughs and many more are directly the result of advancements in chips as predicted by Moore's Law.

Moore's Law is an amazing story of how technological progress came to affect our everyday lives and will affect our children's lives for many generations to come. But its history is far richer than the development of semiconductors, which to some extent, explains why Moore's Law was so readily accepted. This history also explains why there has been an insatiable demand for more powerful computers no matter what people have thought to the contrary.

The quest to store, retrieve, process, and communicate information is one task that makes humans different. No known animal uses tools to store, retrieve, and process information. Moreover the social and technological progress of the human race can be directly traced to this attribute.

Man's earliest attempts to store, retrieve, and process information date back to prehistoric times when humans first carved images in stone walls. Then in ancient times, Sumerian clay tokens developed as a way to track purchases and assets. By 3000 B.C. this early accounting tool had developed into the first complete system of writing on clay tablets. Ironically, these were the first silicon based storage technologies and would be abandoned by 2000 B.C. when the Egyptians developed papyrus based writing materials. It would take almost four millennia before silicon would stage a comeback as the base material, with the main addition being the ability to process stored information. In 105 A.D. a Chinese court official named Ts'ai Lun invented wood-based paper. It wasn't until around 1436 that Johann Gutenberg invented the movable type printing press so that books could be reproduced cost effectively in volume. The first large book was the Gutenberg Bible, published in 1456. So something akin to Moore's Law occurred, as Gutenberg's innovation enabled progressing from printing single pages to entire books in 20 years. At the same time, resolution also improved, allowing finer type as well as image storage. Yet, this was primarily a storage mechanism. It would take at least another 400 years before retrieval would be an issue. In 1876, Melvil Dewey published his classification system that enabled libraries to store and retrieve all the books that were being made by that time. Alan Turing's "Turing Machine", first described in 1936, was the step that would make the transformation from books to computers. So Moore's Law can be seen to have a social significance that reaches back more than five millennia.

Moore's law is also indelibly linked to the history of our industry and the economic benefits that it has provided over the years. Carver Mead was the first to call the relationship "Moore's Law." Moore's observations about semiconductor technology are not without precedent. As early as 1887, Karl Marx, in predicting the coming importance of science and technology in the twentieth century, noted that for every question science answered, it created two new ones; and that the answers were generated at minimal cost in proportion to the productivity gains made. More important was Marx's observation that investments in science and engineering led to technology, which paid off in a way that grew economies, not just military might.

It was this exponential growth of scientific 'answers' that led to these developments, as well as, to the invention of the transistor in 1947 -- and ultimately the integrated circuit in 1958. The IC developed rapidly, leading to Moore's observation that became known as a law – and in-turn, launched the information revolution.

In 1964, *Electronics* magazine asked Moore, then at Fairchild Semiconductor, to write about what trends he thought would be important in the semiconductor industry over the next ten years for its 35th anniversary issue. Integrated circuits (ICs) were relatively new. Many designers didn't see a use for them and worse, some still argued over whether transistors would replace tubes. A few even saw integrated circuits as a threat: if the system could be integrated into an IC, who would need system designers?

The article, titled "Cramming more components into integrated circuits," was published by Electronics magazine in its April 19, 1965 issue. This issue's contents exemplify how so few really understood the importance of the integrated circuit. Ahead of it was the cover article by RCA's legendary David Sarnoff who, facing retirement, reminisced about "Electronics' first 35 years" with a look ahead. Behind this were several more articles with Moore's paper buried on page 114. Electronics magazine was the most respected publication covering its field. Today, the magazine is defunct, not surviving the Moore's Law.

Moore's paper proved so long lasting because it was more than just a prediction. The paper provided the basis for understanding how and why integrated circuits would transform the industry. Moore considered user benefits, technology trends, and the economics of manufacturing in his assessment. Thus he had described the basic business model for the semiconductor industry - - a business model that lasted through the end of the millennium.

From a user perspective, his major points in favor of ICs were that they had proven to be reliable; they lowered system costs; and often improved performance. He concluded, "Thus a foundation has been constructed for integrated electronics to pervade all of electronics." From a manufacturing perspective, Moore's major points in favor of ICs were that integration levels could be systematically increased based on continuous improvements in largely existing manufacturing technology. He saw improvements in lithography as the key driver.

From an economics perspective Moore recognized the business import of these manufacturing trends and wrote, "Reduced cost is one of the big attractions of integrated electronics, and the cost advantage continues to increase as the technology evolves toward the production of larger and larger circuit functions on a single semiconductor substrate. For simple circuits, the cost per component is nearly inversely proportional to the number of components, the result of the equivalent package containing more components."

The essential economic statement of Moore's law is that the evolution of technology brings more components and thus greater functionality for the same cost. Computing power improves essentially for free; driving productivity in the economy; thus fueling demand for more semiconductors. This is why the growth in transistor production has been so explosive. Lower cost of production has led to an amazing ability to not only produce transistors on a massive scale, but to consume them as well.

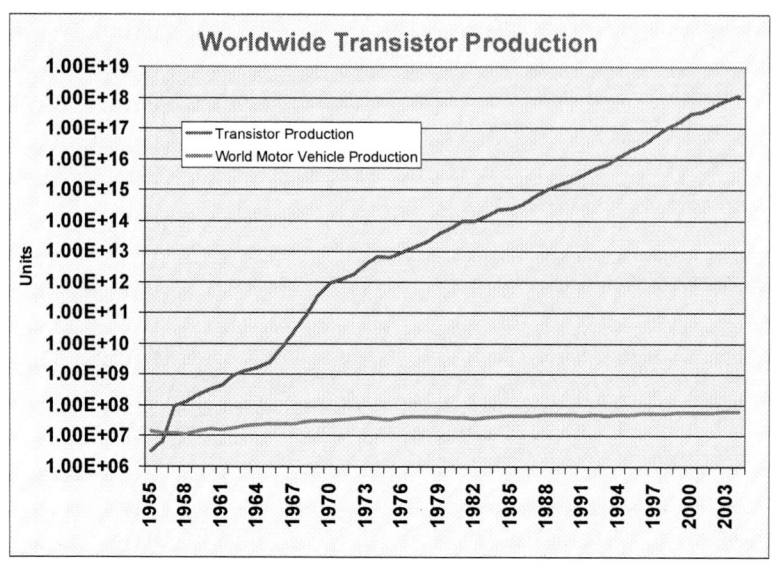

The economic value of Moore's Law is that it has been a powerful deflationary force in the world's macro-economy. Inflation is a measure of price changes without any qualitative change. So if price per function is declining, it is deflationary. This effect has never been fully accounted for in government statistics. The decline in price per bit has been stunning:

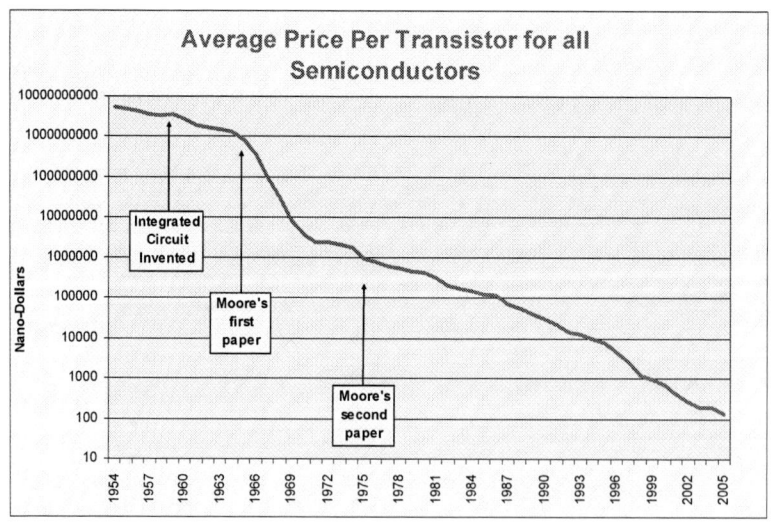

In 1954, five years before the IC was invented, the average selling price of a transistor was $5.52. Fifty years later, in 2004, this had dropped to 191 nano-dollars (a billionth of a dollar). *If the semiconductor were fully adjusted for inflation, its size in 2004 would have been 6 million-trillion dollars.* That is many orders of magnitude greater than Gross World Product! So it is hard to understate the long term economic impact of the semiconductor industry. Much of this impact has come directly to America, because it has been the world's leader in semiconductors.

So what makes Moore's Law work? There are three primary technical factors that make Moore's Law possible: reductions in feature size, increased yield, and increased packing density. The first two are largely driven by improvements in manufacturing and the latter largely by improvements in design methodology.

Reductions in feature sizes have made the largest contributions by far, accounting for roughly half of the gains since 1976. Feature sizes are reduced by improvements in lithography. Transistors can be made smaller and hence more can be packed into a given area.

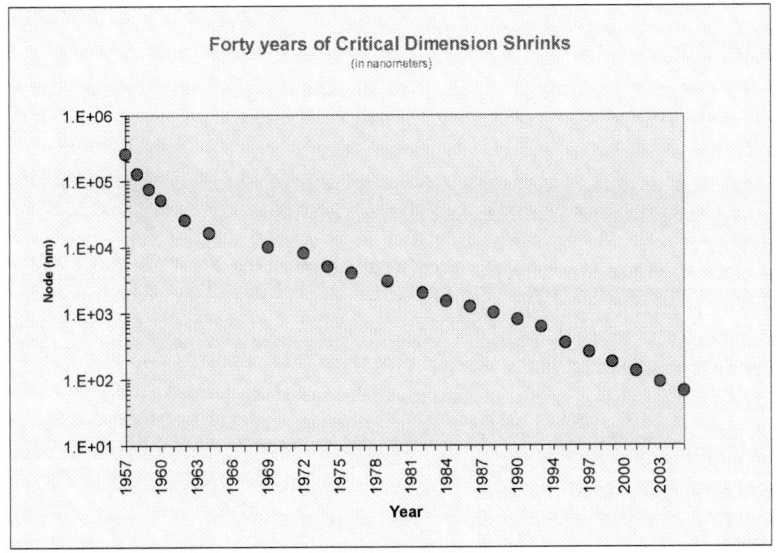

These gains have come from new lithography tools; resist processing tools and materials; as well as etch tools. Lithography tools were not always the most costly tool in the factory. The camel's hair brush, first used in 1957 to paint on hot wax for the mesa transistors, cost little more than a dime. But since that time prices have escalated rapidly, increasing roughly an order-of-magnitude every decade and a half. The industry passed the $10M mark in 2003 and some tools now cost as much as $20M.

Over the decades, these cost increases have been consistently pointed to as a threat to the continuance of Moore's Law. It is testimony to the power of this law that these costs can be absorbed with little effect.

Wafer size is another area that Moore addressed in his papers. He highlighted the economic benefits that would be systematically gained with each progression in wafer size. He saw wafers growing ever larger as a way to keep the cost of making a transistor ever smaller and even at one point predicted a six foot wafer. While the six foot wafer never came to pass, his predictions about growing wafer sizes and their continual economic benefits did come true.

As with the various technologies above, the economic benefits to be gained with another wafer size increase has been the subject of intense debate with each new generation, since the four inch wafer debuted in the early eighties. Yet, we have continued to bring ever larger wafer sizes to production since then. Today, we are faced with yet another wafer size: 450mm. Few think it will come to pass at this stage, even though it is in the ITRS roadmap. But the same was true for 300mm, roughly ten years ago. Today, there are few who question the viability of 300mm. The real issue for our industry with 450mm is not to ask if it will come to pass, but to ask how we can work together efficiently to ensure that it brings the same economic gains as other wafer size generations.

At some point the effect of these technologies translating into high costs will cause Moore's Law to cease. This is why the spotlight is always on cost and how to defray them.

The idea of Moore's Law meeting Moore's Wall and the show stopping, or the contrary belief that there will be unending prosperity in the 21st Century buoyed by Moore's Law, have been recurring themes in the media and technical community since the mid-seventies. I have built my career, in part, by predicting that end of Moore's Law was not coming anytime soon. Many others have lost theirs over the past thirty years by predicting its demise due to physical limits. The reason is that I have always had faith in the ability of the brightest minds in science and technology to come up with the ideas needed to overcome these limits. But I am growing concerned.

The costs of the research to keep Moore's clock ticking are rising with each node. I fear the day that it becomes too expensive for the private sector to afford and the clock stops. In part because of the many conveniences, but mostly because of the dramatic effect it has had in driving America's productivity and thus, its leadership in the global economy. When Moore's clock stops the consequences to the economy should be obvious.

What will America do as a nation when Moore's Law has beat its last heart beat; when it no longer delivers its productivity gains and anti-inflationary effects? How will we pay for ever rising health care costs? What will happen if America's economy falls behind and is no longer the global leader? Other nations recognize the importance of semiconductors at the public level and are investing heavily. These are important questions for legislators to consider.

As Gordon commented on his law a few years back, "No exponential lasts forever. But forever can be postponed." Let's invest to postpone it.

To learn more, see "THE ECONOMIC IMPLICATIONS OF MOORE'S LAW" (1).

ACKNOWLEDGMENTS

This article is an updated and abridged version of "The Economic Implications of Moore's Law," from the book "High Dielectric Constant Materials," Springer Series in Advanced Microelectronics, Volume 16, Springer-Verlag, New York, 2005; excerpted with permission.

Present and Future of Si-based Transistor Technology for Memories

Donggun Park and Byung-Il Ryu

Semiconductor R&D Center, Samsung Electronics
San24 Nongseodong Kiheung, Yongin, Korea, 449-711, dgpark@samsung.com

Nanotechnology is named as a promising research area for the semiconductor technology to be extended or replaced by it. Since the silicon based devices such as NAND Flash memory, DRAM, SRAM, and CPUs are scaled down below 100 nm entering into the nanoscale regime late 1990s, in the middle of 2005, we are already facing the appearance of sub 50 nm technology in production. In this paper, we introduce recent technology development activities in Samsung to realize the sub 50 nm technologies into the devices such as SRAM, DRAM, and Flash memories as well as the high performance logic devices. RCAT (Recessed Cell Array Transistor), PiFET (Partially-insulated MOSFET) and FinFET, McFET (Multi channel FET), MBCFET (Multi-Bridge Channel MOSFET), and Twin Silicon Nanowire MOSFET (TSNWFET) are the 3-dimensional structure CMOS transistors to be introduced for nanoscale applications.

Introduction

The requirement of memory, one of the fundamental elements, for the rapid IT industry growth has been dramatically increased as shown in Fig.1. The device development speed has been accelerated to meet the market demand since sub 100 nm technology has been used for mass production in 2002. With the achievement of this technology trend, the digital equipment industry grows fast and we are being able to enjoy the plentitude of almost 16 gigabytes memory capacity as illustrated in Fig. 2.

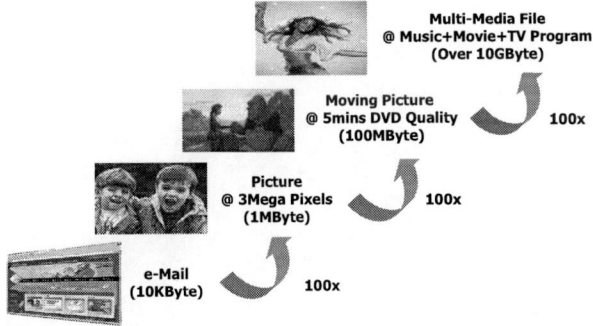

Figure 1. Information density increases by 100 times in sequence driving explosive demand for semiconductors.

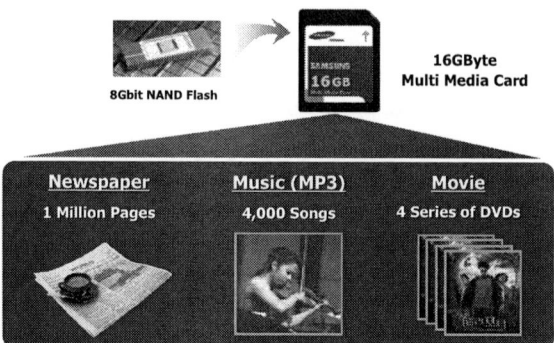

Figure 2. With 16 gigabytes of memory, we can enjoy up to 4 series of DVD titles or 4000 songs with MP3 quality.

However, transistor, the basic element for Si based devices, is facing a barrier in the use of conventional planar CMOS technology due to its physical limit to the scaling. In this paper, we introduce various non-planar transistor development activities in Samsung to be implemented into sub 50 nm devices. This development is to extend the use of Si based transistor down to sub 20 nm technology nodes. It would act as a bridge for the future devices to enter into the nanotechnology era as illustrated in Fig.3.

As the transistor gate length scales down below 45 nm, short channel effects such as DIBL (Drain Induced Barrier Lowering),Vt (Threshold voltage) roll-off, and short HCL (Hot Carrier Lifetime), are crucial to the device fabrication. Many efforts to extend the use of planar transistor, such as strained Si, high-k dielectric, and elevated S/D (Source/Drain) are introduced. However, planar transistor structure is predicted to come to the end around 30nm technology node (1).

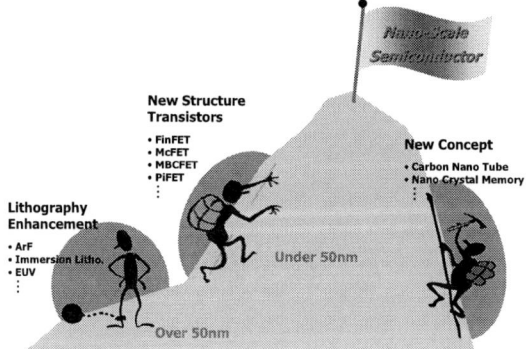

Figure 3. For the nanoscale semiconductor fabrication, new material, structure, concept are required in addition to fine patterning technologies.

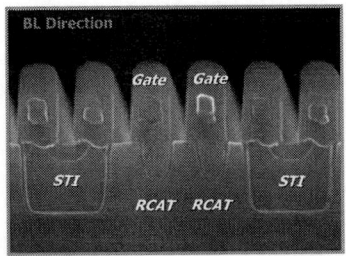

Figure 4. To overcome the scaling limit in DRAM cell transistor below 100nm, the channel region is recessed enlarging the gate length.

To overcome the scaling limit of planar Si CMOS transistor, 3 dimensional transistor structures have been introduced. Among them are RCAT (2), TSM (Twin SONOS Flash Memory) cell transistor (3,4), PiFET (5,6), FinFET (Fin shape FET) (7-9), McFET (10,11), MBCFET (12-15), and TSNWFET (16).

RCAT is a quasi-planar MOSFET enlarging the gate length. PiFET is a quasi-SOI transistor to control the channel and S/D depth effectively without demerits of SOI MOSFET. FinFET is the first highly manufacturable 3 dimensional non-planar transistor with fin shape channels. Various application examples of FinFET to memory devices have been reported (17-22). McFET is a further advanced form of FinFET having twin fin channels without using any sophisticated lithography tools. All the non-planar transistors show excellent electrical characteristics even though there are still some difficult tasks to be implemented to conventional CMOS processes. MBCFET that has two 20 nm thin Si bodies with surrounding gate shows the best transistor performance exceeding the ITRS (International Technology Roadmap for Semiconductors) prediction. From now on we will introduce these newly developed silicon based nanoscale CMOS transistors in detail.

Nanoscale Planar MOSFETs

In this chapter, we show planar MOSFET structures to be extended down to 50 nm technology node.

RCAT (Recessed Cell Array Transistor)

The newly developed recess channel array transistor of DRAM cell is shown in Fig. 4. Recessing the transistor channel, the gate length in small area is effectively enlarged so that not only the Vt distribution of RCAT is improved over the planar transistor but the drain induced barrier lowering (DIBL) is also reduced. This RCAT technology makes possible the DRAM scaling down to 50 nm technology node.

TSM (Twin SONOS Memory) Cell Transistor

While data storage of 2 bits per cell is popular in Flash memories, SONOS type Flash memory is an ultimate shape of Flash memory without floating gate. However, in

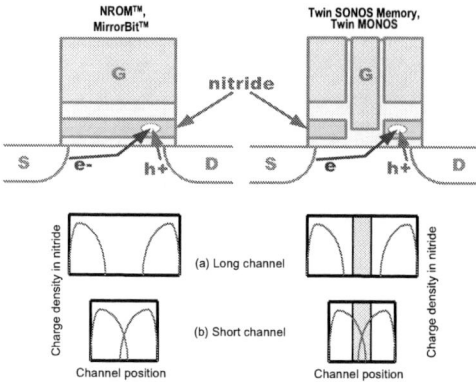

Figure 5. Non-volatile SONOS type Flash memory is facing the scaling limit due to the charge interaction along the gate. Twin SONOS memory cell transistor can be a breakthrough technology to overcome the limit.

NOR type SONOS memory, program is performed with CHI (Channel Hot Carrier Injection) with high drain voltage to enhance the program speed. For this purpose, the gate length should be large enough so that the transistor punchthrough voltage is larger than the program voltage. In addition, since the 2 bits/cell operation requires large enough separation of the programmed charges at the source and drain region, charge interaction in the nitride, charge trapping layer replacing the floating gate limits the gate length scale as illustrated in Fig.5. ITRS (International Technology Roadmap of Semiconductor) 2003 predicts the SONOS Flash Memory devices cannot be scaled down below 140 nm of the gate length even at the 20 nm NVM technology. Fig.6 shows our newly developed TSM cell transistor having physically separated nitride storage nodes at the source and drain regions and the gate oxide thickness at the transistor center is so thin that the transistor can be scaled down without punchthrough problem down to the total gate length of 80 nm. With this scheme, the excellent data retention time has been obtained as shown in Fig.7.

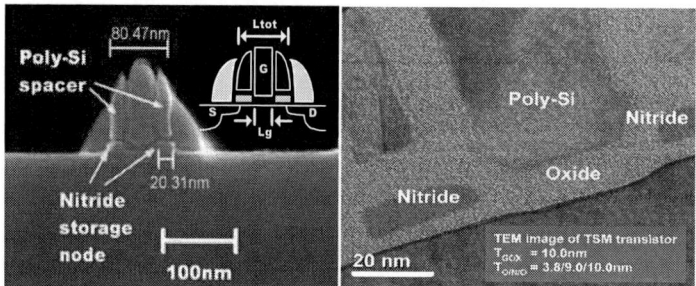

Figure 6. Using TSM cell transistor, SONOS type Flash memory can be scaled down below 80 nm without any charge interaction issues having better gate controllability.

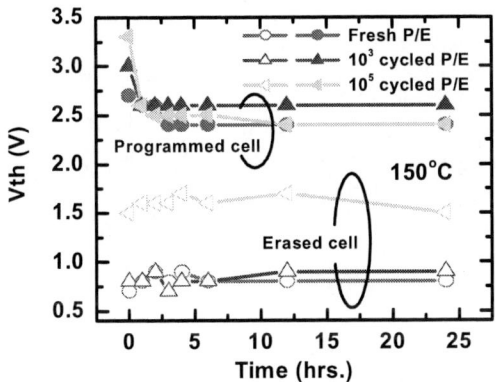

Figure 7. TSM cell shows reasonable program and erase Vth as well as excellent data retention time more than 24 hrs at 150°C.

PiFET (Partially insulated MOSFET)

As an effort to extend the planar MOSFET structure below 100 nm, SOI (Silicon On Insulator) wafer had been introduced. Using SOI wafer with thin silicon on SiO_2 layer, the channel Si region is thinned down to eliminate bulk punchthrough leakage path, while the shallow junction depth is obtained by the self-limited dopant diffusion as well. In this section, we demonstrate a PiCAT (PiFET Cell Array Transistor) structure that has PiOX, a buried oxide layer, to block the junction leakage current path physically. Using this PiCAT structure, improved DIBL characteristic, self-limited shallow junction, and reduced junction capacitance owing to its structural benefit have been obtained.

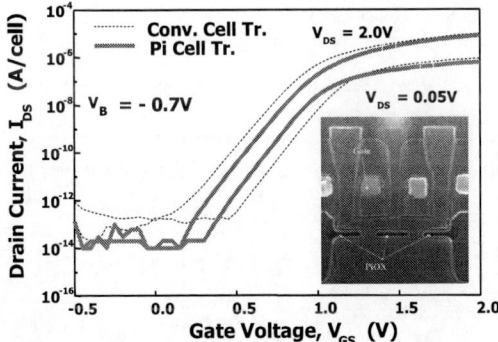

Figure 8. Self-limiting junction depth of piFET helps the PiCAT to show improved DIBL over the conventional ones.

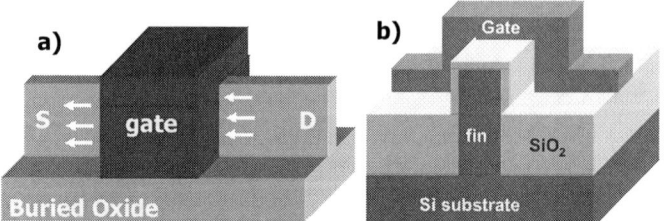

Figure 9. a) FinFET fabricated on SOI (Silicon On Insulator) waer, b) Body-Tied FinFET on Bulk Si. Body-Tied FinFET has several benefits such as eliminating floating-body, better heat dissipation, low cost, etc over FinFET on SOI.

The cross-sectional SEM picture of PiCAT for the fully integrated 80 nm 512M DRAM and the transistor characteristics are shown in Fig. 8. Using this unique process technology, the Si channel is tied to the silicon substrate so that the floating body problem of the SOI MOSFET is avoided.

3D MOSFETs for sub 50 nm Scaling

In this chapter, we show several 3 dimensional MOSFET structures that would be useful for the technology node below 50 nm down to 10 nm.

FinFETs (Fin shape MOSFETs)

Various FinFET structures, as illustrated in Fig. 9-a such as Omega-gate (23), Tri-gate (24), Dual-Gate UTB (Ultra-Thin Body), are introduced for the application below 50 nm technology node due to its good SCE immunity resulted from the excellent gate controllability with thin Si on Oxide (SOI).
We had introduced body-tied FinFETs fabricated on bulk Si instead of SOI wafer with the benefits of low wafer cost, good heat dissipation characteristic, and the elimination of floating body effect. Fig. 9-b shows schematic illustration of the body-tied FinFET.

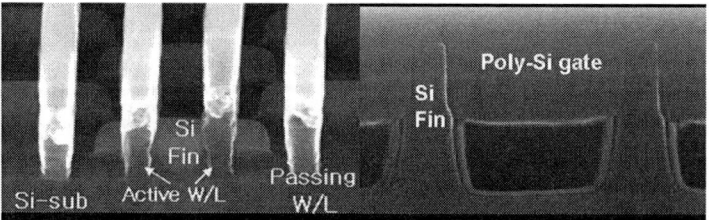

Figure 10. Body-Tied FinFET is adopted for DRAM cell array transistor application.

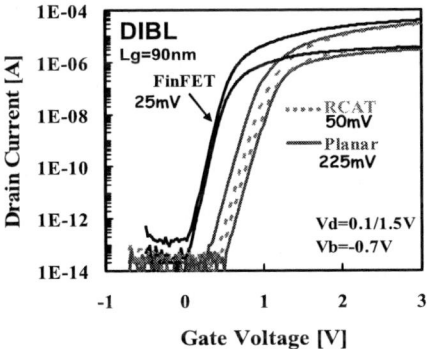

Figure 11. Due to the excellent short channel effect of FinFET, full-processed 512M 90 nm DRAM FinFET cell array transistor shows drain induced barrier lowering as low as 25 mV/V.

Fig. 10 shows the application of FinFET to fully working 90 nm 512M DRAM. Cell array transistors are made by FinFET, while the peripheral circuitries are still made by conventional planar transistors for convenience of analysis. In Fig. 11, we can clearly see that the body-tied FinFET DRAM cell transistor has superior transistor characteristics such as DIBL, subthreshold swing, and current drivability, over RCAT and planar transistors.

FinFET application to NOR Flash memory is shown in Fig. 12. With round fin, uniform tunnel oxide is formed around the fin. Fig. 13 shows that the major benefit of using FinFET to NOR cell transistor, punchthrough margin, is achieved with FinFET that is crucial to hot carrier program. The NOR Flash memory fabricated with FinFET shows programmed state of 7.2 V under the bias condition of Vgs = 10 V, Vds = 4.0 V, and Vb = -0.5V. Erase is performed at Vgb = 16 V for 10 msec. Under this program/erase conditions, Vth window (sensing margin) more than 3 V, endurance characteristics up to 10^5 program/erase cycling, 4.4 V sensing margin after 10^5 cycle endurance are achieved. In addition, data retention of programmed Vth shift of only 0.6 V at 300 °C for 12 hours is obtained. In spite of the above excellent performance of FinFET, FinFET cell has an inherent demerit of lower coupling ratio than planar one. However, the elongated channel

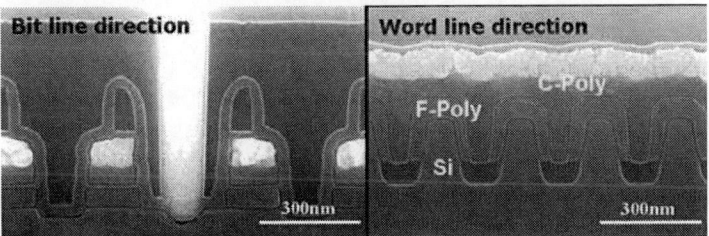

Figure 12. FinFET NOR Flash cell transistor

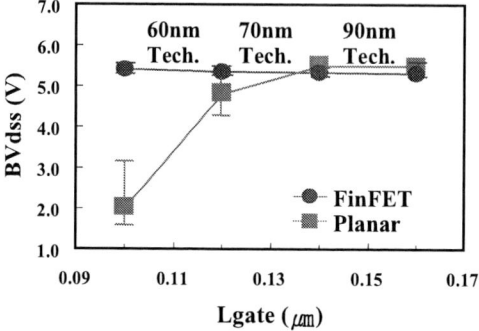

Figure 13. FinFET NOR Flash cell transistor is free from the transistor punchthrough down to 60 nm technology node that is expected to use 100 nm gate length.

width has a significant benefit on the erase Vth distribution. The coupling ratio variation, that is one of the key factors of erase Vth variation, can be dramatically reduced due to the almost same effective active width along the scaling in FinFET structure. As the cell scales down, FinFET would be an essential transistor for NOR flash. This feature would surpass the demerit of coupling ratio lowering.

FinFET SONOS Flash memory that is believed as a successor of floating gate NAND Flash is shown in Fig. 14. Combining FinFET and SONOS scheme, the issues such as interference between floating gates and worst on-cell current can be eliminated effectively. Fabricated FinFET SONOS cells show very uniform Vth distribution compared to floating gate flash cells that is smaller than 0.7 V. This result is owing to insensitivity of active width variation by widening effective channel width with tall fin structure and no coupling interference nature of SONOS device. Vth distributions of initial, programmed, and erased cells on chip are excellent without abnormal tail bits after program and erase operation. Table 1 shows the comparison of several FinFET SONOS schemes showing the excellence of body-tied FinFET SONOS.

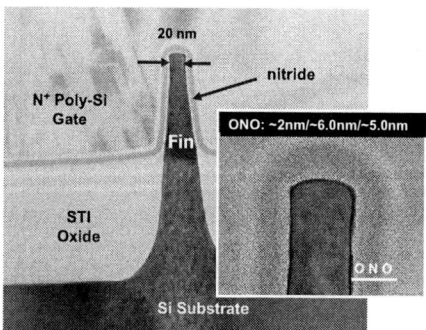

Figure 14. FinFET SONOS Flash cell transistor

TABLE I. Comparison of FinFET SONOS. SONOS FinFET fabricated on Bulk Si shows the best performance among the FinFET SONOS memories.

	IEDM, 2003 [1]	VLSI, 2004 [2]	IEDM, 2004
Structure	**SOI FinFET**	**SOI FinFET**	**Bulk FinFET**
[1] P. Xuan, *et al*, IEDM 2003 [2] M. Specht, *et al*, VLSI 2004			
Gate length	350 nm	30 nm	50 nm
Fin Width	30 nm	20 nm	30 nm
ONO	3nm/6.8nm/4.8nm	3nm/4nm/4.8nm	~2nm/~6nm/~5nm
ΔV_{TH}	~2.0V	~1.0V	>4.5V
Program speed	~2.5V @ 10ms/10V	–	4.1V @1µs/12V
Erase speed	2.3V @ 10ms/-12V	–	4.0V @50µs/-12V
Retention	>1.4V @ 10yrs	>1.1V @ 10yrs	>2.4V @ 10yrs
Endurance	> 10^5	> 10^3	> 10^4

McFET (Multi channel FET)

As an extended technology of FinFET, we successfully developed McFET that has, using conventional CMOS silicon processes, twin fins doubling the fin numbers in the same area as shown in Fig. 15. McFET is fabricated without any patterning limit, while it is hard to have narrower pitch than the design rule due to the patterning limit of lithography tools. In addition, since McFET has the thin fin only at the channel regime without hurting the source and drain contact active area, any parasitic resistance issues do not arise. Using McFET structure, drive current is increased 4~5 times with excellent short channel effect immunity. Fig. 16 shows that both of the n-ch and p-ch MOSFETs with McFET scheme are excellent in the transistor characteristics such as DIBL (Drain Induced Barrier Lowering), SS (Subthreshold Swing), and drive current. In addition, using mid-gap work function TiN metal gate to the thin FinFET, symmetric threshold voltage is obtained in both n-ch and p-ch MOSFETs. Fig. 17 shows that the McFET characteristics are uniform across the 8 inch wafer representing the uniform thin fin formation. We applied the McFET scheme to the 80 nm high performance 6T SRAM cell to successfully obtain the excellent Static Noise Margin of 310 mV and 350 mV at 0.8 V

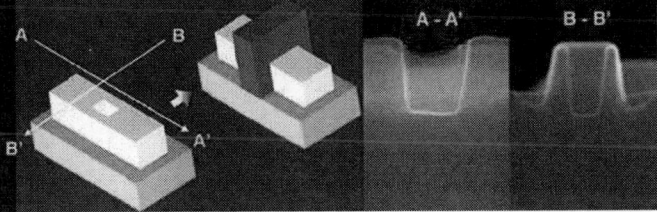

Fig. 15 Schematic view and SEM images of McFET

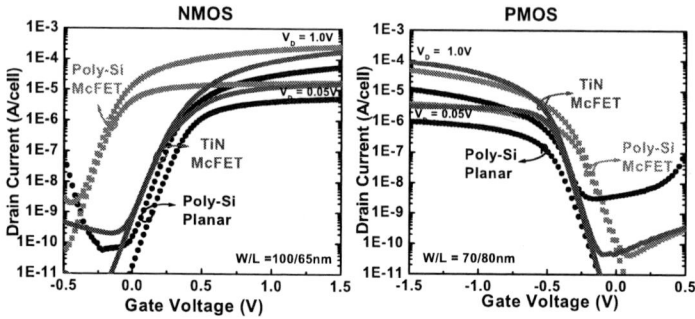

Fig. 16 n-ch and p-ch McFETs show excellent DIBL, subthreshold swing, and large drive current for both N⁺poly-Si and TiN metal gates. Using TiN gate to thin Si-body McFET, symmetric low enough Vtn and Vtp are obtained.

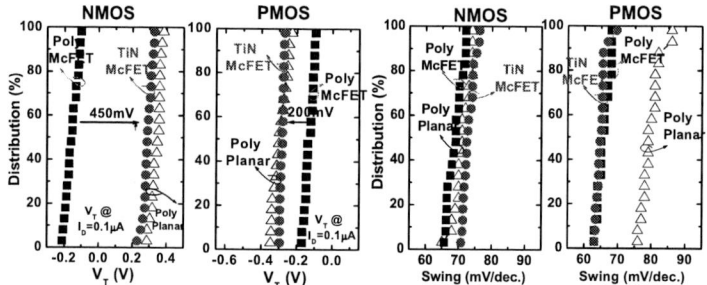

Fig. 17 TiN gate McFET shows optimal threshold voltages for both n-ch and p-ch MOSFETs as well as low and uniform subthreshold swing.

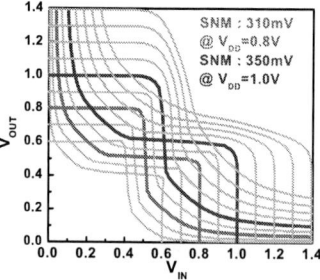

Fig. 18 The operation of SRAM cell inverter made by n-ch and p-ch McFETs show extremely large static noise margin of 310 mV even at Vdd = 0.8 V.

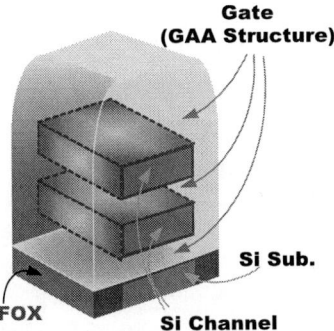

Gate (GAA Structure)

Si Sub.

FOX

Si Channel

Fig. 19 Schematic diagrams of MBCFET. Multiple silicon channels are stacked to enhance the current drivability. Gate is surrounding the channel Si bridges resulting in excellent gate controllability.

and 1.0 V, respectively as shown in Fig. 18. This SNM is superior to the one of planar SRAM cell that is about 170 mV.

MBCFET (Multi-Bridge Channel MOSFET)

In this section, we introduce a novel MBCFET fabricated on bulk silicon substrate. MBCFET have several benefits originated from its vertically stacked surrounding gate structure as shown in Fig. 19. First of all, the current improvement far exceeds the gate capacitance increase due to the mobility enhancement that is resulted from the inversion charge volume inversion and reduced vertical electric field in the channel region. Secondly, it has an excellent immunity to the short-channel effect down to 30 nm gate length with 2 nm thick gate oxide and without any halo implantation, that suppresses off-leakage current and enables the MBCFET to operate at low Vdd and low Vt, resulting in mobility enhancement. Thirdly, its area efficiency is excellent stacking the channels vertically, that makes it possible to increase the driving current without any junction capacitance increment. Lastly, all of its gate lengths are same. It cannot be realized without our newly developed damascene gate process.

Similar to other multi-channel transistors, however, our early developed MBCFET also had the threshold voltage control issue due to the ultra thin silicon bridge thickness causing very low threshold voltage with large off-leakage current. While gate work function engineering is required for CMOS application, to simplify and improve the manufacturability of the dual-metal gate process, we introduce further developed a highly manufacturable CMOS MBCFET process including a simple elevated flat source/drain and single-metal gate process that is shown in Fig. 20. The multi-bridge-channels are completely surrounded by the gate, while typical gate oxide and Si channel thicknesses are 2.2nm and 17nm, respectively. Fig. 21 shows that the MBCFET with two 17nm thick silicon bridges has excellent symmetry of n-ch and p-ch MBCFETs so that CMOSFET is realized with the single metal gate at the thin Si bridge MBCFETs. In addition, the n-ch

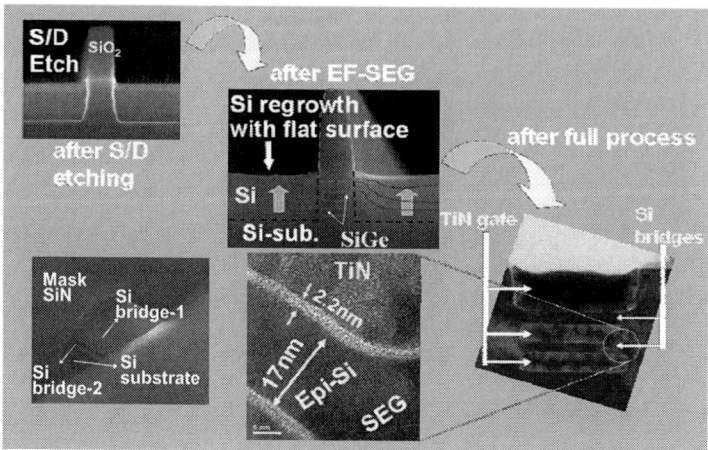

Fig. 20 SEM images of MBCFET. TiN gate is surrounding the two silicon channel bridges. Si channel thickness is 17nm and gate oxide is 2.2nm thick SiO_2.

and p-ch MBCFETs show excellent transistor characteristics of subthreshold swing less than 70 mV/dec and DIBL less than 35 mV/V even at the small gate length of 25 nm and 30nm for n-ch and p-ch MBCFETs, respectively. The drive current was 2.65 mA/um and 1.44 mA/um at $V_{DD} =1.0$ V for n-ch and p-ch MOSFETs, respectively. These are the MOSFET driving currents far exceeding the expectation of ITRS roadmap2003.

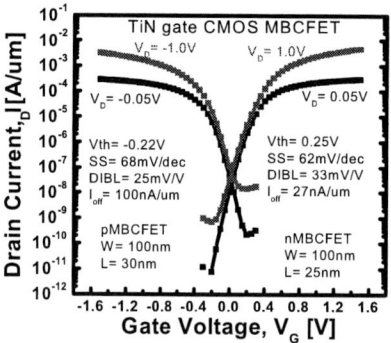

Fig. 21 Symmetric threshold voltage is achieved with mid-gap TiN gate. DIBL and SW are extremely low for both n-ch and p-ch MBCFETs even at the gate length of 25nm.

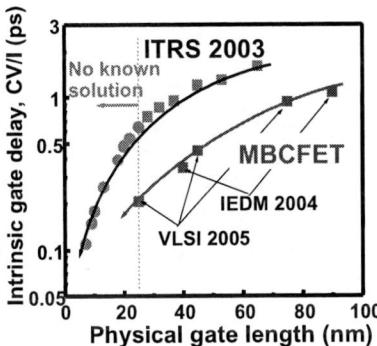

Fig. 22 MBCFET is a breakthrough device to overcome the scaling limit far exceeding the ITRS expectation.

Fig. 22 compares the propagation delay prediction of an inverter to the ITRS roadmap prediction. Even with the gate length of 50 nm, we could meet the goal that is predicted without any known solution to achieve. Moreover, with 25 nm gate length, the requirement of propagation delay for 2013 is achieved with our MBCFET. This result is obtained due to the mobility enhancement effect of surrounding gate and the inversion charge volume inversion effect of ultra-thin Si channel scheme of the MBCFET as shown in Fig. 23.

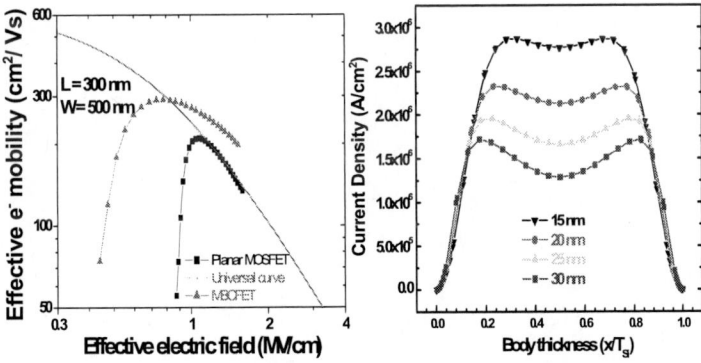

Fig. 23 Using thin silicon bridges and surrounding gate in MBCFET, benefits such as volume inversion and mobility enhancement are realized enhancing the circuit performance.

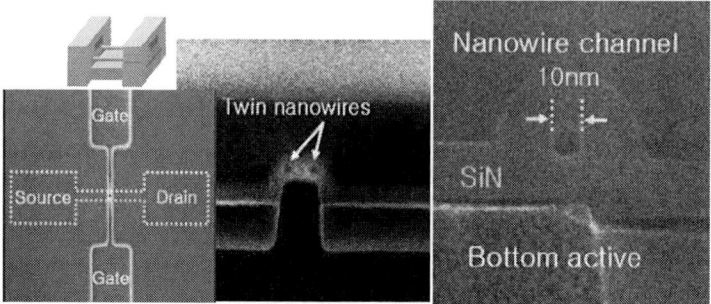

Fig. 24 Top view and cross sectional SEM images of Si nanowire with d=10nm and damascene-gate Lg=30nm.

Twin Silicon Nanowire MOSFET (TSNWFET)

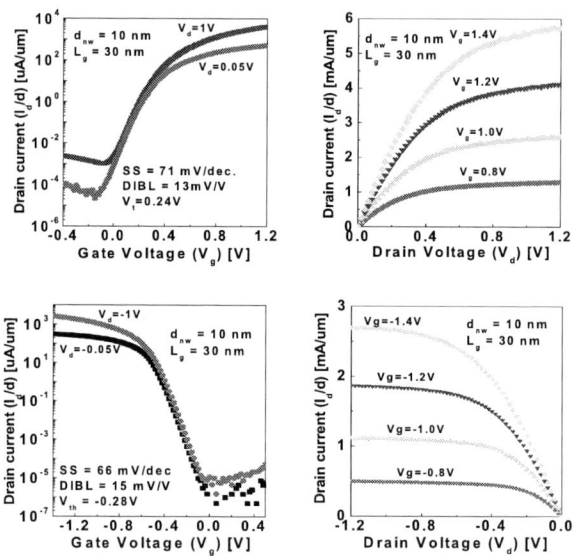

Fig. 22 Excellent SS and DIBL for both n-ch and p-ch TSNWFETs with symmetric Vtn = 0.24 V and Vtp = -0.28 V using mid-gap TiN gate and thin Si nanowire of d = 10 nm.

As transistor scaling approaches to the end of technology roadmap, various new-concept transistors have been reported as a candidate mentioned above. Si nanowire transistor (SNWT) has been focused among those new transistors due to its improved transport property and CMOS process compatibility (25-27). Fig. 24 shows our newly developed high-performance gate-all-around (GAA) twin SNWFET (TSNWFET) fabricated on bulk Si wafer with conventional CMOS processes eliminating the process difficulties and unexpectedly low current drivability of previous works (16).

We could confirm that the gate-all-around (GAA) the silicon nanowire channel fabricated on bulk Si wafer using self-aligned damascene-gate process with TiN metal gate let us achieve the increased transistor drive current as well as excellent subthreshold swing and DIBL with negligible threshold voltage roll-off down to 30nm gate length. This highly manufacturable MOSFETs using silicon nanowire of 5nm radius, namely TSNWFETs show 2.64 mA/μm and 1.11 mA/μm at Vd=Vg=1V, Lg=30nm for n-ch and p-ch, respectively, while the transistor width is counted d=10nm as shown in Fig. 25. These achievements far exceed the ITRS roadmap requirements without any help of emerging research materials.

Summary

We introduce newly developed nanoscale CMOS transistors based on silicon technology to overcome the scaling limits such as area, physics, lithography, etc. For the scaling od planar transistors down to 50 nm, RCAT, PiFET, and Twin SONOS Memory cell transistors are developed. As the scaling of transistors is required further below 50 nm, 3 dimensional structure transistors such as FinFET, McFET, MBCFET, and TSNWFET are newly developed to overcome the physical scaling limits. Using these technologies, with nanotechnology implementation, we believe that the Si based CMOS transistor can be scaled down to 10nm with manufacturability and reliability.

Acknowledgments

Authors would like to thank our research engineers in Device Research Team of Semiconductor R&D Center of Samsung Electronics for their excellent works to realize various future nanoscale CMOS transistors for memory and logic application.

References

1. Panel Discussion, Will Planar CMOS End? When? Why? *Symposium on VLSI Technology* (2005)
2. J.Y. Kim, et al., *Symposium on VLSI Technology*, p. 11 (2003)
3. Y.K. Lee, et al., *Solid State Devices and Materials*, p. 252 (2004)
4. B.Y. Choi, et al., *Symposium on VLSI Technology*, p. 118 (2005)
5. C.W. Oh, et al., *ESSDERC*, p. 233 (2004)
6. K.-H. Yeo, et al., *Symposium on VLSI Technology*, p. 30 (2004)
7. D. Hisamoto, et al., *Trans. on Electron Devices*, **38**, 1419 (1991)
8. B. Yu; et al., *International Electron Devices Meeting*, p. 251 (2002)
9. Y.-K. Choi, et al., *International Electron Devices Meeting*, p177, p. 679 (2003)

10. S.M. Kim, et al., *International Electron Devices Meeting,* p. 639 (2004)
11. S.M. Kim, et al., *Symposium on VLSI Technology*, p. 196 (2005)
12. S.-Y. Lee, et al., *Trans. on Nanotechnology*, Vol.2, p. 253 (2003)
13. S.-Y. Lee, et al., *Symposium on VLSI Technology*, p. 200 (2004)
14. E.J. Yoon, et al., *International Electron Devices Meeting*, p. 627 (2004)
15. S.-Y. Lee, et al., *Symposium on VLSI Technology*, p. 154 (2005)
16. S.D. Suk, et al., *International Electron Devices Meeting*, p. 735 (2005)
17. C.W. Oh, et al., *International Electron Devices Meeting*, p. 893 (2004)
18. C.H. Lee, et al., *Symposium on VLSI Technology*, p. 130 (2004)
19. C. Lee, et al., *International Electron Devices Meeting*, p. 61 (2004)
20. S.K. Sung, et al., *Silicon Nanoelectronics Workshop*, p. 102 (2005)
21. T.Y. Kim, et al., *Silicon Nanoelectronics Workshop*, p. 98 (2005)
22. E.S. Cho, et al., *ESSDERC*, p. 289 (2004)
23. F.-L. Yang, et al., *International Electron Devices Meeting*, p. 255 (2002)
24. B. Doyle, et al., *Symposium on VLSI Technology*, p. 133 (2003)
25. F.-L. Yang, *et al., Symp. on VLSI Tech.,* p.196 (2004)
26. Y. Cui, *et al., Nano Letter*, **3**, 149 (2003)
27. J. Wang, *et al., International Electron Devices Meeting*, p. 695 (2003)

ECS Transactions, 2 (2) 27-31 (2006)
10.1149/1.2195646, copyright The Electrochemical Society

Front-End Process Technology for hp65/45 CMOS Devices

Yasuo Nara, Fumio Ootsuka, and Kunio Nakamura

Semiconductor Leading Edge Technologies, Inc. (Selete)
16-1, Onogawa, Tsukuba, Ibaraki 305-8569, Japan

In order to obtain high performance CMOS devices, introduction of new technologies into the front-end fabrication process are required and therefore so-called "technology boosters" such as strained channel, metal gate, high-k gate dielectrics, thin body SOI, and multi-gate transistor, are proposed so far. Among these technologies, gate stack technology is common key issue for scaled CMOS devices. In this proceeding, we will discuss on high-k gate dielectrics and metal gate technology, and recent achievements of these technologies are reviewed.

Introduction

In order to realize high-performance, low-power, and low-cost CMOS devices, shrinkage of device size is essential but not sufficient, because the simple scaling is reaching the limit. Therefore, in addition to scaling the device size, introduction of new technologies such as strained channel, metal gate, high-k gate dielectrics, thin body SOI, and multi-gate transistor (FinFET) is extensively studied. Among these technologies, gate stack technology (combination of gate dielectrics and gate electrode) is quite important because scaling of MOS structure is indispensable and common issue for any type of MOSFETs. In this proceeding, key issues and achievements of gate stack technology are addressed.

For gate dielectrics, its EOT (equivalent oxide thickness) should be scaled to 0.9-1.6nm for hp65 node and 0.7-1.3nm for hp45 node (target EOT depends on the device type; high performance or power) [1]. However, thickness scaling of conventional SiO_2-based dielectrics is facing a severe power consumption problem due to large gate leakage current by direct tunneling. Therefore, realization of high-k gate dielectrics is strongly required. As a result of recent intensive investigation of high-k gate dielectrics, Hf-based oxide is thought to be the most promising material for high-k gate dielectrics for hp65. Application of this particular material to the further scaled devices, such as the hp45 node and beyond, is the most straightforward from the fabrication point of view.

Metal gate technology is required for eliminating issues associated with conventional poly-Si gate such as gate depletion, boron penetration, and Fermi-level-pinning [2,3]. Fully silicided (FUSI) gate technology is one of the promising methods as conventional integration scheme is easily applied [4,5]. But further investigation is still needed for suppressing the Fermi-level-pinning effect and obtaining wide range tuning of work function of gate material. Dual metal gate technology, in which the work functions of NMOS and PMOS are independently determined, is ultimately required [6,7].

27

High-k gate dielectrics for hp65/45 and beyond

HfSiON film has superior characteristics as a high-k gate dielectric such as high thermal stability, high carrier mobility, and tolerance to boron penetration. Therefore, HfSiON is thought to be the most promising material for hp65 node [8-11]. Nitrogen atoms incorporated into HfSiO film effectively suppress the crystallization, phase separation, and boron penetration after high temperature source/drain activation process. High quality interfacial SiO_2 layer is also important for good electrical characteristics.

Poly-Si/HfSiON transistors are integrated for hp65 node low stand-by power devices with EOT of 1.2nm as shown in Fig.1 [11]. Symmetrical Vth values (+/- 0.2V) have been obtained at gate length of 50nm for NMOS and PMOS transistors (Fig.2.). This was accomplished by optimizing the channel impurity profile using low thermal budget process with flash lamp annealing and low temperature RTA. On current (Ion) of 350uA/um and 150uA/um for NMOS and PMOS was respectively obtained with low off-current (Ioff) of 20pA/um at 1.1V supply voltage. Operation of SRAM with 1.11um^2 cell area has been successfully demonstrated (Fig.3). Lower PMOS Vth was essential to obtain higher bit yield as shown in Fig.4.

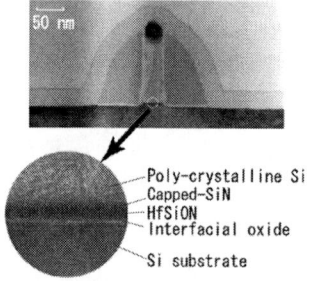

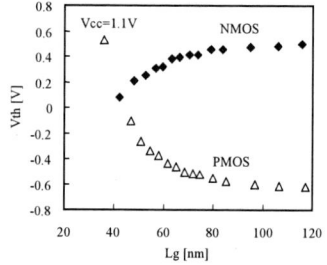

Fig.1 Cross sectional TEM image of Poly-Si/capped-SiN/HfSiON gate stack.

Fig.2 Gate length dependence of threshold voltage for Poly-Si/capped-SiN/HfSiON n- and p-MOSFETs.

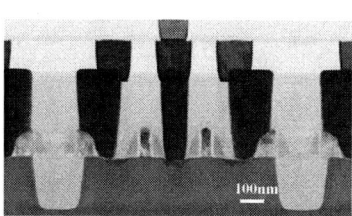

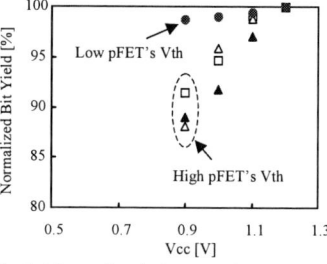

Fig.3 Cross sectional TEM image of SRAM cell array. Cross sectional image of p-MOSFETs is shown in this figure.

Fig.4 Normalized SRAM bit-yield vs. supply voltage.

Further EOT reduction to less than 0.9nm with HfSiON/interfacial layer structure has been achieved by carefully designing the HfSiON formation process. After interfacial layer and HfSiO deposition process, there are some thermal and nitridation process, such as post-deposition-annealing, plasma nitridation, and post-nitridation-annealing. It is found that these post-HfSiO deposition processes are critical factors for the thickness and film quality of thin interfacial layer [12,13]. For example, during the plasma nitridation of HfSiO film, oxygen atoms in the Si-O or Hf-O bonds in the HfSiO film were recoiled by nitrogen atoms. Some of the recoiled oxygen atoms diffused out from the HfSiON/interfacial layer, leading to a decrease in the thickness of the structure. Other recoiled oxygen atoms diffused to the Si substrate, resulting in the growth of the interfacial layer, or remained in HfSiON film as interstitial oxygen, also resulting in interfacial layer growth during post-nitridation-annealing. By combining the optimized process flow of plasma nitridation and high temperature post-nitridation-annealing at 1050°C, an EOT of 0.81nm with a leakage current density of $0.74A/cm^2$ at $Vg = 0.7V$ has been successfully achieved as shown in Fig.5. Figure 6 shows the carrier mobilities for Poly-Si gate/HfSiON dielectrics gate stack with EOT of 0.86nm. High carrier mobility (almost 90% of SiO_2 gate dielectrics) was obtained. These values satisfy the specification for hp45 low-operating-power device and HfSiON will be possibly applied to further scaled technology node.

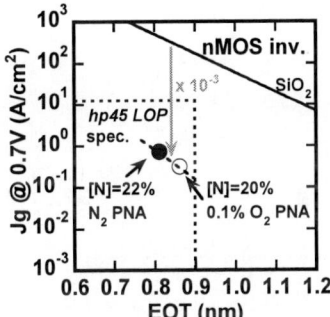

Fig.5 Leakage current density (Jg) vs. EOT relationship for HfSiON gate dielectrics. Two different fabrication conditions (22% nitrogen concentration with nitrogen post-nitridation-annealing (PNA) and 20% nitrogen concentration with slightly O_2 added PNA) are shown. hp45 LOP spec. was satisfied for both conditions.

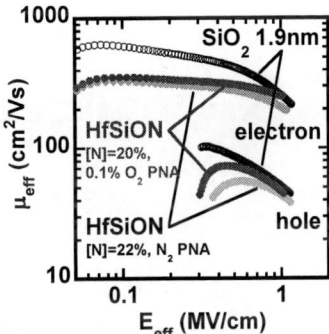

Fig.6 Effective carrier mobilities for Poly-Si/HfSiON gate stack with EOT of 0.86nm.

Thermally stable TaSi/HfSiON gate stack

TaSi metal gate technology has been applied to ultra thin HfSiON gate dielectrics and thermal stability of TaSi/HfSiON gate stack has been investigated [13]. After 1000C spike annealing, EOT of 0.9nm was obtained. EOT increase during annealing was as low as 0.05nm. Figure 7 shows effective electron mobility of TaSi/HfSiON gate stack. Electron mobility at 0.8MV/cm was 264cm^2/Vs, which is approximately 90% of SiO$_2$ gate dielectrics. Figure 8 shows the n-channel MOSFETs characteristics with TaSi metal gate and HfSiON gate dielectrics. Similar to p+ poly-Si/HfSiON, flat band voltage was negatively shifted and, as a result, Vth close to n+ poly-Si/HfSiON (0.28V @gate length=1um) was obtained for the MOSFETs. Thanks to high electron mobility and elimination of gate depletion, 30% higher on-current compared with poly-Si/HfSiON was obtained as shown in Fig.8.

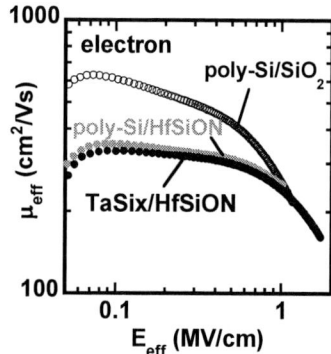

Fig.7 Effective electron mobility of TaSi/HfSiON n-MOSFETs.

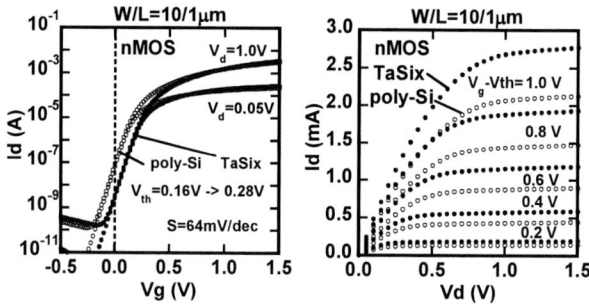

Fig.8 Transistor characteristics of TaSi/HfSiON n-MOSFETs compared with poly-Si/HfSiON n-MOSFETs. Due to high electron mobility and elimination of gate depletion, 30% higher on-current was obtained with TaSi MOSFETs.

Conclusions

As a key front-end process technology for scaled CMOS, gate stack with high-k gate dielectrics and metal gate has been reviewed. HfSiON gate dielectrics will be promising for hp65 node and also be a potential material for hp45 and beyond. Thermally stable TaSi/HfSiON gate stack suitable for gate-first integration of NMOS was demonstrated. However, stability of EOT and effective work function of dual metal gate technology are major remaining issues and further extensive studies are needed.

Acknowledgement

The authors would like to thank all the present and former members of front-end program of Selete for their cooperation and contribution.

References

[1] The International Technology Roadmap for Semiconductors, 2004 Update.
[2] C. Hobbs et. al., VLSI Tech. Symposium, p.9, 2003.
[3] K. Shiraishi et. al., VLSI Tech. Symposium, p.108, 2004.
[4] B. Tavel et. al., Tech. Digest of IEDM, p.825, 2001.
[5] A. Lauwers et. al., Tech. Digest of IEDM, p.661, 2005.
[6] S. B. Samavedam, et al., Tech. Digest of IEDM, p.433, 2002.
[7] Z. B. Zhang et. al., VLSI Tech. Symposium, p.50, 2005.
[8] S. Inumiya et al., VLSI Tech. Symposium, p.18, 2003.
[9] K. Sekine et al., Tech. Dig. IEDM, p.103, 2003.
[10] Y. Tamura et al., VLSI Tech. Symposium, p.210, 2004.
[11] A. Mineji et al., Tech. Dig. IEDM, p.927, 2004.
[12] S. Inumiya et. al., Ext. Abst. of SSDM, p.10, 2005.
[13] S. Inumiya et. al., Tech. Dig. IEDM, p.27, 2005.

32

Trends in European R&D - Advanced Process Control Down to Atomic Scale for Micro- and Nanotechnologies

L. Pfitzner, M. Schellenberger, R. Oechsner, G. Roeder, M. Pfeffer

Fraunhofer Institute Integrated Systems and Device Technology, Schottkystrasse 10, 91058 Erlangen, Germany

Corresponding author contact: lothar.pfitzner@iisb.fraunhofer.de, +49 9131 761 110

Manufacturing of semiconductor devices continues to follow Moore´s Law. Key challenge of future manufacturing of micro- and nanotechnologies will be a much tighter process control. This paper will describe some trends in advanced process control (APC). Definitions of APC, current status of basic principles, elements and future requirements will be given. Three methods of advanced process control are presented, which have to be adapted to further challenges in micro- and nanotechnologies: integrated metrology, run-to-run control (feed-forward / feedback control) and fault detection and classification (FDC). Integrated process and metrology tools have to provide relevant data. The data may be tool data, process data and/or wafer data. An engine analyzes the data for SPC, application of run-to-run control loops and fault detection in order to initiate necessary control actions. Examples from 20 years of own R&D embedded in European activities related to these focal fields will be given.

Introduction

The business model of today's micro- and nano-electronics is relying on shrinking pattern dimensions, on increasing complexity resulting in increasing functionality, on improvements in electrical performance, all combined with lower manufacturing cost per bit or per transistor. Gordon Moore stated in 1965 already (1): "... the complexity for minimum component costs has increased at a rate of roughly a factor two per year. Certainly over the short term this rate can be expected to continue, if not to increase. Over the longer term, the rate of increase is a bit more uncertain, although there is no reason to believe it will not remain nearly constant ..."

Moore's Law still continues today, as can be depicted from the International Technology Roadmap for Semiconductors (ITRS) (2). Major contributions to the decrease of costs in the manufacturing of integrated circuits are (3),(4),(5):

- Feature size: The reduction of critical dimensions achieved by continuous improvements in lithography and structuring of the patterns typically contributed in the past with approximately 12% to 14% to the historical level of cost reduction of 25% to 30% per year.
- Wafer size: Technology improvements in crystal growth have allowed an increase in wafer size. Today, leading edge production of integrated circuits can deploy 300 mm wafer diameters The discussion and standardization procedures of a further increase towards 450mm wafer diameter have just started. The estimated

contribution per year to the overall cost reduction was considered to be 3% to 5%, however, dropping to a smaller percentage in the future due to the heavy increase in fab costs.

- Yield improvement: A contribution of typically 2% was achieved in the past by improving the yield. Today, typically production yields are 90 % and even more in a mature production. Thus, yield enhancement will have minor impact on the reduction of costs. However, huge efforts will have to be necessary to maintain such high yields, when shrinking geometries and higher complexity require significant increases in processing quality.

- Equipment productivity: The most important part of further cost reduction of the production of integrated circuits has to be achieved by much higher equipment productivity. Today's level of mechanical automation is high. Major contributions to an increase in equipment productivity beyond the current level of less than 8% to a level of 10% to 15% will have to involve improvements in equipment control and in process control, the latter one on the level of the individual equipment level as well as on the fab level. Central elements of these improvements are known as "Advanced Process Control" and will be described in details of the principle and of several applications in this paper.

Necessary research and development has been recognized and encouraged within several research projects funded by the European Commission. Combining the strength of European metrology activities, they allowed advanced equipment developments especially in optical technologies, and early introduction of 300mm manufacturing technology in leading edge production of integrated circuits.

Following a short introduction into basic elements and principles of advanced process control, some examples and results are given.

Basic Elements and Principles of Advanced Process Control

Definitions

A uniform definition of APC and related terms is still missing. Many companies have their individual understanding and terms of advanced process control: with some definitions, APC comprises just the methodologies of fault detection, fault classification and run-to-run control, while explicitly excluding "standard methods" such as statistical process control. Other definitions combine all process control and process monitoring capabilities under the term APC, but metrology and sensors are left out, even though they form the basis of information for control routines.

The terminology of APC provided here includes metrology and sensors as well as statistical process control and data management as building blocks. This is a proven approach which allows for a consistent and holistic view on the complex control and monitoring loops that are to be established when utilizing advanced process control.

Advanced Process Control (APC)

Advanced process control is the manufacturing discipline for applying control strategies and/or employing analysis and computation mechanisms to recommend optimized machine settings and detect faults and determine their cause (SEMI E133). The final goal of APC is to initiate control actions in case of process deviations or faults in order to compensate the drifts or at least to stop further misprocessing. Thus, appropriate analysis and computation mechanisms have to be implemented. These mechanisms rely on quality data from the tools, processes and wafers. These data may come from process tools directly or from additional or integrated metrology and sensors. Data as well as control actions are distributed via a well defined communication infrastructure.

Figure 1 illustrates the basic elements of APC which will be defined and explained further on.

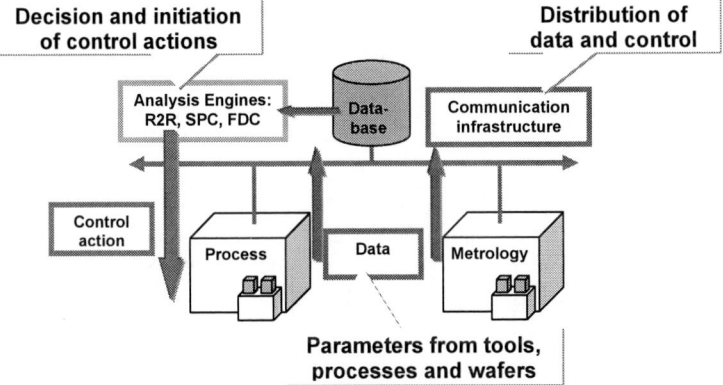

Figure 1: Basic elements of APC

APC Analysis Engines

Analysis engines are the most important part of an APC infrastructure. Based on the data derived from the processes, tools and wafers, decisions are taken and control actions are initiated. The most common analysis engines are depicted in
Figure 2.

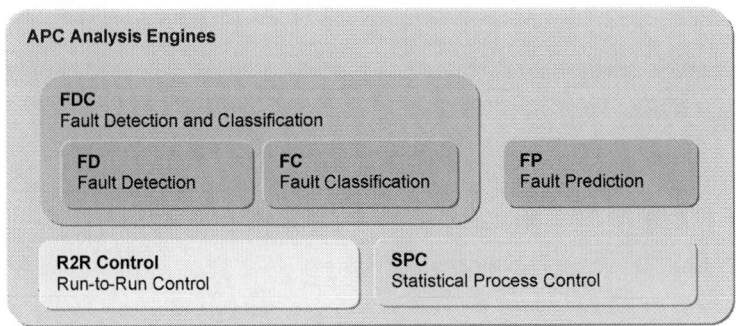

Figure 2: Common APC analysis engines

Fault Detection and Classification (FDC) actually covers two functional APC engines: Fault Detection (FD) comprises methods to detect anomalies and faults. This is done by monitoring specific data coming from a tool or a process, which have been identified in advance to be correlated to an expected fault. After the detection of a fault, its cause is determined by a Fault Classification engine (FC).

Another APC engine handling faults is Fault Prediction (FP): again, tool data or process data are being analyzed for variations or specific patterns, yet not to detect a fault that already occurred, but to predict anomalies and faults in advance.

Run-to-Run Control is the technique of modifying recipe parameters or the selection of control parameters between runs to improve processing performance. A "run" can be a batch, lot, or an individual wafer (SEMI E133).

Statistical Process Control is the technique of using statistical methods to analyze process or product metrics to take appropriate actions to achieve and maintain a state of statistical control and continuously improve the process capability (SEMI E133).

Metrology

As described before, all control actions are based on data about process equipment, processes and wafers. Data is collected by the process tools directly, by stand-alone metrology tools or integrated metrology systems and sensors.

Dependent on the location of a measurement system or sensor, three main groups of metrology equipment can be defined:

- Offline Metrology: The metrology tool is located apart from the process equipment. Those metrology tools are often referred to as stand-alone metrology tools. Currently, this is the most frequent case of using metrology.
- Inline Metrology: In this case, the metrology tool is combined with the process equipment. This can be achieved in various ways, e.g. by attaching a sensor directly to the process tool, by using a dedicated metrology module in a cluster system, or by mounting the metrology system on a FOUP adapter.
- *In situ* metrology: The metrology tool or sensor is directly integrated into the process chamber of process equipment. This is the most complex approach, as the process chamber and the metrology system have to be matched carefully in a way that the necessary data can be obtained without interfering with the process.

The latter two groups are commonly referred to as "integrated metrology", as opposed to "stand-alone metrology" covered by the first group.

Integrated Metrology as Basis for APC Control Actions

In order to achieve a short response time in case of an error or a mis-processed wafer, it is important to get relevant data from the equipment, the process or from the wafer as soon as possible (6). This implies that measurement methods have to be applied as near as possible to the process itself in order to reduce the time between processes, measurement and corrective actions (figure 3).

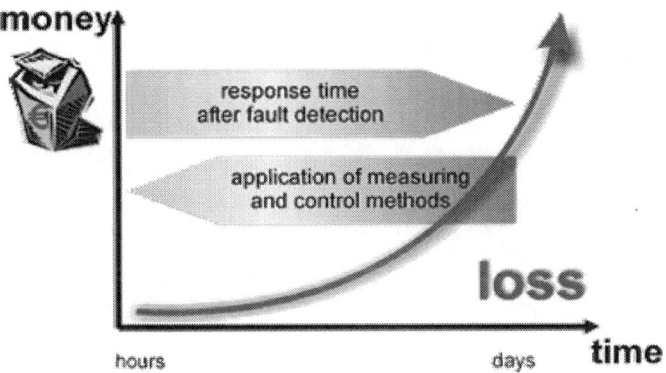

Figure 3: The necessity for a short response time

Looking again at the three main groups of metrology defined above, they can be classified according to their ability in providing a short interval between a process and the measurement:

- Offline Metrology: In this case, a long time elapses between process and measurement. Normally, a batch has to be processed completely before being sent to a metrology tool. In the worst case, this whole batch may be lost in case of systematic errors.
- Inline Metrology: This methodology enables a measurement immediately before and after processing. According to the boundary conditions, even every single wafer may be measured. This allows for a quick reaction on errors or deviations that may arise during processing.
- *In situ* metrology: In this case, the measurement is accomplished during processing. By definition, every single wafer is measured. This kind of metrology even allows for real-time control actions.

Run-to-Run Control

Run-to-run control is one of the most important APC methods. In general, it means that recipes or control parameters are automatically modified between runs to improve processing performance. The goal of the automated control is to reduce variations in the process outputs. Adjustments of process and recipe parameters have always to be made by process engineers if shifts or drifts occur. The emphasis is on automated control actions. A run-to-run control engine is analyzing available metrology data and automatically starts control activities based on well-defined boundary conditions, control models and algorithms if required. A "run" may be a batch, a lot, or an individual wafer.

The heart of a run-to-run application is the run-to-run controller. It should be able to compensate various disturbances, such as process drifts, process shifts (step disturbances) due to maintenance or other factors like model or sensor errors, etc. Moreover, it should be able to deal with limitations, bounds (upper and lower limits), cost requirements, multiple targets and time delays that are often encountered in real processes.

Run-to-run control is a model-based process control. The adjustment of the process parameters requires a process model which describes the influence or correlation between the input parameter and the output results of the process. Based on that model, the controller is able to calculate required corrections of recipe parameters to reduce deviations from the target value. The used model may be a physical model, a semi-physical model or an empirical model. The control algorithm comprises the collection and evaluation of measurement data from the metrology tool or a data base, decision upon control actions and the provision of changed recipe parameters for the process tool, if required.

Run-to-run control consists of feed-forward and feedback control. This means you can either control the process, subsequently following the metrology step by using the measurement results for controlling the next process step (feed forward control), or you can feed back the results by controlling the process step before (feedback control), see also figure 4. Often you use both methods at the same time – feed-forward and feedback. A further modification is that you can feed the information to more than one subsequent and preceding process step. Sometimes, a run-to-run control loop involves more than one processing equipment, metrology tool and sensor.

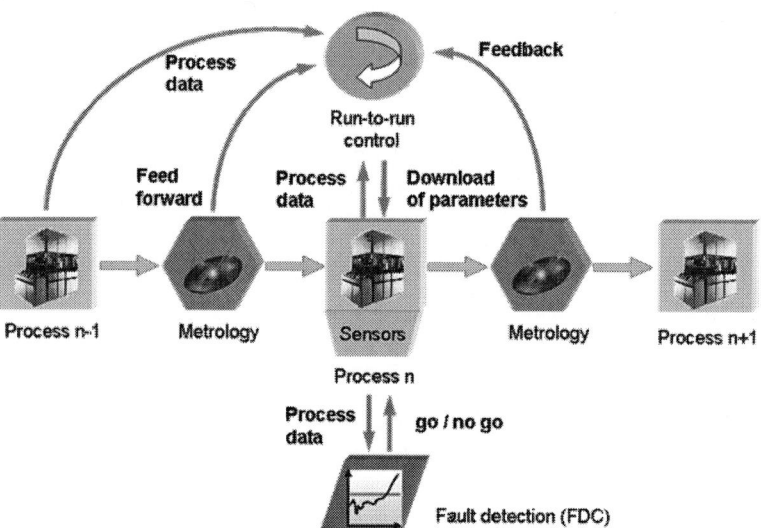

Figure 4: Application of APC capabilities: The process flow is indicated from left to right. Feedback and feed-forward control uses metrology data from preceding or subsequent measurements for recipe parameter adjustment. FDC analyzes process and equipment data e.g. for making a 'go' or 'no go' decision.

Typically, the results of wafer processing at one equipment will be passed on to a subsequent manufacturing step (feed-forward) and used to influence the future processing of that same wafer. Another application is to collect processing data at a processing step and pass the data back (feedback) to the previous processing step to improve processing of other wafers. The whole information gained, which means process as well as metrology data, is used by a run-to-run controller to adjust the recipe or control parameter of the previous or subsequent processing step. Run-to-run control is performed according to process models as mentioned above. A single application can easily accommodate multiple feed-forward and feedback loops.

The basic set-up of run-to-run control and also the mechanisms of feed-forward and feedback can be found in the picture above (Fig. 4).

Fault Detection and Classification

Fault detection and classification (FDC) is also one component of APC. Fault detection and classification is a model-based detection and classification of equipment and/or process problems. It provides measures to decide if a tool operates in an acceptable region. It may deliver criteria for "go" or "no go" decisions. "Go" means the wafer, lot or batch can be processed further, "no go" means the equipment conditions or the resulted process does not fulfill the specifications and can not be accepted. Rework is required or the wafer, lot or batch is scrap. Sometimes, FDC is also correlated with tool health.

Figure 4 also shows how APC components may be combined with run-to-run control applications and the integration of sensors and metrology, and how they interact to provide both run-to-run control and fault detection and classification capabilities.

The data is collected from processing equipment or from sensors or metrology tools and analyzed using an idealized mathematical, semi-physical or empirical process model. The results of this analysis can be used to detect when an equipment fault has occurred (or is likely to occur) and to determine the type of the fault. The type of a fault is correlated to its cause. It is important to detect faults at the moment when they occur, classify them and initiate appropriate correction activities. Figure 3 gives an idea why it is necessary to detect faults as soon as possible. It shows the schematic correlation between the time needed for the detection of failures (response time), and the loss of money due to the increase of wafer value and wasted additional process steps, if scrap is produced. By detection of failures at the tool level (based on the integration of sensors and metrology), the response time to detect faults can be drastically reduced.

Results of fault detection and classification can also be used for fault prediction. Variations in equipment and process data are being monitored and analyzed for prediction of anomalies in following process steps. This is a look ahead to detect tendencies and avoid detectable faults before they occur.

The methods for data analysis are very different. There are univariate as well as multivariate analysis methods needed. In case of a univariate analysis, only one quality characteristic is controlled, and in case of a multivariate analysis, multiple quality characteristics are analyzed and checked. The application of techniques like principle component analyses (PCA) is often used for fault classification. Even if multivariate techniques are used to link faults to multiple variables, univariate analysis capabilities can be very successful. Univariate methods bring the advantage that the application is much easier understandable and applicable.

FDC can be used with yield analysis tools at factory level. Yield management systems can use and coordinate results from FDC to identify causes of yield loss. Specific software tools are used to link yield data to equipment data and process data, and to investigate correlations between input and output parameters. Yield data, FDC data, and logistic data as well as data from infrastructure, maintenance and supply chain can be combined for data mining.

Examples

Control of Layer Thickness in Furnaces

Especially in high-temperature processing of thin films or film stacks (e.g. oxide-nitride-oxide stacks), ellipsometric *in situ* layer thickness control helps to shorten the time for process ramp-up and to efficiently improve the process stability. An implementation of such an *in situ* ellipsometry in production furnaces has been carried out (7). The major goal of this work was the examination of the possibilities for the use of *in situ* spectroscopic ellipsometry in high-temperature processing steps in semiconductor manufacturing. For *in situ* measurements at high temperatures, an exact knowledge of the spectral refractive indices of the measured substrate and layer materials is of crucial importance. Since there is still a lack of optical data as a function of temperature reported in literature, these optical data of the relevant materials were determined at high temperatures to make *in situ* measurement possible. Furthermore, by integrating this technology into a furnace, it has been demonstrated that these modifications of the equipment do not affect the processing results and that the measurement data are accurate enough to enable *in situ* process control and end-point detection.

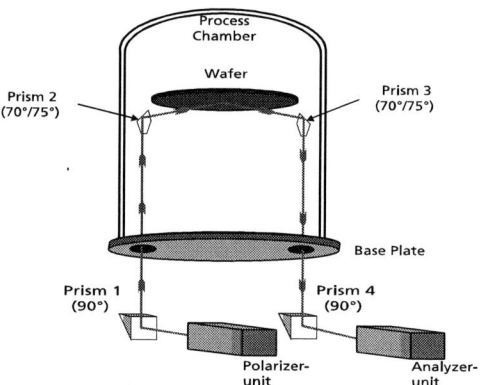

Figure 5: Integration of a spectroscopic ellipsometer in a vertical furnace system

The integration of the spectroscopic ellipsometer was successfully carried out in a vertical furnace for chemical vapor deposition (CVD) and thermal oxidation. Modifications of the furnace and furnace geometry were restricted as far as possible, to proof the rapid integration into an existing industrial equipment at minor costs. In figure 5 the ellipsometer arrangement is shown. The two ellipsometer heads are placed side by side and are mechanically coupled to the base plate of the furnace. Base plate and wafer carrier form a mechanical unit which moves vertically with the boat loader. The light beam of the ellipsometer is guided through the base plate into the furnace tube and directed onto the wafer by quartz glass prisms operating in total internal reflection mode, as can be depicted from figure 6. This arrangement introduces a well-defined additional phase shift in the polarization state of the light, which can be calculated and subtracted from the measured phase shift. As shown in figure 5, the ellipsometer beam must be deflected four times in the ellipsometer arrangement described. The measurements performed showed no significant negative influence from the additional reflections to the measurement signal.

Figure 6:   The figure shows the location of the two 70° prisms in the furnace near the boat (some wafers in the boat removed so that both prisms are visible)

The system was used in a first step to determine the optical reference data for crystalline silicon, silicon oxide and silicon nitride as a function of temperature. These data were implemented in the refractive index library of the *in situ* ellipsometer. There they can be used for end-point detection and monitoring of layer composition which was done in the second step. The measured refractive indices of crystalline silicon for different temperatures are shown in figure 7.

For end-point detection and *in situ* layer thickness control, the ellipsometer software has been adapted to support measurements with the additional optical components (prisms). The measurement modules of the software were extended to provide the necessary correction during the measurements itself. Automation was also necessary to achieve a remotely controlled ellipsometer operation by the furnace controller.

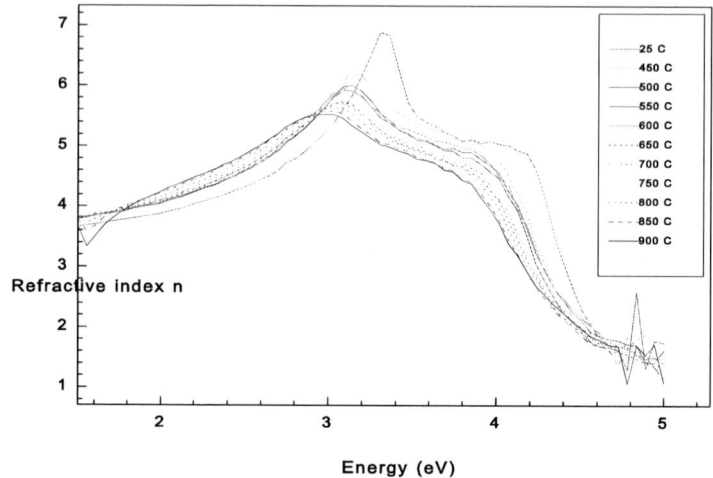

Figure 7: Measured refractive indices of crystalline silicon for different temperatures

The extension of the refractive index library of the *in situ* ellipsometer with the high-temperature data enabled *in situ* monitoring of high-temperature processes by spectroscopic ellipsometry for the first time. Examples of *in situ* ellipsometry applications were thermal oxidation (e.g. gate oxidation), end-point detection of the processing, real-time control of integrated multilayer processing and BPSG control (boron phosphorus silicon glass).

Figure 8 gives an example of a dry thermal oxidation process at 900°C. The process temperature, as measured by the bottom profile thermocouple, and the oxygen flow during the process are also shown in the figure. The oxide thickness data were calculated by the *in situ* spectroscopic ellipsometer with the refractive index data for 900°C. The decrease of oxide thickness during the temperature ramp-down from 900°C to 600°C, which is shown in the oxide growth curve, is an artifact. It is caused by still using the 900°C refractive index data during the ramp-down and withdrawal steps for the thickness calculation. This demonstrates the importance of knowing the accurate optical data as a function of the temperature for *in situ* process monitoring.

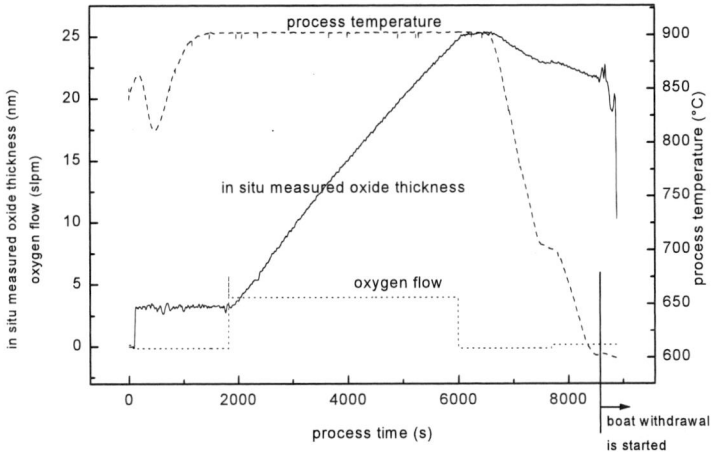

Figure 8: Example of a 25 nm dry thermal oxidation process at 900°C.

Here, for the first time, a layer thickness sensor has been successfully integrated into a vertical furnace, resulting in increased process stability, rapid process optimization and significant cost reduction. The arrangement of the *in situ* spectroscopic ellipsometer enabled post process measurements on selected wafers of the batch during the unloading sequence. *In situ* as well as post process data were used by the furnace controller for immediate and automated correction of parameter settings.

Particle Control in Furnaces

Yield-endangering particle contamination is one of the challenges modern semiconductor manufacturing is facing, and is more critical with each new technology node. *In situ* monitoring of particles (ISPM) generated in vacuum or liquid processes is well known in semiconductor manufacturing. Particles from coatings may flake from wafer edges, wafer surfaces, or the boat during loading or unloading wafers in process furnaces. This can particularly be caused by insufficiently adjusted handling components. Since this is frequently not detected until completion of processing, particle contamination may already have spread into the furnace and onto other process wafers. A cost-effective particle detection system which is capable of detecting particle contamination in the furnace handling area as soon as it arises, was developed (8). The concept is based on the detection of generated particles carried by the airflow inside the loading station. The system allows the detection of particles generated by scratching wafers at the boat due to misalignment of the handling system, the detection of particles generated by bursting off the boat due to increased layer thickness on the quartz ware, the optimization of quartz ware recycling, the immediate correction of equipment faults, the avoidance of yield-endangering further processing and finally the reduction of equipment downtime.

The generation of particles due to misaligned handling systems and their transport inside the loading station was investigated in a setup simulating the conditions inside the loading station of a vertical production furnace. Inside the flow tunnel misalignments between a boat model and a wafer handler were performed. The particle contamination of wafer placement with a well-aligned handling system is negligible. In contrast, the number of particles generated by scratching wafers with different layer materials, commonly used in semiconductor manufacturing, is significant. A substantial part of these particles may be found on the surface of the processed product wafers.

Due to the rotating boat handler of the furnace (see figure 9), particle detection directly at the quartz bar of the boat is impossible due to a potential collision between the rotating boat and the particle sensor in case of a system failure. It was decided to attach the sensor to the leeward wall. The particle detection probability in such a distance of the particle generation is small. But the class 1 or better clean room conditions inside the loading station and the large number of generated particles make the monitoring of the handling system possible. Based on these investigations, a *HandMon*-ISPM prototype for the LPCVD nitride vertical furnace in a 300 mm pilot production was designed and integrated.

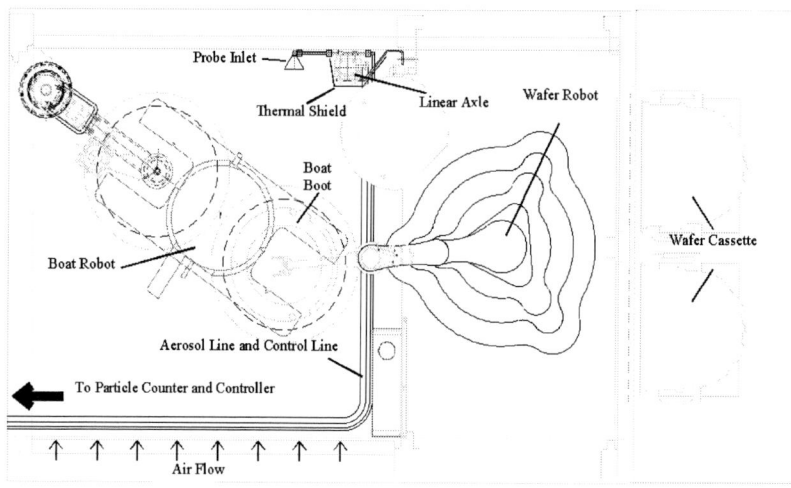

Figure 9: Prototype of an ISPM system (Handling Monitoring - *In Situ* Particle Monitoring system)

A vertical linear axle moves the probe inlet relatively to the slot currently loaded or unloaded by the wafer robot, where the actual slot is estimated by intercepting the communication between the furnace and the robot controller. The aerosol is provided to a particle counter, commonly used for clean room monitoring. To ensure safe operation, several safety features are implemented into the *in situ* particle monitoring system. It is protected against temperature excursion by means of monitoring the boat rotation, boat and aerosol temperature. If necessary, the aerosol line can be locked by a shut-off valve and the flexible tubing inside the loading station can be hidden behind the thermal shield, which is protecting the axle. In case of a power failure or disconnection of signal lines, the system sets itself into an idle state to guarantee safe operation of the furnace. The system controller is connected to the fab network and the ISPM data are sent to the fab wide APC-software for further analysis.

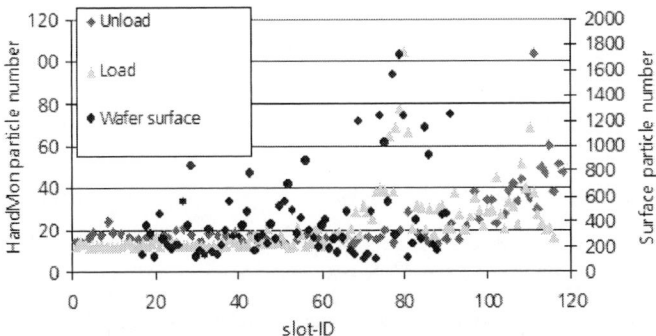

Figure 10: Particle detection by the *HandMon*-ISPM through misalignment of the wafer handling system

After integrating *HandMon*-ISPM into a vertical furnace, first zero measurements without wafer handling were performed to estimate the particle contamination inside the minienvironment of the loading station. After analyzing several loading and unloading sequences of the 2 boats, the handling system was characterized (see figure 10). The rising particle counts in the upper part of the boat due to handling activities originate from the specific wafer handler design. Particle test wafers have been processed in the lower, middle and upper part of the boat and a correlation with *HandMon*-data was found. With the help of *HandMon*-ISPM measurement data the alignment of the handling system was optimized.

Up to this point, the *HandMon*-ISPM measurement data as well as the wafer surface particle tests were in an acceptable range. When loading new wafers a particle contamination due to a misaligned boat was recognized by the *HandMon* system. There-fore, all product wafers were measured by a surface scanner. An increased number of particles were found on the wafer surface as well. With the usual particle contamination control strategy this problem might be undetected. By means of the fast fault detection yield-endangering further processing of product wafers can be avoided.

A significant part of equipment downtime is caused by handling and particle problems. Wafer surface particle tests showed another contamination problem. In contrast, the *HandMon* particle counts were in an acceptable range. The problem was traced to the process tube during processing. With the help of the ISPM system the root cause was narrowed down and estimated earlier. Expensive equipment downtime was reduced and a rise in productivity was achieved.

A complete life cycle of a new quartz ware was monitored. The quartz ware of the furnace is changed routinely after reaching a nitride layer thickness of 15 μm on the boats. During a complete life cycle neither the *HandMon* particle counts nor the wafer surface particle tests showed any significant rise with increasing layer thickness. The next step was the expansion of the quartz ware life time, allowing an optimization of the quartz ware recycling cycles with additional economical benefit.

In summary, the benefits from this type of integrated metrology tool included the continuous monitoring of the minienvironment, the analysis of wafer handling character-istics by monitoring loading and unloading sequences, a support of the control of the alignment of the handling system and a fast detection of misaligned handling systems, the prevention of further contamination and yield-endangering processing, the reduction of equipment downtime by narrowing down the root cause of a particle contamination, the optimization of quartz ware recycling cycles by correlation of ISPM system and wafer surface particle data and finally saving of particle test wafers by a reliable ISPM application and stable processes.

Phi-Scatterometry for Integrated Line width Control in DRAM Manufacturing

A cost-effective scatterometry method has been developed, which is suited for integrated line width control and which is a supplement to conventional SEM's. Those CD-SEM's are stand-alone tools with high costs for investment and maintenance. In addition, their throughput is relatively small which means that for an economic production, only few of all processed wafers can be controlled.

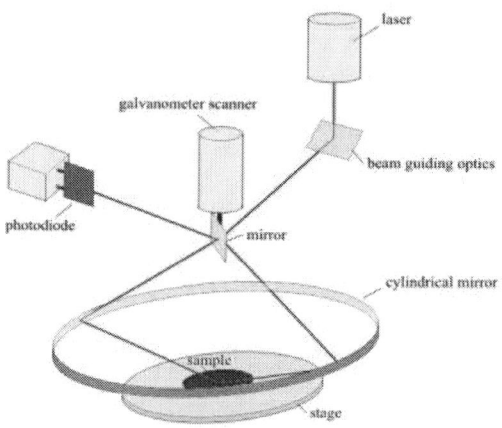

Figure 11: Outline of measurement system

The "phi-scatterometry" procedure described here is carried out directly on periodic functional patterns instead of using additional test structures. In scatterometric applications, the angle of incidence is mainly used as measurement parameter (9). However, it can be verified (10) that the azimuth angle gives similar performance. This permits the construction of a set-up being appropriate for low-cost, integrated CD-control. Different azimuth angles of the measurement beam are realized by a combination of a rotating and a cylindrical mirror, see figure 11 above (11).

The measurement system (12) was designed for easy hook-up integration into production tools. Based on a single-wafer FOUP adapter it can be docked on any 300-mm load port and, therefore, takes use of existing automated and costly wafer-handling systems (figure 12). This allows flexible, rapid and cost effective in-line measurements without downtimes and modifications of productions tools. Easy function tests of metrology and manufacturing equipment are enabled.

Figure12: Open CD-control system with FOUP adapter on manufacturing equipment in production line

There are no long-term simulations of diffraction effects required, as it is the case with other scatterometric approaches. Instead, the measurement results are evaluated by neural networks performing classifications of pattern parameters. Thereby, a fast fault detection and immediate process control is enabled. The measurement principle was verified with DRAM patterns having trenches in two dimensions (see Figure 13).

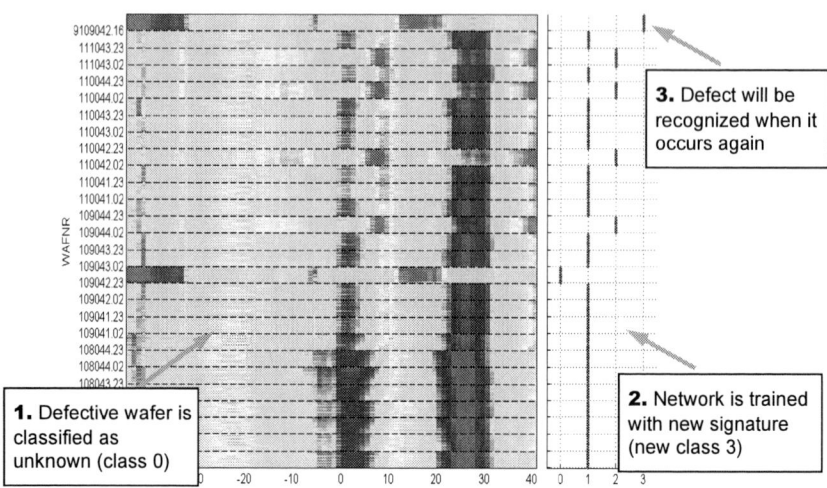

Figure 13: Test of the classification system after training with 6 wafers

Optimization and control of plasma processing

During manufacturing of Integrated Circuits, plasma processes are frequently applied for the deposition and etching of layers. In order to reliably achieve the required process results, the plasma process reactors have to be maintained at a well defined process state. However, either in plasma deposition as well as in plasma etching steps unwanted film formation may occur on the reactor walls. In plasma deposition, film formation should only proceed on the wafer surface but typically unwanted film formation on the reactor wall may not be completely suppressed. In plasma etching, especially in oxide etching, polymer forming gases are added to the etching gases. On the wafer surface polymer film formation is adjusted such that on the sidewall of the etched pattern no lateral etching occurs. On the bottom surface of the pattern, which is exposed to direct ion bombardment, etching rapidly proceeds and, hence, highly directional pattern transfer is achieved. On the reactor walls which are not subject to ion bombardment, typically strong film formation occurs.

Since the unwanted film formation may lead to particle formation and defects on the wafers or may influence the complex plasma process chemistry, plasma cleaning of the reactors is applied periodically to maximize the time between time-consuming wet chamber cleaning. Two APC measures were developed to optimize reactor cleaning for the reduction of non-productive clean times. These were the development, implementation, and test of an advanced endpoint detection method for the cleaning of process chambers in deposition systems and the development of a sensor for the measurement of chamber wall depositions in semiconductor manufacturing equipment.

Advanced endpoint detection

Endpoint detection methods are frequently applied to achieve improved plasma processes for reactor chamber cleaning. However, often the starting conditions for chamber cleaning are not well defined and a large variety of layer materials with different layer thickness may be present on the reactor walls, which results in complex cleaning processes with vastly different clean durations. For the optimization of these cleaning processes, novel methods for endpoint detection have to be applied which enable precise endpoint determination in spite of the starting conditions.

A novel advanced endpoint detection method applying optical emission spectroscopy as non-intrusive measurement technique was developed for application in chamber dry-cleaning processes on a plasma enhanced CVD system (PECVD) for the deposition of dielectric layers (13),(14),(15),(16). An RF-discharge in the deposition reactor was used for plasma generation applying NF_3 as etching gas. The cleaning processes consisted of two steps. The first step was performed to preferably clean the electrode surfaces whereas the second step was performed to preferably clean the reactor walls. Figure 14 shows an optical emission spectrum for the first clean step. From the emission spectrum typical emission lines of the reactant atomic fluorine as well as the reaction products atomic oxygen and hydrogen can be extracted. The new endpoint detection method is based on a specific algorithm for the evaluation of these emission line intensities. With this algorithm, an auto scaled sigmoid-curve versus clean time is obtained for all clean processes independent of the starting condition (figure 15). For this sigmoid-curve the process endpoint can be clearly determined using a well-defined threshold value. The sigmoid-curve is analyzed in real-time during the clean processes and after reaching the inflection point the endpoint time can be determined. The endpoint algorithm was implemented and tested in a 300mm pilot-production environment. It could be proven that the developed algorithms enabled precise endpoint control even for complex sequences in the preceding deposition step and a more precise adjustment of the clean duration compared to standard endpoint detection procedures.

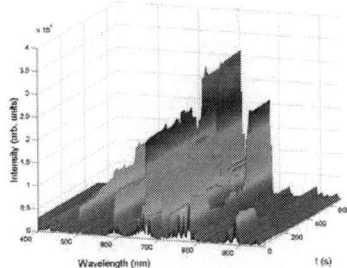 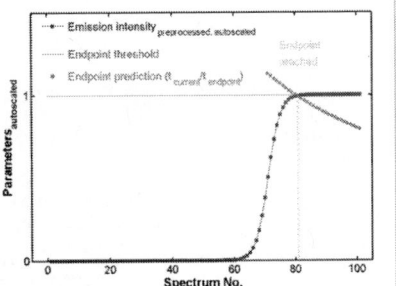

Figure 14: Optical emission spectrum of the first clean step.

Figure15: Principle of the endpoint detection method.

Chamber wall sensor

Either wet or plasma cleaning steps is typically performed at fixed time intervals, e.g. after pre-determined RF hours obtained on a plasma system. However, it would be more desirable to perform the cleaning process dependent on the condition of the chamber wall. To address this topic, a chamber wall sensor, which is capable of measuring the film thickness deposited on the chamber wall, was developed (17),(18). The major requirements for this sensor were cost-effectiveness and compatibility for integration into the chamber wall and the harsh plasma process atmosphere. The sensor is based on the principle of micro dielectrometry using interdigitated electrode structures which may be fabricated applying standard processes for micro systems fabrication (figure 16).

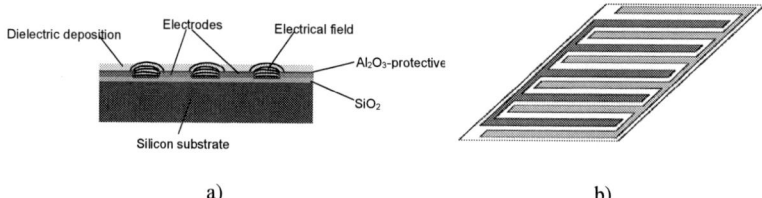

a)   b)

Figure 16: Principle of micro dielectrometry (a) using interdigitated electrode structure (b).

The deposited film will change the capacitance of the capacitor formed between the interdigitated electrode structures. Using a calibration curve of the capacitance versus the deposited film thickness, the film formation process on chamber walls can be monitored directly.

The chamber wall sensor was fabricated and integrated into an oxide etch reactor, replacing a standard viewport. Figure 17 shows the chamber wall sensor with the integrated evaluation circuitry. The sensor was successfully tested in a 300mm pilot-production environment. Figure 18 shows the deposited film thickness measured by the chamber wall sensor versus the number of processed 300mm wafers. Based on this thickness evaluation, the duration for chamber leaning cycles can be optimized.

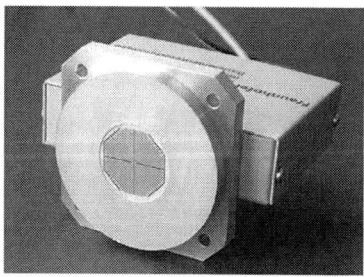

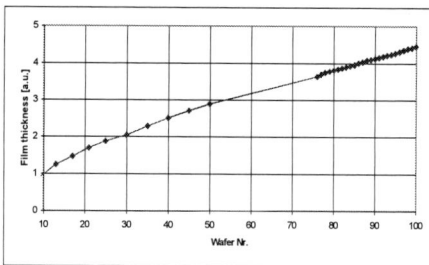

Figure 17: Chamber wall sensor with the integrated evaluation circuitry.

Figure 18: Film thickness measured by the chamber wall sensor vs. the number of processed 300mm wafers.

European Projects

Within a project with the Acronym FAB 2000, funded by the European Commission the Framework Programme III (ESPRIT) in the 90ies of the last century, the goal was to assess and demonstrate the economical feasibility of a dedicated ASIC minifab that is characterized as an extremely flexible, ultra short cycle time manufacturing line producing state-of-the-art CMOS and BiCMOS products.

The work included software simulation for single wafer lots and significant consideration of the methodology for integrated manufacture including automated process monitoring. Considerations towards advanced process control included the development of smart sensors. The integration of an ellipsometer into a vertical furnace for real time control an ONO process, i.e. an oxidation process followed by a nitride deposition and a subsequent final oxide, allowed a reduction of the processing time of more than 20 %, resulting in an according increase in throughput and productivity in an ASIC Fab. In addition, the influence of the thermal mass of heavily differing load sizes between 10 wafers and more than 100 wafers in the furnace on the resulting layer thicknesses grown could be minimized.

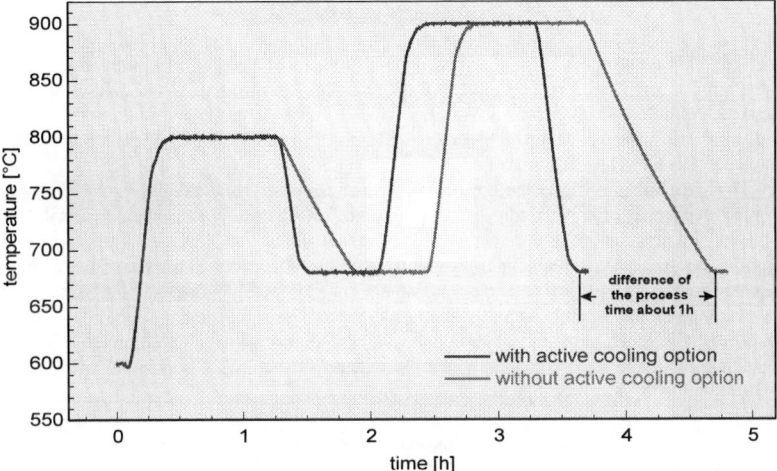

Figure 19: Throughput improvements of a furnace used for an ONO (oxide-nitride-oxide) process using integrated metrology and an active cooling option

In the frame of the establishment of the first 300mm pilot production facility, a national German project funded the collaboration of equipment manufacturers, clean room manufacturers, metrology manufacturers, semiconductor manufacturers and research institutions. Included in this national project were several sub-projects dealing with aspects of advanced process control. Part of the work described above and final test and evaluation were performed in the 300 mm pilot production line.

More recently, APC related activities were started in NanoCMOS, one of the first so-called Integrated Projects funded by the European Commission. The project in general focuses on the research and technology activities necessary to develop the 45nm, 32nm and below CMOS technologies. From these technology nodes it will be mandatory to introduce revolutionary changes in the materials, process modules, device and metallization architectures and all related characterization, test, modeling and simulation technologies, advanced equipment and process control technologies, to keep the scaling trends viable and make all future IST applications possible. Within this ongoing project NanoCMOS, a study of best suited processes and according sensors for integrated metrology was performed. It revealed a broad range of key applications for the patterning processes, for metal layers, for gate oxide and gate structures in general and for new materials, especially for new high-k and low-k materials. Beyond optical inspection for macro and micro defects, thickness and uniformity control, especially aspects of composition, grain size, crystallographic texture and porosity were investigated.

Results from the NanoCMOS projects were used for the development of new tools. A second *Integrated Project* funded by the European Commission with the project acronym SEANET including more than 25 participating companies just started to provide according sensing devices and integrated metrology for integration into semiconductor manufacturing equipment, for process integration and subsequent assessment in pilot production and in high volume production.

Conclusions

The continuation of Moore's law into future technology nodes will deploy a maximum of advanced process control. Basic elements and principles were described in details. Examples of successful integration of APC were given, the results proved the high value with respect to further development into future technology nodes and revealed the throughput and cost benefits.

The principles of advanced process control are well understood, however, the implementation is often suffering from the absence of well suited sensors and metrology systems. In most cases, the sensing devices are available in principle, but not in an appropriate status for direct integration and adoption. Therefore it is essential to bring together competencies of sensing and measuring, of semiconductor processing, of software development and of application know-how. Especially the creation of dedicated standards for mechanical integration as well as for software integration is a huge challenge, which has to be tackled by all participating structures, a perception which has led and will lead to further joint R&D activities in Europe.

Acknowledgments

The authors would like to express their thanks for parts of the work used in this paper to Norbert Benesch, Ralph Trunk, Rudi Berger and Wolfgang Lehnert and especially in commemoration of Claus Schneider (V). Results of this paper were partly obtained in joint research projects with Centrotherm, Sopra, Leica, Atmel, austriamicrosystems, Infineon, Freescale, Philips, SC300, STMicroelectronics and X-Fab. This work was partly funded by the European Commission within the project FAB 2000, Project Reference: 8413 (Framework Program 3), NanoCMOS No. 0507587 and SEANET No. 0279280, partly funded by the German BMBF, Federal Ministry for Education and Research within its funding program for the 300mm technology, and partly funded by BayStMWIVT, the Bavarian Ministry of Economic Affairs, Infrastructure, Transport and Technology.

References

1. Moore, G. E.: Cramming more components onto integrated circuits. Electronics, Vol. 28, Nr. 8, April 19, 1965.

2. International Technology Roadmap for Semiconductrors (ITRS). 2005 Edition, to be published by ITRS/Semiconductor Industry Association, 2706 Montopolis Drive, Austin, Texas 78741.

3. Ed Ward, Ron Horwath, Hamilton Hayes, Bob Bracamonte, Solid State Technology 11, 1999.

4. Pfitzner, L., Kuecher, P.: A Roadmap towards Cost Efficient 300 mm Equipment. Material Science in Semiconductor Processing Vol 5/4-5, p.321 - 331, Elsevier Science Publishing Co LtD., Amsterdam, 2003

5. Pfitzner, L. Oechsner, R. Schneider, C. Ryssel, H. Riemer, M. von Podewils, M.: Novel process control strategies for 300mm semiconductor production. Microelectronic Engieneering 45, 247 (1999)

6. Schneider, C., Pfitzner, L., Ryssel H.: Integrated metrology – An enabler for advanced process control. Proc. SPIE, 4406, 118 (2001)

7. Berger, R., Schneider, C., Lehnert, W., Pfitzner, L., Ryssel H.: Advanced process control system for vertical furnaces (invited paper). Proc. SPIE 2876, 16 (1996)

8. Trunk, R.; et al: *HandMon*-ISPM: Handling Monitoring in a Loading Station of a Furnace; Advanced Semiconductor Manufacturing Conference 2002, ASMC 2002, Boston, USA, April 30 - May 2, 2002, p. 113 – 118

9. Raymond, C. J.; Murane, M. R.; Prins, S. L. et al.: Multiparameter grating metrology using optical scatterometry; Vac. Science and Techn. B, Vol. 15, No. 2, Feb. 1997, pp.361.

10. Benesch, N.; Hettwer, A.; Schneider, C. et al.: Phi-Scatterometry for On-line Process Control; AEC/APC XII, Sept. 23, 2000, Lake Tahoe, pp. 715.

11. Benesch, N.; Schneider, C.; Pfitzner, L.: Gerät zur schnellen Messung winkelabhängiger Beugungseffekte an feinststrukturierten Oberflächen; Dt. Patent- und Markenamt, DE19914696.9, München, 2000.

12. Benesch, N.; Hettwer, A.; Schneider, C. et al.: Application of Phi-Scatterometry for Process Control in DRAM Manufacturing; 2nd European AEC/APC Conference, April 18, 2001, Dresden, Germany.

13. Roeder, G.; v. Andrian-Werburg, M.; Schneider C.; Becher S.; John, P.; Vatel, O.; Tampe, A.; Tegeder, V.: Advanced Process Control for *In Situ* - Chamber Clean Processes. Advanced Equipment Control / Advanced Process Control (AEC/APC) Workshop Europe - Dresden, Germany, March 30 - 31, 2000.

14. Roeder, G.; v. Andrian-Werburg, M.; Schneider C.; Becher S.; John, P.; Vatel, O.; Tampe, A.; Tegeder, V.: Advanced Endpoint Detection for *In Situ* - Chamber Clean Processes. European IEEE/SEMI Manufacturing Conference, Munich, Germany, April 3, 2000.

15. Roeder, G.; v. Andrian-Werburg, M.; Tschaftary, T.; Schneider, C. ; Pfitzner, L.; Ryssel, H.; John, P.; Tegeder, V.: In - Production Monitoring and Control of *In Situ* - Chamber Clean Processes. AEC/APC Symposium XII, Lake Tahoe, Nevada, September 23-28, 2000.

16. Roeder, G.; v. Andrian-Werburg, M.; Tschaftary, T.; Schneider, C. ; Pfitzner, L.; Ryssel, H.; John, P.; Tegeder, V.: Development, implementation, and test of a novel endpoint detection method for the cleaning of process chambers in deposition systems. 2nd European AEC/APC Conference, Dresden, Germany, April 18-20, 2001.

17. Ziegler, J.; Waller, R.; Roeder, G.; Schneider, C.: Development of sensors for the observation of chamber wall depositions, AEC/APC Symposium XI, Vail, Colorado, 12.-16. September 1999.

18. Waller, R.; Schneider, C.; Pfitzner, L.; Ryssel, H.; Marx, E.; Schneider, T.: Development of sensors for the measurement of chamber wall depositions in semiconductor manufacturing equipment. AEC/APC Symposium XIII, Banff, Alberta, Canada, 6.-11. October 2001.

SESSION 2

300 mm

56

300 mm Session Summary

Dieter Gilles
Siltronic, AG
Burghausen, Germany

Walter Huber
SUMCO USA
Freemont, CA

The 300 mm session provides a comprehensive and insightful overview about the achievements of the silicon materials industry, the collaboration and definition of industry standards that need to precede a successful transition in wafer diameter and the successful utilization of SOI wafers in high volume CMOS fabs. The semiconductor industry continued its relentless pursuit of Moore's law achieving high volume production of transistors with critical dimensions < 100nm. On the other hand the materials industry has successfully transitioned to manufacturing 300 mm wafers. The convergence of the deep sub-micron device technologies and the introduction and ramp of 300 mm wafers provided unique technical challenges for the silicon wafer manufacturers. The shrinking device structures and in-line process monitors continued to become more sensitive to crystallographic defects. The materials industry had to develop the appropriate crystal pulling hardware and large scale production capability of eliminating extended defects in CZ crystals. A good portion of the 300 mm session is dedicated to the modeling and study of defect formation in CZ crystals.

The continuous shrinking of the critical dimensions of the devices created the demand for silicon wafers with near surface regions that are almost void of any defects or tailored to specific applications. Such demands lead to the introduction of highly tuned CZ crystal processes, high-temperature annealing of 300 mm wafers, and further scaling of epitaxial processes. The requirements of complex and advanced logic designs ushered in the industrial use of 300 mm SOI wafers. The shrinking depth of focus in optical lithography produced the demand for double-side polished wafers and its unique requirement for edge handling and control of backside cleanliness. The second half of the 300 mm session addresses the issues of tuning the wafers strength for thermal processing, the continuous improvement of SOI wafer quality, and the industry lessons learned from the 300 mm transition and presenting an outlook for the next diameter (450 mm) transition of the industry.

The theory by V.V. Voronkov about the generation of intrinsic point defects in CZ crystals and their relationship to crystal pulling parameters provided the theoretical tools to the industry in controlling the type and density of bulk defects on an industrial scale. The first paper of the 300 mm session is an invited talk by V.V. Voronkov et al. This paper deals with the quantification of the diffusivities (Dv, Di) and the equilibrium concentrations (Cv, Ci) of point defects such as vacancies and silicon self-interstitials, which are of crucial importance for simulating and modeling the phenomena of defect formation during crystal growth and subsequent wafer processing.

The talk by M.S. Kukarni addresses the complexity of modeling and the quantification of the formation of micro-defects, and presents an overview of models that are available for quantifying the defect dynamics in CZ material. Over the years, a few mathematical approaches have emerged, like the application of the Fokker-Planck equation or by using a so called lump model, which approximates the micro-defect population by another equivalent population of identical micro-defects. However, the complexity still limits these models to one dimensional calculations.

A joint paper by the University of Pennsylvania and Siltronic describes an accurate model for defect agglomeration in silicon by taking into account the entropic contributions. A key parameter for void formation is the free-energy formation of vacancy clusters as a function of size and temperature. In previous models the surface energy was estimated from the silicon (111) values because large voids are bounded by these planes. However, this lead to a large underestimation of the void formation temperature during the CZ crystal growth. The authors found that the ground state configuration of small clusters at higher temperatures is a collection of extended configurations that are even mobile. With the presented model an excellent quantitative agreement with experimental values is achieved, and an aggregation temperature of 1070-1110 °C is predicted, which is 70-100 °C higher than for previous models.

H. Tu and his coworkers report a new doping process for achieving high concentrations of As in silicon single crystals and their ability to increase the interstitial oxygen content.

W. Sugimura et al investigated the void formation and the kinetics of oxygen precipitation in heavily boron and arsenic doped silicon. In the boron doped material no grown-in defects were observed for concentrations < $9.7x10^{18}/cm^3$, while in arsenic doped crystals the void density peaks at As concentrations of $1.4x10^{19}/cm^3$ and then rapidly drops off with further increase of As doping. The nucleation rate for oxygen precipitates is accelerated in heavily-doped boron doped wafers that is explained by the enhanced diffusivity of oxygen, while the diffusion of oxygen is not affected by the heavy As doping. On the other hand, a retardation of the nucleation rate is observed with heavy As doping, which is explained by the change in the formation energy of interstitials in silicon.

The emergence of high-temperature anneals with shorter and shorter anneal times and automatic wafer handling resurrected the interest of studying the brittle fracture of silicon wafers. The invited talk by T. Ono et al. addresses the relationship between crack size and wafer breakage stress. The intrinsic breakage stress of a silicon wafer reaches 700 – 1200 MPa but drastically decreases to 100 MPa for crack sizes ranging from 10 – 115 microns and then becomes independent of larger crack sizes. At room temperature, no impact of boron and nitrogen doping, and oxygen precipitates on brittle wafer breakage is revealed. On the other hand, boron concentrations above $8 \times 10^{18}/cm^3$ are very effective in suppressing wafer slip, and tuning the oxygen precipitate size and density can be applied to reducing the propagation of slip.

P. Gupta et al model the impact of wire saw slicing on the wafer shape based on applying some fundamental physical principles to the dynamics of the cutting process. The shape of a sliced wafer is predominately determined by non-uniform local thermal expansion of the silicon ingot in the vicinity of the cutting wire. This phenomena is modeled with a 3D finite element mesh where slicing is simulated by "deactivating" elements from the mesh. The predicted results are in good agreement with the observed wafer shape.

S. Kramer and co-workers from Sematech discuss the lessons and conclusions derived from the transition to 300 mm wafers. Their experience calls for a clear upfront consensus and strategy by the industry for the next diameter conversion. Rigorous analysis of the economics and market needs to be combined with open discussions of business conditions: understanding the development and R&D cycles, costs of equipment and material suppliers, modeling of fab operations and productivity. This needs to go hand-in-hand with developing global standards for materials, equipment, and fab automation.

M. Watanabe et al. look at the process of standardizing relevant parameters for 450 mm wafers. Besides the diameter, the wafer thickness is of great importance and high priority for standardization. Wafer thickness can have an impact on sip generation, breakage, gravitational sag, and cost. Simulations of thermal stress during Flash lamp annealing revealed very little sensitivity to diameter and thickness, indicating no overwhelming reason for increasing wafer thickness. On the other hand, the type of wafer support during thermal processing and wafer handling also play a significant role in defining the 450 mm wafer thickness.

IBM pioneered the use of SOI in high volume manufacturing of advanced MPU's. G. Pfeiffer et al share data about the quality and performance of 300 mm SOI wafers. The use of thin SOI wafers has added the requirement for controlling the thickness uniformity of the superficial silicon layer, the elimination of voids or small Si holes (divots) and sub-surface metallic contamination because the traditional bulk gettering strategies cannot be directly applied. Process changes in the manufacture of SOI have reduced the within-a-wafer Si thickness uniformity to ± 2-3 nm. The detection of divots, which are killer defects with typically tens of microns in size, is more of a challenge than large voids because automatic surface inspection tools cannot differentiate them from localized light scatterers (LLS's). However, 95% of 300 mm SOI wafers have divot density below levels of $0.04–0.07/cm^2$. Additionally, flat metal concentration profiles of Ni and Cu, which typically pile up at the BOX/silicon interfaces, as observed by SIMS profiling, are significantly reduced compared to 200 mm SOI wafers. Overall the quality of 300 mm SOI has reached or exceeded the level of quality of 200 mm SOI.

60

ECS Transactions, 2 (2) 61-75 (2006)
10.1149/1.2195649, copyright The Electrochemical Society

Parameters of Intrinsic Point Defects in Silicon Based on Crystal Growth, Wafer Processing, Self- and Metal- Diffusion

V.V.Voronkov[a] and R.Falster[b]

[a] MEMC Electoronic Materials, via Nazionale 59, I-39012 Merano BZ, Italy

[b] MEMC Electoronic Materials, viale Gherzi 31, I-28100 Novara, Italy

The vacancies and self-interstitials in silicon are involved, in a straightforward way, in various phenomena such as formation of grown-in microdefects, diffusion of metals (Au, Zn), self-diffusion and installation of vacancy depth profiles into wafers by Rapid Thermal Annealing. The available data is sufficient to deduce the diffusivities and equilibrium concentrations of the intrinsic point defects leaving only one parameter (the migration energy of self-interstitials) not specified. The diffusivities are high while the equilibrium concentrations are remarkably low.

Introduction

The basic parameters of vacancies and self-interstitials in silicon – the diffusivities D_V and D_I, the equilibrium concentrations C_{Ve} and C_{Ie} – are of a crucial importance to understand and simulate the effects that occur during wafer processing and crystal growth. In spite of numerous efforts, there is still a considerable discrepancy (sometimes, by orders of magnitude) between the parameter values used in different works. The safest approach, to specify reliable values for the parameters, is to use only the phenomena where the intrinsic point defects are involved in a clear unambiguous way.

The list of the relevant phenomena starts in the field of crystal growth where the pure silicon material is initially in its virgin state, with only point defects present. The survived (incorporated) intrinsic point defects turn out to be either vacancies or self-interstitials – but not both simultaneously. At a later stage of crystal cooling, the survived defects are agglomerated into observable "microdefects" - or perhaps "nanodefects". The properties of the grown-in microdefects (especially, of the grown-in voids produced by vacancy agglomeration) tell a lot about the properties of the source point defects. Although mostly semi-quantitative estimates are gathered from the crystal growth data - implying very high values for the diffusivities and very low values for the equilibrium concentrations - these conclusions are indispensable for a proper treatment of the data coming form the other sources of information.

Particularly, the notion of high D_V, D_I and low C_{Ve}, C_{Ie} is important to simulate properly the in-diffusion profiles of metals (Au, Zn) that are controlled by the out-flux of the kicked-out self-interstitials, and - to a less extent - by the in-flux of vacancies. These fluxes limit a conversion rate of in-diffusing interstitial metal species M_i into the dominant substitutional form M_s. In this way, the "transport capacity" of self-interstitials, $P_I = D_I C_{Ie}$, can be deduced more reliably than it was done before (although a correction

61

to the previously deduced values of P_I is not large). The vacancy transport capacity, $P_V = D_V\,C_{Ve}$, can be only roughly estimated from the metal profiles.

A lack of information on P_V is compensated by the data on the self-diffusivity available in a wide temperature range. The self-diffusivity is definitely composed of the two comparable contributions, that of self-interstitials and that of vacancies. Therefore – with a specified P_I – the value of P_V can be extracted more reliably than it could be done considering only metal diffusion.

With both P_I and P_V specified, it is much easier to analyze the vacancy profiles installed into wafers by Rapid Thermal Annealing (RTA) and monitored by diffusion of Pt or Au at relatively low T. At low T, the only operating mechanism to create substitutional metal species is by interaction of M_i with the already existing vacancy species. The relevant vacancy species are not free, but trapped vacancies. In Czochralski-grown (CZ) material, the major trap is represented by the oxygen impurity. The traps serve simply to fix the vacancy profile developed during a quench – otherwise the fast-diffusing free vacancies would be completely lost by diffusion to the wafer surface. The installed vacancy profiles provide another strong piece of evidence in favor of a very low value of C_{Ve} (as well as of C_{Ie}) and of very high diffusivities D_V and D_I. The vacancy diffusivity at the RTA temperature can be deduced while D_I can be only estimated – to be higher than D_V.

The procedure outlined above will be now described in more details.

Crystal Growth Data

The type of initially incorporated point defect (either vacancy or self-interstitial) has long been known to be controlled by the V/G ratio (1,2). Here V is the growth rate and G is the axial temperature gradient in the vicinity of the crystal-melt interface.

Vacancies are incorporated if V/G exceeds a certain critical value, $(V/G)_{cr}$, and the incorporated concentration C_V is an increasing function of V/G. It is zero for the critical ratio and saturates at larger V/G. The saturated value is equal to the concentration difference $\Delta C = C_{Ve} - C_{Ie}$ taken at the melting point (T_m). The kind of vacancy agglomerates depends on C_V. At larger C_V (larger V/G) voids are produced. By Transmission Electron Microscopy (TEM), the voids are of octahedral shape and of a diameter typically around 100 nm (3-5). At low C_V the dominant vacancy-type microdefects in Czochralski-grown (CZ) crystals are oxide particles formed by vacancy-assisted oxygen agglomeration (2,6). Spatially, the particles are often present as a narrow band (P-band) that gives rise, upon oxidation anneal, to a well-known Ring of Stacking Faults (OSF ring).

Self-interstitials are incorporated into a growing crystal at V/G < $(V/G)_{cr}$, and the concentration C_I increases upon reducing V/G. If C_I is not too low, the produced microdefects are interstitial-type dislocation loops initially known under the name of A-swirl-defects (7). At lower C_I, small globular agglomerates (known as B-swirl-defects) are produced.

Since V/G varies through the crystal body both in radial and axial directions, banded microdefect patterns are often developed, represented by a vacancy region of voids surrounded by a marginal P-band (OSF ring), and by an adjacent interstitial region of loops (A-defects) surrounded by a marginal B-band. In reality, the banded structure found in the vicinity of the V/I boundary (that separates the vacancy-dominated region from the self-interstitial-dominated one) includes more bands (6,8), due to peculiarities of defect clustering at very low C_V and C_I. Such banded microdefect patterns are used to delineate resulting V/I boundaries and to determine the critical V/G ratio.

Concentration Scale by the Value of $(V/G)_{cr}$

The "V/G rule" is based on a fast recombination of vacancies and self-interstitials, in a temperature range close to the melting point. Strong concentration gradients, induced by recombination, result in axial in-diffusing fluxes of the defects, from the interface into the bulk. The competition of these fluxes (scaled in proportion to G) and the convection fluxes (scaled in proportion to V) means that the type and concentration of the surviving (initially incorporated) defects depends on the V/G ratio. The value of $(V/G)_{cr}$ is thus expressed directly (1,9) through the diffusivities and equilibrium concentrations taken at T_m:

$$(V/G)_{cr} = (1/kT_m^2)\ [P_I\ (E - \varepsilon_I) - P_V\ (E - \varepsilon_V)]\ /\ \Delta C . \qquad [1]$$

Here $P_V = D_V C_{Ve}$ is the "vacancy transport capacity", and $P_I = D_I C_{Ie}$ is the self-interstitial transport capacity. The energy E is equal to $(E_V + E_I)\ /\ 2$ – the average of the two formation energies, E_V and E_I. Finally, ε_V is the drift energy that defines the velocity of the uphill drift of vacancies along the temperature gradient (ε_I is a similar quantity for self-interstitials). The transport capacities are approximately known from metal and self-diffusion (10,11), and E is roughly 4 eV. The drift energies, although not known for certain, are unlikely to exceed E. The experimental value of $(V/G)_{cr}$ is (0.16 ± 0.04) mm^2/minK (2). In spite of some uncertainty in the parameters, Eq.[1] is useful to estimate the concentration difference $\Delta C = C_{Ve} - C_{Ie}$ at the melting point. The resulting value is roughly 1.5×10^{14} cm^{-3} or less. Since it is unlikely for the two concentrations, C_{Ve} and C_{Ie}, to be almost identical, the estimated concentration difference sets a scale for C_{Ve} expected to be in a range 3×10^{14} to 10^{15} cm^{-3}. The value of C_{Ie} is smaller by ΔC.

Concentration Scale by the Total Volume of Voids

The initially incorporated vacancy concentration C_V – if not too low - is almost completely spent on production of voids (2). Therefore, the measured total volume of voids can be used to determine C_V. By the Laser Scattering Tomography, both the void density and the volumes of individual voids (calibrated by TEM data) are obtained (12). The measured dependence of C_V on V/G gives the saturated concentration ΔC which is about 1.6×10^{14} cm^{-3}, in agreement with the previous estimate.

Vacancy diffusivity by the Void Properties

The voids in CZ crystals are produced typically at around 1100°C (2,12,13). The void production is definitely assisted (14) by oxygen that can be adsorbed at the void walls thus reducing the surface energy. Another possibility is that first not voids but globular oxide particles are nucleated, by a joint clustering of vacancies and oxygen. Very short after that, the particles are converted into voids by cavitation, due to a huge negative pressure in oxide induced by the vacancy super-saturation (14). In both scenarios of oxygen-assisted void production, the calculated void density is insensitive to the oxygen concentration C_{ox}. The effect of oxygen is just to facilitate nucleation and thus to shift the nucleation temperature to a higher T. The calculated void density N_v is strongly affected by the assumed value of D_V and by the cooling rate $q = - dT/dt$ (both taken at the void production temperature - at about 1100°C). For diffusion-limited vacancy consumption by voids, N_v is scaled as $q^{3/2}$ (2), well in accord with reported experimental data (12). Therefore, $D_V(1100°C)$ can be deduced using the reported value of N_v (around 1.5×10^6 cm^{-3} at $q = 2$ K/min). The result is almost the same for both scenarios of oxygen-assisted void production: the deduced D_V is around 5×10^{-5} cm^2/s (14,15). Such a high diffusivity is further evidence in favor of very low C_{Ve} since the two parameters, D_V and C_{Ve}, are coupled by a known value of the vacancy transport capacity, $P_V = D_V\, C_{Ve}$. The previously reported P_V (11) is not far from the refined value of P_V to be deduced later: it is about 5×10^7 cm^{-1}s^{-1} at 1100°C. Accordingly, $C_{Ve}(1100°C)$ is about 10^{12} cm^{-3}.

Estimate of D_V by Void Striations. It was mentioned above that although void production is oxygen-assisted, the void density N_v is insensitive to C_{ox} both by computations (14,15) and by measurements (16). At the same time in crystals of strong oxygen striations, the voids were found to follow the oxygen profile: N_v is distinctly higher in oxygen-rich layers (16,17). This effect clearly indicates a high vacancy diffusivity. Indeed, the voids are first formed in the oxygen-rich layers. The void suppression found in the adjacent oxygen-lean layers means that these layers become depleted of vacancies, by diffusion to the voids that already exist in the oxygen-rich layers. Such a depletion can be efficient only if D_V is sufficiently high, since the vacancies should diffuse by a half-period of the striations (by a distance of 0.5 mm), within a characteristic time of some minutes. The resulting estimate for D_V is similar to the value extracted above from the absolute value of N_v.

Metal Diffusion

There are some metal impurities (zinc, gold, platinum) that are dissolved mainly in the substitutional form (M_s) but have an appreciable interstitial component (M_i). Accumulation of M_s species in a wafer, during an in-diffusion anneal, proceeds through in-diffusion of fast M_i species that are converted into the dominant M_s form by filling vacancies and by kicking-out self-interstitials. Therefore, the produced concentration of the M_s species is limited by a supply of vacancies from the surface, and by removal of kicked-out self-interstitials to the surface. After a sufficiently long diffusion time, the equilibrium with respect to the two reactions, $M_i + V \rightarrow M_s$, and $M_i \rightarrow M_s + I$, is maintained throughout the sample depth. The profiles of C_I and C_V then follow those of the two metal species

$$C_V = (C_s / C_i)\, C_{Ve}\, R_{is} , \qquad\qquad C_I = (C_i / C_s)\, C_{Ie} / R_{is} , \qquad [2]$$

where $R_{is} = C_{ie} / C_{se}$ is the ratio of the interstitial and substitutional metal solubilities.

The dynamic equations (18) for the two metal profiles, $C_i(z,t)$ and $C_s(z,t)$, contain all the parameters of the intrinsic point defects. However, due to very low equilibrium concentrations C_{Ve} and C_{Ie}, in comparison to C_s, only two parameters are relevant: the transport capacities P_I and P_V. Indeed, the out-flux of I and the in-flux of V – that are responsible for accumulation of the M_s species - are controlled by D_IC_I and D_VC_V, respectively. According to Eq.[2], these products are expressed through P_I and P_V. The problem also includes metal parameters: the solubility C_{se}, the diffusivity D_i of M_i, the solubility ratio R_{is}. The parameters C_{se} and D_iC_{ie} (the metal transport capacity) are known from other measurements while a small ratio R_{is} has only a minor impact on the computed profiles. With only two basic fitting parameters, the reliability of the deduced values is essentially increased. Yet it turns out that even just two fitting parameters are too many for both to be deduced reliably. The self-interstitials play the dominant role in metal in-diffusion since the wafer bulk is supersaturated with I species and undersaturated with V species, according to Eq.[2], taking into account that the bulk value of C_s is normally well below C_{se}. For this reason only $P_I(T)$ can be deduced by a fitting procedure with a good confidence while the deduced values of a less important parameter P_V show such a strong scatter that only a rough estimate of $P_V(T)$ is possible (18). Yet it is essential to take into account the vacancy flux while simulating the metal in-diffusion. This is clear from comparison of the in-flux of V and out-flux of I taken right at the wafer surface ($z = 0$). From Eq.[2], the gradients $\partial C_V/\partial z$ and $\partial C_I/\partial z$ are expressed, and then the ratio of the two diffusion fluxes is found, to be equal to P_V/P_I. It means that the two fluxes are comparable as far as P_V and P_I have similar values. Therefore the amount of M_s species created by in-diffusion of vacancies is comparable to that created by out-diffusion of self-interstitials. The difference between the two contributions is that the in-coming vacancies are consumed in the near-surface region thus affecting mostly the near-surface part of the metal depth profile. The bulk part of the profile is controlled mostly by the out-going flux of self-interstitials.

Metal profiles are available for Zn, Au, Pt. For a reliable determination of the P_I and P_V parameters, several profiles of a sequence of diffusion times are required. Profile series of a good quality are available only for Zn and Au. A series of profiles reported for Pt (19) is of a questionable reliability since the Pt solubility, by this work, is in a severe contradiction to that obtained by the Neutron Activation Analysis (NAA) in ref.(20). For this reason, we will take into consideration only the profiles reported for Zn and Au.

Simulation of Zn Diffusion Profiles

In-diffusion of Zn was performed (10) at five temperatures: 870, 942, 1021, 1115 and 1208°C. At each T, a series of evolving depth profiles of Zn was obtained, for different diffusion time t and sample thickness d. The profile shape, plotted in dependence of the normalized depth, z/d, evolves in dependence of the "effective time" t_{eff} proportional to t/d^2 (18). The hole concentration p(z) - controlled by the Zn_s acceptors - was traced by Spreading Resistance Profiling. The conversion of p into C_s presents a problem since Zn_s is a deep (double) acceptor; normally it exists in neutral and single-negative charge states. The concentration ratio of these two charge states is dependent, apart from the position of the first acceptor level of Zn_s, also on the degeneracy factors of the two charge states which are not definite. The expression for C_s through p contains (if C_s is not too low) a factor of indefinite accuracy. However, this factor is excluded from the normalized concentration $S_s = C_s/C_{se}$ that can be thus well defined by the spreading resistance data. In the original work (10) the profiles for $T \geq 1000$°C were simulated altogether neglecting the vacancy contribution. The profiles at $T < 1000$°C were, on the contrary, simulated under assumption of a high C_{Ve}. This approach is not self-consistent. Neither it is consistent with a low value of C_{Ve} based on the crystal growth results discussed above.

Therefore a refined simulation is necessary to deduce the parameters P_I and P_V. Since the role of vacancies is less important than that of self-interstitials, the refined values for P_I do not deviate much from the previously obtained values (10).

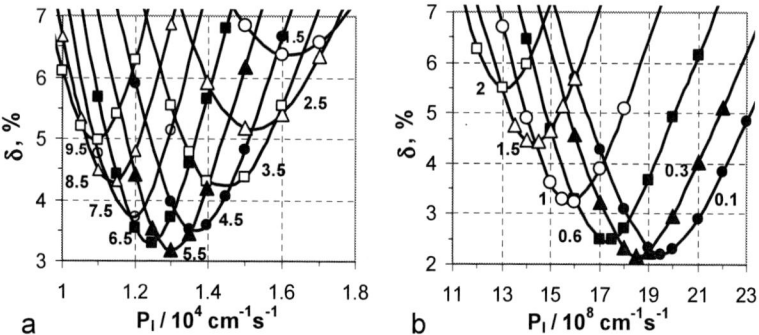

Figure 1. Deviation of the computed middle-wafer zinc concentrations, C_s/C_{se}, from the experimental values, in dependence of P_I. Each curve corresponds to a particular value of P_V/P_I indicated at the curve. The diffusion temperature is 870°C (a) and 1208°C (b).

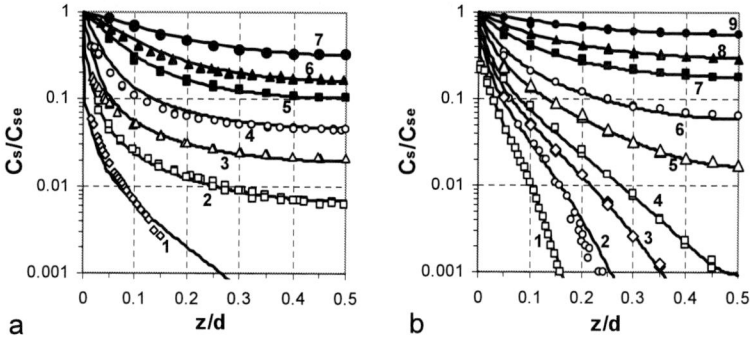

Figure 2. Normalized depth profiles of Zn_s computed with the best-fit parameters (solid lines) compared to the measured profiles (numbered in the order of increasing effective time, t/d^2). The diffusion temperature is 870°C (a) and 1208°C (b).

In simulations, it is convenient to fix several representative values for the P_V/P_I ratio, and for each ratio to examine different values of P_I. At the first step, only five profiles of the longest t_{eff} were considered – since in this case the relevant parameters are just P_I and P_V. The criterion of the fit quality was the relative mean square deviation (δ) of the computed middle wafer concentration (C_s/C_{se} at $z = d/2$) from the measured values, for the selected 5 profiles. An example of computed $\delta(P_I)$ curves is shown in Fig.1, for the lowest (870°C) and the highest (1208°C) diffusion temperature. At every P_V/P_I, a sharp minimum in $\delta(P_I)$ is reached. This minimum is the lowest for some interval of the P_V/P_I ratio that is around 5.5 at the lower T and around 0.3 at the higher T.

At the second step, the profiles for shorter t_{eff} were also included. The already specified best-fit values for P_I and P_V were used for the simulation, and the R_{is} parameter of Zn was adjusted to improve the fit for shorter-time profiles. These profiles are quite sensitive to R_{is} but not sensitive at all to the diffusivities D_I and D_V - as far as these were assumed to be higher than 5×10^{-6} cm^2/s (which is the case, according to the crystal growth data).

With the best-fit parameters - that minimize the deviation δ at the wafer middle - the complete depth profiles of Zn_s are also well reproduced. It is demonstrated in Fig.2, for the same two diffusion temperatures, 870 and 1208°C.

Simulation of Gold Diffusion Profiles

For gold, a series of profiles (for several values of t_{eff}) are available for 3 diffusion temperatures 1000, 1098 and 1200°C (21-23). Here the profiles of the total Au concentration ($C = C_s + C_i$) were monitored by NAA and normalized by the total solubility $C_e = C_{se} + C_{ie}$. In the original publications, the data were analyzed under some simplifying assumptions. For the present numerical simulation, we do not need any such simplifications, and the gold data were thus re-analyzed using the procedure just discussed for zinc. The $\delta(P_I)$ dependence was computed for various values of P_V/P_I, to find the best-fit parameters. For gold, about 5 profiles at each T are available, all for relatively long t_{eff}, and all of them were used to find δ. A peculiarity of the present case is that a small contribution of Au_i is essential, at least for the profiles of the lowest middle-wafer concentration C_s. The parameter R_{is} was therefore fitted, for every selected combination of P_I and P_V/P_I. The fitted values of R_{is} were about 0.01 or less, but even such small values provided a definite improvement of the fit. A series of the $\delta(P_I)$ curves, for various P_V/P_I, looks similar to those for Zn. Also the total depth profiles of C/C_e are well described with the best-fit parameters, similar to Fig.2 for Zn.

Temperature Dependence of P_I and P_V/P_I

The best-fit values of P_V/P_I, at each diffusion temperature of Zn or Au, are within an interval that is not always narrow; more than that, this interval changes with T not quite regularly (Fig.3a). Accordingly, the best-fit value of P_I, at each T, is specified within some error bar (Fig.3b). These error bars are however sufficiently narrow to draw a reliable Arrhenius line and thus to define $P_I(T)$. This function (together with other temperature-dependent properties of the intrinsic point defects, to be deduced later) is specified in the Table I. The activation energy for P_I is remarkably high, close to 5 eV.

Regarding the capacity ratio P_V/P_I, only semi-quantitative conclusions can be drawn from the metal diffusion data. This ratio is basically in the order of 1 around 1000°C; on average, it is a decreasing function of T. The solid line in Fig.3a follows from the self-diffusivity data discussed below.

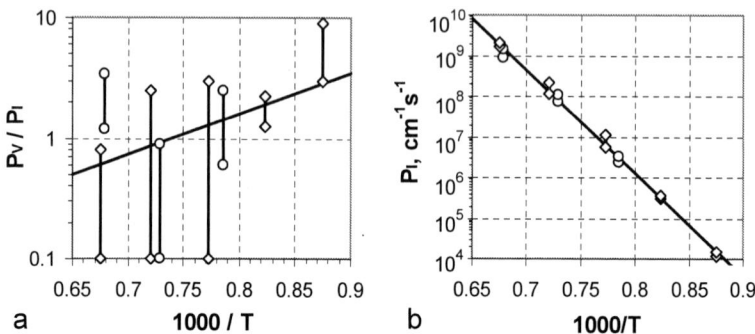

a   b

Figure 3. The values of the P_V/P_I ratio (a) and of the transport capacity P_I (b) deduced from the metal diffusion data (Zn-based bars: rhombs, Au-based bars: circles).

Table I. The transport capacities (P_V, P_I), diffusivities (D_V, D_I) and equilibrium concentrations (C_{Ve}, C_{Ie}) of the intrinsic point defects represented by Arrhenius-type functions.

Quantity	Prefactor	Activation energy, eV
P_V, cm^{-1}s^{-1}	2.4×10^{24}	4.5
D_V, cm^2/s	0.002	0.38
C_{Ve}, cm^{-3}	1.2×10^{27}	4.12
P_I, cm^{-1}s^{-1}	2.55×10^{26}	5.04
D_I, cm^2/s	0.004	0.3
C_{Ie}, cm^{-3}	6.4×10^{28}	4.74

Vacancy Transport Capacity by Self-Diffusion

The self-diffusion coefficient D_{sd} was reported by many authors who used different techniques, with a remarkable discrepancy between the results. The most reliable values were obtained, in a wide temperature range, using isotope heterostructures (11). Under the equilibrium conditions, the diffusivity D_{sd} is expressed through the transport capacities of the self-interstitials and vacancies :

$$D_{sd} = (P_I^* + P_V^*) / \rho \quad , \qquad\qquad [3]$$

where ρ is the lattice site density (5×10^{22} cm^{-3}). The quantity P_I^* is a modified self-interstitial transport capacity, $f_I P_I$, that contains a so called correlation factor (f_I) for the tracer jumps. Similarly, $P_V^* = f_V P_V$. The vacancy migrates in a simple way, by jumping to one of the four neighboring sites, and the factor f_V is then equal to 0.5 (24). For self-interstitials, the path of migration is not definite. A self-interstitial just located in an interstice can migrate by simple jumps into the 4 neighboring interstices. Another way is by kicking out one of the 4 lattice neighbors. Still more complicated way of migration occurs for an interstitialcy configuration of the I species (two silicon atoms residing around one lattice site). The value of f_I is therefore not known.

The measured dependence $D_{sd}(T)$ can be well described by an Arrhenius line with a single activation energy, 4.75 eV (11). This energy is smaller than the activation energy $E_{PI} = 5.04$ eV for the self-interstitial component P_I^*. It can be thus concluded that (i) the vacancy contribution P_V^* in Eq.[3] has a value comparable to P_I^*, (ii) the activation energy E_{PV} for P_V^* is essentially smaller than the apparent activation energy of 4.75 eV (since the apparent activation energy is intermediate between E_{PI} and E_{PV}). A separation of D_{sd} into the two components, with a specified $E_{PI} = 5.04$ eV (but with the prefactor in P_I^* still to be fitted), is not unique. Almost any value (less than 4.75 eV) can be assumed for E_{PV} with a result of nearly the same averaged deviation of the experimental points from the calculated D_{sd}.

This uncertainty is removed if the data for self-diffusion under oxidation or nitridation conditions (25,26) are used. The $P_I^* + P_V^*$ sum in Eq.[3] is then replaced with $S_I P_I^* + S_V P_V^*$. The supersaturation ratios S_I and S_V were estimated (25) by simultaneous study of diffusion of phosphorus (that migrates predominantly via self-interstitials) and antimony (that migrates predominantly via vacancies). The phosphorus diffusivity is scaled in proportion to S_I, and that of antimony – in proportion to S_V. Under oxidation ($S_I > 1$, $S_V < 1$), both the self-diffusion and phosphorus diffusion were enhanced; the antimony diffusion was retarded. Under nitridation ($S_V > 1$, $S_I < 1$), both self-diffusion and antimony diffusion were enhanced; the phosphorus diffusion was retarded. These data (obtained at 1000 and 1100°C) imply that the two components of the self-diffusion, P_I^* and P_V^*, are almost identical in this temperature range. It is also in accord with the data on P_V/P_I ratio by the metal diffusion (Fig.3a). With this constraint ($P_V^*/P_I^* \approx 1$ at T ≈ 1050°C), acceptable values for the activation energy E_{PV} of the vacancy contribution are limited to quite a narrow range 4.3 to 4.5 eV. Using the deduced value of P_I^* and the previously specified value of P_I, the correlation factor $f_I = P_I^*/P_I$ is calculated to be 0.5 to 0.6 (depending on the assumed E_{PV}).

We prefer a higher value (4.5 eV) for E_{PV}, from the above-mentioned interval (4.3 to 4.5 eV) because this number allows for a better fit of the vacancy profiles installed into wafers by RTA (next section). By the deduced vacancy contribution $P_V^*(T)$ - with $E_{PV} = 4.5$ eV - the vacancy transport capacity is defined: $P_V = P_V^* / f_V$. This result is included into the Table I.

Simulation of Vacancy Profiles in RTA Wafers

A Rapid Thermal Annealing (RTA) of thin wafers at T > 1150°C was shown to install a vacancy depth profile of inverted-U shape (27). The vacancy concentration increases upon increasing the RTA duration, but already after several seconds it is not changed any longer. This important result shows that the wafer, in the course of RTA, is fast saturated with the intrinsic point defects, due to very high diffusivities. The uniform profiles ($C_V = C_{Ve}$, $C_I = C_{Ie}$) are created during RTA. In the course of a subsequent quench, the profiles evolve due to annihilation of V and I defects, and due to a simultaneous out-diffusion. The dominance of vacancies in quenched wafers suggests that $C_{Ve} > C_{Ie}$ at the RTA temperature - just like at T_m.

Installed vacancy profiles can be used to control the depth profiles of subsequently formed oxide precipitates, since oxygen precipitation is strongly enhanced in the presence of the vacancy species (28). In the Magic Denuded Zone® (MDZ®) wafer, a near-surface vacancy depletion results in a precipitate-free near-surface region. The wafer bulk becomes densely populated with the oxide precipitates, due to vacancy-enhanced nucleation during heat treatment (27). MDZ® wafers provide an efficient implementation of internal gettering.

The installation of vacancy profiles into MDZ® wafers is partially the result of the trapping of the free vacancies by oxygen. Free vacancies V dominate initially, at the RTA temperature, but upon lowering T, they are eventually trapped into VO and VO_2 (6, 29); at low T, the VO_2 form is the dominant one. For oxygen concentrations typical of CZ silicon, the characteristic "binding temperature" T_b that separates the higher-T range of V dominance from the lower-T range of VO_2 dominance was estimated to be around 1050°C (29). Therefore, efficient vacancy out-diffusion from the wafer, in the course of a post-RTA quench, is limited to the temperature interval from T_{RTA} to T_b. Below T_b, the vacancy out-diffusion becomes suppressed, due to a strongly reduced fraction of mobile V species. The trapping effect is most important to preserve some vacancies (in the VO_2 form) – otherwise the vacancies would be lost almost completely by out-diffusion, due to a very high diffusivity D_V.

The concentration of quenched-in vacancy species (VO_2) was measured by diffusion of Pt at around 730°C (27). At this – relatively low - temperature, the metal in-diffusion is qualitatively different from that discussed above. The transport capacities P_I and P_V are now very low, and the substitutional metal M_s is created mostly by filling the existing vacancy species: $M_i + VO_2 \rightarrow M_s + O_2$. The metal solubility C_{se} well exceeds the equilibrium concentration of VO_2 (for specified oxygen concentration), and almost all VO_2 species are replaced with M_s. The depth profile of M_s is then measured by DLTS, and in this way the vacancy depth profile is obtained. The obtained vacancy concentration should not depend on the kind of metal used to "count" the vacancy species. It was confirmed by measuring sister wafers, using either Pt or Au.

Another way to judge of the vacancy depth profile is to induce oxygen precipitates (e.g. by annealing 800 + 1000°C) and to monitor the depth profile of the precipitate density $N_p(z)$ that follows the vacancy depth profile (27,30). However, the translation of N_p into the vacancy concentration is not unambiguous since the measured density N_p is often scattered significantly.

Recently new RTA experiments with carefully monitored temperature-time profiles have been performed. In the present study, RTA processing was performed at 1240°C for 10s (in an Ar ambient), with different characteristic cooling rates, 100, 40 and 20 K/s and known full cooling profiles T(t). It is the dependence of the profile shape on the cooling rate that is most useful to deduce the defect parameters by simulation. One wafer was subjected to a shorter RTA (2 s) with subsequent fast quench (100 K/s). The dependence of the vacancy profile on the RTA duration is also a valuable source of information. The vacancy depth profiles were monitored by Au diffusion at 730°C for 3 h (the diffusion/DLTS test was performed by Prof. E.Yakimov to whom the present authors are grateful).

The simulated $C_V(z)$ and $C_I(z)$ profiles developed in a wafer after a full cycle (ramp-up to T_{RTA}, holding at T_{RTA} and ramp-down) are controlled by diffusion of the V and I species and by their recombination. The equilibrium boundary condition for C_V and C_I at the wafer surfaces was assumed. The dynamic equations are simple

$$\partial C_V/\partial t = D_V \, \partial^2 C_V/\partial z^2 - J \,, \quad \partial C_I/\partial t = D_I \, \partial^2 C_I/\partial z^2 - J \,. \qquad [4]$$

The net recombination rate, $J = K \, (C_I \, C_V - C_{Ie} \, C_{Ve})$, contains the kinetic recombination coefficient K which is another fitting parameter. If the recombination were diffusion-limited, K would be equal to $4 \pi R \, (D_I + D_V)$ (where R is the capture radius conventionally adopted to be 5×10^{-8} cm), but in reality K can be much smaller. The simulations were performed assuming various representative values of K at T_{RTA}. The activation energy for K has a very slight impact on the computed profiles; this energy was estimated by the absolute value of $K(T_{RTA})$ using a tentative prefactor 3×10^{-9} cm^3/s (resulting from the diffusion-limited expression for K). The activation energies for D_V and D_I were assumed to be small; in this case they have only a small impact on the computed profiles.

For $K < 10^{-14}$ cm^3/s the recombination is negligible in the time scale of ramp-down. Then the self-interstitials (fast diffusers) almost completely out-diffuse from the wafer. The vacancy species are frozen-in, due to vacancy binding at $T < T_b$. For the fastest quench (100 K/s), the middle-wafer concentration is close to the initial value of C_{Ve}.

For $K > 10^{-12}$ cm^3/s the recombination is almost instantaneous thus supporting $C_I C_V$ product at the equilibrium value of $C_{Ie} C_{Ve}$. In this case the self-interstitials are almost completely annihilated by the vacancies. The middle-wafer vacancy concentration (for the fastest quench) is close to the initial difference $C_{Ve} - C_{Ie}$.

In either case, the measured middle-wafer vacancy concentration (close to 10^{13} cm^{-3}) is a strong evidence in favor of a low equilibrium concentration of vacancies at T_{RTA}. In addition, the near-surface regions of reduced vacancy concentration (about 150 microns wide) clearly indicate to a high vacancy diffusivity. The complete profiles were already presented (18); here we reproduce only the two profiles of the fastest quench (Fig.4).

Since the transport capacities, $P_I = D_I C_{Ie}$ and $P_V = D_V C_{Ve}$, are already specified, there are basically only two fitting parameters – either two concentrations C_{Ie}, C_{Ve} or two diffusivities D_I, D_V, for every assumed value of K. In simulation of 3 profiles for 3 different cooling rates, the best-fit vacancy diffusivity (at T_{RTA}) was found to be only moderately sensitive to K, and quite a narrow range for D_V is specified, $(1 \text{ to } 1.7) \times 10^{-4}$ cm^2/s. The best-fit value of D_I is less definite although significantly higher than D_V. A reasonably good fit to all the available profiles (the 3 profiles of

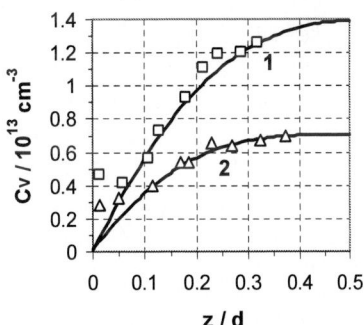

Figure 4. Vacancy depth profiles induced by RTA 1240°C followed by a fast quench, 100 K/s, and monitored by Au diffusion. Curves 1 and 2 are for the duration of 10 and 2 s, respectively. Solid lines: simulation with the best-fit parameters.

different cooling rates, and the 2 profiles of different RTA duration) was achieved only under assumption of small K. The best-fit computed profiles for slow recombination (K = 3×10^{-15} cm^3/s) correspond to D_V = 1.7×10^{-4} cm^2/s and D_I = 7×10^{-4} cm^2/s (and thus C_{Ve} = 1.5×10^{13} cm^{-3}, C_{Ie} = 6×10^{12} cm^{-3}). These numbers refer to T = T_{RTA} = 1240°C.

The value of K = 3×10^{-15} cm^3/s is small considering a very short time-scale of a post-RTA quench. For the point defect incorporation that occurs in the vicinity of the crystal-melt interface, this value of K is on the contrary effectively high: the equilibration time between the V and I defects, τ = (K (C_{Ve} + C_{Ie}))$^{-1}$, is then about 0.5 s. Actually τ is still much shorter since K(T_m) is expected to be considerably larger than K(T_{RTA}). In effect, the equilibrium between I and V is fast supported within the near-interface region of point defect incorporation.

Temperature Dependence of D_V and C_{Ve}

Two independent high-T values for D_V have been obtained above: $D_V \approx 5\times10^{-5}$ cm^2/s at 1100°C (based on the void density produced by diffusion-limited vacancy clustering) and $D_V \approx 1.5\times10^{-4}$ cm^2/s at 1240°C (based on simulation of vacancy profiles installed into wafers by RTA). At very low T, the vacancy diffusivity can be deduced from the lifetime of radiation-induced vacancies reported long ago (31). The vacancies disappear because they are trapped by oxygen atoms, to produce so called A-centers. If diffusion-limited, the trapping time is 1/(4 π R D_V C_{ox}) where C_{ox} is the oxygen concentration and R is the capture radius (adopted to be 5×10^{-8} cm). The deduced low-T values of D_V, together with the high-T values, are well described by a single Arrhenius line, with the activation (migration) energy of about 0.38 eV. The prefactor in D_V can be estimated only approximately, due to the point scatter (an uncertainty in the prefactor is estimated to be about ±30%). The parameters of the Arrhenius-type function D_V(T) are included into the Table I.

The equilibrium vacancy concentration is immediately specified through D_V and P_V as C_{Ve} = P_V/D_V. This result is also included into the Table I.

Parameters at the Melting Point

The transport capacities at T_m, are calculated according to the deduced Arrhenius-type expressions (presented in the Table I): P_V = 8.6×10^{10} cm^{-1}s^{-1} and P_I = 2.2×10^{11} cm^{-1}s^{-1}. These refined numbers will be now used in Eq.[1]. For E, a realistic value is about 4.4 eV. The concentration difference, ΔC, is approximately 1.5×10^{14} cm^{-3} as estimated from the total volume of voids. If the self-interstitial drift energy ε_I is neglected, the calculated (V/G)$_{cr}$ is consistent with the experimental critical ratio only if the term P_V (E - ε_V) is much less than P_I E . In other words, the vacancy uphill drift should be appreciable (the drift energy ε_V should be comparable to E).

The equilibrium vacancy concentration at T_m, according to the deduced Arrhenius dependence (presented in Table I) is close to 6×10^{14} cm^{-3}. The equilibrium self-interstitial concentration is smaller by ΔC, and therefore close to 4.5×10^{14} cm^{-3}. Finally, D_I at T_m is found as P_I/C_{Ie}; the number is close to 5×10^{-4} cm^2/s. All the four parameters (D_I, D_V, C_{Ve}, C_{Ie}) are thus specified at T_m. These melting-point parameters are especially important for the incorporation stage of the intrinsic point defects during crystal growth. The obtained numbers are of course approximate. An uncertainty is roughly 30 to 40% for the diffusivities and the concentrations. The values of the transport capacities, P_I and P_V, seem to be more accurate.

Activation Energies

To specify the temperature dependence of the parameters, we need 4 activation energies. The vacancy-related energies are already specified: the formation energy E_V is 4.12 eV, the migration energy $E_{Vm} = 0.38$ eV. For self-interstitials, only the sum $E_I + E_{Im}$ (which is the activation energy E_{PI} of the transport capacity P_I) is known to be 5.04 eV. The only remaining problem is how to split E_{PI} into E_I and E_{Im}, and accordingly to split P_I into D_I and C_{Ie}.

Since a high value of D_I at 1240°C - comparable to $D_I(T_m)$ - follows from simulations of the vacancy profiles in the previous section, we can conclude that E_{Im} is relatively small, most likely less than 0.4 eV. A more definite number is difficult to define since the temperature of 1240°C is too close to T_m. For the sake of completeness we temporarily adopt $E_{Im} = 0.3$ eV. This number is to be refined using some new data. With this (still somewhat uncertain) value for E_{Im}, the $D_I(T)$ function, as well as $C_{Ie}(T) = P_I / D_I$, is specified and inserted into the summary table (Table I).

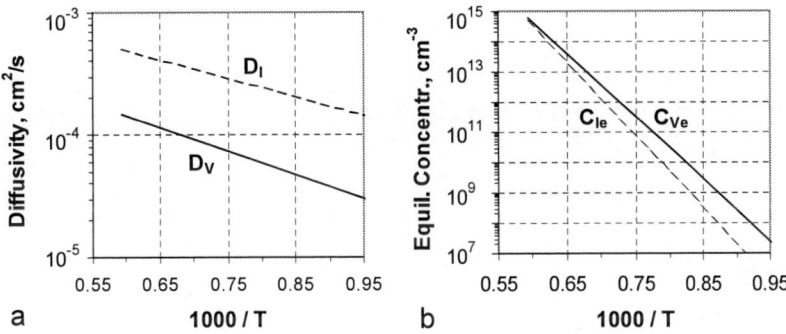

Figure 5. Deduced temperature dependence of the diffusivities (a) and the equilibrium concentrations (b) of vacancies and self-interstitials.

The Arrhenius lines for the diffusivity of vacancies (D_V) and self-interstitials (D_I) are shown in Fig.5a, and those for the equilibrium concentrations (C_{Ve} and C_{Ie}) – in Fig.5b. At the left, the lines terminate at the melting point. While the vacancy properties are well defined, the values for self-interstitial are still tentative, and therefore shown by broken lines. Yet the melting-point values of both D_I and C_{Ie} are thought to be specified reasonably well.

Summary

Using experimental data from various fields – where vacancies and self-interstitials are involved directly - it is possible to define almost all the parameters of these intrinsic point defects. Agglomeration of point defects into grown-in microdefects provides a general concept of low equilibrium concentrations (C_{Ve}, C_{Ie}) and high diffusivities (D_V, D_I) of the intrinsic point defects. This concept is important to analyze properly the data on Zn and Au diffusion and to deduce a refined value for the self-interstitial transport capacity, $P_I = D_I\, C_{Ie}$. Next, also the transport capacity of vacancies ($P_V = D_V\, C_{Ve}$) was extracted from the self-diffusion data. The vacancy diffusivity D_V at high T can be deduced from the properties of grown-in voids, and by simulating the vacancy depth profiles installed into wafers by Rapid Thermal Annealing. Together with low-T values of D_V based on the radiation experiments, it allows to specify the $D_V(T)$ function. With a known concentration difference $C_{Ve} - C_{Ie}$ at the melting point, the only not yet specified parameter is the migration energy of self-interstitials. This energy is however thought to be rather low.

An additional source of information on the parameters of the intrinsic point defects is provided by the banded patterns of grown-in microdefects (32-34). However it is difficult to deduce many fitting parameters using only such patterns. A list of fitting parameters now includes two more – the two drift energies ε_I and ε_V. A proper approach is to use the approximate values of the basic parameters $D_V(T)$, $D_I(T)$, $C_{Ve}(T)$ and $C_{Ie}(T)$ deduced in the present study as a basis for simulation of the defect dynamics (diffusion and clustering) and for predicting the microdefect patterns. In principle, refined values of the parameters – adjusted to the best reproduction of the observed patterns – could be obtained. It is not an easy task though, since the temperature field $T(r,z)$ in a growing crystal - affected by a complicated thermal environment - should be specified precisely.

References

1. V.V.Voronkov, *J.Crystal Growth*, **59**, 625 (1982).
2. V.V.Voronkov and R.Falster, *J.Crystal Growth*, **194**, 76 (1998).
3. M.Kato, T.Yoshida, Y.Ikeda and Y.Kitagawara, *Jpn. J. Appl. Phys.*, **35**, 5597 (1996).
4. M.Nishimura, S.Yoshino, H.Motoura, S.Shimura, T.Mchedlidze and T.Hikone, *J. Electrochem. Soc.*, **143**, L243 (1996).
5. T.Ueki, M.Itsumi and T.Takeda, *Jpn. J. Appl. Phys.*, **37**, 1667 (1998).
6. V.V.Voronkov and R.Falster, *J. Crystal Growth*, **204**, 462 (1999).
7. A.J.R de Kock, *Philips Res. Repts, Suppl.* **1**, 1 (1973).
8. R.Falster, V.V.Voronkov, J.C.Holzer, S.Markgraf, S.McQuaid and L.Mule'Stagno, in *Semiconductor Silicon/1998*, H.R.Huff, H.Tsuya and U.Gosele, Editors, PV 98-1, p.468, The Electrochemical Society Proceedings Series, Pennington, NJ (1998).
9. V.V.Voronkov and R.Falster, *J.Appl.Phys.*, **86**, 5975 (1999).
10. H.Bracht, N.A.Stolwijk and H.Mehrer, *Phys.Rev.*, **B52**, 16542 (1995).
11. H.Bracht, E.E.Haller and R.Clark-Felps, *Phys.Rev.Letters*, **81**, 393 (1998).
12. T.Saishoji, K.Nakamura, N.Nakajima, T.Yokoyama, F.Ishikawa and J.Tomioka, in *High Purity Silicon V*, C.L.Claeys, P.Rai-Choudhury, M.Watanabe, P.Stallhofer and H.J.Dawson, Editors, PV 98-13, p.28, The Electrochemical Society Proceedings Series (1998).

13. N.I.Puzanov and A.M.Eidenzon, *Semicond. Sci. Technol.*, **7**, 406 (1992).
14. V.V.Voronkov, *Materials Science and Engineering*, **B56**, 69 (1999).
15. V.V.Voronkov and R.Falster, *J.Crystal Growth*, **273**, 412 (2005).
16. K.Nakamura, T.Saishoji and J.Tomioka, in *High Purity Silicon VIII*, C.Claeys, M.Watanabe, R.Falster and P.Stallhofer, Editors, PV 2004-5, p.237, The Electrochemical Society Proceedings Series, Pennington, NJ (2004).
17. S.Umeno, S.Sadamitsu, H.Murakami, M.Hourai, S.Sumita and T.Shigematsu, *Jpn. J. Appl. Phys.*, **32**, L699 (1993).
18. V.V.Voronkov and R.Falster, *Solid State Phenomena*, **108-109**, 1 (2005).
19. S.Mantovani, F.Nava, C.Nobili and G.Ottaviani, *Phys.Rev.*, **B33**, 5536 (1986).
20. J.Hauber, W.Frank and N.A.Stolwijk, in *Mater. Science Forum*, **38-41**, 707 (1989).
21. N.A.Stolwijk, B.Schuster, J.Holzl, H.Mehrer and W.Frank, *Physica*, **116B**, 335 (1983).
22. N.A.Stolwijk, B.Schuster and J.Holzl, *Appl.Phys.*, **A33**, 133 (1984).
23. J.Hauber, N.A.Stolwijk, L.Tapfer, H.Mehrer and W.Frank, *J.Phys.C: Solid State Phys.*, **19**, 5817 (1986).
24. K.Compaan and Y.Haven, *Trans. Faraday Soc.*, **52**, 786 (1956).
25. A.Ural, P.B.Griffin and J.D.Plummer, *Appl.Phys.Lett.*, **73**, 1706 (1998).
26. A.Ural, P.B.Griffin and J.D.Plummer, *Phys.Rev.Lett.*, **83**, 3454 (1999).
27. R.Falster, M.Pagani, D.Gambaro, M.Cornara, M.Olmo, G.Ferrero, P.Pichler and M.Jacob, *Solid State Phenomena*, **57-58**, 129 (1997).
28. V.V.Voronkov and R.Falster, *J. Appl. Phys.*, **91**, 5802 (2002).
29. V.V.Voronkov and R.Falster, *J.Electrochem.Soc.*, **149**, G167 (2002).
30. M.Akatsuka, M.Okui and K.Sueoka, *Nuclear Instruments and Methods in Physics Research*, **B 186**, 46 (2002).
31. G.D.Watkins, *J. Phys. Soc. Japan*, **18**, Suppl. II, 22 (1963).
32. T.Sinno, R.A.Brown, E.Dornberger and W. von Ammon, *J.Electrochem.Soc.*, **145**, 303 (1998).
33. K.Nakamura, T.Saishoji and J.Tomioka, in *Semiconductor Silicon/2002*, H:R.Huff, L.Fabry and S.Kishino, Editors, PV 2002-2, p.554, The Electrochemical Society Proceedings Series, Pennington, NJ (2002).
34. V.V.Voronkov and R.Falster, in High Purity Silicon VII, C.L.Claeys, M.Watanabe, P.Rai-Choudhury and P.Stallhofer, Editors, PV 2002-20, p.16, The Electrochemical Society Proceedings Series, Pennington, NJ (2002).

76

ECS Transactions, 2 (2) 77-88 (2006)
10.1149/1.2195650, copyright The Electrochemical Society

An Atomically Accurate Model For Point Defect Aggregation In Silicon

T. Sinno[a], W. Haeckl[b], and W. von Ammon[b]

[a] Department of Chemical and Biomolecular Engineering, University of Pennsylvania,
Philadelphia, Pennsylvania 19104, USA
[b] Siltronic AG, Burghausen, D84479, GERMANY

An atomically accurate model for vacancy aggregation is presented. The model is based on two recently developed elements: (1) an accurate set of point defect properties derived from extensive model regression to multiple types of experimental data, and (2) a new description of point defect clusters at elevated temperature, which includes previously neglected entropic contributions. The latter are found to substantially alter both the cluster thermodynamics and the cluster structures, leading to modified clustering dynamics during crystal growth and wafer annealing. The model is tested extensively against experimental measurements of void densities, sizes, and aggregation temperatures, and is found to provide excellent agreement with experimental data across a wide range of crystal growth conditions.

Introduction

Vacancy aggregation during crystal growth of silicon continues to be of technological interest for multiple reasons. Most importantly, voids, which appear in both Czochralski (CZ) and float-zone (FZ) silicon single-crystals as ~100 nm octahedral cavities (1), are known to detrimentally impact the electrical quality of polished wafers used in the fabrication of DRAM memory devices (2,3). Although various strategies have been devised for reducing void sizes and densities formed during crystal growth, a fully predictive process model is required to optimize these approaches. Vacancy aggregation in crystalline silicon also serves as an ideal system for testing new multiscale modeling approaches for solid-state homogeneous nucleation. It is a single component system that can be accurately described with empirical interatomic potentials that are required for large-scale atomistic simulations (which, as will be shown, can often be instrumental in formulating accurate continuum models). Moreover, ample experimental data is available that can be used to test models at all length scales, ranging from measurements of single void dimensions to average void densities across macroscopic distances.

In this paper we present several elements that are brought together to build a process scale model for vacancy aggregation that we believe addresses all the important phenomena required to **quantitatively** predict void size distributions, densities, and evolution kinetics as a function of crystal growth operating conditions. While the basic theory governing point defect evolution during single-crystal growth and wafer post-processing is well established, the quantitative prediction of microdefect distribution is still questionable, particularly if such models are to be employed in lieu of experiments.

The first element of the void formation model presented in this paper is a detailed rate-equation system that explicitly accounts for single vacancies and self-interstitials, vacancy-oxygen complexes, and vacancy clusters up to arbitrary sizes. The latter are represented by a hybrid discrete-continuous model, which is described in previous publications (4,5,6,7) and briefly reviewed in the following section. The second element is an accurate set of point defect thermophysical properties, namely diffusion coefficients and equilibrium concentrations. These have been obtained in previous work (8,9) using a multi-model, multi-dataset regression analysis that included experimental information from crystal growth, metal diffusion, and wafer annealing experiments. The last element is a thermodynamic and structural model for describing vacancy clusters. This model is based on recent atomistic simulation results that highlight the importance of cluster configurational entropy at high temperature, which has been ignored in previous studies. The combination of these three elements is shown to provide a fully quantitative representation of void evolution during crystal growth and wafer annealing.

Defect Aggregation Model

The collective rate equation system to describe point defect aggregation is given by three sub-components that are described below. As discussed in detail in refs. (7,10), vacancy clusters are represented by a hybrid approach in which the first few sizes are each modeled by a single rate equation, and the remainder is collectively represented by a single Fokker-Planck equation.

Single Point Defects and Oxygen:

$$\frac{\partial C_V}{\partial t} + V\frac{\partial C_V}{\partial z} = D_V\frac{\partial^2 C_V}{\partial z^2} - k_{IV}\left(C_I C_V - C_I^{eq} C_V^{eq}\right) - k_{OV}\left(C_O C_V - C_O^{eq} C_V^{eq}\frac{C_{OV}}{C_{OV}^{eq}}\right)$$
$$- J_2^1 - \sum_{j=3}^{N_{disc}} J_j^1 + \int_{N_{disc}}^{N_{max}-1} n\frac{\partial I^1}{\partial n}dn \qquad [1]$$

$$\frac{\partial C_I}{\partial t} + V\frac{\partial C_I}{\partial z} = D_I\frac{\partial^2 C_I}{\partial z^2} - k_{IV}\left(C_I C_V - C_I^{eq} C_V^{eq}\right) \qquad [2]$$

$$\frac{\partial C_O}{\partial t} + V\frac{\partial C_O}{\partial z} = D_O\frac{\partial^2 C_O}{\partial z^2} - k_{OV}\left(C_O C_V - C_O^{eq} C_V^{eq}\frac{C_{OV}}{C_{OV}^{eq}}\right)$$
$$- k_{O_2V}\left(C_O C_{OV} - C_O^{eq} C_{OV}^{eq}\frac{C_{O2V}}{C_{O2V}^{eq}}\right) \qquad [3]$$

Discrete clusters:

$$\frac{\partial C_i}{\partial t} + V\frac{\partial C_i}{\partial z} = \sum_{j=1}^{i-1}\left[J_i^j - J_{i+j}^j\right] - J_{2i}^i - \sum_{j\geq i}^{N_{max}-i} J_{i+j}^i + \frac{1}{i}\int_{N_{disc}}^{N_{max}-1} n\frac{\partial I^i}{\partial n}dn, \ 1 < i \leq N_d \qquad [4]$$

78

$$\frac{\partial C_i}{\partial t}+V\frac{\partial C_i}{\partial z}=\sum_{j=1}^{N_d}\left[J_i^j-J_{i+j}^j\right], \quad N_d<i\le N_{disc}-1 \tag{5}$$

Continuous clusters:

$$\frac{\partial f(n,z)}{\partial t}+V\frac{\partial f(n,z)}{\partial z}=-\frac{\partial}{\partial n}\left[\left(A(n)-\frac{1}{2}\frac{\partial B(n)}{\partial n}\right)f(n)-\frac{1}{2}B(n)\frac{\partial f(n,z)}{\partial n}\right],$$

$$N_{disc}+1\le n<N_{max}-1 \tag{6}$$

The following definitions apply: C_I, C_V, and C_O are the concentrations of self-interstitials, vacancies and interstitial oxygen, respectively. The loss of vacancies to oxygen complexes is due to the sequence of reactions, $V+O\leftrightarrow OV+O\leftrightarrow O_2V$, although these reactions are not explicitly considered in the simulations performed here. All quantities with the superscript "eq" represent equilibrium concentrations. C_i and $f(n)$ are the concentrations of discrete and continuous vacancy clusters, where n is a continuous size variable for large clusters. N_d is the largest mobile cluster, N_{disc} is the largest discrete cluster, and N_{max} is the largest overall vacancy cluster considered in the model. J_{i+j}^j in eqs. [1], [4], and [5], represents the net forward flux at size i due to the reaction with a mobile cluster of size j to create a cluster of size $i+j$, and is given by

$$J_{i+j}^j=g(i,j)C_i-d(i+j,j)C_{i+j} \tag{7}$$

where the growth and dissolution rates, $g(i,j)$ and $d(i+j,j)$, respectively, are defined by the reversible reaction, $i+j\underset{d(i+j,j)}{\overset{g(i,j)}{\leftrightarrow}}(i+j)$. The continuous analog of the net forward flux is denoted as I and is given by the quantity in square brackets in eq. [6]. Thus, I^1 in the last term of eq. [1] is the net flux of single vacancies into continuous clusters. The terms $A(n)$ and $B(n)$ in eq. [6] represent the drift and diffusion components of the continuous flux and are given by

$$A(i)=\int w(i+j;i)j\,dj$$
$$B(i)=\int w(i+j;i)j^2\,dj \tag{8}$$

where $w(i+j;i)$ represents the transition probability from size i to size $(i+j)$. Eq. [8] is a generalization of the transition probabilities given in ref. (11) that include transitions due to cluster-cluster interactions.

The model defined by eqs. [1]-[8] requires several boundary conditions. All mobile native species (i.e. excluding oxygen) are set to their local equilibrium values at all crystal surfaces, while cluster concentrations are set to zero (equilibrium concentrations of clusters are extremely small). The matching condition at the interface between the continuous and discrete representation of vacancy clusters is taken to be

$C_{N_{match}} = f(N_{match})$ for all spatial coordinates. Finally, a no-flux condition is imposed at the largest cluster size so that $I(n = N_{match}) = 0$.

Thermophysical Property Estimation

The model represented by eqs. [1]-[8] requires the specification of several parameters related to the diffusivities and equilibrium concentrations of point defects, aggregation and dissolution rates of clusters, and a model for the free energies and capture zones of vacancy clusters as a function of size and temperature. The self-interstitial and vacancy properties used here have been regressed using a rigorous inverse modeling approach based on data from several experimental systems (8,9). These include the response of the OSF-ring and IV-boundary position to changes in the pull-rate and temperature profile during CZ crystal growth, the behavior of oxygen precipitation, and specifically the thickness of the surface denuded zone, in wafers subjected to various Rapid Thermal Annealing (RTA) treatments, and the concentration profiles of in-diffused zinc atoms in wafers as a function of time and annealing conditions. Importantly, these experimental data span a large temperature range, and exhibit varying sensitivities to the different point defect properties. A central result of this multi-model multi-dataset regression study is that while none of the experimental systems alone is sufficient to uniquely specify the point defect properties, collectively they can establish tight bounds. The parametric dependencies of each experimental system are shown in Fig. 1.

	Zinc Diffusion	DZ Depth	IVB	Crystal Growth OSF-R	Voids
$D_I(T)$	▦	▦	▦	▦	▦
$D_V(T)$	▦	▦	▦	▦	▦
$C_I^{eq}(T)$	▦	▦	▦	▦	
$C_V^{eq}(T)$		▦		▦	▦
C_V^*		▦		▦	
$\{C_{ZnS}/C_{ZnI}\}(T)$	▦				
T^O_{nuc} & T^{OV}_{bind}		▦		▦	

Figure 1. Parametric dependencies of various experimental systems used to extract property information. C_V^* is the threshold vacancy needed for oxide precipitation to proceed; C_{ZnS}/C_{ZnI} is the equilibrium ratio of substitutional-to-interstitial zinc atoms; T^O_{nuc} is the temperature at which oxide precipitation proceeds and T^{OV}_{bind} is the temperature at which vacancies are bound into oxygen-vacancy complexes – see ref. (9)

TABLE I. Point Defect Properties

Parameter	Value	Unit
$D_I(T)$	$0.237\exp\left(-\dfrac{0.937}{kT}\right)$	cm^2/s
$D_V(T)$	$7.87\times10^{-4}\exp\left(-\dfrac{0.457}{kT}\right)$	cm^2/s
$C_I^{eq}(T)$	$2.97\times10^{23}\exp(7.67)\exp\left(-\dfrac{4.0}{kT}\right)$	cm^{-3}
$C_V^{eq}(T)$	$4.97\times10^{22}\exp(7.6)\exp\left(-\dfrac{3.7}{kT}\right)$	cm^{-3}

Vacancy Cluster Thermodynamics

In general, the coalescence rate between two vacancy clusters is a function of (i) cluster mobility, (ii) cluster size (r_{cap} represents the capture radius), and (iii) cluster thermodynamics. The free energy of formation for vacancy clusters, in particular, is a critically important parameter in aggregation simulators. It appears in the expression for the cluster coalescence rate,

$$g(i,j) = k_r(i,j)C_{j,V}(r_{cap},i),\qquad [9]$$

where $k_r(i,j)$ is given by

$$k_r(i,j) = \frac{4\pi}{\delta}(D_i + D_j)r_{cap}^2(i,j)\exp\left(-\frac{\Delta G^B_{(i)+(j)\to(i+j)}}{kT}\right).\qquad [10]$$

In eq. [10], the free energy barrier associated with the coalescence event, $(i)+(j)\to(i+j)$, is given by

$$\Delta G^B_{i+j\to(i+j)} = G_{i+j}^f - G_i^f - G_j^f - kT\cdot\ln(\frac{\Omega_2}{\Omega_1}) = 0,\qquad [11]$$

where G_i^f is the formation free energy of a cluster of size i and Ω_1 and Ω_2 are the initial and final number of distinct ways of placing a distribution of clusters within a lattice (12).

Unfortunately, the vacancy cluster formation free energy is difficult to extract from experimental data and here we demonstrate a new approach for estimating this complex property from atomistic simulations. Previous void formation models have relied on cryogenic measurements of the Si(111) surface free energy to provide an approximate value ($\sim$1.24 J/m^2 (13,14)) for vacancy cluster energies because experimental obser-vations of voids in silicon show highly faceted octahedral structures that are almost exclusively aligned along (111) planes (1). Moreover, theoretical work has shown that the so-called Hexagonal Ring Cluster (HRC) morphology, in which six-vacancy rings are the building blocks for larger clusters, is the energetic ground state configuration (15,16). The HRC morphology naturally leads to octahedral geometry in the large size limit (17).

While the HRC (octahedral) morphology is a good representation of large voids at low temperatures, we have recently shown that it is not adequate for all sizes and temperatures (18). The reason is that entropic effects, namely vibrational and configurational, become important at elevated temperature and override the energetic favorability of the ground state. A crystal system containing a defect can exist in any number of (local minimum energy) configurations. While only one of these local minima corresponds to the global minimum energy configuration, other, higher energy states, can be entropically favored. This situation is shown for the 6-vacancy cluster in Fig. 2, which denotes the probability of encountering a configuration as a function of the configuration's formation energy. The data was generated using long molecular dynamics (MD) simulations at 1600 K based on the EDIP potential for silicon (19); details can be found in ref. (18). Example configurations at different locations in the distribution also are shown in Fig. 2 and highlight the fact that the most likely configurations are more disordered and extended than the energetic ground state; in fact, the ground state HRC structure is never observed at high temperature. The disordered structures near the peak of the distribution (approx. 11.8 eV) are favored because they possess larger configurational and vibrational entropy than the HRC structure.

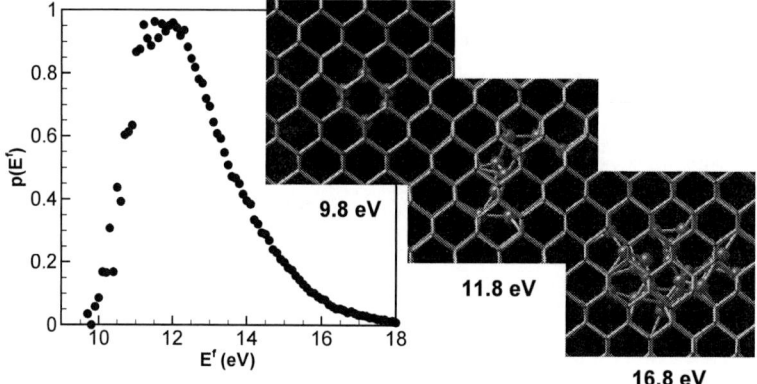

Figure 2. Probability distribution function for the formation energies of the 6-vacancy cluster obtained by EDIP MD simulation at 1600 K. Also shown are sample configurations; large red spheres denote atoms that are displaced by more than 10% of a bond length from their equilibrium positions.

Once the probability distribution function for a given cluster size is determined, it can be used to compute the free energy of formation as a function of temperature by applying the relationship $G^f(n) \sim -k_B T \int p(E^f) dE^f$, where the integral is taken over all possible values of the formation energy. A summary of cluster free energies of formation as a function of temperature and size is shown in Fig. 3. The free energies are expressed in terms of effective surface free energies, which are defined as $\sigma(n,T) = G(n,T)/A_{sph}$, where A_{sph} is the surface area of a sphere with volume equivalent to the sum of the volumes of the vacancies in the cluster. Also shown in Fig. 3 are the surface free energies of HRC clusters (upper surface). As expected, the two estimates converge at low temperature when entropic contributions are small. It is notable that at low temperature the predicted surface energy (for both cases) is in excellent agreement with the experimentally measured values of about 1.24 J/m^2 at 77 K (13,14). However, at elevated temperature, the surface free energies diverge strongly for all sizes considered, and particularly for small clusters. With the new approach, the effective surface free energy decreases to a value of about 0.85 J/m^2 at the melting temperature. The divergence between the two results is entirely attributed to the cluster configurational entropy because the vibrational portion is included in both estimates. Note that the onset of aggregation during crystal growth is driven primarily by small clusters at high temperature, and therefore the deviation between the standard HRC model and our new model is likely to be significant in the context of the predicted void size distributions.

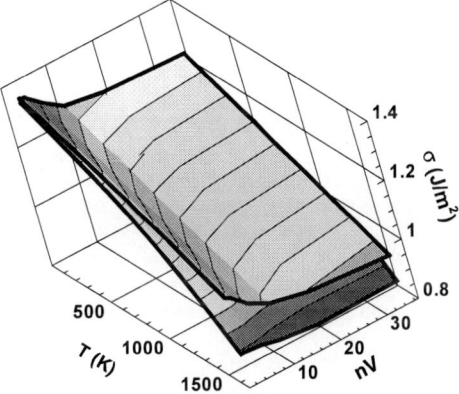

Figure 3. Temperature and size dependence of effective surface free energy (σ) for vacancy clusters. Lower surface (σ_{FULL}): current results including configurational entropy, Upper surface (σ_{HRC}): HRC calculations with vibrational entropy only.

A contour map of the relative deviation between the two free energy surfaces is shown in Fig. 4(a). The associated physical phenomena in each part of the "phase diagram" are highlighted in Fig. 4(b). The deviation is maximal for small clusters at high temperature, and as mentioned above, the new thermodynamic model is expected to play a large role in the onset of aggregation during crystal growth. The persistent deviation for larger clusters at high temperature arises from cluster surface melting; i.e. the new model accounts for the fact that voids are essentially internal surfaces, which locally melt at temperatures below the bulk melting point. Finally, no deviation is apparent for larger clusters at low temperature, and excellent agreement with experimental measurements of the Si(111) surface free energy is obtained with both models.

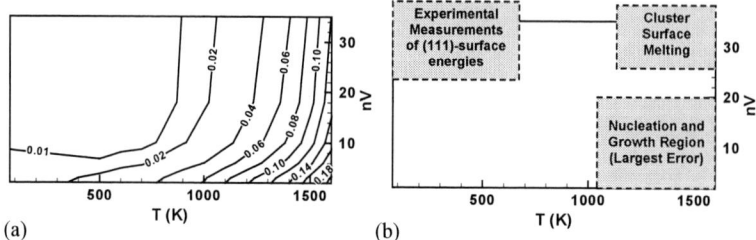

(a)

(b)

Figure 4. (a) Relative error, $(\sigma_{HRC} - \sigma_{FULL})/\sigma_{FULL}$, between current ($\sigma_{FULL}$) and HRC ($\sigma_{HRC}$) surface free energies as a function of temperature and vacancy cluster size; (b) "Phase diagram" showing the various aspects that are influenced by the new surface free energy model.

Vacancy Cluster Size and Structure

The effect of the cluster configurational entropy is not limited to the cluster thermodynamics. As shown in Fig. 2, the most likely structures at high temperature are disordered and extended and therefore are characterized by larger capture zones. An effective capture radius for each cluster size in the interval $1<n<500$ was calculated using MD simulations. In each of these simulations, a single vacancy cluster was allowed to sample various configurations. Each configuration was characterized by tagging all atoms in the vicinity of the cluster that were displaced from their perfect lattice sites by more than a critical distance, β, which was taken to be 0.08 Å based on previous calculations of the vacancy-vacancy interaction distance (17). The cluster volume was then taken as the sum of all tagged atoms. An average effective capture radius was then derived by assuming spherical cluster geometry so that

$$R_{avg} = \left\langle \left(\frac{3V(\beta)}{4\pi} \right)^{\frac{1}{3}} \right\rangle,$$ [12]

where $V(\beta)$ is the volume of all atoms displaced by at least β from their equilibrium positions. A comparison between the cluster sizes predicted by this model and that predicted by assuming that the clusters are compact spheres at 1600 K is shown in Fig. 5. A constant shift is observed for clusters larger than about 100 vacancies, but the effect is strongest for small clusters which are most influenced by "entropic expansion". Given that the agglomeration rate is proportional to the square of the capture radius (eq. [10]), this effect is at least as important as the thermodynamic one.

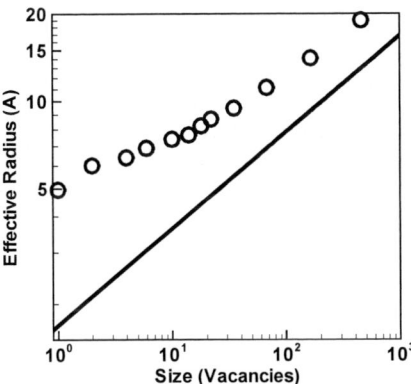

Figure 5. Effective Cluster capture radius as a function of number of vacancies for $\beta = 0.08\,\text{Å}$ (circles). Compact sphere model – solid line. T=1600 K.

Results and Discussion

The model elements discussed in the previous sections were combined in a simulator and applied to multiple crystal growth cases. Several metrics were used to compare the model predictions to experimental data including the total cluster density (both volumetric and surface measurements), average cluster size, and aggregation temperature, which the temperature at which measurable void formation occurs during crystal growth. The total volumetric density is defined as

$$F_{3D}(n_{min}) = \int_{n_{min}}^{\infty} f(n)dn,$$ [13]

where $n_{min} \sim 1 \times 10^5$ vacancies (corresponding to ~30 nm diameter) is the minimum cluster size included in the count. This limit is chosen based on the experimental observation limit, although its precise value does not influence the computed void density for most crystal growth conditions. The surface density is extracted indirectly from gate-oxide integrity (GOI) measurements taken across a wafer surface. It is computed from the volumetric size distribution by

$$F_{2D} = \int F_{3D}(n)D(n)dn,$$ [14]

where $D(n)$ is the effective diameter of a cluster containing n vacancies. Note that GOI only provides a relative measure of the void surface density, i.e. the GOI is proportional to the surface void density but cannot be used to directly estimate the actual value.

A comparison of model predictions and experimental measurements of void densities is shown for several crystals in Figs. 6(a) and (b). For each crystal, two-dimensional crystal geometries and thermal fields were computed separately using large-scale integrated heat transfer models that have been verified by comparison to experimental data; this information was used as input into the void simulator. The void size distribution experimental data considered here are subdivided into two sets. In the first set, (Fig. 5(a)) volumetric densities were measured for seven CZ crystals in which the pull rate and the thermal history were varied; these crystals (both 200 MM and 300 MM diameter crystals are represented) have been analyzed in previous work (20). The void densities (and in some cases the void sizes) were measured using repeated SC1 cleanings and light scattering. In the second set (Fig. 6(b)) five CZ crystals were analyzed using GOI measurements of test capacitor structures grown on the surfaces of wafers sliced from the as-grown ingots.

Clearly, the agreement between the current model predictions for the void density (squares) and experimental data (circles) is quantitatively excellent for both sets of crystals. Note that in Fig. 6(b), crystal 4 was used as a reference to obtain the proportionality constant between GOI and surface void density. The average void size predictions in the inset of Fig. 6(a) also are in excellent agreement with experimental measurements, although this data is somewhat less reliable because it is difficult to characterize large numbers of voids and obtain a statistically converged estimate.

Also shown are the predictions of a model based on HRC void thermodynamics and structures (triangles). While the agreement with the experimental data is also quite good in this case, there are certain crystal growth conditions that are not well modeled (e.g. crystals 1 and 5 in Fig. 6(b)). In particular, the model is seen to fail for fast cooling rate and low V/G environments, which is a significant shortcoming given the importance of these operating condition windows for producing smaller, less visible defects ().

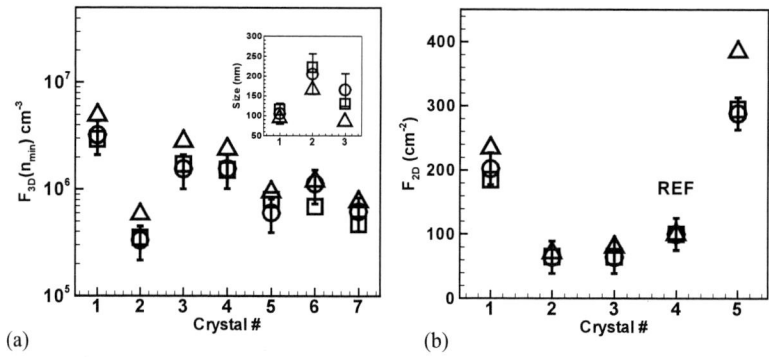

(a) (b)

Figure 6. (a) Comparison between predicted (squares) and experimental (circles) absolute volumetric densities for crystals in Set 1. Inset: Predicted (squares) and experimental (circles) void sizes based on a spherical geometry assumption. (b) Comparison between predicted (squares) and experimental (circles) relative surface densities for crystals in Set 2. Error bars are based on 25% measurement uncertainty.

Finally, the aggregation temperatures predicted by the new vacancy cluster model and the HRC-based model were computed. The temperature range obtained for the new model is approximately 1050-1110 °C, which is again in excellent agreement with experimental measurements. Note that for lower V/G conditions (e.g. V/G~V/G$_{crit}$), the aggregation temperature can decrease significantly below this range; simulations under such operating conditions will be considered in future work. On the other hand, the model based on compact void structures predicts an aggregation temperature range of about 950-1020 °C, which is almost 100 °C too low (7). This shortcoming has been a long-standing issue in previous modeling efforts and *ad hoc* approaches have been used to artificially lower the void surface energy.

Conclusions

An atomistically detailed continuum model for void formation during CZ silicon crystal growth was presented. The model is parameterized using a combination of multiple-model regression to experimental data for the point defect properties and atomistic simulation for vacancy cluster properties. These properties have been shown to provide a quantitative representation of a diverse set of point defect-related phenomena, including, metal diffusion, crystal growth, and oxide precipitation. The atomistic simulations of vacancy cluster properties demonstrate, for the first time, that the vacancy cluster configurational entropy is an important contribution to the cluster thermodynamics and substantially alters both the cluster formation free energies and structural characteristics. Interestingly, the source of the cluster configurational entropy is almost entirely due to off-lattice relaxations in the surrounding matrix; as a result, it is not possible to calculate this entropic effect with lattice-based approaches.

At high temperatures, where nucleation and growth are important, the effect of configurational entropy is to stabilize cluster configurations that are more disordered and spatially extended than the ground state structures (compact, octahedral) that are usually used as a basis for modeling void properties. As the crystal cools during crystal growth and entropic effects diminish, the cluster properties gradually return to values in excellent agreement with experimental measurements. The overall effect of new entropic source is to reduce the effective void surface energy and increase the effective capture radius at high temperatures. These two effects lead to an earlier onset of aggregation and automatically lead to a substantial increase in the void nucleation temperature, which is in excellent quantitative agreement with experimental measurements.

The new void aggregation model presented in this paper therefore is able to provide quantitative prediction of several quantities: (i) void densities, (ii) void sizes, and (iii) the void aggregation temperature during CZ growth for a wide variety of crystal growth conditions. This broad capability is achieved with no parameter fitting because all the point defect properties were previously obtained by model regression to multiple other experimental data sources.

Acknowledgments

We gratefully acknowledge financial support for this work from Siltronic AG and the National Science Foundation.

References

1. M. Itsumi, H. Akiya, T. Ueki, M. Tomita and M. Yamawaki, *J. Appl. Phys.*, **78**, 5984 (1995).
2. E. Dornberger, D. Temmler and W. von Ammon. *J. Electrochem. Soc.*, **149**, G226 (2002).
3. J. Eaglesham, P. A. Stolk, H.-J. Gossmann and J. M. Poate. *Appl. Phys. Lett.*, **65**, 2305 (1994).
4. R. A. Hartzell, H. F. Schaake and R. G. Massey, *Mater. Res. Soc. Symp. Proc.*, **36**, 217 (1985).
5. J. Esfandyari, C. Schmeiser, S. Senkader, G. Hobler and B. Murphy, *J. Electrochem. Soc.*, **143**, 995 (1996).
6. M. Schrems, T. Brabec, M. Budil, H. Potzl, E. Guerrero, D. Huber and P. Pongratz, *Mat. Sci. Eng.*, **B4**, 393 (1989).
7. T. Sinno and R. A. Brown, *J. Electrochem. Soc.*, **146**, 2300 (1999).
8. T. Frewen, T. Sinno, E. Dornberger, R. Hoelzl, W. von Ammon and H. Bracht, *J. Electrochem. Soc.*, **150**, G673 (2003).
9. T. A. Frewen, T. Sinno, W. Haeckl and W. von Ammon, *Comp. & Chem. Eng.*, **29**, 713 (2004).
10. T. A. Frewen, S. S. Kapur, W. Haeckl, W. von Ammon and T. Sinno, *J. Crystal Growth*, **279**, 258 (2005).
11. H. Risken, *The Fokker-Planck Equation: Methods of Solution and Applications.* Springer-Verlag. Berlin (1989).
12. J. L. Katz and H. Wiedersich, *J. Chem. Phys.*, **55**, 1414 (1971).
13. R. J. Jaccodine, *J. Electrochem. Soc.*, **110**, 524 (1963).
14. J. J. Gilman, *J. Appl. Phys.*, **31**, 3208 (1960).
15. A. Bongiorno, L. Colombo, F. Cargnoni, C. Gatti and M. Rosati, *Europhys. Lett.*, **50**, 608 (2000).
16. M. Prasad and T. Sinno, *Appl. Phys. Lett.*, **80**, 1951 (2002).
17. M. Prasad and T. Sinno, *Phys. Rev. B*, **68**, 045206 (2003).
18. S. Kapur, M. Prasad, J. C. Crocker and T. Sinno, *Phys. Rev. B*, **72**, 014119 (2005).
19. M. Z. Bazant, E. Kaxiras and J. F. Justo, *Phys. Rev. B*, **56**, 8542 (1997).
20. E. Dornberger, Ph.D. Thesis, Universite Catholique de Louvain, Belgium (1998).

ECS Transactions, 2 (2) 89-94 (2006)
10.1149/1.2195651, copyright The Electrochemical Society

Growth Technologies for 300 mm Arsenic Heavily Doped Silicon Crystals

Hailing Tu, Qigang Zhou, Guohu Zhang, Xiaolin Dai, Zhiqiang Wu, Taotao Jia

National Engineering Research Center for Semiconductor Materials,
General Research Institute for Nonferrous Metals
GRINM Semiconductor Materials Co., Ltd.
No.2 Xin Jie Kou Wai St., Beijing100088, China

Arsenic heavily doped silicon substrates are vital to the power devices and power integrated circuits. The present work focuses on the series of growth technologies for 300 mm silicon crystals heavily doped with arsenic. The 22" hot zone has been employed and growth parameters have been optimized to improve the yield of the as-grown ingots. The systemic processes have been applied for the human safety and environmental concerns, which guarantee the exhaust gas and water have met the requirement for the national laws of the environmental protection.

Introduction

As-doped silicon polished wafer is a key substrate material for IC industry. Enlarging wafer diameter has become the main developing trend. At present, for commercial usage, As-doped silicon polished wafers of 150 mm and 200 mm in diameter are widely used. The growth of larger diameter As-doped silicon single crystal ingots will meet the requirement of the devices which need N-type and low resistivity substrates. This experiment is focused on the growth of 300 mm arsenic heavily doped silicon ingot on 22" hot zone. The properties of As-doped ingots and lightly doped ingots are compared.

Crystal Growth and Measurements

a) The ingots were grown in a KX-150 crystal puller with 22" hot zone that was specially designed and suited for 90 kg charge. The graphite heat shield was used to optimize the thermal profile. Boron content and oxygen concentration were measured by secondary ion mass spectroscopy (SIMS) and gas fusion analysis (GFA) respectively. Oxidation induced stacking fault (OSF) and flow pattern defect (FPD) (1) were detected by Secco (2) and Wright (3) etch and Vacancy/interstitial (V/I) boundary was observed by Cu decoration (4, 5).
b) Under the same growing condition, the 308 mm boron-doped (P-type, lightly doped) ingot and arsenic heavily doped ingot <100> orientation were grown respectively.
c) Doping method: Gas doping.
d) The conditions of body growth are listed in Table 1.

Table I. Crystal Growth Conditions and Parameters

Average Growth Rate (mm/min)	Seed Rotation (rpm)	Crucible Rotation (rpm)	Tank Pressure (Torr)	Distance between Heat Shield and Melt Surface (mm)	CUSP Magnet Field (upper/lower, A)
0.58	5-12	6-12	15-45	45	260/220

From adding arsenic to start of body growth, it took 26 hours. The diameter was 306.2-306.7 mm. The body length was 249 mm and the weight was 76.2 kg.

e) Inspection sites and method: several slugs of wafer samples (1~3 mm thick) from different part of ingot were cut off, and the following data were inspected: arsenic concentration and resistivity radial gradient (RRG), total oxygen content, radial oxygen variation (ROV), boron content, dislocation density, OSF, FPD, V/I boundary. The samples were taken from 0 mm, 120 mm and 249 mm respectively on body. In order to compare reasonably, the total oxygen content was inspected for both boron and arsenic doped ingot. Arsenic concentration was obtained according to F723-82. The detailed inspection method, reference standard and testing equipments are listed in Table 2.

Table II. Inspection Methods, Standards and Testing Equipments

As concentration & RRG	Total Oxygen & ROV	Boron	Dislocation Density	OSF	FPD	V/I
ASTM F84-84a ASTM F723-82	GFA	SIMS	ASTM F47-84	ASTM F522	Schimmel/Secco Etching	Copper decoration

Results and Discussion

i. The arsenic concentration in the body part is (8.48-11.4) E18 atom/cm^3.

ii. RRG is <5% (Plan-B).

iii. OSF ring is not observed on the sample surface, and OSF density is <5 pcs/wafer.

iv. The total oxygen of As-doped ingot is 10-25% less than the lightly doped ingot. ROV is <15%.

v. V/I boundary was not found on slugs cut from As-doped crystal, which was obviously different from the lightly doped ingot, even with the same growth parameter.

vi. By SIMS measurement, the boron concentration of ingot tail slug was 4.9E13 atom/cm^3.

The ideal hot zone size for 300 mm ingot growth should be 28 inches, 32 inches, or even larger. The ratio of crystal/crucible diameters should be >2.33. In this experiment, it was only a test to pull large diameter ingot with 22" hot zone, the ratio of ingot/crucible diameters is only 1.83, which can decrease the crystal resistivity along the body length rapidly. By changing pressure in the puller, the axial resistivity distribution was controlled (see Figure 1). The escape of arsenic atoms mainly happened during doping (about 25%). The evaporated quantity of arsenic during body growth was mainly determined by hot zone structure and pressure in the puller. Generally, when heat shield is used, more arsenic escapes from puller. If the inner pressure of the puller is higher, the less arsenic escapes.

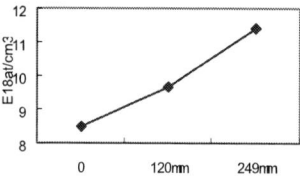

Figure 1. Axial arsenic concentration

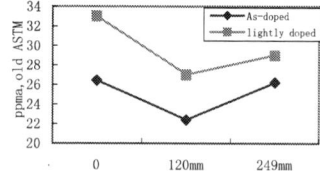

Figure 2. Oxygen content comparison

From oxygen data (Figure 2), we can see that in As-doped ingot, the oxygen content is 10-25% lower than that in the lightly doped ingot. This result is in coincidence with the analysis of Ryuta etc (6). With the application of 22" hot zone, the oxygen content can be increased effectively in the As-doped ingot, and the mechanical strength of polished wafers can also be increased, at the same time. The yield loss caused by breakage can be decreased in wafer processing. Therefore, the internal gettering effect can be strengthened in IC process.

Regarding FPD, OSF and Cu decoration test result, FPD has been found, but OSF and V/I boundary have not been observed. This result is in conformance with Porrini's analysis (7). Because large quantity of N-type dopant was added, the ξ_0 (V/G) value was decreased and the voids defect was easily formed. As the ξ_{real} value is much higher than ξ_0 (8-10) value, the "L-band" and "P-band" (OSF) (8, 9) have not been found. In addition, the inspection on FPD showed more FPD defects at the tail. This is related to ingot growth condition. In this test, the heat shield was used and the ingot tail was grown very fast. Figure 3 & 4 show pictures of the As-doped ingot FPD and Cu decoration test result. In Figure 3a and 4a, the striations are caused by distribution of impurities. Figure 5 is the lightly doped ingot inspection result. There is an 85 mm radius V/I boundary, and less FPD defects appeared. The influence factors for defects formation are: hot zone properties, shape of melt-body interface, the distance between silicon liquid surface and heat shield, growth rate, etc.

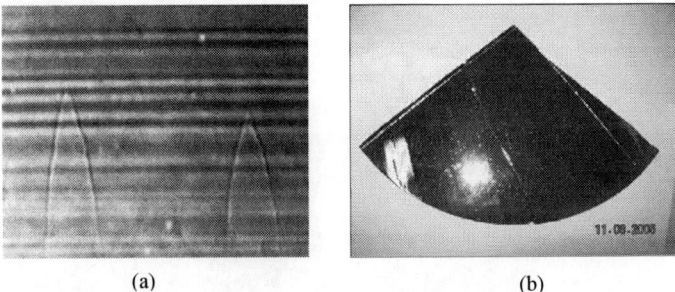

(a)   (b)

Figure 3. FPD (Figure 3a) and Cu decoration (Figure 3b) at 120 mm body length of As-doped crystal.

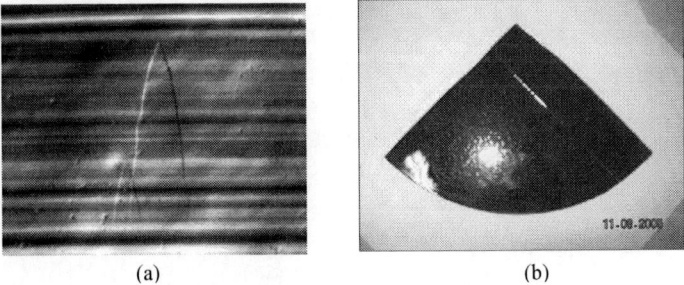

(a)   (b)

Figure 4. FPD (Figure 4a) and Cu decoration (Figure 4b) at 249 mm body length of As-doped crystal.

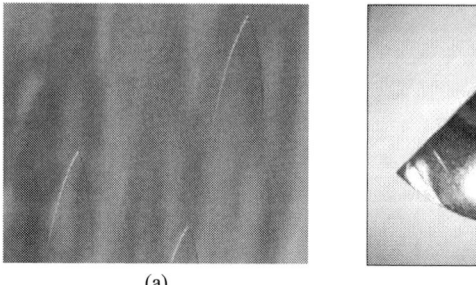 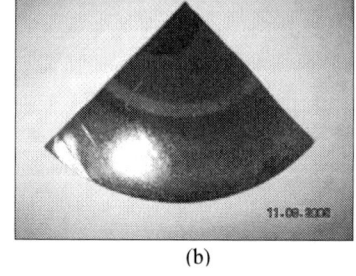

(a)   (b)

Figure 5. FPD (Figure 5a) and Cu decoration (Figure 5b) for lightly doped crystal.

After taking special measure, the boron content in the ingot can be controlled within the acceptable range.
With MCZ Kayex-150 type puller and 22" hot zone, the 300 mm As-doped and <100> single-crystal ingot was successfully grown. The arsenic concentration, body oxygen content, dislocation density, OSF and FPD etc. were inspected. Figure 6 shows the single crystal body shape.

Figure 6. Appearance of As-doped Si crystal

Environmental Protection

Arsenic compound are harmful to human beings and environment. In this experiment, a special technology for safety production and waste treatment is developed. It includes arsenic storage, arsenic doping process, cleaning puller process, waste treatment, and operator protection, etc. The details are as follows:

i.   Arsenic storage: The arsenic is kept in a special safe box. When operators use it, the record will be made, and confirmed frequently.

ii. When arsenic is weighed and doped into the puller, operators wear special clothes used only one time, so as to avoid touching arsenic powder.

iii. When cleaning puller chamber, the humid-absorbing system is used. It is a kind operation system of internal cycle with that the arsenic-contained waste water and gas can not leak out. All the wastes are collected and sent to professional factory for treatment.

iv. This is the flow chart of the technology as shown in Figure 7.

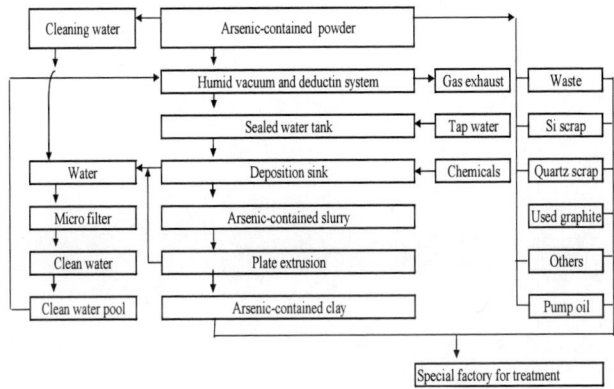

Figure 7. The systematic processes for the arsenic-contained waste treatment

By using this technology, the As-doped ingot production has been approved by Beijing Environment Bureau. There is no waste water produced. And its exhaust gas is up to the National Protection Standard for Industry Exhaust Gas and can be exhausted directly.

Summary

300 mm silicon crystal ingots have been grown in a Kayex-150 crystal puller with specially designed 22" hot zone. The graphite heat shield has been applied. The longitudinal arsenic concentration distribution is from 8.48 to 11.4×10^{18} atom/cm^3 and oxygen concentration distribution is from 22.4 to 26.4 ppma (Old ASTM), and the radial oxygen vibration is from 8.3 to 12.3%. The properties of heavily As-doped and lightly B-doped ingots are compared.

References

1. H. Yamagishi, I. Fusegawa, N. Fujimaki and M. Katayama, Semicond. Sci. Technol. **7**, A135 (1992).
2. F. Secco d'Aragona, J. Electrochem. Soc., **119**, 948 (1972).
3. Magarate Wright Jenkins, J. Electrochem. Soc., **124**, 757 (1977).
4. T. Abe, H. Harada, and J. Chikawa, in Oxygen, Carbon, Hydrogen, and Nitrogen in Crystalline Silicon, J. C. Mikkelsen, Jr., S. J. Pearton, J. W. Corbett, and S. J. Pennycook, Editors, Vol.59, p.537, Materials Research Society, Pittsburgh (1986).
5. A.J.R. de Kock, Philips Res. Rept. Suppl., **1**, 1 (1973).
6. J. Ryuta, E. Morita, T. Tanaka, Y. Shimanuki, Jap. Journal of Appl. Phys., 29, L1947 (1990).
7. M. Porrini, Cryst. Res. Technol. **40**, 1054 (2005).
8. V. V. Voronkov and R. Falster, J. Cryst. Growth, **204**, 462 (1999).
9. V. V. Voronkov and R. Falster, ECS, **149**, G167 (2002).
10. R. Falster, V. V. Voronkov, J. Holzer, S. Markgraf, S. McQuaid and L. Mule'Stagno, in Semiconductor Silicon/1998 (8th International Symposium), H. R. Huff, H. Tsuya, U. Gösele, Editors, PV 98-1, p. 468, The Electrochemical Soc., San Diego, CA (1998).

ECS Transactions, 2 (2) 95-107 (2006)
10.1149/1.2195652, copyright The Electrochemical Society

DEFECT FORMATION BEHAVIORS IN HEAVILY DOPED CZOCHRALSKI SILICON

Wataru Sugimura[a], Toshiaki Ono[a], Shigeru Umeno[a], Masataka Hourai[a] and Koji Sueoka[b]

[a] Advanced Technology Development Department, SUMCO Corporation,
Kouhoku, Kishima-gun, Saga 849-0597, Japan
[b] Department of System Engineering Okayama Prefectual University,
111 Kuboki, Soja, Okayama, 719-1197, Japan

To reveal a difference of defect formation behaviors, i.e. grown-in void formation during crystal growth and oxide precipitation in n- and p-type silicon, we have investigated by using heavily boron- and arsenic-doped silicon crystals. The density of void defects in heavily boron doped silicon was decreased with a shrinking OSF-ring, but in arsenic doped silicon were increased with resistivities below 3.3mΩcm. On the other hand, for oxygen precipitation, the nucleation rate in boron doped silicon was enhanced with increasing resistivities, while decreased by one tenth in reference to lightly doped silicon for resistivities up to 4.4mΩcm in arsenic doped silicon. These contrastive phenomena between n- and p-type cannot be explained with a growth model of precipitates by an accelerated diffusion of oxygen in silicon. We believed that the nucleation rate of oxide precipitates related to a dependence of point defects on fermi level closely.

INTRODUCTION

The effect of heavily boron doping on defect formation behaviors in silicon crystal, i.e., grown-in defect formation or nucleation rate of oxide precipitates has been widely studied(1-11) due to its importance on a semiconductor industry. For grow-in defect formation in heavily boron doped crystals, OSF-ring shrunk and is dominated by a region of interstitial silicon(I-rich)(12). On the other hand, heavily antimony doped silicon has an increased density of void defects with concentrations up to 4×10^{18}atoms/cm^3(11), its concentration is maximum in the case of antimony doped silicon from considering atomic size. The difference in the grow-in defect formation mechanism by the type of dopant was proposed by several models. One is a model in which the thermal equilibrium concentration of a point defect has changed because of the addition of the dopant atom(13). Another model that considers the stability of point defects for each concentration of dopant, assumes that the formation energy of point defects is influenced by dopant concentration(14). For the kinetics of oxide precipitation, the nucleation rate of boron doped silicon accelerated with decreasing resitivities but was suppressed in antimony or phosphorus doped silicon(10). These phenomena have been verified by many experimental results, but no model has explained the difference of defects formation between p- and n-type silicon clearly. In this paper, the dopant concentration dependence of the defect formation was statistically investigated using boron and arsenic doped silicon with a wide range of impurity concentration. It is considered that a validity of a defect formation model is verifiable by using arsenic as a dopant because it can be

95

added in high concentration because of equal atomic size to silicon. Also boron can be added in high concentrations to silicon.

EXPERIMENTAL

Boron doped 200mm diameter CZ silicon wafers with concentrations ranging from 1.34×10^{15} to 2.0×10^{20} atoms/cm^3, and arsenic doped 150mm wafers with 1.7×10^{15} to 1.6×10^{19} atoms/cm^3 were prepared in this study. For arsenic doped crystals, several types of hot-zones were used. The initial oxygen concentration of the boron doped wafers were 11.0 to 12.2×10^{17} atoms/cm^3, and 14.0 to 16.0×10^{17} atoms/cm^3, respectively, as determined by using the old ASTM standard. Grown-in void defects known as aggregate supersaturated vacancies were measured by light scattering method, with the commercially available tool MO-601. The location of crystal-originated particles(COPs) on the wafer surface were determined using a laser particle counter(KLA-Tencor Surfscan 6220) and the structure of COPs were observed using an AFM(Seiko Instruments SPA360).

In order to understand the difference of oxygen nuclei formation between p- and n-type silicon, samples were subjected to an nucleation heat treatment (Temperature: 600-900°C/Time: 1-96 hours) in an O$_2$ atmosphere after an Epi-process, which acts a dissolution treatment of grown-in oxygen precipitates. Then, the density of micro defects was characterized by Wright etching the sample for 2min after a growth anneal of 16 hours at 1000°C in an O$_2$ atmosphere. The nucleation rate was determined by using the measured density of micro defects(BMD) of each annealing time. The oxygen depth profiles due to out-diffusion from the surface were measured by secondary ion mass spectroscopy(SIMS), in a CAMECA IMS $3f$, using a Cs$^+$ primary ion beam having an energy of 14.5 keV, after heat treatment(Temperature: 700-1050°C/Time: 1-64 hours) in a N2 ambient. From these profiles, a diffusion constant of oxygen in silicon was estimated.

RESULTS

Grow-in defect formation in heavily doped silicon

Figure 1 shows LSTD Maps of boron doped wafers, when measured by MO601, with dopant concentration from 5.1×10^{18} to 9.7×10^{18} atoms/cm^3. As shown in Fig.1, the density of void defects decreased with increasing concentration of boron, and no defects were observed for 9.7×10^{18} atoms/cm^3. These results suggest that the diameter of an OSF-ring has shrunk with increasing concentration of boron.

Figure 2(a) shows MO601 LSTD maps of arsenic doped wafers, and (b) shows a location of the OSF-ring using an X-ray topography after a growth heat treatment 16 hours at 1100°C in an O$_2$ atmosphere. The OSF-ring is located at the edge of a wafer for every arsenic concentration. As shown in Fig. 2(a), the LSTD density was increasing with dopant concentration up to 1.4×10^{19} atoms/cm^3, but dropping off rapidly after that concentration of arsenic. These results suggest that a mechanism of void formation was dependent on the concentration of arsenic when 1.8×10^{19} atoms/cm^3.

Figure 3 shows the arsenic concentration dependence of LSTD defect density for crystals that were grown using four different types of hot-zones(A,B,C and D) to change the V/G conditions. It has been reported that the density of grow-in defects change with

V/G(15). As shown in Fig. 3, the defect density decreased rapidly with arsenic concentrations higher than 1.8×10^{19} atoms/cm^3 independent of the hot-zone configuration. From this result, it is confirmed that a decrease in void defects has strongly depends on arsenic concentration.

Figure 4 shows AFM image of COPs in an arsenic doped wafer with a concentration of 1.4×10^{19} atoms/cm^3. As shown in Fig 4, no difference in COP feature was observed in heavily arsenic doped wafer as compared with lightly doped wafers. From this result, there is no effect of heavily arsenic doping on the structure of COPs.

To reveal a mechanism of void formation in arsenic doped silicon, information of the total volume of void defects at each arsenic concentration is very important for understanding the process of void formation. Here, the size distribution and defect density in arsenic doped silicon were measured by using MO601. Results are shown in Figures 5. From these results, two tendencies were revealed for arsenic concentrations ranging from 1.5×10^{15} atoms/cm^3 to 2.7×10^{19} atoms/cm^3. The total volume of void defects increased with dopant concentration up to 1.4×10^{19} atoms/cm^3, and decreased above that concentration, as shown in Fig 5. These results suggest that isolated vacancies might exist in heavily arsenic doped silicon concentrations higher than 1.4×10^{19} atoms/cm^3 since all vacancies are not consumed by void formation.

Oxygen nuclei formation in heavily doped silicon

Figures 6 shows annealing time dependence of bulk micro defects(BMD) density as a function of heat treatment temperature in boron doped wafer, each boron concentration is (a)1.34×10^{15} atoms/cm^3, (b)4.1×10^{18} atoms/cm^3, (c)1.13×10^{19} atoms/cm^3 and (d)2.0×10^{20} atoms/cm^3, respectively. After an epitaxial process, these wafers were subjected to a 2step heat treatment(600-800°C/4hr+1000°C16hr). The results of the BMD density after Wright etching were plotted in a graph as shown in Fig. 6. Since the epitaxial process involves a high temperature process, nuclei generated during crystal growth shrink and the BMD density are lower as a whole. In fact, there no precipitates were generated after a 16 hour anneal at 1000°C16hr. As shown in Fig. 6, the nuclei formation rate goes up with increasing of boron concentration, a nuclei formation occurred at extremely short times at more than 1.0×10^{19} atoms/cm^3 of boron concentration. An effect of promoting nuclei formation in heavily boron doped silicon was confirmed and it is revealed that the nuclei formation occurred at high temperatures.

Figures 7 shows the time dependence of BMD density in the case of arsenic doped wafers, each concentration is (a)1.7×10^{15} atoms/cm^3, (b)1.7×10^{18} atoms/cm^3, and (c)1.6×10^{19} atoms/cm^3. These wafers were subjected to a 2step heat treatment(600-750°C/4hr+1000°C16hr). As shown in Fig 7, the nuclei formation was controlled by the heavily arsenic doping. Furthermore, an incubation time of more than 15 hours was observed at 650℃ temperature for 1.6×10^{19} atoms/cm^3 of arsenic. These results suggest that heavily arsenic doping has an influence at the initial process of nuclei formation.

The oxygen diffusion constant in heavily doped silicon

Figure 8 shows the SIMS depth profile of oxygen concentration for 1.7×10^{15} atoms/cm^3 of arsenic concentration annealed at 900°C for 4 hours. The vertical axis indicates the oxygen concentration normalized by initial oxygen concentrations. To determine the diffusivity of oxygen in silicon, the oxygen depth profile was fitted to the error function equation. Here, normalized oxygen, C_{os} is given by

$$C_{os}(x)=C_{sn}+(1-C_{sn})erf(x/2*(Dt)^{1/2}) \qquad [1]$$

where C_{sn} is the normalized surface oxygen concentration, x is the depth, D and t are diffusivity of oxygen and annealing time respectively. Fitting was done by varying the surface oxygen concentration:C_{sn} and the oxygen diffusivity:D. The best fitting or error function to SIMS data is also shown in Fig. 8. By this fitting work, obtained the value of D at each temperature in boron- and arsenic-doped silicon were plotted in Figures 9.

Figures 9 shows the oxygen diffusion constant in (a)boron doped wafer and (b)arsenic doped wafer as a function of the dopant concentration. The solid line represents the result of Mikkelsen et al(1982)(16). As shown in Figure 9(a), the oxygen diffusion constant in heavily boron-doped wafer was observed to be higher, as reported by Sueoka et.al(8). However, no deviation of the diffusion constant from the Mikkelsen data were observed in arsenic doped wafer. These results suggested an accelerated diffusion of oxygen in heavily boron-doped silicon for concentrations higher than 3.8×10^{18}atoms/cm^3. This phenomena was considered to have caused acceleration of nuclei formation.

DISCUSSION

In this section, we will first discuss about that behavior of void formation in heavily doped silicon. Experimental results are compared with calculation results using a size effect model that considers the equilibium concentration of point defect. Later we will turn to the discussion of nucleation in heavily boron or arsenic doped silicon. A mechanism of nuclei formation in heavily doped silicon is discussed using a classical nuclei formation theory.

Behavior of void formation

Size effect model is explained by an interstitial silicon providing relief to a tensile stress around impurities when the size of the dopant is smaller than that of a silicon atom. On the other hand, when the size of an impurity atom is bigger than that of a silicon atom, the concentration of vacancies is increased.

Kikuchi et al. reported(13), a formula for the vacancy equilibrium concentration for impurity concentrations N,

$$[V]/[V]_0=exp[16Q\Omega N/(kT)] \qquad [2]$$

where $[V]_0$ is the equilibrium concentration of vacancy in undoped silicon, Q is the potential energy change, and Ω is the potential wall volume. By using the experimental size distribution and defect density, the vacancy concentration changes were estimated to be about 4.5 times at 1.0×10^{19}atoms/cm^3 of arsenic. But calculation result by Eq.[2] considering an appropriate value of Q were only 1.1 times because the size of an arsenic atom is equal to that of silicon atom.

On the other hand, Sueoka et al. proposed(14) a decreasing formation energy as vacancy increase in n-type silicon with dopant concentrations more than 1.0×10^{19}atoms/cm^3. However, the experimental results revealed a weak trend with dopant concentration. The behavior of void formation in heavily arsenic doped silicon cannot explain only by the above model. These results may suggest that isolated vacancies do not contribute to void formation in heavily arsenic doped silicon. In fact, Ranki *et al.* reported that the vacancies are mainly isolated from impurities at high temperature in heavily arsenic doped silicon based on positron annihilation measurements(17). Investigating a stability of monovacansy at high temperature is a key point to reveal the behaviors of void formation in heavily arsenic doped silicon.

Nuclei formation of oxygen precipitation

In this section, we will describe a calculation method for the nucleation rate(J) using a classic nucleation theory at first. Then, the nucleation rate was determined from experiment results of the annealing time dependence of BMD density. After that, the difference of two results was discussed by using diffusion constant of oxygen.

According to the classic nuclei formation theory, supersaturated oxygen will form a cluster with their continuous size distribution determined by temperature as expressed in the lower formula by using a cluster of an oxygen atom as a heterogeneous nuclei.

$$N(r)=[Oi]exp[-\Delta G(r)/kT] \hspace{3cm} [3]$$

$$\Delta G(r)=(4\pi/3)r^3\Delta G_v+4\pi r^2\sigma \hspace{3cm} [4]$$

Here, [Oi] is the concentration of interstitial oxygen, $\Delta G(r)$ is the formation energy of nuclei, ΔG_v is the free energy per unit volume and σ is the interface energy. A nucleation rate(J) is written by,

$$J=ZN(r_c)\omega_c \hspace{3cm} [5]$$

$$r_c=-2\sigma/\Delta G(r_c) \hspace{3cm} [6]$$

where Z is the Zerdovich factor, ω_c is the adhesion frequency of an oxygen atom to a critical nucleus and r_c is a critical radius. Using the diffusion constant D, ω_c can be written as

$$\omega_c=4\pi r^2[Oi]D/A \hspace{3cm} [7]$$

where A is a distance between sites that can be trap an interstitial oxygen atom. For heat treatments of limited time, the equilibrium distribution given by Eq.[3] is not realized, but the stationary density N'_c can be expressed as

$$N'_c=N(r_c)Z \hspace{3cm} [8]$$

After all, the nucleation rate J is expressed below,

$$J=[Oi]Zexp[-\{(4\pi/3)r^3\Delta G_v+4\pi r^2\sigma\}/kT]*4\pi r^2[Oi]D/A \hspace{2cm} [9]$$

Then, the nucleation rate was calculated from the experimental annealing time dependence of the BMD density in boron or arsenic doped wafers for a wide range of dopant concentrations. In order to calculate the nucleation rate in a stationary state, the initial nuclei formation during the annealing process was eliminated in the calcuration. The nucleation rate J_{ex} can be written as:

$$J_{ex}=(d-d_i)/(t-t_i) \qquad [10]$$

where d is the increasing density of oxide precipitates per unit time, d_i is the initial increasing density, t is the annealing time for nuclei formation and t_i is the initial annealing time for nuclei generation. Using Eq.[10], we determined the nucleation rate at temperatures from 600°C to 800°C. The temperature dependence of the nucleation rate as a function of the boron and arsenic concentration was plotted in Figures 10. As shown in Fig. 10, differences in nucleation rates were observed between boron doped and arsenic doped silicon. The effects of enhancement and suppression of nuclei formation were confirmed. In arsenic doped wafer, the nucleation rate was supressed for arsenic concentrations higher than 1.7×10^{18} atoms/cm^3. Although arsenic doped silicon had a higher oxygen concentration as compared to boron doped wafers, the suppression effect of the oxygen precipitation has been demonstrated.

Furthermore, calculation results from Eq.[9] are plotted in Fig. 10. In order to fit the experimental results, several cases were calculated by changing the value of the oxygen diffusion constant. In particular, the Zerdovich factor of Eq.[9] was determined by fitting the experimental results of the nucleation rate in intrinsic silicon. In the case of heavily boron doped wafer, a 50 times and 200 times higher diffusion constant were used in Eq.[9], and a tenth of the oxygen diffusion constant was used for arsenic-doped wafers. If only homogeneous nucleation occurred in heavily boron doped wafers, a 200 times higher diffusion constant of oxygen was necessary to fit the experimental results. But the SIMS depth profiles in Fig.9, only show an enhancement of 10 times for the diffusion constant at 1000°C. These results suggest that the enhancement of the nucleation kinetics in heavily boron-doped wafers cannot be explained only by homogeneous nucleation, but heterogeneous nucleation needs be taken into account for heavily boron doped wafers.

On the other hand, only a tenth of the intrinsic diffusion constant for oxygen in heavily arsenic-doped wafers was necessary to fit the experimental results. But no changes of the oxygen diffusion constant with arsenic doping were observed, as shown in Fig.10(b). This fact suggests that factors other than oxygen diffusion contribute to controlling the nuclei formation. Recently, Sueoka et al. reported(14) that precipitation in n-type silicon can be attributed to the increasing formation enegy for interstitial silicon. In general, silicon self-interstitials are known to be generated during oxygen nucleation and precipitate growth according to the following reaction

$$2Si + 2Oi = SiO_2 + I_{si} \qquad [11]$$

where Isi is the interstitial silicon. By considering an equilibrium state in Eq.[11], increasing formation energy of interstitial silicon drives the reaction to the right. Therefore, suppressing nucleation in heavily doped n-type silicon. The nucleation rate of oxide precipitation necessary to be considered a charge state of point defects influenced to formation energy with precipitation process.

CONCLUSION

In summary, the effect of heavy doping on grown-in defect formation and oxygen precipitation in CZ silicon wafers has been studied for concentrations ranging from 1.34

$\times 10^{15}$ to 2.0×10^{20} atoms/cm^3 of boron doped, and 1.7×10^{15} to 1.6×10^{19} atoms/cm^3 of arsenic doped. In heavily boron doped silicon, the grown-void defect density decreased with shrinking the OSF-ring, and no grown-in defects were observed in wafers more than 1×10^{19} atoms/cm^3 of boron. On the other hand, density of void defects in arsenic doped silicon increased with dopant concentrations up to 1.4×10^{19} atoms/cm^3. However, both size and density decreased with increasing arsenic concentrations higher more than this values. The behavior in arsenic-doped wafers did not show an agreement with the size effect model in case of very high dopant concentrations. Estimating the total volume of vacancies from experimental resulted in 4 times values as compared with that of the calculation results by this model. This suggests that isolated vacancies that do not contribute to defect formation exist in high concentration arsenic-doped silicon.

On the other hand, the nucleation rate of oxygen in heavily doped silicon was determined, and compared with the calculated results of the classic nucleation theory. Just considering a change in diffusion constant of oxygen could not fully explain the nucleation rate by only homogeneous nucleation, but also heterogeneous nucleation needs to by taken account in heavily boron doped wafers. Furthermore, the nucleation kinetics in heavily arsenic-doped silicon could not be explained by a reduced oxygen diffusion constant. It is necessary to consider the effect of charge state of point defects to understand the nucleation behavior in heavily arsenic-doped silicon.

References

1. H.Tsuya, Y.Kondo and M.Kanamori, *Jpn.J.Appl.Phys.*,**22**,L16(1983).
2. S.Matsumoto, I.Ishihara, H.Kaneko, H.Harada and T.Abe, *Appl.Phys.Lett.*, **46**,957(1985).
3. K.Wada and N.Inoue, in *Semiconductor Silicon*(1986), H.R.Huff, T.Abe and B.O.Kolbesen, Editors, P778, The Electrochem. Soc., Pennington, NJ(1986).
4. S.Hahn, F.A.Ponce, W.A.Tiller, V.Stojanoff, D.A.P.Bulla and W.E.Castro, Jr.,*J.Appl.Phys.*,**64**, 4454(1988).
5. E.Asayama, T.Ono, M.Hourai, M.Sano and H.Tsuya, in *Advanced Science and Technology of Silicon Materials*, M.Umeno, Editor, P.276, The 145th Committee of the JSPS(1996).
6. T.Ono, E.Asayama, H.Horie, M.Hourai, M.Sano, H.Tsuya and K.Nakai, *Jpn.J.Appl. Phys.*, **36**,L249(1997).
7. T.Ono, E.Asayama, H.Horie, M.Hourai, K.Sueoka, H.Tsuya, and G.A.Rozgonyi, in Semiconductor Silicon(1998), H.R.Huff, U.Gosele and Tsuya, Editors, p1113, The Electrochem., Pennington, NJ(1998).
8. K.Sueoka, M.Yonemura, M.Akatsuka and H.Katahama, Electrochem. Soc. Proc.**99-1**, p253(1999).
9. T.Ono, G.A.Rozgonyi, E.Asayama, H.Horie, H.Tsuya and K.Sueoka, Appl.Phys.Lett.,**74**,3648(1999).
10. H.Tsuya, Y.Kondo and M.Kanamori, *Jpn.J.Appl.Phys.***22**, L16(1983).
11. T.Ono, H.Horie, M.Miyazaki, H.Tsuya and G.A.Rozgonyi, Soc. Proc.**99-1**, p300(1999).
12. M.Shuren, D.Gräf, U.Lambert and P.Wagner : *High Purity Silicon IV*, p132, Electrochemical Society(1996).
13. M.Kikuchi. K.Tanahashi and N.Inoue, Electrochem. Soc. Proc.**99-1**, p491(1999).
14. K.Sueoka, S.Shiba and S.Fukutani, Proceedings of The 4th Internatinal Symposium on Advanced Science of Silicon Materials, Kona, Hawaii,pp75(2004).

15. M.Hourai, E.Kajita, T.Nagashima, H.Fujiwara, S.Umeno, S.Sadamitsu, S.Miki and T.Shigematsu : in Materials Science Forum Vols.**196-201**,p1713(1995).
16. J.C.Mikkelsen, Appl. Phys. Lett.**40**, 336(1982).
17. V.Ranki and K.Saarinen, Phys.Rev.Lett, **93**,255502(2004).

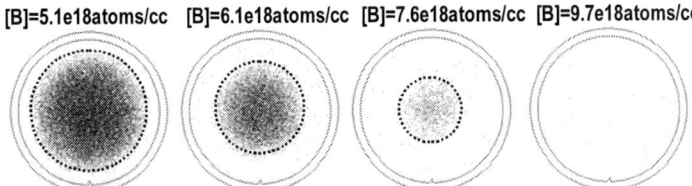

Fig 1. LSTD Maps of boron doped-wafer by, with boron concentrations ranging from 5.1 $\times 10^{18}$ to 9.7×10^{18} atoms/cm^3. Inside dash circle shows the void region.

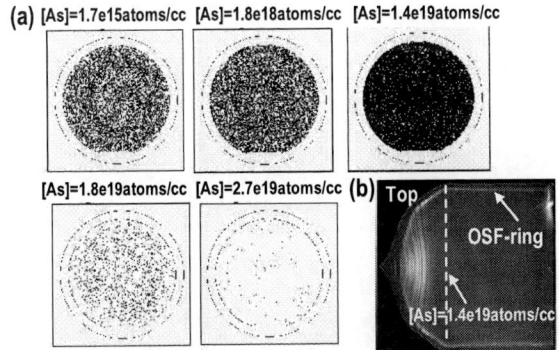

Fig 2. MO601 LSTD Maps of arsenic doped wafer, (a) showing concentration dependence of LSTD density and (b) is a XRT image of a wafer heat treated at 1100°C for 16 hours showing the location of the OSF-ring.

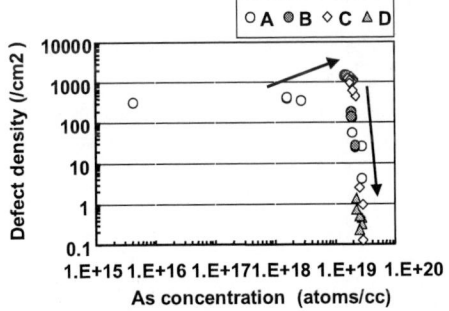

Fig.3 Arsenic concentration dependence of defect density of crystals were prepared using different 4 types of hot-zones(A,B,C and D).

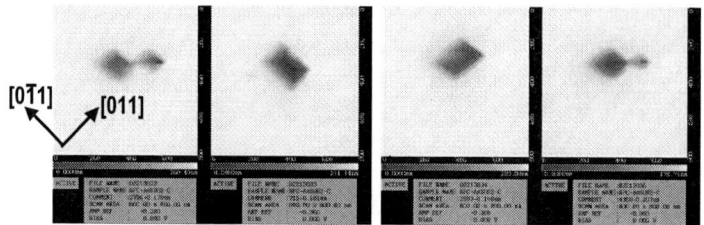

Fig 4. AFM images of COPs in arsenic doped wafers. These results are for 1.4×10^{19} atoms/cm^3 of arsenic.

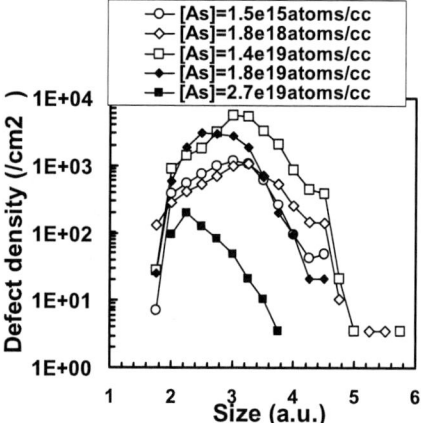

Fig.5 Defect size distribution and density in arsenic-doped wafers when measured by MO601, range of arsenic concentration is 1.5×10^{15} atoms/cm^3 to 2.7×10^{19} atoms/cm^3.

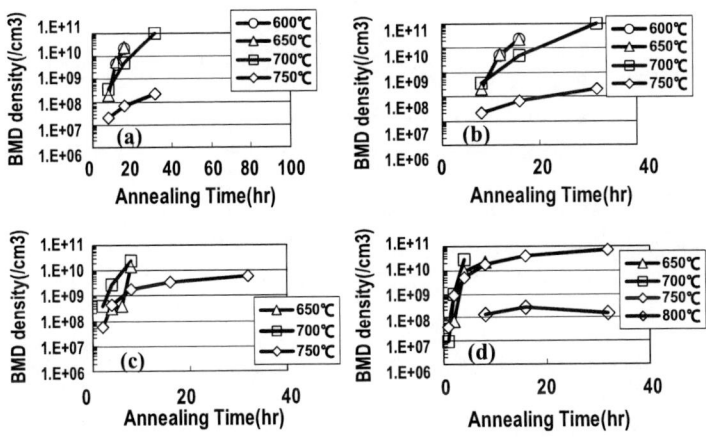

Fig.6 Annealing time dependence of BMD density in boron-doped wafers, concentration is (a)1.34×10^{15}atoms/cm^3, (b)4.1×10^{18}atoms/cm^3, (c)1.13×10^{19}atoms/cm^3 and (d)2.0×10^{20}atoms/cm^3.

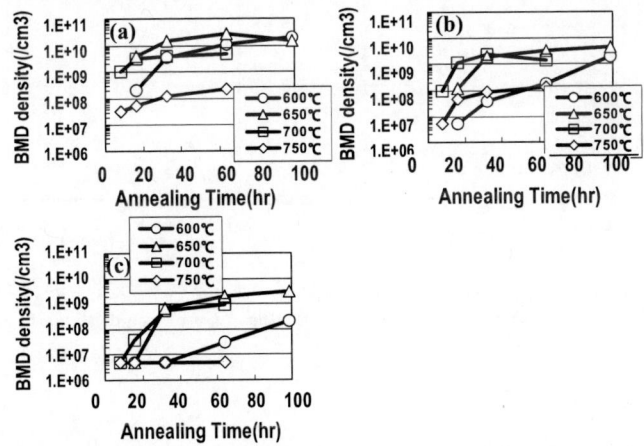

Fig.7 Annealing time dependence of BMD density in arsenic-doped wafers, concentration is (a)1.7×10^{15}atoms/cm^3, (b)1.7×10^{18}atoms/cm^3 and (c)1.6×10^{19}atoms/cm^3.

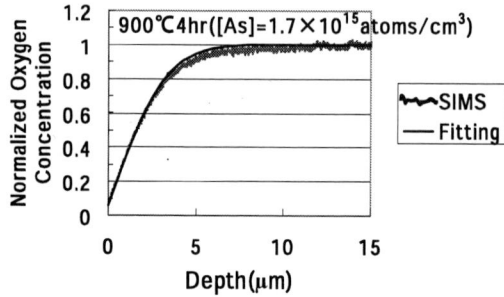

Fig.8 Depth profile of the normalized oxygen and calculated fitting curve for arsenic-doped wafer annealed at 900°C for 4 hours. Arsenic concentration is 1.7×10^{15} atoms/cm³.

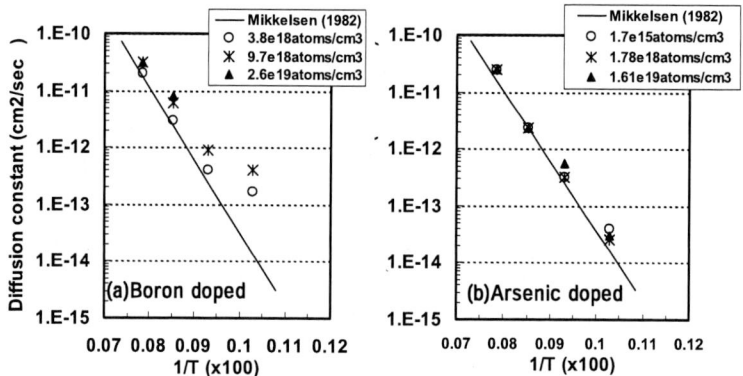

Fig.9 Impurity concentration dependence of the oxygen diffusion constant in (a)boron doped wafers and (b) arsenic doped wafers, calculated from oxygen depth profiles by using SIMS. The solid line represents the results of Mikkelsen *et al*(1982).

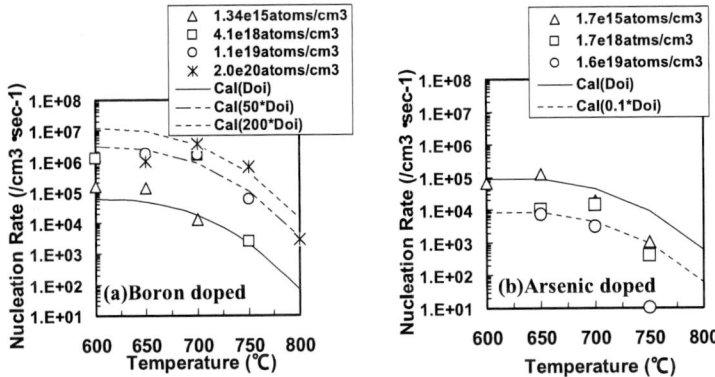

Fig.10 The symbols represent the experimentally determined Nucleation rate J in (a)boron doped wafers and (b)arsenic doped wafers. The solid and dash line show the calculated values obtained the classic nuclei formation theory using various values for the diffusion constant of oxygen(Doi).

108

ECS Transactions, 2 (2) 109-122 (2006)
10.1149/1.2195653, copyright The Electrochemical Society

WAFER STRENGTH AND SLIP GENERATION BEHAVIOR
IN 300mm WAFERS

Toshiaki Ono, Wataru Sugimura, Takayuki Kihara, and Masataka Hourai

Advanced Technology Development Sect., SUMCO Corporation
2201 Kouhoku-cho, Kishima-gun, Saga 849-05, Japan

Dependence of mechanical strength of large diameter wafers as a
function of impurity concentration and density of oxide
precipitates has been studied in terms of brittle fracture and slip
dislocation propagation. Fracture toughness does not depend on a
type of impurity, defects and its concentration. The crack size
induced on a wafer is a more sensitive factor to actual breakage
stress. However, the breakage stress without crack is
approximately 1GPa. Dislocations generation and propagation are
suppressed drastically when the boron concentration exceeds
$8 \times 10^{18}/cm^3$, while the oxygen has a little impact on slip length in
the range of MPa. Oxygen precipitation of appropriate "Density
and Size" is effective to suppress slip propagation, although large
precipitate become an origin of wafer softening.

INTRODUCTION

The sub-micron semiconductor process is required to utilize shorter annealing times
in the order of msec / μsec, because of ultra-shallow junctions (1,2,3). Achieving uniform
temperature control during such short rapid anneals will be more difficult with increasing
wafer diameter. Therefore, issues regarding the mechanical strength of wafers are
becoming even more critical in future semiconductor processes. The stress induced in
silicon wafers might be about hundreds of MPa during such an advanced rapid annealing.
The stress release via dislocation motion and propagation in silicon crystal might be
difficult, since the annealing times are extremely short, and the strain rate is large.
Therefore, brittle fracture of silicon wafer will be more critical issue for next generation
device processing.

On the other hand, of particular current importance to wafer strength is also the
behavior of slip dislocations. Recent device process is becoming more sensitive to the
slip dislocations because of the decreasing dimensions of the components in integrated
circuits. The slip dislocations have a tendency to generate at a wafer support area and
propagate during the high temperature annealing, i.e., the plastic deformation occurs at
localized areas of silicon wafers. The slip dislocation motion make a surface relief on the
wafer, therefore, the local plastic deformation is a cause for misalignment during
photolithography. Both the stress due to bending and thermal become lager in 300mm
wafers, issues regarding the mechanical strength of wafers are becoming even more
critical.

In the present paper, effects of impurities (oxygen, boron, etc.) and oxide
precipitates on the mechanical properties of silicon wafers are discussed. The brittle
fracture is discussed based on the fracture toughness and the actual stress for brittle

fracture with inducing an artificial crack. The brittle fracture stress was determined as a breakage stress by using the three-point bending tests at room temperature. The slip dislocation behavior is also evaluated by mean of X-ray topography. Thermal stress was induced in large diameter wafers by using vertical and rapid thermal annealing (RTA) furnaces.

BRITTLE FRACTURE

In order for the wafer fracture to occur, stress intensity factor (K_I) at the point of crack must be larger than the fracture toughness (K_IC). Stress intensity factor with tensile stress σ can be written as

$$K_1 = \sigma\sqrt{\pi a_1} \times f \qquad [1]$$

where a_1 is the crack size, f the factor depending on the shape of sample and crack. Therefore, wafer breakage not only depended on the stress, but crack shape or morphology is also important.

Fracture toughness is the material constant that can be determined by the indentation fracture (IF) method, utilizing a Vickers hardness tester (4). The cracks are created around an indent with a loading P on the wafer surface, K_IC can be determined as

$$K_1C = 0.0026E^{\frac{1}{2}}P^{\frac{1}{2}} \times c \times a^{\frac{-3}{2}} \qquad [2]$$

where E is the Young modulus, c and a are half of an indent size and crack, respectively. The materials with lager K_IC show a stronger resistance to the brittle fracture. The optical micrograph of an indent on a (100) wafer surface at room temperature is shown in Fig.1. The cracks are introduced from the edge of the indent along four <011> directions.

Fracture Toughness Dependence on Impurity and Bulk Defects

200mm, (100) CZ wafers with various types of impurities and concentrations were used in this study. The load for the Vickers indenter was set to 200g. The fracture toughness measured by IF method is shown in Fig.2. All samples had almost the same K_IC of 0.6MPa·m$^{1/2}$, indicating that oxygen, boron, nitrogen and carbon had no impact on brittle fractures at room temperature. Tanaka et al. have calculated that fracture toughness reaches a minimum value of 0.73MPa·m$^{1/2}$ for a {110} crack plane, although their experimental value obtained by IF method is about 1.1MPa·m$^{1/2}$ (5).

The temperature dependence of K_IC for p-(10Ω cm) and p+(9mΩ cm) samples are presented in Fig.3. In these experiments the IF method was carried out with a heated sample holder. The fracture toughness was constant from room temperature to 450°C, however, K_IC increased at 600 °C for both p- and p+ samples. K_IC of heavily boron doped sample of 9mΩ cm was lower than the lightly doped sample of 10Ω cm at 600 °C. The increase of K_IC is considered to be related to brittle-ductility transformations (BDT). Tanaka et al. have explained with detailed observations by transmission electron microscopy that the K_IC increase in BDT region is due to the stress release via dislocation formation and motion at the point of cracks (6). It also has been reported that the activation energy of BDT is equal to that of the dislocation motion (7). The differences of K_IC values between p- and p+ samples at 600 °C readily explained the lower dislocation mobility and / or higher critical stress for dislocation generation in heavily boron doped samples. Dislocation behavior in heavily boron doped wafers is mentioned in later chapters.

Breakage Stress Dependence on Crack Size and Impurity

The crack can originate from a scratch or chip that are sometime introduced during device processing, however, the wafer seldom brake during a the thermal process. Since the breakage stress depends on both crack size and morphology, as mentioned above; artificial cracks were induced by a Vickers hardness tester to control the cracks size with the same morphology. The crack size could be controlled from 10 to 300μm by changing the indent load, see Fig.4(a). After the indent at the periphery of a conventional mirror-polished 200mm, CZ wafer (Oi=1.2×10^{17}atoms/cc, ρ=10 Ω cm) with various loads, the breakage stress was measured by using a three-point bending tester. The wafer configuration on the bending tester is shown in Fig.4(b). The upper rod press speed was constant at 5mm/min.

Figure 5 shows the breakage stress dependence on the cracks size. The intrinsic breakage stress of the wafers without cracks was 700-1200MPa. The breakage stress degreased drastically when the wafers had a crack with the size from 10 to 115μm, however, it was almost constant, about 100MPa, when the crack size was over 140μm. The fragments sizes of broken wafers also depended on the crack size. The wafer broke only into two or three prices when the crack size was larger than 90μm, 10mm square sizes with cracks of 30-80μm, and 5mm square pieces when the crack size was less than 20μm, respectively. In order to understand the effects of impurity doping and oxide precipitate on the breakage stress, 60 and 270μm cracks were induced on various 200mm CZ silicon wafers. These wafers had various impurity (boron, nitrogen) concentrations and oxygen precipitates. Breakage stress of all samples were about 150 MPa and 100MPa for the crack size at 270μm and 60μm, respectively. It did not depend on the type of impurity and bulk micro-defects, as shown in Fig.6. These experimental results suggest that wafer breakage during thermal processing it is important to suppress/minimize the size of a crack, thereby increasing the stress for the brittle fracture. On the other hand internal structures, impurity and bulk defects, have no influence on wafer breakage.

SLIP DISLOCATION

The shear stress for inducing slip dislocation in an "ideal diamond crystal" was calculated by Tyson (8). The "ideal" means that crystal has a perfect arrangement of atoms and flat surface. His calculations results by considering the interaction potential between two neighboring atoms show the ideal strength is about 14GPa. However, the propagation of slip dislocations at a high temperature annealing process (>1000 °C) are observed, although the thermal stress, τ_{th}, is anticipated to be in the order of MPa. The slip dislocations usually are observed at the contact point with the wafer carrier and the handling scratch from wafer annealing process. Sumino et al suggested that mechanical shock or chemical reaction at the surface act as the preferential generation centers of dislocations under stress at a high temperature(9). Moreover, an amorphous region accompanying a dislocated crystalline region were observed after scratching the surface with a light load at room temperature (9,10).

It is known that flow stress, τ_a, depends on dislocation density(N), temperature(T), and internal structures(S). In order to prevent the wafer deformation, $\tau_a > \tau_{th}$ must be satisfied. The reduction of thermal stress is useful to suppress the slip dislocation propagation, although increasing τ_a that is accomplished by the reduction of N and lower T, is also effective. The internal structure S is dependent on impurity concentration and crystal defects in wafers, therefore, there are some possibilities for optimization by increasing τ_a.

Impurity Effect of Oxygen, Boron and Nitrogen

We examined the impurity effect on slip dislocation propagation by using 200mm vertical and RTA furnace. In the vertical furnace wafers were supported by four points of a SiC coated boat with 6.75mm wafer spacing, and pushed into the furnace at 900 °C. After insertion into the furnace, wafers were ramped from 900 °C to 1150 °C at a rate of 10 °C/min, and isothermally annealed at 1150 °C for 30min. The ramp down rate from 1150 °C to 900 °C was 3 °C/min, and pulled out at 900 °C. The compressive stress was induced at the wafer periphery, because of a temperature difference during the ramp up process. The stress at the wafer periphery was estimated about 4MPa by numerical simulation (11). In case of RTA annealing, the wafers was held at 600 °C, and then heated from 600 °C to1250 °C with a ramp rate of 50 °C/min. In the process chamber, wafers were supported by 3 pins. The stress value is unclear; however, this thermal process must have a higher strain rate (ε) as compared with vertical furnace annealing. After thermal stress tests boat or pin marks with slip dislocations are clearly seen by X-ray topography, highlighted by the circled regions in Fig. 7.

In Figure 8, slip dislocation length measured by X-ray topography are plotted against the impurity concentration of (a) oxygen, (b) boron, and (c) nitrogen. The higher oxygen concentration wafers had a higher slip tolerance during the vertical furnace annealing, but the slip length was almost the same after RTA annealing, as shown in Fig.8(a). Immobilization of dislocations by dissolved impurities in silicon crystals has been reported by Sumino and Yonenaga (9). The released stress was measured by means of in situ X-ray topography below 750°C. Their detail experimental and theoretical investigations demonstrated that dislocation immobilization is due to the oxygen precipitation at the dislocation core, not expected to develop "Cottrell atmosphere" around dislocation. Cottrell atmosphere are known as the disproportionately distribution of solute atoms near dislocations due to a interaction between solute atoms and dislocation in impure crystal (12). Therefore, the sufficient time to getter oxygen by dislocation is unexpected during RTA annealing. However, heavily boron doping in silicon was very effective in suppressing the slip dislocation propagation for both annealing in the vertical and RTA furnaces. No expansion of slip dislocations from boat and pin marks were observed for boron concentrations in excess of 8×10^{18}/cm^3, as shown in Fig.8(b). These results suggested that critical stress for dislocation generation is increasing with increasing the boron concentration, although Yonenaga et al. have reported that boron concentration of 8×10^{18}/cm^3 is same as in un-doped CZ-silicon (13). On the other hand a weaker dependence of impurity concentration was observed in nitrogen-doped wafers, as shown in Fig.8(c). Although, it has been reported that a nitrogen concentration of 5.4×10^{15}atoms/cc, immobilized dislocations in FZ silicon crystals effectively (14). The solid solubility of nitrogen is about 5×10^{15}atoms/cc. Our experimental results indicate that nitrogen doping with concentrations below solid solubility has little impact on slip propagation under the stress in the range of MPa.

Precipitate Dispersion Hardening

CZ silicon wafers contain interstitial oxygen atoms which are incorporated during crystal growth and are supersaturated during subsequent heat-treatments, therefore, oxygen precipitation reaction readily occurs. These oxide precipitates are believed being a cause of precipitation-softening in silicon wafers. Sumino et al. have reported that oxygen precipitation in silicon results in a decrease in mechanical strength because of punched out dislocations originating from the precipitate (9). Furthermore, Sueoka et al. reported that slip dislocations are generated when the oxide precipitate is larger than 200nm, and is independent of precipitate density (15,16,17). On the other hand, Yasutake et al showed that the small but high density oxide precipitates became an obstacle for dislocation motion (18).

Figure 9 shows the slip length dependence on precipitate density in wafers. 200mm, CZ (100) wafers (Oi=1.35×10^{17}atoms/cc, ρ =12 Ω cm) were subjected to a two step heat treatment, 700°C /4 to 16hr + 1050 °C /0.5 to 16hr, to create various sizes and densities of oxide precipitates. For these anneals, both heating and cooling rate were 3°C /min to avoid propagation of the slip dislocation. After precipitation annealing, the thermal stress was induced by the same procedure as mention in the previous section. The maximum length of slip dislocation at the boat marks were decreased with increasing oxide precipitate density, indicating that the obstacle for dislocation motion is controlled by the spacing of precipitates. A slip length of less than 5mm was observed in samples with an oxide precipitate density > 5×10^9/cm^3.

Resistance force due to oxide precipitates can be estimated by the "Orowan mechanism". The Orowan stress τ_0 ,act as the resistance force for dislocation motion, described as (19),

$$\tau_0 = \frac{Eb}{L} \qquad\qquad [3]$$

, here b is the Burgers vector of slip dislocation 1/2<110>, and L the space between oxide precipitates. Using this simple model, τ_0 = 4.4MPa is obtained for an oxide precipitate density of 5×10^9/cm^3 , which readily explains the experimental results in Fig.9.

On the other hand, precipitates are known to degrade the mechanical strength of wafers (15,16). Since oxide precipitates cause a localized compressive stress in a silicon matrix, the precipitates generate the punch out dislocations and motion as a slip in wafers with thermally compressive stressed regions. Therefore, it is suggested that the critical size for precipitate softening is also important for controlling the wafer strength. The optimum precipitate size and density for suppressing slip dislocation is the shaded region in Fig.10. The critical size, s_{cri}, for precipitate softening was determined to be approximately 150 nm, and critical density, d_{cri} , for obtaining τ_0 lager than τ_{th} was 5×10^9/cm^3. It is concluded that precipitation of appropriate "Density and Size" prevents dislocation motion and is effective in suppressing slip propagation. Note that this optimization results were obtained from a thermally induced stress of 4MPa. If the thermal stress is increasing more, s_{cri} is to become smaller, and d_{cri} higher, therefore, understanding of thermal stress during actual device processing plays a key role in a practical use of precipitate dispersion hardening.

Slip Tolerance of 300mm Wafers

We investigated the tolerance for slip dislocations using various types of 300mm wafers, as shown in Table I. Thermal stress was applied with an annealing in a vertical furnace with the following heat cycle. The wafers were loaded with a 9.5mm pitch onto SiC coated boats (4 points support), and pushed into the furnace at 700 °C. After insertion, the wafers were ramped up from 700 °C to 1100 °C at a rate of 3, 8 °C /min, and isothermally annealed at 1100 °C for 60min. The ramp down rate from 1100 °C to 700 °C was 1.5 °C /min, and pulled from furnace at 700 °C. The compressive stress was induced at the wafers periphery because of the temperature gradient during the ramping process. A summary of the slip length determined by X-ray observations and oxide precipitate density after a 1000°C/ 16hr anneal are presented in Table I.

Table I : List of 300mm wafers used and slip length determined by X-ray topography

Samples	Oxygen concentration ($\times 10^{17}$atoms/cc)	Nitrogen doped	Oxide precipitate density (/cm^3)	Slip length (3°C/min)	Slip length (8°C/min)
A (p-)	10.0	Non-doping	< 1e8	15.5	22
B (p/p-)	11.6	Doping	2.8e8	13.5	17.5
C (p/p-)	12.1	Doping	1.32e9	12.5	16.5
D (p/p-)	14.4	Doping	2.5e9	11	13
E (p/p+)	13.1	Non-doping	< 1e8	6	8
F (p/p+ +)	13.0	Non-doping	< 1e8	0	5

Figure 11 shows the dependence of oxygen concentration on the slip propagation length. The close symbols denote polished p- wafers and p/p- Epi, while open symbols are p/p++ Epi wafers. In Fig.11, the dashed line provides the linking between the lowest oxygen samples A and highest samples D, which indicates the oxygen influence on dislocation motion. Note that the slip lengths in higher oxide precipitates samples B, C, and D, are shorter compared with the dashed line, suggests that oxide precipitates become an obstacle for dislocation motion during the 300mm thermal process, although its density is not enough to suppress slip dislocation motions completely. It is also apparent that heavily boron doped p/p++ has a stronger slip resistance than oxide precipitates.

SUMMARY

The brittle fracture toughness and slip dislocation behavior in large diameter wafers were examined in terms of impurity and oxide precipitates in silicon. At the room temperature, crack size and morphology are play a key role in brittle fracture. However, heavily boron doping impacts fracture toughness above 500°C, perhaps due to the higher critical stress for dislocation generation. Actual intrinsic wafer breakage stress is about 1GPa, which is likely to be larger than thermal stress in advanced wafer annealing processes. The size and morphology of cracks induced in wafers are a sensitive factor for preventing the brittle fracture during the thermal processes. For the stronger slip resistance, heavily boron doping is more effective than oxygen even in thermal processes for 300mm wafers. Oxygen Precipitation of appropriate "Density and Size" prevents dislocation motion and is also effective in suppressing slip propagation. The suitable internal structure of a wafer is depends on the thermal stress value, therefore, understanding of temperature and stress at the slip sensitive step is important.

ACKNOWLEDGMENTS

The authors would like to thank Dr. K. Higashida of Kyushu University for many valuable discussions.

REFERENCES

1. A. Shima, Y. Wang, S. Talwar and A. Hiraiwa, in *Symposium on VLSI Technology 2004, Digest of Technical Papers*, IEEE, p493
2. P. J. Timans, W. Lerch, J. Niess, T. Huelsmann and P. Schmid, *Proc. Solid State Technology*, p35 (2004)
3. S. Talwar, D. Markle and M. Thompson, *Proc. Solid State Technology*, p83(2003)
4. K. Niihara, J. *Mater. Sci. Lett.* , **2**, 221 (1983)
5. M. Tanaka, K. Higashida, H. Nakashima, H. Takagi and M. Fujiwara, *Mater. Trans.* , **44**, 681 (2003)
6. M. Tanaka, K. Higashida, T. Kishikawa, andT. Morikawa, *Mater. Transa.* , **43**, 2169 (2003)
7. P. B. Hirsch and S. G. Roberts, *Acta. Meter.*, **44**, 2361 (1996)
8. W. R. Tyson, Phil. Mag., **14**, 925 (1966)
9. K. Sumino, in *Oxygen in Silicon*, F. Shimura, Editor, p449, Academic Press Inc.(1994)
10. K. Minowa and K. Sumino, *Phys. Rev. Lett.* , **69**, 320 (1992)
11. M. Akatsuka, K. Sueoke, H. Katahama and N. Adachi, *J. Electrochem. Soc.* , **146**, 2683 (1999)
12. J. P. Hirth and J. Lothe, in *Theory of Dislocations*, p650, John Wiley and Sons Inc. (1982)
13. I. Yonenaga, T. Taishi, X. Huang, and K. Hoshikawa, *J. Appl. Phys.* , **89**, 5788 (2001)
14. K. Sumino and M. Imai, *Philos. Mag.*, **A47**, 753 (1983)
15. K. Sueoka, M. Akatsuka, H. Katahama, and N. Adachi, *J. Electrochem. Soc.* , **144**, 1111 (1997)
16. K. Sueoka, M. Akatsuka, H. Katahama, and N. Adach, *Jpn. J. Appl. Phys.* , **36**, 7095 (1997)
17. M. Akatsuka, K. Sueoka, H. Katahama, Y. Koike, and S. Sadamitsu, *Jpn. J. Appl. Phys.* , **37**, 4663 (1998)
18. K. Yasutake, M.Umeno and H. Kawabe, *Appl. Phy. Lett.* , **37**, 789 (1980)
19. E. Orowan, *Symposium on Internal Stress*, p451, Institute of Metals, London (1948)

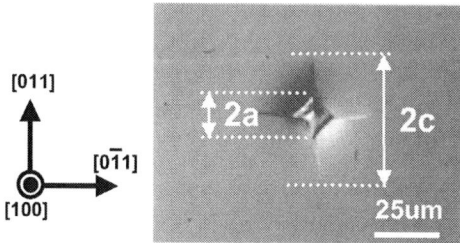

Fig. 1 Optical micrograph of an indent on (100) wafers.
{011} cracks are introduced from the edge of the indent.

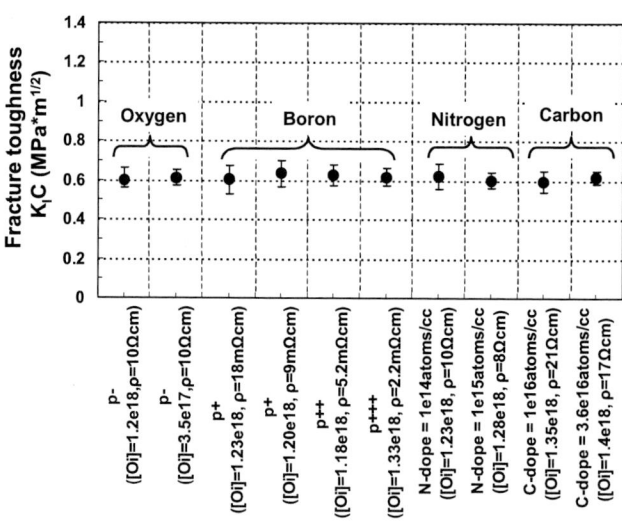

Fig. 2   Room Temperature fracture toughness dependence on oxygen, boron, nitrogen and carbon concentration.

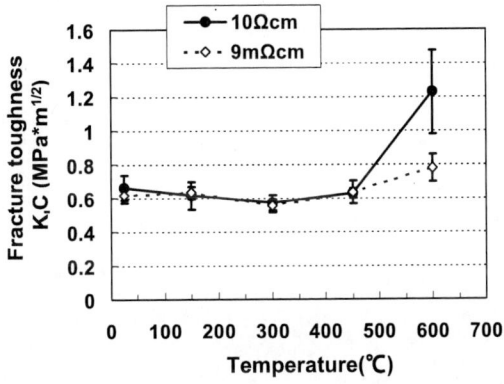

Fig.3 Temperature dependence of fracture toughness for p- (10 Ω cm) and p+ (9m Ω cm) wafers.

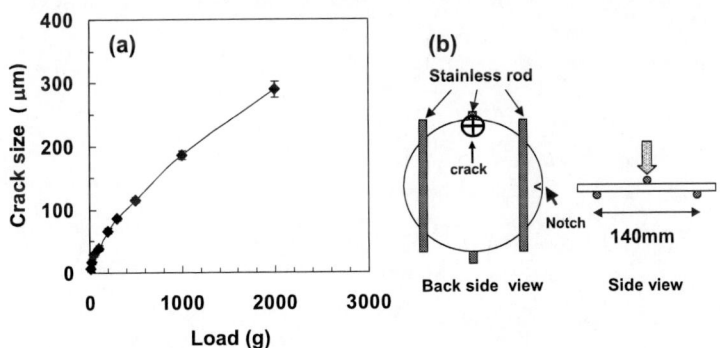

Fig. 4   Dependence of crack size on the indent load (a), and the schematic wafer configuration in the three point bending machine (b).

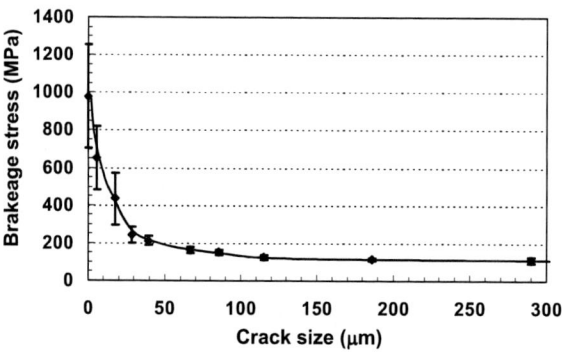

Fig. 5 Breakage stress dependence on crack size. The artificial cracks were induced by indent.

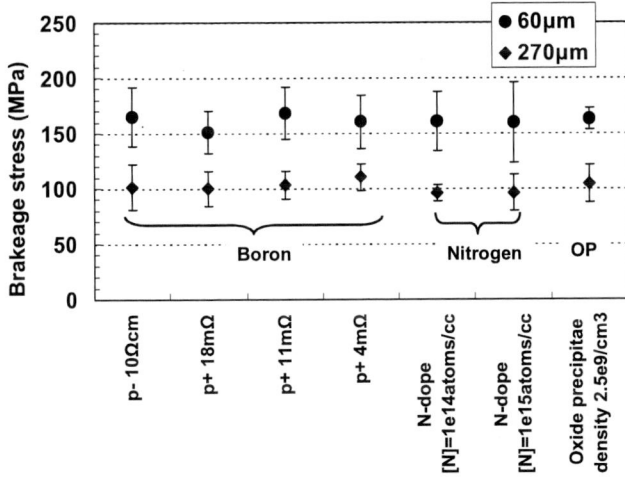

Fig.6 Impurity doping and oxide precipitation dependence of the breakage stress.

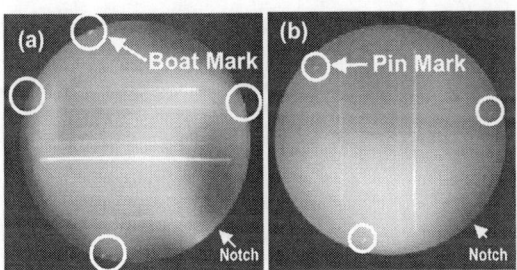

Fig. 7 X-ray topography images after a thermal stress test using
(a) Vertical furnace and (b) RTA furnace.

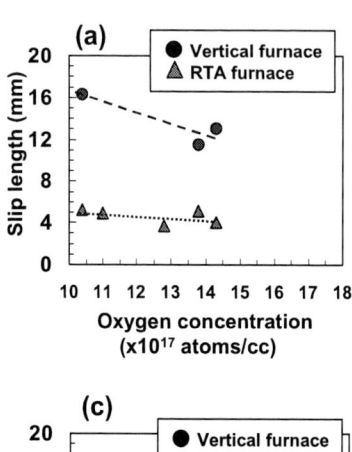
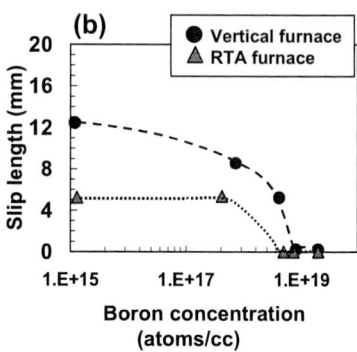
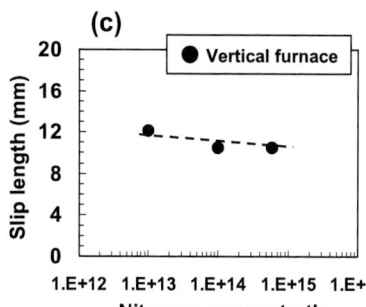

Fig. 8   Slip length dependence on the impurity concentration of (a) oxygen, (b)boron and (c) nitrogen.

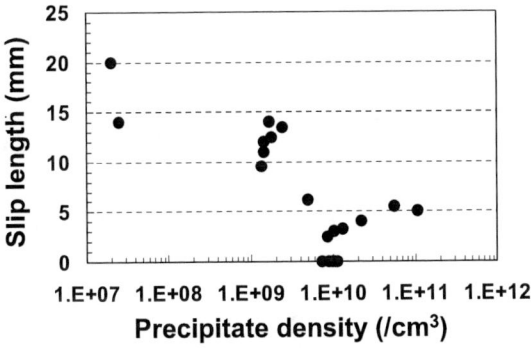

Fig. 9 Slip length dependence on precipitate density after a thermal stress test using a vertical furnace.

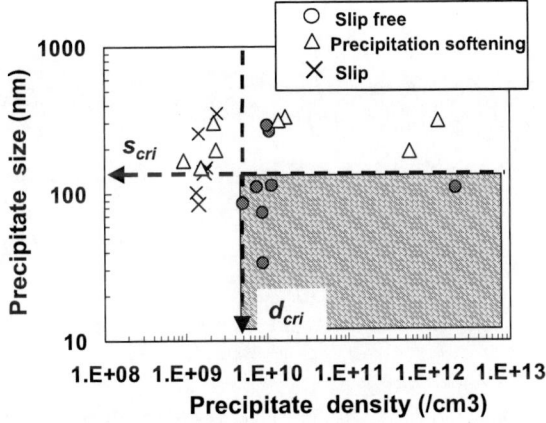

Fig. 10 The appropriate "Density and Size" region (shaded) for preventing dislocation propagation due to the thermal stress induced by the heat treatment. In the region higher S_{cri}, oxide precipitates become an origin of slip. In the region lower d_{cri}, slip dislocation propagate due to poor resistance force τ_0.

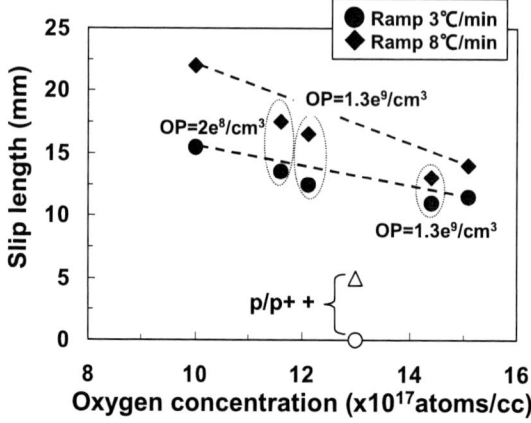

Fig. 11 Slip length in 300mm wafers after a thermal stress test.
Note that OP means oxide precipitate density.

ECS Transactions, 2 (2) 123-134 (2006)
10.1149/1.2195654, copyright The Electrochemical Society

Warp of Silicon Wafers Produced from WireSaw Slicing: Modeling, Simulation, and Experiments

Puneet Gupta and Milind S. Kulkarni

Quantitative Silicon Research, MEMC Electronic Materials Inc., St. Peters, MO 63376, USA

The slicing of wafers from a silicon ingot by the wiresaw process involves the cutting action of the fine slurry particles trapped between the wire and the ingot surface being cut. The heat generated from the cutting action of the slurry is partly convected away by the fast moving slurry at the point of cut. The remainder of the heat leads to differential thermal expansion of the ingot depending on the local conditions at the instant and location of cut. It is conjectured that the non-uniform heat generation and partition is responsible for the warp observed in wafers as cut by the wiresaw process. In this study, a coupled thermal and structural finite element based approach is utilized to model and simulate warp of sliced wafers. Within the constraints and assumptions of the coupled FE model, excellent comparison of simulation results with experimental warp data is observed.

Introduction

The capacity to simultaneously cut multiple thin slices from large bodies or workpieces with minimal kerf loss makes the modern wiresaw process the preferred technology for meeting the slicing requirements of both the semiconductor as well as the photovoltaic industries [1]. In the silicon wafer processing industry, the wiresaw technology is popularly applied in the production of wafers of small thicknesses (typically in the range of $800\,\mu m$ - $900\,\mu m$) from large cylindrical bodies known as ingots. This slicing technology employs the cutting action of free-floating slurry particles in a process fluid (mineral-oil/glycol). As seen from Figure 1 [2], this slurry is introduced on an assembly mimicking multiple wires placed parallel to one another, which carries it into the ingot cutting-zone by a periodic reciprocating motion. The ingot to be sliced is pressed against this reciprocating assembly of wires and is progressively sliced by the cutting action of the slurry particles by a combination of rolling, indenting, cutting, and scratching mechanism [3].

The current slicing technology in its industrially practiced form has been arrived at primarily through empirical means with the prime focus on cutting as fast as possible, and little is reported about the fundamental mechanisms that lead to the surface features observed on the sliced wafers. Therefore, with the requirements for better surface finish getting tighter by the day, it is becoming increasingly difficult to produce wafers with

123

ever decreasing warp without controlling the process based on more fundamental understanding.

The warp of sliced wafers is predominantly produced by the non-uniform local thermal expansion of the ingot in the vicinity of the location of the cut during the slicing cycle, by the heat generated from the cutting action. Yamada and co-workers [4] recently proposed a framework based on domain deactivation for modeling the process of kerf loss during the slicing process. This study is concerned with the development of modeling and simulation capabilities that can relate the process variables to the observed warp of sliced silicon wafers produced from wiresaw slicing.

Model Development

The fluid dynamic interaction of the cutting slurry with the ingot being sliced is an extremely complex process and a detailed model of the slurry fluid dynamics and its coupling with the ingot structural and thermal model is not pursued in this study. Instead, a simplified description of the slurry flow over the ingot surface, as discussed below, is adopted and interfaced with the detailed finite element model of the ingot. The numerical implementation of the developed model is formalized in the framework of coupled thermal and structural 3D Finite Element analysis using the commercial software MARC from MSC.Software Corporation. Novel approaches for approximating the heat-transfer resulting from the slurry flow over the ingot and for relating the power consumption to rate of heat-generation have been incorporated into the Finite Element model, where slicing is simulated by "deactivating" elements from the computational domain [4]. The solution scheme involves sequential transient thermal and quasi-static structural calculations. The wafer warp is evaluated from total axial displacement frozen at the instant of cut.

Thermal Model

Figure 2 shows the schematic of slurry flow at two different instants of the cut. In the first half of the cut when the wire-web is below the mid-point of the ingot circular face, the slurry upon impact with the ingot surface and due to geometric restrictions creates a small recirculating slurry pool, and then falls below into the collection tank. However, when the wire-web is above the mid-point of the ingot circular face, the slurry has the freedom to splash upwards due to the absence of the geometric restriction encountered when wire-web is below the ingot mid-point. The capturing of this distinction in slurry flow is important for modeling heat loss to the slurry from the ingot at different instances during the cut.

The governing equation for temperature distribution within the ingot is the standard transient conduction equation as shown below. Here, ρ is the density of the ingot, C_P is its specific heat at constant pressure, k its thermal conductivity and T its temperature. $\dot{Q}$ is the heat source flux term arising from the cutting action of the slurry, and the method of its evaluation is described subsequently.

$$\rho C_P \frac{\partial T}{\partial t} = \nabla \cdot (k \nabla T) + \dot{Q} \tag{1}$$

The initial temperature of the ingot is set to be equal to the ambient temperature, while the boundary conditions where the ingot is losing heat to the slurry, excluding the cutting surface, or the ambient are the Robin type boundary conditions and modeled in terms of film heat transfer coefficients.

$$-k \nabla T \big|_{Boundary} = h_i (T - T_{i,\infty}) \big|_{Boundary} \qquad i \equiv "air", "slurry" \tag{2}$$

In the above equation, h_i is the film heat-transfer coefficient that is predominantly governed by forced convective flow of the slurry over the ingot surface and by natural convection at locations where heat-loss from the ingot surface is to the ambient air. On the other hand, $T_{i,\infty}$ is the temperature of the phase to which the ingot boundary is losing heat (either "air" or "slurry"). In the model simulations, the ambient air temperature is assumed constant, while the ambient slurry temperature is linearly and dynamically interpolated between the *measured* inlet and outlet slurry temperatures from the experiments based on the current vertical location of the wire-web with respect to the vertical location of the wire-web at the start of the cut, which is at the bottom of the ingot.

The film heat transfer coefficient for heat loss from the ingot to the slurry due to its flow over the ingot surface is evaluated from the following correlation for Nusselt number [5] by assuming that the heat loss mechanism is similar to that encountered for forced-convection of a liquid over a flat-plate.

$$h_{sl} = \frac{Nu_{sl} k_{sl}}{L_{C\_Ingot}}; \qquad\qquad L_{C\_ingot} = 0.5 \pi D_{Ingot}$$

$$Nu_{sl} = 0.664 \, Re_{sl}^{1/2} \, Pr_{sl}^{1/3}; \qquad Re_{sl} = \frac{L_{C\_Ingot} \, \rho_{sl} \, v_{sl}}{\mu_{sl}}; \tag{3}$$

$$Pr_{sl} = \frac{\mu_{sl} C_{P\_sl}}{k_{sl}}; \quad v_{sl} = \frac{\rho_{sl} \, g \, \delta_{sl}^2}{3 \mu_{sl}}; \quad \delta_{sl} = \left(\frac{3 Q_{sl} \, \mu_{sl}}{\rho_{sl} \, L_{Ingot}} \right)^{1/3}$$

On the other hand, the film heat transfer coefficient for heat loss from the ingot to the ambient air is primarily a result of natural-convection heat-transfer and is evaluated from the following correlation [6].

$$h_{air} = \frac{Nu_{air} k_{air}}{L_{C\_Ingot}} ; \qquad\qquad L_{C\_ingot} = 0.5 \pi D_{Ingot}$$

$$\sqrt{Nu_{air}} = \sqrt{Nu_0} + \left[\frac{\dfrac{Gr_{air} \, \mathrm{Pr}_{air}}{300}}{\left\{ 1 + \left(\dfrac{0.5}{\mathrm{Pr}_{air}} \right)^{9/16} \right\}^{16/9}} \right]^{1/6} ;$$

$$Nu_0 = 0.36 \quad \textit{for cylinders} \tag{4}$$

$$Gr_{air} = \frac{L_{C\_Ingot}^3 \, g \rho_{air}^2 \, \beta_{air} \left| T_s - T_{air,\infty} \right|}{\mu_{air}} ;$$

$$\mathrm{Pr}_{air} = \frac{\mu_{air} C_{P\_air}}{k_{air}} ; \qquad \beta_{air} = \frac{2}{T_s + T_{air,\infty}}$$

In addition to the above two routes for heat loss from the ingot to the surrounding fluids, an additional source of cooling the ingot via slurry flow through a channel in the ingot-holder needs to be accounted for. The cross-section of the channel is rectangular and its layout in the ingot-holder is "U-shaped". The heat loss to the slurry flowing through the ingot-holder channel is again modeled by a film heat transfer coefficient estimated from the following correlation for Nusselt number [7].

$$h_{chan} = \frac{Nu_{chan} k_{sl}}{D_{h\_chan}} ; \qquad D_{h\_chan} = \frac{4 W_{chan} H_{chan}}{2 \left(W_{chan} + H_{chan} \right)}$$

$$Nu_{chan} = 1.86 \left(\mathrm{Re}_{chan} \, \mathrm{Pr}_{chan} \, \frac{D_{h\_chan}}{L_{chan}} \right)^{1/3} \left(\frac{\mu_{sl}}{\mu_{sl\_w}} \right)^{0.14} ;$$

$$\mathrm{Re}_{chan} = \frac{D_{h\_chan} \, v_{chan} \rho_{sl}}{\mu_{sl}} ; \qquad \mathrm{Pr}_{chan} = \frac{\mu_{sl} C_{P\_sl}}{k_{sl}} ; \tag{5}$$

$$v_{chan} = \frac{Q_{chan}}{W_{chan} H_{chan}}$$

The heat loss from the ingot to the slurry by the first two mechanisms is due to the fluids contacting the ingot surface. There is therefore one more mechanism for heat loss from the ingot to the slurry, and that is from the grooves (representing kerf loss) that are machined into the ingot. There is a slurry film trapped in between the reciprocating wire and the groove surface, details of which are provided elsewhere [1]. However, since there is a continuous movement of the interface between the slurry film and the surfaces of the corresponding groove due to material removal from the ingot as a result of the cutting action, an effective Stefan type boundary condition describes the heat loss from the ingot to the slurry. The Stefan boundary condition essentially balances the sum of conductive and convective heat flux on the ingot-side with the sum of conductive and convective heat flux on the slurry side. The material removal due to cutting can be effectively thought of as a phase change (ingot-dissolution) process where the material cut away representing a phase-change from the ingot to a constituting slurry particle. In the absence of phase change, the convective part of the heat balance is trivially zero, and the Stefan condition reduces to a balance of conductive fluxes on the two sides of the interface.

$$\begin{Bmatrix} -k_{Ingot} \nabla_n T_{Ingot} \big|_{Interface} + \\ v_{Interface} U_{Si,Ingot} \end{Bmatrix} = \begin{Bmatrix} -k_{sl} \nabla_n T_{sl} \big|_{Interface} + \\ v_{Interface} U_{Si,Particle} \end{Bmatrix}$$

$Or,$

$$-k_{Ingot} \nabla_n T_{Ingot} \big|_{Interface} = \begin{bmatrix} -k_{sl} \nabla_n T_{sl} \big|_{Interface} + \\ v_{Interface} \left(U_{Si,Particle} - U_{Si,Ingot} \right) \end{bmatrix}$$

$Or,$

$$\underbrace{-k_{Ingot} \nabla T_{Ingot} \big|_{Interface}}_{FE\ Calculation} = \begin{Bmatrix} \underbrace{h_{sl,Int} \left(T_{Ingot} \big|_{Interface} - T_{slurry,\infty} \right)}_{Effective\ film\ heat-transfer} + \\ \underbrace{v_{Interface} \left(-\Delta U_{Si,Ingot-Particle} \right)}_{Effective\ Heat\ Source-Flux} \end{Bmatrix}$$

$where,$

$$h_{sl,Int} = \frac{k_{sl}}{\delta_{sl,Int}} \tag{6}$$

In the above equation, $-\Delta U_{Si,Ingot-Particle}$ is the change in the effective specific internal energy of silicon as it changes its form due to the cutting action of the wires, from being part of the ingot-crystal to being part of the slurry as a silicon particle. Alternatively, it can be effectively thought of as the energy released as heat when part of silicon in the ingot becomes a free particle floating in the slurry. This ad-hoc specific internal energy change can be estimated from the power consumption measurements of the wiresaw machine, and is discussed below briefly. The heat-source flux is therefore readily evaluated by the multiplication of the user-specified ingot feed-rate into the wire-web ($v_{Interface}$) with $-\Delta U_{Si,Ingot-Particle}$.

The other terms in Equation (6) are temperature gradients in the ingot and the slurry on either side of the moving interface. Since in this study, we are not explicitly modeling the flow and temperature fields of slurry, the temperature gradient on the slurry side is modeled in terms of an effective slurry thickness ($\delta_{sl,Int}$) between the ingot-groove and the cutting wire, which is of the order of $100\,\mu m$ - $300\,\mu m$ [1]. Thus, the conductive heat loss to the slurry from the ingot at the interface is modeled in terms of an effective film heat-transfer coefficient ($h_{sl, Int}$). As for the conductive heat flux on the ingot-side, it is automatically calculated as part of the overall coupled thermal and structural finite element calculations.

Estimation of the heat source flux term. The heat-source flux term is evaluated based on the power consumption profile of a wiresaw machine during the cutting cycle. Figure 3 shows a typical power consumption profile during the wiresaw slicing cycle. Upon integration of the total power consumed with time, one obtains the total energy consumed in slicing the ingot. However, part of this total energy is consumed for driving the bearings that carry the wire-guides through which the wires travel for producing the reciprocating motion of the wire-web. Hence, only a fraction of the total energy consumed during the slicing process is directed towards cutting the ingot, which is essentially dissipated as heat at the location of the cut. Thus, $-\Delta U_{Si, Ingot-Particle}$ can be obtained as:

$$-\Delta U_{Si, Ingot-Particle} = \frac{\beta \int P(t)dt}{N_k \dfrac{\pi}{4} D_{Ingot}^2 t_{kerf}}$$

where,

$P(t)$ $\equiv$ *Power consumption profile*   (7)

N_k $\equiv$ *Number of cuts*

t_{kerf} $\equiv$ *thickness of each kerf − loss*

β $\equiv$ *Fraction of energy used for cutting*

Structural Model

The structural part of the calculations is carried out assuming elastic and equilibrium conditions. The static (equilibrium) assumption is valid given the total time scale of the cutting cycle. Similarly, the elastic assumption is fairly valid since on the scale of the ingot, deformations are primarily caused by the thermal expansion of the ingot, and are in the sub-millimeter range. Thus, given that it is a small displacement and small deformation problem, the assumption of elastic equilibrium is justified. The differential equations describing the static equilibrium are:

$$[\partial]^T \{\sigma\} + \{F\} = \{0\}$$

where,

$\{F\}$ $\equiv$ *External Force Field* $= \begin{bmatrix} F_x & F_y & F_z \end{bmatrix}^T$ (8)

$\{\sigma\}$ $\equiv$ *Stress − Tensor*

$[\partial]^T$ $\equiv$ *Gradient operator* [8]

The external force field has two sources, one is the gravitational force and other is the force due to the thermal expansion of the ingot due to the heat generated by the cutting action of the slurry particles. The relationship of the stress-tensor to the strains and that of the strains to the displacements is well-documented [8].

Simulations via Finite Element Method

The coupled thermal and structural model of the wiresaw process described above was solved in the framework of finite elements using the commercial Finite-Element software MARC from MSC.Software Corporation. The geometry of the computational model as well as the mesh for the finite-element calculations was created in the software pre-processor MENTAT. Figure 4 shows the layout of the modeled geometry and the resulting mesh using 8-noded hexahedral brick elements. To reduce the size of the computational model, a geometric approximation in the ingot axial direction was invoked where instead of simulating the actual wafer thickness, a 9 times larger thickness was used. Similarly, instead of the actual kerf-loss thickness, and 9 times larger kerf-loss thickness was modeled. With this approximation, while the actual size of the geometry was still captured, the size of the computational model was approximately one-tenth of what would result if the true wafer and kerf thicknesses were resolved. Since a model with full resolution of the kerf-loss thickness was beyond the scope of the study and intractable with available computational resources, it is difficult to quantify the impact of this assumption. It is, however, conjectured that the aforementioned approximation should still be able to provide valuable insights at the ingot scale, and is indirectly justified based on the reasonably good comparison of the experimental data with simulations as presented below.

Computational Details

Based on the knowledge of the imposed experimental profile of the ingot feed-rate into the wire-web ($v_{Interface}$), the location of the wire-web with respect to the ingot (as a function of time during the slicing cycle) was explicitly calculated *a priori*, and interfaced with the finite-element software via a user defined subroutine. Through another set of user-subroutines, an elaborate scheme of dynamically finding appropriate element "faces" was developed, coded and interfaced with the FE software for imposing the heat-source flux and the moving-boundary heat-loss at the cutting location, and the heat loss to the slurry and the ambient. In addition, based on the relative location of the wire-web with the ingot, the elements representing kerf-loss below the wire-web were "deactivated" using a third set of sub-routines. The "deactivation" of the elements (as also described elsewhere [4]) below the wire-web simulated the loss of ingot material to the slurry due to cutting. This method of selective deactivation of the computational domain is capable of simulating not only the loss in the overall stiffness of the ingot due to loss of material, but also the loss of ingot material available for conductive heat transport within the ingot.

For evaluating the wafer shape from the computational results, again a user-defined subroutine was developed that selectively recorded the local axial displacements of the ingot for a few sliced surfaces along the ingot length. The selection criterion was the maximum axial nodal displacement during the time-period for which the wire-web remained within the geometric bounds of the element to which the node belonged. In other words, the nodal displacements in the vicinity of the cut were dynamically recorded and frozen to yield the displacement data from which the wafer shape can be estimated [4]. To evaluate the wafer shape and subsequently the wafer-warp, parameters of the equation describing a best-fit reference plane through a given sliced surface were calculated by the least-squared method that minimized the sum of the squares of differences between the computed axial displacements and those predicted by the equation of the best-fit reference plane. Subsequently, the reported 2D wafer-shape contours were evaluated as the local difference in the computed axial displacements and those predicted by the equation of the best-fit reference plane. Lastly, the wafer warp number for each of the sliced surfaces was evaluated based on the definition of warp as discussed by Yamada and co-workers [4]. All the processing of the data from the finite-element calculations for evaluating the best-fit reference planes, evaluation of wafer-warp and contour plotting of wafer shapes was accomplished via another commercial software MATLAB.

Results

Figure 5a shows the predicted wafer-shape with respect to the best-fit reference plane for a wafer of 300-mm diameter as sliced from a silicon ingot by the wiresaw process, while Figure 5b shows the measured wafer shape of the equivalent wafer from experiments. The wafer for this comparison was chosen on of the ends of the ingot where the measured wafer warp is the highest. From these two figures, it can be seen that the main characteristic features of the measured wafer shape are well captured in the simulated results. Additionally, Figure 5c shows the comparison of the one-dimensional wafer shape profile along the vertical chord through the wafer center from measurements with that from simulations. Again, an excellent agreement between the two is observed along with a similar strong agreement between the computed and measured warps. It is, however, to be noted that a couple of model parameters were tuned to get good agreement between the predictions and measurements. The primary among them are parameter β, which is the fraction of the total energy consumed in the process that is used towards cutting and the slurry film thickness at the interface between the wire and the groove surface. However, given that there could be large uncertainties in some of the material properties, and their variations with temperatures, as well as the crude approximations made regarding slurry flow and splashing over the ingot surface and the associated heat transfer coefficients, the qualitative features of the wafer shapes are well captured by the simulations. Therefore, the ability to capture the behavior of an extremely complex machining process by tuning a small set of parameters can indeed be considered the outcome of a reasonably accurate description of the important process physics in the developed model.

Summary and Conclusions

A 3D finite-element methodology has been successfully used to simulate the important process physics relevant to an industrial wiresaw process. This includes a simultaneous solution of the partial differential equations describing the coupled thermal (transient) and structural (static) process physics. For an accurate prediction of the wafer warp and shape, it is imperative that a very consistent set of modeling approximations in a 3D model representing many details of the actual geometry is implemented. Given that the predicted results agree well with experimental data, it is evident that the developed methodology can be successfully applied for improvement of the industrial wiresaw process.

Acknowledgements

The authors gratefully acknowledge Carlo Zavattari and Roland Vandamme for support of the experimental part of this work, and Milind Bhagavat for preliminary discussions on wiresaw modeling.

References

1. "Elasto-Hydrodynamic Interaction in the Free Abrasive Wafer Slicing Using a Wiresaw: Modeling and Finite Element Analysis", M. Bhagavat, V. Prasad and I. Kao, *J. Tribology*, **122**, 394 (2000)
2. "Development of Fixed-Abrasive-Grain Wire Saw with Less Cutting Loss", J. Sugawara, H. Hara and A. Mizoguchi, *SEI Technical Review*, **58**, 7 (2004)
3. "Free Abrasive Machining in Slicing Brittle Materials with Wiresaw", F. Yang and I. Kao, *J. Electronic Packaging*, **123**, 254 (2001)
4. "Warpage Analysis of Silicon Wafer in Ingot Slicing by Wire-Saw Machine", T. Yamada, F. Kinai, T. Ichikawa, A. Yokoyama, M. Fukunaga and T. Ohshita, in *Materials Processing and Design: Modeling, Simulation and Applications, NUMIFORM 2004*, S. Ghosh, J. C. Castro and J. K. Lee, PV 712-2, p. 1459, American Institute of Physics Conference Proceedings, Columbus, OH (2004)
5. "Der Warmeaustausch zwischen festen Korpern und Flussigkeiten mit kleiner Reibung und kleiner Warmeleitung", E. Pohlhausen, *Z.Agnew. Mat. Mech.*, **1**, 115 (1921)
6. "Free convection around immersed bodies", S. W. Churchill, in *Heat Exchanger Design Handbook 2*, C. F. Beaton, D. Butterworth, M. Morris and J. Taborek, Editors, Ch. 2.5.7, Begel House, New York (2002)
7. E. N. Sieder and G. E. Tate, *Ind. Eng. Chem.*, **28**, 1429 (1936)
8. R. D. Cook, D. S. Malkus, M. E. Plesha and R. J. Witt, *Concepts and Applications of Finite Element Analysis*, Ch. 3,John Wiley & Sons, Inc., New York (2002)

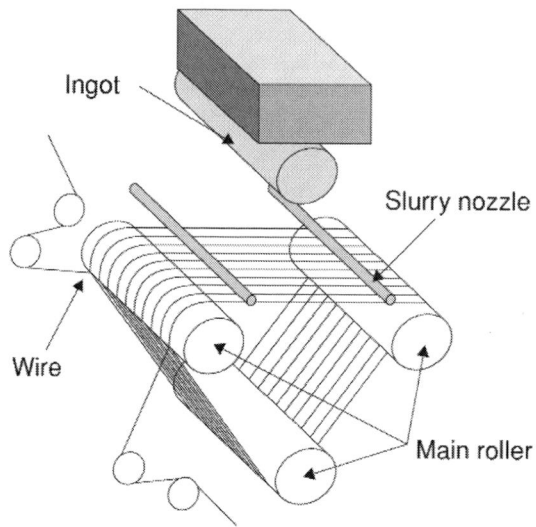

Figure 1. Schematic of a typical wiresaw slicing process [2]

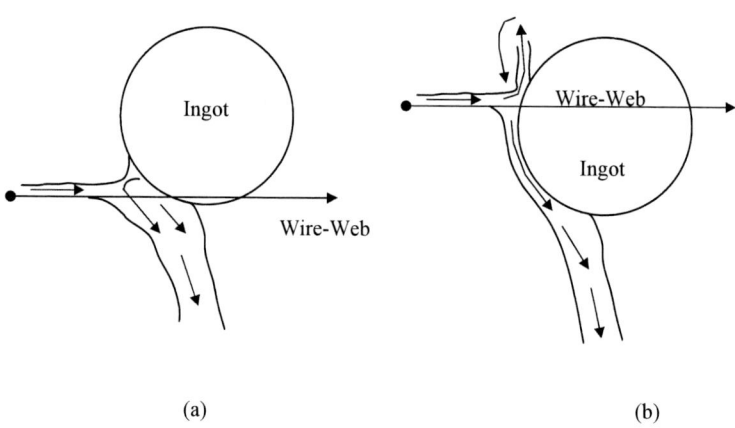

(a)   (b)

Figure 2. Schematic of slurry flow. Wire-web a) below Ingot mid-point b) above Ingot mid-point

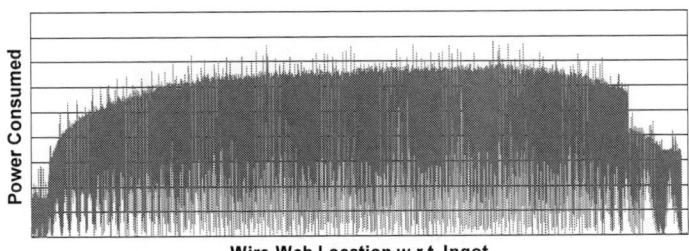

Figure 3. Typical power consumption profile of a wiresaw during the cutting cycle

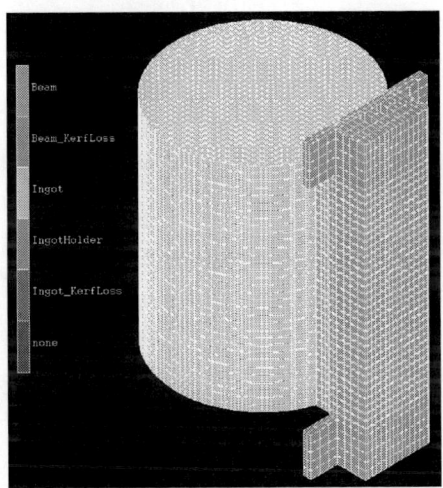

Figure 4. Finite Element Model of the Wiresaw Process

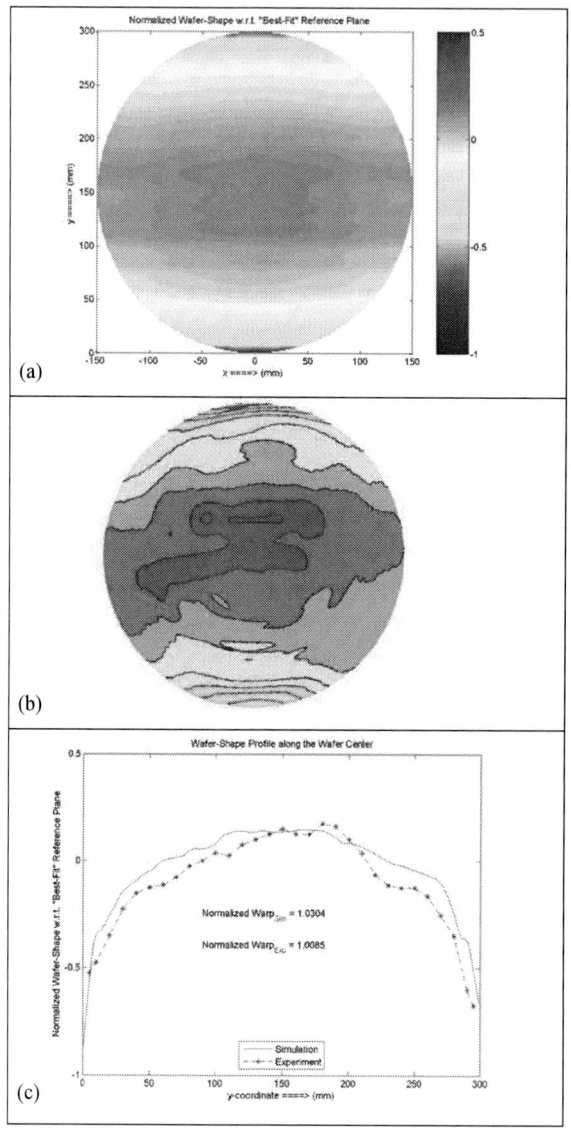

Figure 5. Comparison of simulated and experimental wafer-shapes and wafer-warp.
2D wafer-shape: a) Computed   b) Measured   c) 1D shape profiles

Lessons Learned from the 300mm Transition

J. Draina, D. Fandel, J. Ferrell, and S. Kramer

International SEMATECH Manufacturing Initiative (ISMI)

The 300mm transition required unprecedented levels of industry collaboration, standardization, and investment. This paper reviews the experiences of the 300mm transition and what we have learned, and offers a strategy to comprehend those learnings and improve as we move forward into the 450 mm transition

Consensus on an industry strategy for a next transition must be driven by rigorous economic and market analysis. Modeling based on supplier and device maker data must support evaluation and analysis of alternatives to validate assumptions and tradeoffs, economic and architectural attributes. The strategy should comprehend technology nodes and new or alternate substrate and process technologies.

A vision for opening the architecture of next generation 300mm fab configuration before the transition to the next wafer size occurs enables more flexibility for high mix business models and allows a more cost-effective scale-up from 300mm to 450 mm.

This paper reviews the experiences and learning from wafer size transitions, particularly the 200mm to 300mm transition. The intent of studying lessons learned from the 300mm transition is to make improvements during the 450 mm transition. Thus, this paper also discusses strategies for applying those lessons learned to the next wafer size transition.

Most importantly, and paramount in catalyzing and mobilizing the industry in the next transition, is the indisputable need for a clear consensus and strategy. This strategy should comprehend the needs of different IC maker business models and early consensus on common global semiconductor manufacturing requirements, including flexible fab architecture and operations, new manufacturing technologies and metrics, and support for multiple leading-edge operational modes. This architecture should comprehend wafer size bridging in toolsets and backward application of productivity improvements.

Consensus on an industry strategy for the next transition must be driven by rigorous economic and market analysis. Modeling based on source data must support evaluation and analysis of alternatives to identify program assumptions and tradeoffs as well as economic and architectural attributes.

A vision for opening the architecture of next generation 300mm fab configuration before the transition to the next wafer size enables more flexibility for high mix business models and allows a gradual scale-up from 300mm to 450 mm. This can minimize the impact on equipment design and development burden on suppliers, and minimize the number of configurations to develop, test, and deploy for the next wafer size.

Historical Motivation for Wafer Size Transitions

Wafer size transitions have traditionally been a cornerstone of semiconductor productivity improvement. In addition to technology improvements such as a higher degree of function integration, periodic lithography feature size reductions, and chip design improvements, improved factory integration and manufacturing methods also boost productivity. Major manufacturing paradigms such as wafer size transitions allow the IC manufacturers to build a "differently architectured fab" and take advantage of productivity improvement opportunities, operational efficiencies, and process and control system upgrades. This was especially true in the 300mm transition.

The progression in wafer size has continued since the inception of the industry from about ½ inch (13 mm) in the early 60s to the latest 300mm. Early transitions were driven by individual IC manufacturing companies to provide competitive advantage. As a result, the transition timing and wafer sizes varied. Thus, the transitions were 1 inch; 2 inches, and 2 $^1/_4$ inches; 3 and 3 $^1/_4$ inches; and 4 inches, 5 inches, and 6 inches.

During the 8 inch (200mm) transition, the industry achieved de facto standardization on wafer size. This was an important milestone, paving the way for consensus on the next wafer size transition. The transition cost for 200mm wafers has been estimated to be between $5B and $10B. The transition to 200mm yielded only 5% net productivity gain relative to the 6-inch wafer size, even though the total surface area available per process step was 78% more than the previous wafer size. The major industry drive in 300mm was to avoid those pitfalls since, despite the low productivity gains for 200mm wafers, industry consensus remains that periodic wafer size increases are necessary to keep the industry on the 30 year historical ~30% cost per function reduction productivity path as illustrated in one of the Moore's Law charts in Figure 1.

Historical Productivity Curve

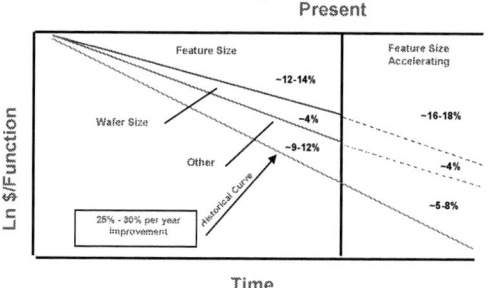

- Cost improvement equivalent to or better than price decline
- Function improvement
- Market growth
- Productivity improvement
    - Continued feature size improvement
    - Transition to next wafer size
    - Factory productivity investment effectiveness

Source: I300I Archives

Figure 1. Historic Productivity Curve

Lessons of the 200mm Transition

Each time a transition is made, the industry builds on the learning and technology advances from the previous transition. In the 200mm transition, the philosophy was to control wafer contamination by keeping only the wafer environment clean. The ensuing advantages were reductions in air handling system cost, more worker-friendly fabs, and reduction in the time spent by workers in garmenting. Early studies showed improved contamination control on wafers and reduced equipment post-maintenance recovery time. Standard mechanical interface (SMIF) pods were an integral part of this system that led to automated handling using load ports, transport systems, and automated stockers for SMIF pod storage and retrieval.

The emergent fab architecture—an improvement over bay and chase system—was a ballroom concept with a support system away from the clean room and into the lower floors and the air handling system on the upper floor. Concepts such as individual process bays (e.g., litho, implant) for optimal travel distances were already being investigated. Reticle handling and storage systems were being developed. Unfortunately, adoption of such systems was sporadic because of the following:
- Lack of industry discussion and consensus on a common strategy
- Lack of industry standards
- Standard development after commercial and proprietary solutions abound
- Lack of consensus on the use of standards
- Development of new systems in parallel with a maturing 200mm transition
- Internal equipment configuration and safety requirements resulting in custom equipment
- Variations in fab architecture (Full and Partial SMIF, Open Cassette & Ball Room, Bay & Chase)
- Evolving control and communication systems

Consensus on a wafer traceability method or standard was not established, but it was recognized that a consensus method would reduce or eliminate non-productive steps in the fab such as laser scribe and post-scribe clean operations. While overall equipment and fab costs were escalating due to technology demands, low equipment maturity and custom design drove an additional cost burden for tool and system enhancements and debug, reducing yield ramps and resulting in high manufacturing development costs.

300mm Transition Cost Targets

Critical decisions such as wafer size and transition timing require extensive modeling and collection of industry source data to support evaluation and analysis of economic and business impacts, emerging and alternative technologies, and fab architecture assumptions and tradeoffs. Modeling efforts such as the static Cost Resource Model (CRM) and dynamic fab operation models were used to show the impact of fab configuration and suggested that the 300mm wafer size could achieve a 30% cost reduction per unit area of silicon processed. However, the modeling effort had some drawbacks:
1) The software available supported only limited modeling compared to the complexities of a major industry transition.

2) Dynamic cost modeling was implemented on an incremental basis and the elements of fabs were not linked into a single thread. Issues of individual elements of fabs, such as stocker logistics issues, were successfully identified.
3) Linkage to fab operating scenarios was not integrated with the operational models even though such controls and communications drove fab logistics.

The 300mm Transition Early Decisions

Early Decisions - Technical

An initial and major question that confronted the industry was the appropriate wafer size and relative wafer thickness for the next transition. This work had already started in the early 90s in cooperation with silicon wafer manufacturers. Two international working groups (US/Europe and Asia/Japan) were established at the Large Wafer Summit meeting in July 1994 to examine the issue of the next wafer size. Wafer sizes considered ranged from 250mm (10 inches) to 350mm (14 inches). The industry consensus to optimize at 300mm was based on the reduced risk with sufficient gain in surface area (better than 2X), optimal number of wafers produced per boule, and minimal risk to slip and warpage. (The 50% increase in wafer diameter and resultant surface area increase of 2.25X has now been incorporated in the *International Technology Roadmap for Semiconductors* [ITRS].)

The 13-wafer and 25-wafer batch sizes represented a fundamental fab logistics and operational decision. The 25-wafer batches loaded in "SMIF like" containers weighing over 8 kg necessitated an automated transport system. The SMIF pod was modified according to SEMATECH member company consensus to a front opening unified pod (FOUP) to reduce separate handling of the cassette and frame and to reduce factory footprint. However, the FOUP design required a large development effort without realizing factory footprint benefits.

Early Decisions – Organizational

A key, overriding message is that the magnitude of a wafer size conversion requires highly representative global participation, coordination, and consensus. To be affordable, the industry must share and leverage research and development resources. The next wafer size transition (300mm) was predicted to cost $15B to $50B and considered far beyond the means of any single company. This was deemed a global industry issue that could be effectively addressed by a coordinated effort using three guiding principles:
1) Transition cost must be contained and the cost shared among all parties
2) Provide guidelines to supplier community
   a. Create a common fab architecture and industry standards
   b. Create an environment friendly fab that is resource efficient
3) Fab investment effectiveness and overall productivity gains must be addressed with lessons learned from 200mm

Consensus was reached at the International Business Partners Meeting in 1994 to create industry-wide cooperation. Industry organizations provided the framework for research, development, and global industry collaboration to drive the 300mm conversion. These included the I300I and SELETE device maker consortia; J300 (a coalition of five Japan associations); trade associations such as SEMI/SEMATECH, SEAJ, JEITA, KSIA, and

SEMI; government-sponsored initiatives such as the European Commission's Semiconductor Equipment Assessment Initiative and JESSI; institutes and national labs such as Fraunhofer, IMEC, and ITRI; and professional societies such as the Ultraclean Society. The Industry Executive Forum addressed strategic issues and global economic trends.

Guiding the 300mm Transition

It was agreed that there would be consortium-to-consortium cooperation to identify and drive the industry to a common, mutually beneficial vision. The 300mm transition rested on twin pillars of cost effectiveness and productivity, which translated into high reliability and throughput and insertion into an automated fab. The consortia provided infrastructure support on equipment evaluation, data and analysis availability, process capability, and test wafer availability, with cross coordination of development.

Silicon wafer thickness was a critical early decision made in consensus with the industry participants. I300I collaborated with silicon suppliers to provide a modified version of the Cost Resource Model and investigate process options to reduce the cost of substrates,. New metrology equipment was driven by both technology generation and 300mm. Defect detection and data baseline with tool-to-tool repeatability and cross-tool calibration were key advanced technology requirements for metrology equipment.

Automated material handling efforts focused on wafer transport, storage, and handling, including equipment interfaces, the interoperability of loadports, wafer handlers, end effectors and FOUPs, and related equipment such as FOUP cleaners, contamination control, and fire retardation. The collaboration produced the Unified Checklist and requirements that led to standards requirements, interoperability guidelines, and assessment tools and methodologies. Performance metrics were refined with the use of fab operational modeling

Unified Performance Metrics addressed individual equipment type, as well as overall fab performance. Assessment of process equipment maturity against these metrics was an indicator of equipment and infrastructure readiness. I300I established a multi-phase Demonstration Test Methodology (DTM) to assess process equipment maturity against these metrics and factory integration criteria. Areas for improvement were identified to enable suppliers to take corrective action and track progress against defined targets. Figure 2 is a snapshot of overall equipment supplier performance based on the DTM from 1997 to 1999. Over 65–70% of I300I's budget in was allocated to equipment evaluations.

Figure of Merit (FoM) Performance

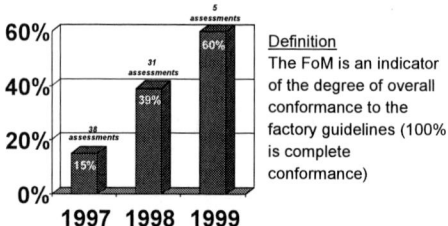

Definition
The FoM is an indicator of the degree of overall conformance to the factory guidelines (100% is complete conformance)

NOTE: All results through 1998 are based upon I300I Factory Guidelines (FGL) Version 3.0.

Figure 2 *I300I Figure of Merit 1997 to 1999   Source: I300I Archives*

Factory Guidelines
The I300I and SELETE consortia collaborated to define a common vision for an automated, highly integrated fab. The initial discussion formed the basis for the first set of 14 guidelines, called "The 14 Points of Light." These guidelines evolved over time to improve the understanding and context of emerging fab configuration and manufacturing as well as to address upcoming issues. The expanded guideline document, I300I/J300 Factory Guidelines, formed one of the bases to drive the formation of a complete set of standards. Guidelines addressed all aspects of fab productivity, logistics, and equipment effectiveness.

CIM Global Joint Guidance
Equipment integration and automation communications and control software collaboration resulted in CIM Global Joint Guidelines (GJG), Base Functionality Software Requirements, and Operational Flowcharts. Topics covered product dispatching, lot and wafer tracking and traceability, transport control, storage, reticle handling, recipe management, data collection, data availability, and logistics control based on production strategy such as hot lots processing and test wafer logistics.

The I300I and J300/Selete Factory and Software Global Joint Guidelines were essential in setting basic requirements for standards; User Requirement Documents (URDs), User System Requirements Documents (USRDs), and Unified Checklists provided another level of detail. SEMI provided the mechanism and forum for transforming these requirements into standards.

Global Supplier Coordination
Device makers unveiled joint guidelines, requirements, and roadmaps in workshops, industry summits, and standards education programs sponsored by SEMI, I300I, and SELETE to deliver a unified message to suppliers about implementation requirements and timing. An assessment of 300mm manufacturing readiness in June 2000 is shown in Figure 3. The variation in aspects of 300mm readiness is the result of evolving factory vision and challenging aspects of automation, control, and integration.

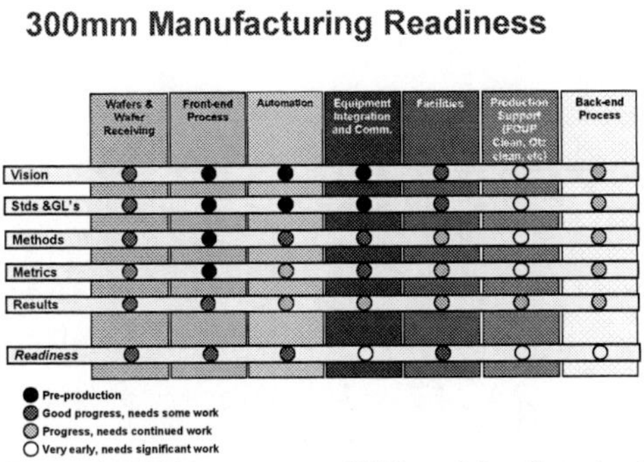

300mm Manufacturing Readiness

Figure 3 I300I Assessment of 300mm Manufacturing Readiness EOY1999
Source: I300I Archives

300mm Industry Standards

As soon as key decisions such as timing and wafer size were determined, the first global workshop for 300mm Factory Integration standards was held in New Orleans in February 1995. A standards roadmap was created and first shown the following July at SEMICON West. This roadmap considered dependencies and timing; for example, silicon wafer thickness was decided before carrier interface standards were finalized.

Early standards needs began with the individual customer implementation visions. Strategic alliances and collaboration among device maker consortia provided a mechanism for sharing visions and identifying common elements, which were documented in the I300I and J300/Selete Factory and Software Global Joint Guidelines, and Unified Checklist. These common elements were transformed into detailed specifications such as URDs and USRDs through negotiation. This negotiation was critical to developing a coherent set of requirements that were universally recognized and translated into industry standards. The standards organizations provided the forum for suppliers and device makers to codify the desired functionality and performance into industry standards. Throughout the 300mm transition, SEMI evolved organizationally to provide more efficient and effective processes and to become a truly global standards organization. Figure 4 shows the status of 300mm-related standards in June 2000. By the end of 2002, the I300I Factory Guidelines endorsed 76 published SEMI standards intended for 300mm fabs and equipment.

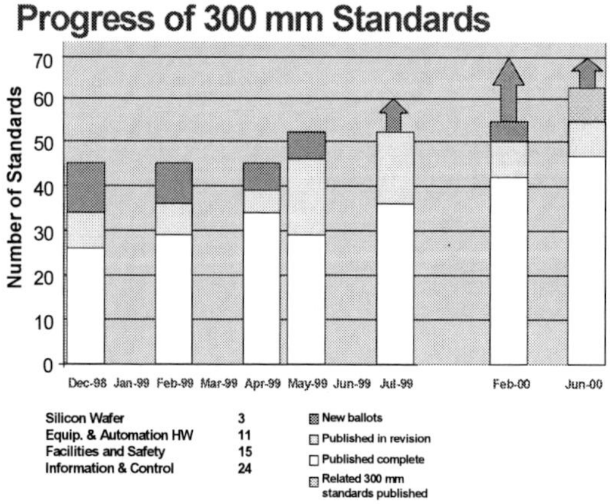

Progress of 300 mm Standards

Number of Standards (y-axis: 0, 10, 20, 30, 40, 50, 60, 70)

Dec-98 Jan-99 Feb-99 Mar-99 Apr-99 May-99 Jun-99 Jul-99 Feb-00 Jun-00

Silicon Wafer	3
Equip. & Automation HW	11
Facilities and Safety	15
Information & Control	24

New ballots
Published in revision
Published complete
Related 300 mm standards published

Source: I300I Archives

Figure 4 Progress in 300mm Standards 1995–2000

300mm Software Conformance Assessment

Implementations conformant to industry standards in general, and software standards in particular, required considerable improvement and progressed slowly. Experiences in implementing the standards and conducting conformance assessments resulted in frequent after-the-fact revisions to the published standards.

In 2001, SEMATECH launched a 300mm equipment software conformance assessment project. This effort had three primary goals:
- Accelerate the maturity and availability of equipment integration and automation software
- Assess the current state of the industry for 300mm standards conformance
- Assist suppliers in standards implementation and problem resolution through interactive discussion and expert guidance

This project clearly illuminated the importance of standardized tests and assessment methodologies. Identification of test requirements and methodologies in parallel with standards development helps to validate the standards and improve implementations. Early prototyping accelerates the availability of technically sound standards and delivered conformance. Establishing objective test criteria, standardized assessment methodologies, and metrics to quantify results accelerates maturation. Common user requirements and interpretations must be factored into the equation. The commercial availability of high quality, stable test tools and testing resources is critical. Close cooperation between suppliers and IC makers in early conformance assessment and improvement significantly accelerates development and improves the likelihood of first-pass success.

The requirements and functions in a standard must be interpreted from the perspective of the end user. Otherwise, costly delays ensue and reprogramming is needed. Final specifications and standards must be available to suppliers as early as possible in the tool development cycle. Failure to engage and educate suppliers early results in lost time and lost opportunities. Suppliers may be forced to commit to non-compliant designs to meet schedules, resulting in disruptive de facto standard implementations, including patent issues, and costly re-dos.

Suppliers that actively participated in the standards process were more successful in implementation. Early development and implementation activities, especially the development of in-house knowledge and expertise, had high leverage and contributed to the ability or success of implementing the new factory automation software standards. The size of the company, long-term relationships, or maturity of the product did not significantly contribute to the ability or success of a company to implement the new software standards.

Proper error-free behavior with adequate recovery capabilities in fully loaded dynamic operation mode is critical to fab success and should be considered as part of the requirements and test development. As 300mm fabs ramped to high volume and fully loaded the equipment, operations revealed implementation errors and aberrant behavior. In some cases, the stress of full loading exposed previously unknown or unobserved defects and incompatibilities in the software, such as software layering and weaknesses in architecture. These issues impacted fab operations and productivity.

Industry Timeline Synchronization

Early understanding of fab operational requirements can provide a blueprint of the decisions that need to be made for an integrated factory. Device maker guidelines help to determine standards requirements, provide supplier implementation guidance, and define systems and performance needed to support fab operation.

Industry roadmaps helped to establish the timing for making key decisions and defining requirements for 300mm fab operations and equipment. Once the physical interfaces were defined, the operational requirements for software and systems to control and manage automation and integration were developed. This development occurred on the trailing end of the industry need. A 300mm software implementation roadmap was established to provide unified device maker evaluation and production-level software need dates and supplier implementation target dates.

Backend, assembly, and die prep were not addressed early in the deliberations to optimize and take full advantage of new technologies and standards. Considerations for high mix and low mix were late in the transition, resulting in architectures and standards that were not flexible enough to support agile manufacturing.

300mm Deployment Lessons Learned

Key Successes

Early prototyping and standardized assessment methodologies were crucial in accelerating the availability of technically sound standards and conformance in equipment. Establishing objective test criteria enabled accelerated maturation. The availability of a high quality, stable test tool was critical to success. Partnering or close cooperation with IC makers in early conformance assessment and improvement significantly improved the likelihood of first-pass success and accelerated development and delivery.

Based on the 200mm experience, the 300mm factory cost was predicted to be $10B to $50B. Industry coordination and collaboration dramatically mitigated that cost. A 90 nm technology node, 300mm fab at 20K wafer starts per month is between $2.5B to $3B. This has come about through the development of industry wide consensus on

1) Common factory architecture guidelines
2) Factory logistics and integration through software, automation, metrics, wafer tracking, and traceability
3) Standards and guidelines for materials, automated material handling, equipment integration and control, facilities, and safety

In 1997 the average time to first wafer start was over 20 months. Current 300 mm fabs are an estimated 18 months to first production wafer. In 1994, a Japan PCS (Productivity Cost Savings) survey reported that 11 variations of 8-inch production cassettes existed. In 300mm, the industry settled on one 25-wafer production FOUP.

A major benefit was enrichment of the factory vision through global cooperation, communication, and the ability to discuss and improve manufacturing effectiveness. It required transcending regional, linguistic, business, and cultural barriers to work together effectively, paving the way for future collaboration.

This global cooperative work circumvented potential derailment caused by early direction and assumptions such as

1) Aggressive transition dates (planned transition date 1997, actual transition 2001)
2) No bridge tools
3) Lack of full industry support and delays in implementation of standards

Frank and open communications enabled the start of industry-wide discussions on business cycles and heightened awareness of the need to maintain the industry's productivity improvement direction as well as ameliorate the intensity of resource requirements by moderating the transition timing. The global cooperation helped to keep transitions requiring productivity improvements on an evolutionary path, allowing for mid-course corrections in terms of visions and needs. Global cooperation has enabled the transition to maintain a steady and deliberate course, recognizing that the transitions took longer than the original aggressive plan.

Areas for Improvement

Implementation of industry standards in general, and software standards in particular, required considerable improvement and progressed slowly. Some suppliers did not anticipate that the 300mm factory automation software standards and user guidance requirements were going to be universally required. For these suppliers, development was delayed until suppliers began to see purchase orders. The expertise and understanding required to implement the standards was inadequate in most supplier engineering ranks when the standards first emerged. Some suppliers did not have sufficient understanding of the factory environment to provide context for their implementation or what it takes to be a well behaved member of a factory tool set. This learning curve created a significant delay.

In areas of high priority for consortia and industry collaboration, backend considerations were overshadowed by emphasis on silicon wafer processing through final passivation. I300I did initiate backend work, especially for wafer-level processes such as test and probers, bumps and wire bond pads, wafer thinning, and tape mounting and dicing and provided guidelines. Although backend suppliers responded well to the guidelines and worked on equipment configurations, several areas were left untended such as ultra thin wafer handling, post-dice pick and chip cassettes, and chip-level traceability.

The ability to evaluate fab logistics in a pre-competitive environment might have provided clearer understanding of requirements, cost sharing for development costs, and a means of early verification and validation. Although individual equipment was evaluated for logistics, control, and communications, no industry pilot facility or test bed existed to understand product logistics and handoff issues.

While IC companies have learned to cooperate in the pre-competitive arena, the equipment supplier community did not organize and achieve a similar level of coordination and collaboration. Without organized sponsorship of pre-competitive supplier discussions, supplier consensus input to the strategy and standards development, and collective dialogue with IC makers, the ability of suppliers and device makers to engage at the right time and level was often sporadic and inadequate.

Top 10 300mm Lessons Learned

#1 A wafer size transition requires global coordination and consensus
#2 Build on learning from previous transitions
#3 Consensus on an industry strategy for a transition must be driven by rigorous analysis of economic, business, and technology impacts
#4 Identify initial, fundamental decisions that need to be made
#5 Establish early consensus on global semiconductor manufacturing requirements
#6 Comprehend bridge tools and backward application in the manufacturing strategy
#7 Form collaborative efforts between device makers to share cost and risk
#8 Early, close cooperation and communication between device makers and suppliers is essential
#9 Accelerate the standards development cycle through early standards prototyping, validation and concurrent conformance test development
#10 Establish and continuously refresh a roadmap and timeline to synchronize priorities and decisions

Going Forward for Next Transition

The Climate Going Forward

Significant changes are already occurring in market forces, business drivers, technology complexity and product configurations, economics, and production centers. Projected average revenue growth is expected to slow down from the previous 16–17% to 11% with some predictions going as low as 6–7%. The industry will remain cyclic; fab cost pressures will continue and perhaps accelerate. Equipment reuse and environmentally benign regulation will be in greater force. The Kyoto protocol is one example of this trend.

At the start of the 300mm transition, the major players were the US, Japan, Europe, Taiwan, and Korea. At present this has expanded to China, Singapore, and Thailand. Within the next few years, more regions will have manufacturing facilities to cater to their expanding markets. Industry fab configuration demands vary from large fabs (40K wafer starts/month), replicated modular fabs (7K wafer starts per module), and requirements for small fabs. This variation in fab configuration may proliferate.

The transition to 300mm as planned at the inception was guided by the current technology trend with only the most rudimentary conception of oncoming "non-incremental" technology changes. For example, the issue of Cu/low-k BEOL logistics was a late consideration and required FOUP isolation and changes in standards to prevent cross-contamination issues. Present interconnect limitations may drive alternative technologies and in the future may drive heterogeneous technology integration. Future fabs will have to contend with the advent of nano-technology. The cost of new part numbers will continue to rise due to design and design verification cost, mask cost, technology complexity, and any other non-recurring engineering costs that might be involved. Product configurations are changing and, in addition to system-on-chip (SOC), acceleration is occurring in system-in-package (SIP) and stacked chips. Application-specific standard products (ASSP) growth is outpacing application-specific IC (ASIC) growth. Also during the 300mm transition, the technical and cost feasibility of a larger silicon wafer size was a serious consideration. Other unexplored or under-explored options are being investigated. These and other alternatives deserve due consideration in the next transition.

The current standards do not cater well to the current need for flexible batch size requirements. Not only is this needed for large part number manufacturers whose batch sizes are often small, but also for small part number manufacturers who need small batch size capability for engineering lots, design verification lots, and specialty parts. While the start of the transition was predicated upon the industry consensus on the batch size, the changing economic and business picture renders it an open issue. Some of the early 300mm decisions, such as switching carrier configuration from SMIF to FOUP, drove major development effort. From 1996 though 1999, the I300I basic thrust was to enable 300mm fabs for maximum production volume. The original plan for 24 product wafers plus one test wafer was modified to 25 product wafers in the FOUPs. It is important to note that while the industry settled on 25-wafer FOUPs for silicon fab operations, the option was kept open for 25- and 13-wafer FOUPs for backend. A single-wafer FOUP and the related interface standards were developed through a supplier initiative with some support from the I300I organization. Early effort was expended on person guided vehicles (PGVs). The cost and benefit of such fundamental, pivotal decisions should be evaluated objectively for the next transitions. Earlier, more focused cooperation with suppliers would likely result in more cost-effective strategies.

Historically, industry-associated manufacturing paradigm shifts with a wafer size transition. While the 300mm transition took on the innovations made during 200mm wafer manufacturing, the industry separated 200mm and 300mm; hence, much of the wisdom accumulated through 200mm was applied only to 300mm. From the inception of the 300mm transition, the number of 200mm fabs doubled and more are being built. Many of the productivity and automation improvements in 300mm fabs were not planned for back application to 200mm fabs. Opportunities to apply new concepts to previous fab generations should be considered to improve overall productivity on an ongoing basis. This will help to relieve schedule pressures for the next transition, allow any needed mid-course corrections, and pace R&D expenditures.

Any future transition must contend with evolving needs and market dynamics. No amount of working together can overcome macroeconomic forces. Device maker differentiation increases with increasing complexity and implementation time variance. Changes in requirements, timing, or business strategies should be continuously refreshed and communicated to allow mid-course corrections.

300mm "Prime"

300 mm lessons learned are being applied and form the basis for next generation fab and next wafer size transition strategies. The concept of 300mm prime is offered as a strategy for a 300mm productivity continuum bridge to the 450mm with an objective to have forward compatibility of the 300mm prime design and architecture in the 450 factory. Meeting this objective would result in minimizing design and architecture changes in the 450mm implementation, giving a high confidence of success when scaling up to the next wafer size. This also gives a benefit to 300mm in continuous productivity improvement while carefully steering the design for a 450mm transition.

As depicted in Figure 5, business model consideration is key to the design transition. An escalating cost of factory and development with NWS drives even harder a need to have common factory equipment and factory architecture. To accommodate a global usage strategy, high and low product mix and high and low volume factory designs need to be considered. A design of this type would make a factory both flexible and nimble in adapting to varied volumes and mixes while maintaining good cycle time. It is recognized, however, that all aspects of a fully transitioned 300mm prime design may not be completely backward compatible to accommodate installation into pre-existing 300mm factories.

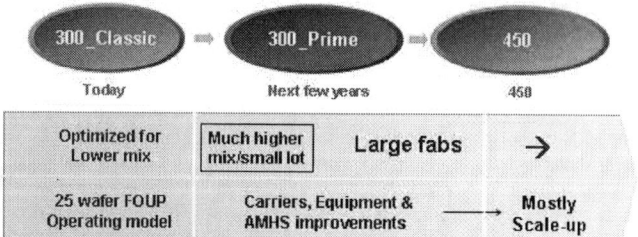

Figure 5 450 mm Productivity Transition Continuum
Source: Dev Pillai, ITRS Factory Integration Technical Working Group presentation at the IRC, Seoul, S. Korea

Areas of focus in 300mm prime evolution may include wafer, carrier, process equipment, factory and facility layout, and data systems integration. Many are interrelated in that changes in one area will likely drive changes in another.

Flexibility in coping with high product mix especially in high volume situations while maintaining good cycle time drives the need to accommodate small or very small lot sizes. Batch equipment such as ion implant, furnace, and wet stations generally need to be fully loaded to maintain optimum uninterrupted throughput and consequently low cycle time. These equipment technologies may need to migrate toward a single-wafer processing tool concept while maintaining equivalent (or better) wafer process performance, cost of ownership, and maintainability.

Small lot size and high volume would stress existing automated material handling system (AMHS) designs into doubling or even quadrupling the number of moves depending on lot size and volume, pushing the number of moves required per hour beyond current wafer transport capabilities. Design energy would need to be expended on evaluating the feasibility and use of high speed transport capabilities or possibly marrying process equipment directly to stockers, which would eliminate much of the carrier travel distance but results in redesign of the equipment footprint, layout, front end interface, etc. Dense packing would also be desired in the latter to optimize factory footprint; however, it will bring additional challenges when shrinking the peripheral support equipment footprint for each tool, i.e., pumps, point-of-use (POU) scrubbers, gas delivery systems, etc.

The above describes the 300mm concept and shows examples of design changes that may be desirable in 300mm allowing for a scale up to the next wafer size. There is no clear line separating 300mm prime design changes vs. "standard" upgrades only in that a fully transitioned 300mm prime design, factory and process equipment, would be used in a scaled up version in 450mm. The early challenge is to identify what areas of standards or design change would be needed in a NWS factory scenario, founded on business model direction and in synch with a 300mm improvement vector. Then appropriate areas of the industry need to plan to make this realty within the timeframes that are needed.

The ITRS Factory Integration Working Group has started investigation and strategic planning for the next wafer size transition. Figure 6 shows an early example of a high-level timeline for standards needed for a 450mm conversion before factory equipment and systems are designed.

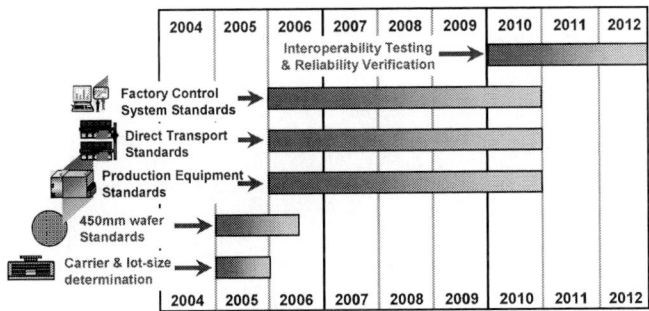

Source: ITRS Factory Integration Working Group, February 2005

Figure 6 ITRS Draft Timeline for Standards to Support a 450mm Conversion

Situation Analysis

The International SEMATECH Manufacturing Initiative (ISMI) has performed a detailed analysis of industry productivity, using the Industry Economic Model (IEM). Interestingly, this tool was not available to plan for, but developed as a result of, the 300mm transition. As such, the IEM now provides a comprehensive look at possible scenarios, with the following observations:

1. A productivity boost of some sort is necessary in the 2010–2012 timeframe; using current ITRS assumptions, that productivity boost is necessary for the industry to stay on the historical productivity curve.

2. Of the several factors that can influence the timing of the next wafer size, the most important is chip density. Consequently, a "roadmap acceleration" can delay, but will not eliminate, the need for a wafer size change.

Declining cost per function has been the *de facto* governing principle of the semiconductor industry for multiple decades. Falling functional cost, or "productivity," is assumed to be responsible for the continuously growing demand for electronics and semiconductor products. It is that success metric that was examined by the ISMI Industry Economic Model (IEM) to evaluate the assumptions contained in the current ITRS. The IEM is a one-of-a-kind tool that integrates at an industry level many semiconductor technology, wafer diameter, and factory and equipment configurations along with many core strategic manufacturing and development planning functions. The IEM logistics, algorithms, and assumptions have been validated and can generate scenarios that not only examine the drivers of past, present, and future productivity, but also assess changes to technology and manufacturing assumptions and the impacts of demand fluctuation and business cycles.

In these studies, the transistor has been used as the surrogate for function, since it is the most measurable metric readily obtainable. The two most common methods used by chip manufacturers to increase the volume of a transistor produced while managing the associated cost is to increase the number of transistors in a die with smaller line widths or the number of die on a wafer with larger wafers. Since both methods increase the number of transistors per wafer area, accelerating the introduction pace and subsequent production ramp of each new technology node, thus providing increased chip densities, can have an effect comparable to increasing the wafer size. Examining the productivity benefits derived from each method with the use of the IEM has provided a useful neutrality rule of thumb; that is, for each year that the technology roadmap is accelerated, the introduction of the next wafer size can be delayed an equivalent year. Consequently, if the introductory pace of technology nodes with historical production ramps can be maintained at two-year intervals for the next few generations (i.e., 65 nm–32 nm), the next wafer size (450 mm) can be delayed until 2015, assuming no change in any other metrics (see results: Table 1).

Technology Node (nm)	180	130	90	65	45	32	22	16
ITRS (3 year)	2000	2002	2004	2007	2010	2013	2016	2019
SMT (2 year)	2000	2002	2004	2006	2008	2010	2013	2016

Wafer Size (mm)	300			450			
ITRS (3 year)	2001			2012			
SMT (2 year)	2001						2015

Table 1 Roadmap Options
Source: ISMI Industry Economic Model

IEM Scenario: Realize the Roadmap
The initial scenario was created to use all the fundamental assumptions in the 2003 ITRS, along with a continuation of recent historical demand trends. As such, the technology introduction pace was set at three-year intervals beginning with the 65 nm node and the density/chip set to their respective product levels. Silicon area was used to establish the industry demand growth trends along with a continuation of historical business cycles in five-year average durations). As depicted by Figure 7, using the silicon area to govern the demand growth and using the ITRS assumptions to describe the physical capabilities

provided some useful observations. Since chip sizes are assumed to continue constant for leading-edge products and decline linearly for lagging/trailing products, the unit output is projected to remain at the historical growth level. The output of transistors, however, is projected to slow by over 20% based on ITRS transistor/cm^2 assumptions. To recoup expected production costs, maintaining reasonable margins would require our customers to pay more per transistors than historic price equilibriums allowed and, assuming long-term market elasticity, retard industry growth.

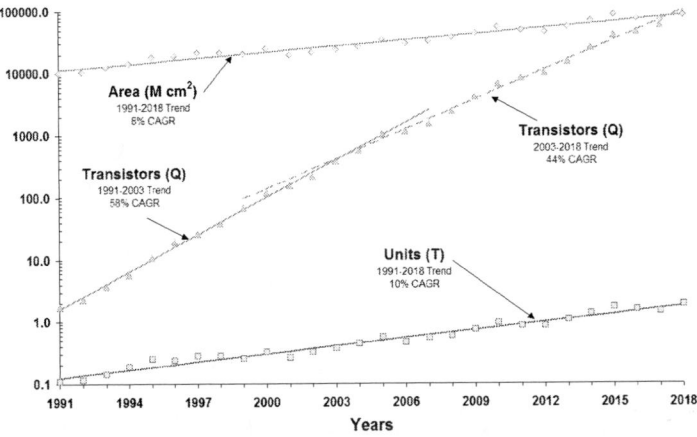

Figure 7 Industry Demand
Source: ISMI Industry Economic Model

The IEM was then executed with and without the next wafer size (see illustration 3) to generate the required fab population based on historical product demand trends, technology lifecycles, and cost-effective fab management principles. As depicted in the illustration, historically the fab population by wafer size grows to around 300 or so, followed by slow attrition. If the 450 mm wafer were not introduced, the required number of 300mm fabs would nearly double the historical norm by the end of the next decade, thereby creating possible environmental and logistic issues. In addition, to operate these additional 300mm fabs vs. less than half of their 450 mm counterparts would add substantially (over $150B cumulative through 2018) to industry production costs, thereby reducing productivity. By 2024, when 450 mm would reach its full production ramp, over $850B could be realized in cumulative savings.

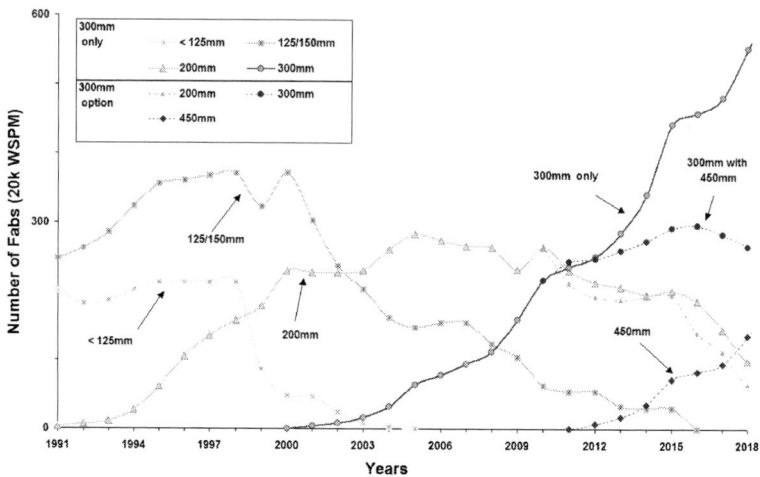

Figure 8 Industry Capacity

Source: ISMI Industry Economic Model

After extrapolating historical manufacturing metrics and supplier values beyond the 65 nm technology node and the 300mm wafer diameter, the overall productivity outlook of the industry was examined and contrasted to the historical trends established over the past dozen years (see Figure 8). As depicted in the illustration, the industry cost/transistor metric permanently crosses over historical trends by 2010. While the 450 mm scenario productivity results do not return the industry to the historical trend, it does show by 2018 (with 135 fabs) a 20% improvement over the 300mm-only scenario, driven by lower manufacturing costs noted previously. By 2024, with continuing growth in 450 mm production, the improvement could grow to 30% over the 300mm-only scenario.

Summary

This paper has covered many of the experiences and learnings from the 200 and 300mm transitions. Our learning from the 300mm transition can help to guide the next transition and avoid costly mistakes. A model for successful global consensus and cooperation exists in the I300I/Selete collaboration and industry standards development. Suppliers should be involved early to define an industry strategy and guidelines and throughout the transition. The approach to wafer size, equipment, and productivity improvement should include bridging between the past and future fabs to mitigate risk. Open, representative industry discussion on technology needs and engagement of organizations such as universities, research centers and national labs must be ensured. Research, modeling, and analysis must be provided with open discussion and awareness of business and economic pressures including better understanding of supplier equipment development cycle.

Most importantly, and paramount in catalyzing and mobilizing the industry for the next transition, is the indisputable need for a clear consensus and strategy that takes into account business and economic factors for the industry, as well as the technical factors.

Acknowledgments
Ashwin Ghatalia, Frank Robertson, Dev Pillai, Jeff Pettinato ITRS Factory Integration Technical Working Group

Referenced Documents[1]
1. J. Ferrell and M. Pratt, *I300I Factory Guidelines*, v. 5.0, I300I, Austin TX (2000)
2. *Global Joint Guidance for 300mm Semiconductor Factories*, v. 1.0, I300I and J300, I300I Archives, Austin TX (1997)
3. K. Gartland and S. Kono, *Automated Material Handling System (AMHS) Framework Document*, v. 1.0, I300I and J300E, I300I Archives, Austin TX (1999)
4. E, Bass and P. Jai, *Metrics for 300mm Automated Material Handling Systems (AMHS) and Production Equipment Interfaces:* Rev 1.0, I300I, I300I Archives, Austin TX (r 1998)
5. R. Goodall, *I300I Guidelines on 300mm Process Tool Mechanical Interfaces for Wafer Lot Delivery, Buffering, and Loading*, Rev. D, I300I, I300I Archives, Austin TX (1996)
6. *Backend Global Joint Guidance for 300mm Semiconductor Factories*, v. 1.0, I300I and J300, I300I Archives, Austin TX (1998)
7. *CIM Global Joint Guidance for 300mm Semiconductor Factories: Release Five*, I300I and J300, I300I Archives, Austin TX (2000)
8. N. Desai and T. Wakabayashi, *Equipment Performance Management User Requirements*, International SEMATECH, I300I Archives, Austin TX (2000)
9. K. Gartland, G. Godding, B. Hodges, N. Muzumbar, S. Hohkibara, S. Kono, M. Honma, and G, Inoue, *Scheduler/Dispatcher User Requirements*, International SEMATECH, I300I Archives, Austin TX (2000)
10. *User System Requirements Document (USRD) for FOUP-to-Loadport Interoperability*, International SEMATECH and J300E, Austin TX (1999)
11. B. Crandell, D. Bloss, G. Crispieri, J.S. Chou, K. Gartland, M. Pratt, P. Cross, S. Kono, and T. Masui, *300mm Software Functionality Requirements Documents, Phase 1*, rev. 2, International SEMATECH and Selete, I300I Archives, Austin TX a (2001)
12. B. Crandell, D. Bloss, G. Crispieri, K. Gartland, L. Pivin, M. Fujita, P. Cross, T. Katsuyama, T. Masui, T. Miki, Y. Kominato, and Y. Ohyama, *300mm Software Functionality Requirements Documents, Phase 2*, rev. 2.0, International SEMATECH and Selete, I300I Archives, Austin TX (2001)
13. G. Crispieri, L. Christal, S. Fulton, *300mm Operational Flowcharts and Scenarios*, v.10, International SEMATECH Manufacturing Initiative (ISMI), Austin TX (2004)
14. *Unified Equipment Performance Metrics for 0.25 micron Technology*, I300I and Selete, I300I Archives, Austin TX (1997)
15. F. Robertson and K.Komiya, *Unified Equipment Performance Metrics for 0.18 micron Technology*, I300I and Selete, I300I Archives, Austin TX (1999)
16. H. Miyatake, K. Kishimoto, and P.Patruno, *Unified Equipment Performance Metrics for 130 nm Technology*, v.2.0, International SEMATECH and Selete I300I Archives, Austin TX (2000)

---

[1] International SEMATECH, ISMI, and I300I documents, are available on the SEMATECH website at http://ismi.sematech.org/standards/guidance/index.htm

17. D. Pillai, *450mm planning presentation to IRC: What we should do in 2006,* presented to the ITRS International Roadmap Committee. Acknowledgement: The ITRS Factory Integration Working Group, Seoul, Korea (2005)

18. Asian New Factory Start-ups, Semiconductor Industry Standard Indicators, International Archives, Austin TX

ECS Transactions, 2 (2) 155-165 (2006)
10.1149/1.2195656, copyright The Electrochemical Society

Discussion on Issues toward 450mm Wafer

M.Watanabe, T.Fukuda, A.Ogura, Y.Kirino, and M.Kohno

Silicon wafer and SOI WG
Semiconductor Manufacturing Technology committee of Japan (SMTJ), Japan
Electronics & Information Technology Industries Association (JEITA)
Kandasurugadai 3-11, Chiyoda-ku,
Tokyo 105-0011 Japan

This reports a present status of recent technical discussions on
450mm wafer. Discussions are focusing on mechanical wafer
specification, especially, wafer thickness based on thermal stress
and other factors that may determine proper wafer thickness.

Introduction

As demand or production of semiconductor devices has been increasing rapidly, 450mm
wafers are expected to be introduced in 2012 according to ITRS (1). Industry Economic
Model of ISMI (2) also predicts that LSI production using 450mm wafers will start in
2012. It is well recognized (1) that the transition to the next generation wafer size
(NGW) needs collaboration among all industry partners. It is necessary, in advance, to
make industry consensus and establish many standards to reduce the cost of wafer size
transition. Discussions of silicon wafer standards need to start in 2005 (3). So that, a
working group has been organized in JEITA and silicon wafer related issues were
discussed toward 450mm era. Silicon wafer and SOI WG in JEITA is to discuss any
technical 450mm wafer related issues. Members are wafer suppliers (Japan Society of
New Metals) and experts from JEITA. Some of them are also active in ITRS and/or
SEMI. The NGW committee organization makes it possible to create a discussion path to
other organization to involve device makers and equipment suppliers. Discussions on
450mm wafer issues are just started and will continue to the next decade based on various
points of view. This reports a present status of technical discussions on 450mm wafer.

450mm Wafer Issues

Crystal Growth and wafer shaping

Growth of 450mm single crystal is a key technology of 450mm era. Pioneering
work (4) by SSi (Super Silicon Crystal Research Institute) was an excellent feasibility
study of 400mm, almost 450mm, single crystal growth, epitaxial growth and wafer
shaping technologies. Discussions among authors (5) are agreed that growing 450mm
single crystal is quite possible. Problems left to be explored are crystal quality and how
to produce 450mm crystal economically to meet the industry requirement. A position
paper (6), "Advantage and Challenges Associates with the Introduction of 450mm
Wafers" is discussing various aspects of the 450mm wafer technology.

Mechanical Wafer

450mm wafer issues have two aspects. One is business issue that is R&D cost and timing of pilot production. The other is wafer technology itself including some wafer specification. First wafer is a so-called mechanical wafer that is used in equipment development. Second one is a test wafer that is for various device process developments and the last one is a device quality wafer. Properties of the mechanical wafer under discussion are,

0. Diameter = 450mm
   Tolerance = ±0.2mm
1. Thickness = 775μm + α
2. Edge shape = TBD
   Edge finish = Polished
3. Notch dimension = Current dimension
4. Front and back surface finish = Polished
5. Flatness
   Site Flatness,
   Nanotopography
6. Others
   Bow and Warp,
   Surface metal and particle

Among these, thickness is very important and first priority to standardize (7). Wafer thickness influences both of device processes and equipments. Wafer cost depends on its thickness. Second priority is edge shape. Current 300mm wafer has variety of edge shapes due to different optimization in the device processes. A single edge shape is preferred for 450mm wafer to reduce wafer production complexity. As for the diameter, 450mm is a given at this moment. Diameter tolerance is a subject of the discussion but its decision can be postponed. Notch dimension and surface finish are same to current 300mm ones since no reason to change them. Flatness and other properties are not discussed since they are not important to the mechanical wafer.

SOI Wafer Structure at 45nm and 32nm Node

A possible SOI wafer structure as starting materials depends on device structures at that technology node. Devices that use SOI substrate today will continue to use similar or advanced SOI substrates because device design based on the SOI advantage will make it difficult to get back to the use of the conventional polished or epitaxial wafer. Use of SOI and strained silicon will rather increase in future. Structure of 450mm SOI wafer will be same to that of 300mm and processes to make it will also be same. Availability of tool set such as ion implanter is a key issue to the development of 450mm SOI wafers.

Wafer Thickness Scaling

Thickness issue should be discussed carefully but quick enough for the equipment development needs. Factors to be taken into consideration are,

- Slip due to thermal stress

- Wafer breakage due to thermal stress or wafer handling
- Sag due to wafer weight
- Wafer cost

Scaling Based on Empirical or Mechanical Property

Empirical Scaling. Wafer thickness increases as wafer diameter increases. However, technical reason to make choice of a particular thickness is not clear at any wafer diameter. Figure 1 in which x-axis is log of diameter show the wafer thickness trend including an extension to 450mm diameter. There seems to exist two slopes. One is in small wafers, 3 inches to 125mm and the other is in large wafers, 125mm to 300mm. Extrapolation of the latter slope to 450mm indicates that 450mm wafer thickness is about 825μm. Figure 1 also shows third order polynomial curve fitting. Extrapolation of this to 450mm indicates that thickness is about 800μm. Wafer thickness of 800μm to 825μm may be a reasonable thickness. It should be noted that 300mm thickness was not decided based on any technical reason and any of above extrapolation does not have any technical reason. It is just a simple empirical extension of the historical diameter trend.

A Scaling Based on Wafer Breakage. Bawa, Petro and Grimes (8) proposed wafer thickness based on wafer breakage. They improperly extended an application of ASTM F394 (9) to determine the 300mm wafer thickness. They use a fracture toughness of as-sawn wafer that has heavily damaged surfaces and it is not appropriate to apply their model to LSI device process that uses polished wafers. Many think thinner wafer is more fragile without experience or data. Wafer breakage will start from a process-induced or a wafer handling-induced crack on front/back surface or wafer periphery if there happens any stress large enough to propagate it to cause fatal breakage. Problem is properties of the induced cracks and how stress concentration factor depends on the wafer thickness. This is open to further study.

Scaling Based on Thermal Stress.

A slip generation due to thermal stress was a subject discussed in the 300mm diameter transition. An experimental data from JEIDA (10) could not show a definite dependence of the slip generation on the wafer thickness. Slip due to the thermal stress can be, in general, thickness sensitive so that it is revisited here again.

Stress Simulation. Flash lamp annealing (FLA) is a most tough process regarding to thermal stress even in 45nm node and beyond. Wafer surface is heated up to 1200°C or higher in msec. to activate ion-implanted dopant. Finite element method simulation of thermal stress distribution during FLA shows interesting results. Wafer diameters studied are 300mm and 450mm. Their thickness is from 725μm to 925μm and wafer is free standing. Figure 2 shows positions on wafer where temperature and stress are discussed below. Figure 3 is temperature change assumed during FLA process. Temperature, T_f, on front surface, A to D, changes as shown in Figure 3. Temperature, T_e, on full round front edge, D to G, is assumed

$$T_e(\theta) = T_b + (T_f - T_b)\cos(\theta)$$
$$T_b = 600°C$$

where θ is an angle from $D(\theta = 0°)$ toward $G(\theta = 90°)$, T_b is back surface temperature. Back surface, A′ to D′, and full round back edge, D′ to G, is kept constant T_b. T_b is 600°C in this simulation. Time dependent temperature distribution and thermal stress during FLA are simulated. Figure 4 (a) and (b) shows thermal stress σ_r on front (a) and back (b) surface. Tensile stress develops on the front surface while compressive stress develops on the back surface. Thermal stress that causes slip is shear stress. Maximum shear stress on the front or the back surface during FLA is shown in Figure 5 (a) and (b) respectively. Although shear stress depends on the position of wafer but it does not change significantly by the wafer thickness, 725µm to 925µm, of 450mm wafer. This means that slip generation due to thermal stress does not depend on the wafer thickness. Wafer breakage will be related, in general, to tensile stress. Maximum compressive stress (note that tensile stress is negative compressive stress in this plot) during FLA is shown in Figure 6. Tensile stress appears only on the front surface. It depends on the wafer thickness, 725µm to 925µm, of 450mm wafer. It varies less than 10%. Figure 6 also shows tensile stress on 200mm 725µm wafer surface and 300mm 775µm wafer surface. Tensile stress of 450mm wafer is slightly, 10% compared to 300mm, larger than 200mm or 300mm wafer. It is interesting that thinner wafer is better for this stress issue. These will mean that slip or breakage does not depend so much on the wafer thickness or diameter and that there seems to be no reason to increase 450mm wafer thickness from current 300mm wafer thickness.

Flash Lamp Annealing Experiment.   Defect generated in bare wafer during FLA was detected by x-ray topography. Figure 7 shows a part of FLA wafer. Repeating FLA for 4 times introduced some defects look like slip but wafer handling might cause them. SOI wafer for which FLA was only once shows some defect on its wafer periphery. Poor thermal conductivity of the buried oxide layer will cause rather easy introduction of defects in the SOI wafer. However, in either case, FLA is not long enough to move dislocation detected as slip because dislocation movement distance is quite limited even at very high temperature (11). In the case of FLA of locally implanted dopant, defect generation, slip or crack, by some FLA conditions is reported (12). So, very short movement, for example, less that 1µm needs to be considered in the defect engineering of the LSI processes. Extending FLA experiments to more realistic situation is the next step. Then, process window regarding to the wafer thickness can be discussed more clearly.

Sag and Stress due to Wafer Weight.

Sag is an important factor to wafer handling and wafer cassette. Sag of 450mm wafer supported by 4 points at its periphery is show in Figure 8 (a). In case of 775µm wafer, sag is 687µm. If this is acceptable, 450mm wafer as thick as 300mm wafer can be used in LSI process. If sag needs to be same as 300mm one that is 136µm, 450mm wafer should be 1744µm thick.   Wafer of this thickness is heavy and will definitely be expensive. It is not a practical choice. Situation is same, as shown in Figure 8 (b), when wafer is supported at 4 points about 70% of wafer radius from the center where supporting stress is minimum. Shear stress of wafer supported at 70% from the center is calculated regarding slip generation by wafer weight. It is shown in Figure 9. If supporting stress of 450mm wafer is required as small as that of 300mm, thickness should be 1744µm as shown in Figure 9. Situation is same to sag case. Stress of 450mm wafer supported by ring at the same position is smaller than that of 300mm supported currently by 3 or 4 points. This means that there would be a solution using an appropriate supporting to make stress comparable to current 300mm batch thermal process.

Edge Shape

Current 300mm wafers are produced with various edge shapes specified by wafer users. They are roughly classified to two typical shapes. One is so-called round type and the other is so-called taper type, as shown in Figure 10. Round type is mechanically more stable during wafer handling. Taper type is more commonly used. To detect wafer in a cassette, taper type is made use of light reflection from its flat apex while round type wafer is detected using transmitted light. SEMI M1 specification is a template that wafer edge shape must fit within the specified region. This makes some ambiguity or some choice of edge shape within the template. Edge of 450mm wafer should be specified by an exact shape with some tolerance. This will guide device makers to use a single edge shape.

Summary

Empirical extrapolation of wafer thickness to 450mm, it is 800μm to 825 μm without rational argument. Thermal stress of FLA by simulation or experiments does not indicate its thickness dependency. Sag of 775μm thick 450mm wafer supported by 4 points at periphery is 687μm. If this sag is acceptable, there is no reason to increase 450mm wafer thickness from 775μm that is current 300mm thickness.

Acknowledgement

Authors thank Mr. Akira Fukuda of Ebara Research Co. Ltd for the stress simulation.

References

1. ITRS (International Technology Roadmap for Semiconductors) 2005
2. J. Ferrell and D. Fandel, "Next Wafer Size Considerations" presented at SEMI Silicon Wafer Committee, March 2nd (2005) also presentation at ITRS 2005 Seoul meeting
3. J. Pettinatto and D. Pillai , *ISSM* 2004 F01 (2004)
4. K. Takada, H. Yamagishi, H. Minami, M. Imai, *Semiconductor Silicon 1998*, edited by H. R. Huff, ECS Proc. **98-1**, 376 (1998)
5. To be published as *JEITA Report* in 2006 March
6. Available from ITRS web site
7. ITRS 2005 Factory Integration Chapter
8. M. S. Bawa, E. F. Petro and H. M. Grimes, Fracture Strength of Large Diameter Silicon Wafers, *Semiconductor International* Nov. p115 (1995)
9. ASTM F 394-78 (Reapproved 1996)
10. *JEIDA Reports* 99-20, 1999 March
11. M. Akatsuka, K. Sueoka, N. Adachi, N. Morimoto, H. Katahama, *Microelectronic Engineering* **56**, 99 (2001)
12. T. Ito, K.Suguro, M. Tamura, T. Taniguchi, Y. Ushiku, T. Iinuma, T. Itani, M. Yoshioka, T. Owada, Y. Imaoka, H. Murayama and T. Kusuda, *IEEE Trans. Semicond Manuf.* **16**, 417 (2003)

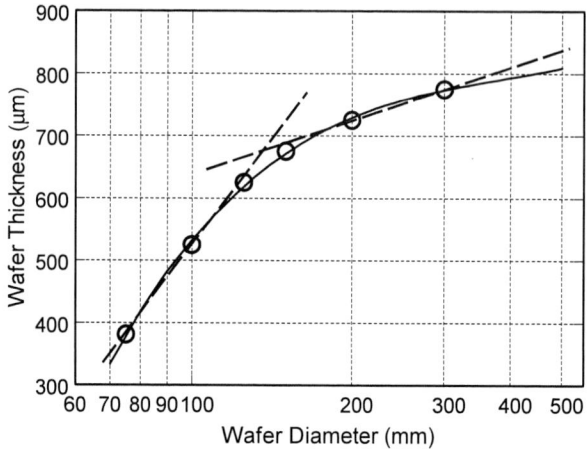

Fig.1 Wafer thickness trend. Thickness plotted vs log (diameter) shows two slopes. Third order polynomial fitting line is also shown.

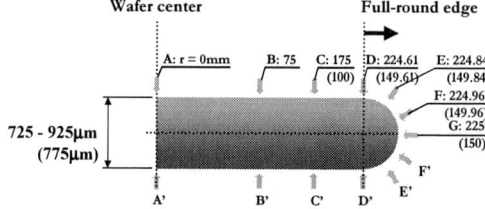

Fig.2 Position of points where temperature and stress are shown in the following figures.

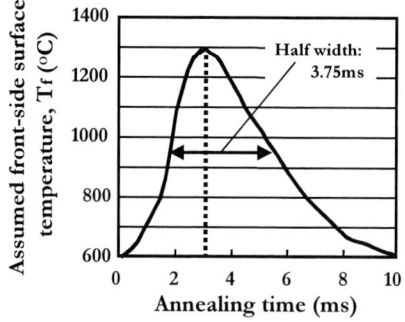

Fig.3 Temperature of wafer front surface during flash lamp anneal used for simulation.

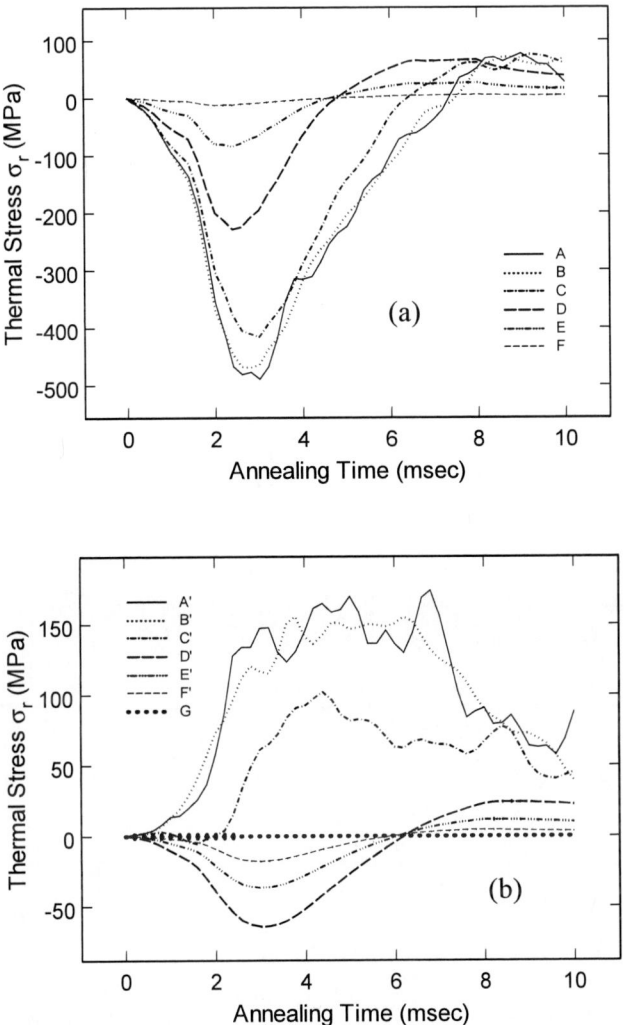

Fig. 4 Compressive thermal stress changes of front surface (a) and back surface (b) of 775μm thick 450mm wafer during FLA. Each position is shown in Fig.2 and temperature is shown in Fig.3.

ECS Transactions, 2 (2) 155-165 (2006)

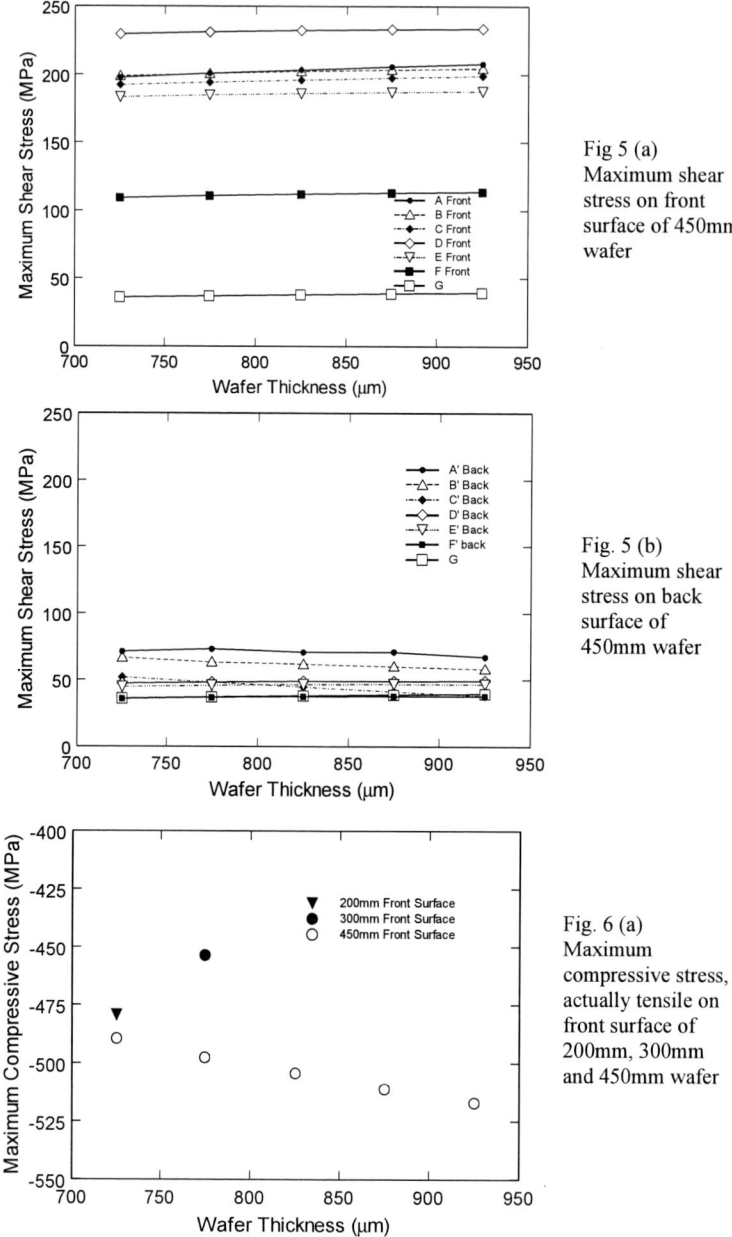

Fig 5 (a)
Maximum shear
stress on front
surface of 450mm
wafer

Fig. 5 (b)
Maximum shear
stress on back
surface of
450mm wafer

Fig. 6 (a)
Maximum
compressive stress,
actually tensile on
front surface of
200mm, 300mm
and 450mm wafer

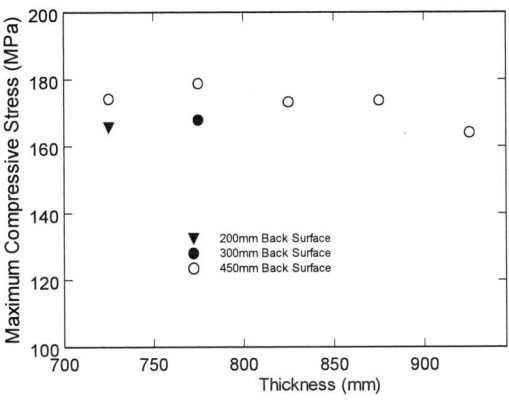

Fig. 6 (b)
Maximum
compressive stress
on back surface of
200mm, 300mm
and 450mm wafer

300mmBulk

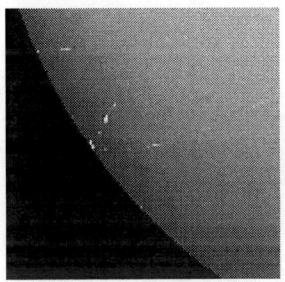 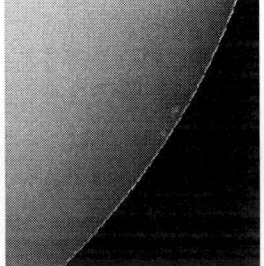

FLA 4times   FLA once

300mmSOI

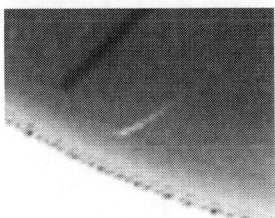

Fig.7   X-ray topographies of flash
annealed wafers. "300mm Bulk" is that
bare wafers were flash lamp annealed
once and 4 times while "300mm SOI"
is that SOI wafer was flash lamp
annealed once by the same process as
bulk anneal. X-ray topography contrast
was enhanced to show some defects at
wafer periphery. These defects may
not be a slip induced by the thermal
stress as discussed in text. Defects by 4
times FLA looks more clear.

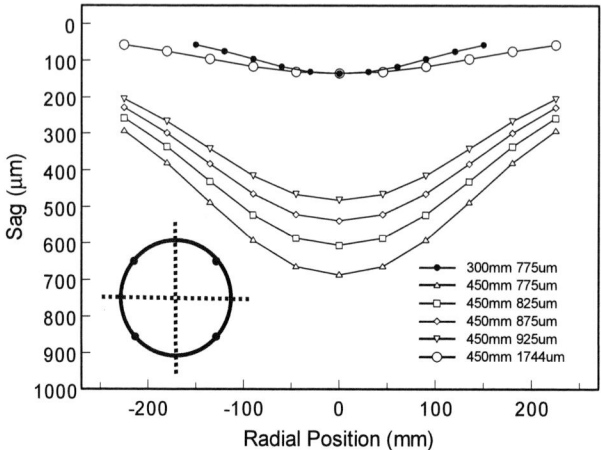

Fig. 8 (a) Bending of wafers supported by 4 points on the wafer periphery. Sag of 450mm wafer with 1744µm thickness is same as current 300mm wafer with 775µm thickness.

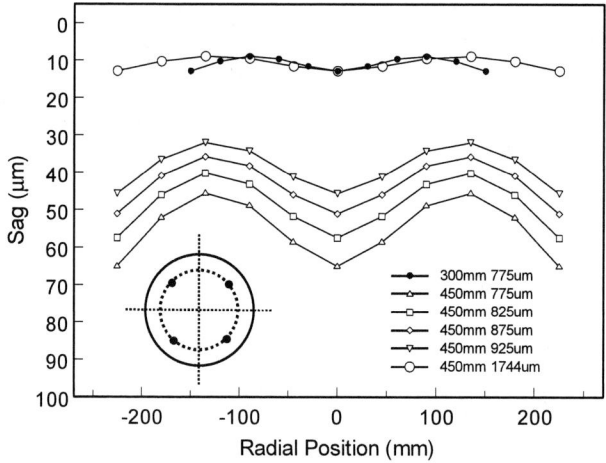

Fig. 8 (b) Bending of wafers supported by 4 points at 70% radius from center. This position is for minimum supporting stress. Sag of 450mm wafer of 1744µm thickness is same as current 300mm wafer of 775µm thickness

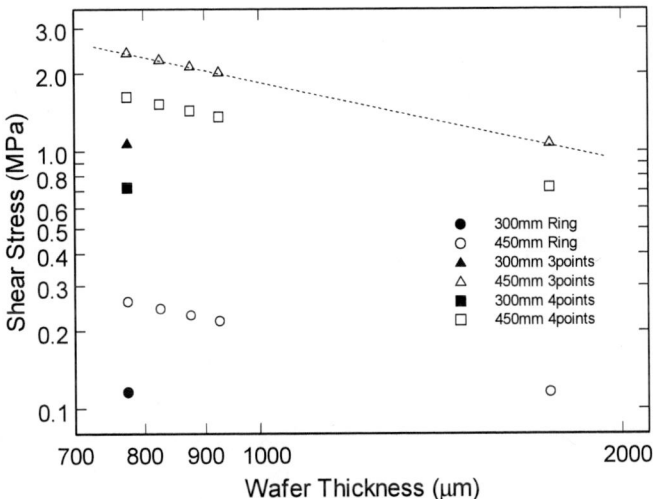

Fig.9 Shear stress by 3, 4 and ring support. Stress of 1744μm thick 450mm wafer is same to current 775μm thick 300mm wafer. Stress of ring supported 775μm thick 450mm wafer is less than 4 points supported 775μm thick 300mm wafer.

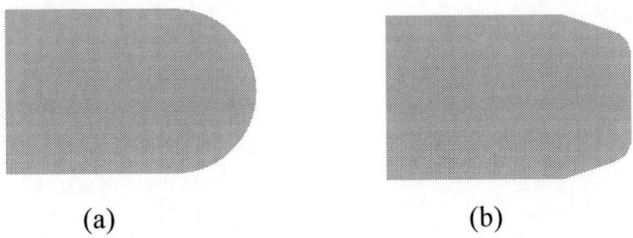

(a)   (b)

Fig.10 two typical edge shapes. One (a) is round type and the other (b) is taper type. Everyone prefers to unify edge shape of 450mm wafer. Either one or, at least, both of them should be standardized with single shape parameter set for each edge shape.

ECS Transactions, 2 (2) 167-181 (2006)
10.1149/1.2195657, copyright The Electrochemical Society

300 MM SOI for High Volume Manufacturing

G. Pfeiffer[a], M. Haag[b], M. Schmidt[b], R. Krause[b], P. Tsai[a] and J.D. Lee[a]

[a] IBM Corp., Hopewell Junction, NY 12533, USA
[b] IBM Procurement Technology Center, Mainz, Germany

300 mm SOI for leading edge CMOS applications has been
moving into high volume manufacturing recently but little
information on the quality of these wafers is available. This paper
reviews key characteristics of 300 mm SOI and how they compare
to the status from several years ago. For physical characteristics,
Si layer thickness and uniformity, surface quality and defects,
flatness, surface roughness, metal contamination and slip are
surveyed. Electron mobility, buried oxide charge and interface
state density data obtained from pseudo MOSFET measurements
are employed to monitor the electrical characteristics of the SOI
structure.

INTRODUCTION

The focus of development and manufacturing of leading edge thin film SOI
applications has recently shifted from 200 to 300 mm diameter wafers. As 300 mm SOI
is moving into high volume manufacturing, it will become the dominant diameter and
with the transition, the focus has been on wafer quality improvements so that high yields
for device fabrication can be attained. An earlier study identified several areas of
concern where 300 mm SOI could not yet meet requirements.(1) This paper describes
current capabilities at the end of 2005 and improvements in the relevant quality
characteristics of 300 mm SOI wafers used for IC manufacturing.

SOI MATERIAL CHARACTERIZATION

Quality improvements of SOI wafers over the last several years are the result of a
collaborative effort between the SOI wafer user and supplier community to develop
specifications and requirements and then to validate improvements in the IC fab. As part
of this effort, a collaborative data exchange platform (2) for wafer technical data has been
implemented as a tool for analyzing wafer – IC fab process interactions. Based on
feedback from this approach, SOI wafer suppliers have achieved quality enhancements
via the development of new wafer manufacturing processes and continuous process
improvement. Key areas have been physical characteristics (Si layer uniformity, flatness,
defectivity and metal contamination levels) and electrical properties of the top Si and
buried oxide (BOX) layer (mobilities, BOX charge and interface state density).

The SOI wafers surveyed for this report were typical p- material used for CMOS
technology with a superficial silicon layer thickness in the range of 50 – 70 nm and a
BOX thickness of 140 – 150 nm. Details of the characterization techniques used to
evaluate the quality of SOI wafers have been described previously (1,3) and, for the
purpose of this paper, we will concentrate on parameters which are critical to improving
supplied quality as well as performance, yield and reliability in manufacturing.

Silicon Thickness and Uniformity

Silicon layer thickness is a critical design parameter for CMOS SOI devices affecting threshold voltage and short channel roll off control so that thickness target and uniformity are primary characteristics of SOI wafers. Typical partially depleted CMOS applications (4) call for Si thickness to be in the 50-70 nm range with a thickness variation of 2-5 nm. According to the International Technology Roadmap for Semiconductors (ITRS) (5), Si thickness will continue to be scaled down for future device generations, driving even more stringent uniformity requirements. Thickness uniformity is typically defined at the wafer level, as the min to max range of points measured on the wafer surface outside the 3 mm edge exclusion. There are two approaches to SOI thickness measurement: Discrete measurement points (10-100 points) using spectroscopic ellipsometry or a full wafer thickness map with several thousand (or even several ten thousand) measurement sites obtained using reflectometry.

Min to max range uniformity measures global thickness variations across a wafer, however, local thickness variations on the length scale of a typical die size can have detrimental impact on the device process (lithography) and performance. It is therefore useful to evaluate SOI uniformity in terms of local sites, similar to the concept of site flatness for Si wafers. Using full wafer thickness map data, a grid of 25 mm x 25 mm squares is overlaid on the wafer surface for a total of 109 local sites. Average Si thickness and its standard deviation are calculated per site, based on approx. 625 measurement points each. A local thickness uniformity metric is essential to quantify areas of poor uniformity on a wafer and allows the user to assess the impact of thickness variations on device process and performance. Figure 1 illustrates the differences between a good wafer and a wafer with poor uniformity characteristics.

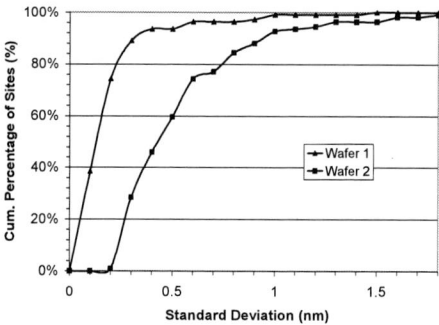

Figure 1. Local uniformity of two SOI wafers based on 25 mm x 25 mm site size. Wafer 1 shows good uniformity whereas wafer 2 has significant local non-uniformity. The thickness variations are visible to the naked eye as a striping pattern and bands of different color across the wafer. In terms of global uniformity, the Si thickness standard deviation for wafer 1 is 0.93 nm and 1.73 nm for wafer 2.

For high volume manufacturing, uniformity requirements in terms of a thickness target and corresponding tolerance cover all points on all wafers in a population so that both thickness variations on each wafer and wafer to wafer (mean) variations have to be included. Due to the critical role of Si uniformity as device parameter and the continuous drive to improve uniformity, the wafer supplier community has engaged in considerable development work over the last several years to improve capabilities. For bonded wafer processes, the key direction has been to replace the CMP finish step (which degrades uniformity) with new surface smoothing techniques. The resulting improvements have been significant, as illustrated in Fig. 2 by the difference between an older process, B1(CMP finish), and two newer processes, B2 and B3 (CMP-free finish). The data show that current uniformity capabilities are better than +/- 3.5 nm and improvement to the +/- 2.5-3.0 nm range in the near future appears within reach. For the case of the SIMOX process, work to improve Si thickness uniformity is focused on reducing implant dose and temperature variations during the oxygen implantation process and refinements to the high temperature anneal process following implantation. Based on this work, uniformity capabilities in the +/- 2.5-3.0 nm range, similar to bonded wafers, have been demonstrated. Further improvements in uniformity may be necessary for future CMOS technologies as the Si layer thickness is further reduced and uniformity requirements become more stringent.

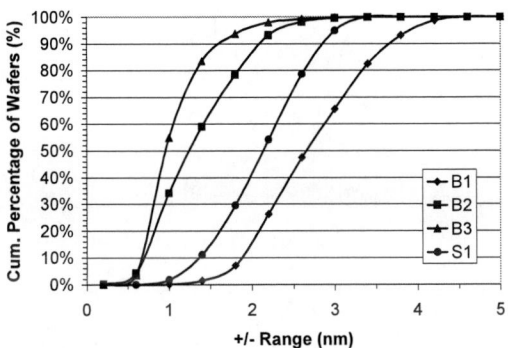

Figure 2. Thickness uniformity of 300 mm SOI in terms of the cumulative percentage of wafers that meet a given +/- range for a specified target thickness. Three different bonded wafer types (B1-B3) and SIMOX (S1) are shown. The data are based on several thousand wafers per category.

Surface Quality and Defects

Defects are a central aspect of wafer quality and the primary focus is on imperfections in the top silicon and BOX layers (see examples shown in Figs. 3-4). These defects can be detected by microscope or automated surface inspection tools for routine inspection of large numbers of wafers. Inspecting SOI wafers with a standard laser scattering surface inspection tool (for example, KLA Tencor SP-1) provides difficulties because interference effects from multiple reflections at the SOI layer interfaces modulate the reflectivity. This phenomenon is illustrated in Fig. 5 where reflectivity is shown as a function of Si layer thickness for a fixed BOX thickness of 145 nm, indicating a series of pronounced maxima and minima. Low reflectivity leads to minimum threshold limitations due to increased background noise for the measurement (6). Furthermore, defect sizing accuracy is impacted by reflectivity variations, causing the apparent size of a defect to increase with the reflectivity of the wafer surface and requiring specialized calibration curves.

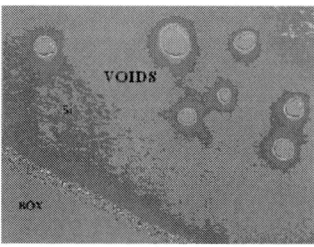

Figure 3. Silicon voids. Optical micrograph at 16X magnification.

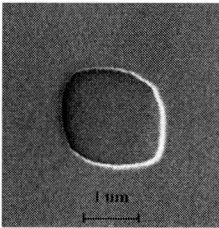

Figure 4. Si hole (divot) defect. The size of this defect is typically one to several microns.

Due to a reflectivity minimum, SOI wafers in the thickness range of 50-60 nm can typically be inspected only with a minimum threshold of 300 nm. For slightly thicker SOI wafers (70 nm range), on the other hand, reflectivity is near a maximum and thresholds down to 160 nm can be achieved. In general, the defect thresholds achievable for SOI are significantly above the critical defect size defined by the 2005 ITRS (5) for current technologies while for bulk Si wafers, thresholds at 90 nm or below have been implemented for some time. Within the constraints imposed by current metrology, the surface quality of 300 mm SOI is excellent as demonstrated by the data shown in Fig. 6. The quality level of production material has improved significantly and now reaches a typical range of 20-30 light point defects (LPDs) per wafer at 160 nm threshold. SOI specific limitations are addressed by a new generation of laser inspection tools (KLA Tencor SP-2) which use UV instead of visible light. The multiple reflection issue is avoided due to the low penetration depth of UV light in Si at a few nm and it has recently been demonstrated that SOI can be inspected at 90 nm threshold, similar to bulk Si wafers. (7)

Laser scattering inspection tools measure the density of surface particles and structural layer defects on a wafer but generally have some difficulty with automated classification of defect types. The ability to classify defects detected by the inspection tool is essential since certain defect types are a much more serious concern from the viewpoint of device process issues and yield impact than others. While large voids and similar types of imperfections in SOI can easily be identified at visual inspection, smaller defects with a size of tens of microns or smaller are a key area where improved classification capabilities are needed. Among the types of small defects are Si hole defects (divots) which are areas, typically a few µm in size, where either the top Si layer or both the Si and BOX layer are missing (see Fig. 4). Si hole defects are major killer defects with high kill ratios (1) and device yield implications so that it is essential for wafer suppliers to have the capability to inspect for this defect type. The desired capability is provided by brightfield inspection tools used in conjunction with SEM defect review to verify the defect type. However, brightfield inspection tools have throughput limitations and are generally available only in IC fab lines so that work has been initiated to develop improved defect classification capabilities for laser scattering inspection tools.(7) Si hole defects are typically captured as part of the area defect category (threshold > 0.5 µm) and the first step was to clarify how the number of area defects (laser scattering) on a wafer correlates to the number of Si hole defects (brightfield inspection). Fig. 7 shows that, on average, the number of area defects exceeds the number of Si hole defects per wafer by 2.5 for a population of bonded wafers. The higher number of laser scattering counts is predominantly due to other non-cleanable defects. While this result is representative of current bonded material, additional work is needed to extend the analysis to other SOI wafer types. Reducing the density of Si hole defects has been an important part of quality improvement efforts. As shown in Figure 8, the Si hole defect density distribution for 300 mm wafers indicates that more than 95 % of the wafers now have a density below the level of $0.04 - 0.07$ cm^{-2} reported earlier.(1) Moreover, approx. 40 % of the wafers have no Si hole defects and this is a significant improvement over the situation from a few years ago. However, further work is needed to reach the target of 100 % of the wafers to be completely free of the defect.

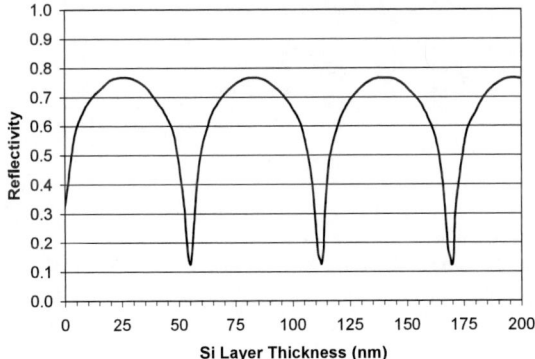

Figure 5: SOI reflectivity as a function of Si layer thickness for a BOX thickness of 145 nm. The calculation is for an incident angle of 70 degrees and a laser wavelength of 488 nm.

Flatness

Site flatness is identified by the ITRS (5) as a critical parameter for wafers due to its impact on the depth of focus for lithography processes and increasingly tighter site flatness requirements have driven substantial capability improvements for 300 mm wafers. Since the device process for SOI is based on the same lithographic requirements, SOI wafers have to meet site flatness requirements equivalent to bulk wafers which are driven by the printed feature size. When considering SOI wafer flatness, it is essential to keep in mind that the starting point for all SOI wafer manufacturing processes is a standard Si wafer and the flatness of the starting wafer (in a bonded wafer process, the base wafer is the key element) to a large extent determines the flatness of the finished SOI wafer. The strategy is therefore to employ a starting wafer with the desired SFQR flatness and avoid degradation during the SOI process. The goal of having SOI wafers available with SFQR flatness equivalent to bulk has been reached as evidenced by wafer types BN1 (bonded) and SM2 (SIMOX) in Fig. 9. Wafer type SM1 is a previous type of SIMOX SOI which included a poly Si layer on the backside of the wafer and the site flatness of the starting wafer for this type was degraded relative to industry standard 300 mm double side polished wafers.

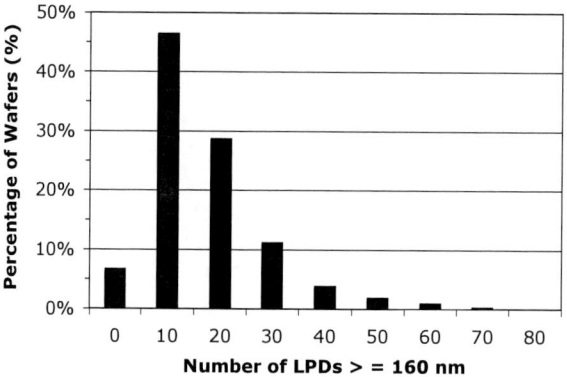

Figure 6. Number of LPDs >= 160 nm per wafer measured by SP-1 on high reflectivity (70 nm Si thickness) 300 mm SOI. The data are for more than 300 wafers measured in sampling mode at incoming inspection. The average value is 17 and the standard deviation is 12.

While we have not observed a noticeable degradation in site flatness for SOI wafers relative to standard polished wafers, there are indications that warp is increased moderately as the result of the SOI process. However, even the increased warp values are still within the desired range and we do not observe a negative impact.

Surface Roughness

Atomic force microscopy (AFM) is the typical approach to evaluate roughness of a wafer surface on the length scale of a few μm to several tens of μm. From the measurement, two relevant parameters are extracted, R_{RMS}, the root mean square average of the measured height variations within the sampling area and R_{MAX}, the vertical distance from the highest peak to the lowest valley in the sample. The concern with high surface roughness (> 0.5 nm R_{RMS}) is gate oxide integrity degradation and significant local Si thickness variations (via high R_{MAX}) which in turn degrade device performance. In addition to measuring roughness parameters, AFM is very useful for detecting and characterizing unusual surface features (pits, holes, mounds, spikes, etc.).

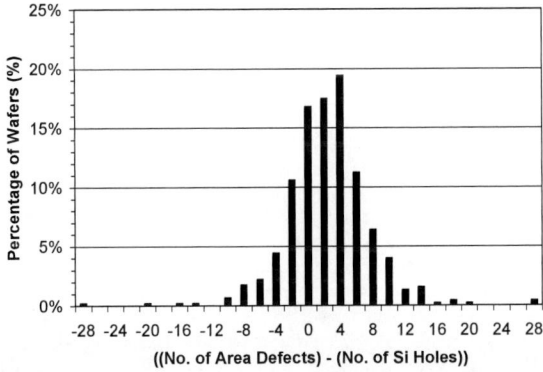

Figure 7. Delta of area defects measured by KLA Tencor SP-2 and Si hole defects by brightfield inspection. The average delta is 2.5 with a standard deviation of 5.4.

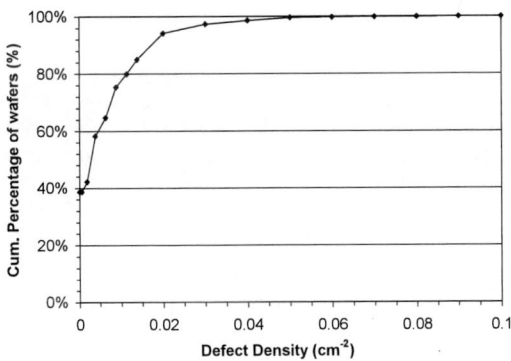

Figure 8. Si hole defect density for 300 mm SOI wafers. Shown is the cumulative percentage of wafers vs. defect density.

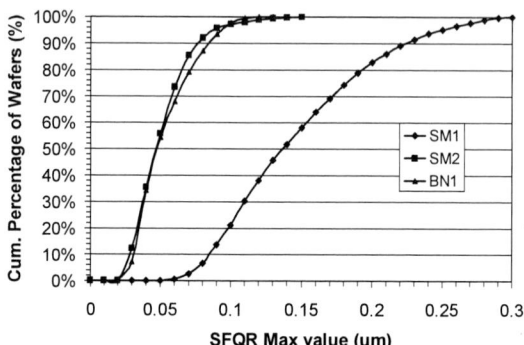

Figure 9. Site flatness (SFQR) distribution for three populations of SOI wafers: SM1 (SIMOX with poly Si backside layer), SM2 (SIMOX without poly Si backside) and BN1 (bonded SOI). Shown is the maximum SFQR site per wafer (25 mm x 25 mm site size with partial sites active).

The target for SOI is a surface roughness close to the surface finish of a polished Si wafer and this goal can be attained easily for wafers finished by a CMP process similar to the one used for bulk Si wafers. However, CMP finished SOI has inferior Si uniformity capabilities and difficulties meeting the uniformity requirements of leading edge technologies (see Fig. 2). Recently developed process variants for bonded wafers which eliminate CMP as well as the SIMOX process produce surface finishes that are slightly different from a typical polished surface. Examples of SOI smoothed without CMP are shown in Fig. 10 as AFM images, indicating some differences in the surface texture (and roughness parameters) between two wafer types. It is noted that the surface roughness of CMP-free SOI is not necessarily uniform and some variation, locally on the length scale of tens of μm to globally across a wafer, is possible.

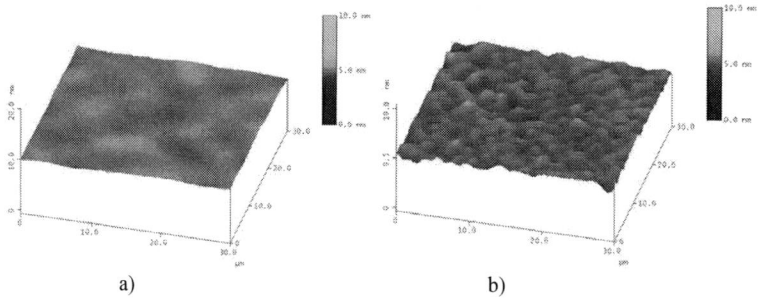

a)   b)

Figure 10. SOI surfaces smoothed without CMP. AFM images (30μm x 30 μm scan area) for wafers processed by two different surface smoothing techniques, a) $R_{RMS} = 0.14$ nm and $R_{MAX} = 1.07$ nm and b) $R_{RMS} = 0.31$ nm and $R_{MAX} = 3.09$ nm.

To capture longer range roughness variations, the scan area has to be large (for example, 30μm x 30 μm) and multiple locations on a wafer should be measured. Overall, much progress has been made over the last several years by the wafer supplier community refining CMP-free surface smoothing processes and additional work is currently in progress. At present, SOI wafers processed this way are approaching the characteristics of a polished surface, reaching the desired range for 0.25 nm R_{RMS} and 2.5 nm for R_{MAX} (30μm x 30 μm scan area) and thereby providing assurance that there is no negative impact on the device process.

Interface roughness at the active layer – BOX interface can be of concern as well because it introduces an additional modulation of the local active layer thickness with the implications discussed earlier. To measure interface roughness, the active layer has to be carefully removed so that the exposed BOX surface can be measured by AFM. High interface roughness is not typically observed and would be indicative of a quality problem with the material.

Metallic Contamination

The BOX layer provides electrical isolation between the top Si layer and the substrate and, additionally, functions as a barrier to transition metal contaminants diffusing from the substrate to the top layer or vice versa.(8-10) This means that contaminants located in the substrate or on the wafer backside have to diffuse through the BOX before they can reach the active area of a device in the top layer. At the same time, contaminants already in the top layer are essentially trapped and cannot be gettered by conventional Si wafer gettering techniques (internal, backside poly Si, etc) in the substrate or at the backside. Diffusion processes through the BOX (8,9) play a stronger role during long high temperature process steps (> 1050°C) employed as part of the SOI wafer manufacturing process. Modern device processes are typically limited to lower temperatures (< 1050°C) with relatively short process times so that the BOX will act as an effective diffusion barrier for transition metals except for fast diffusing Cu (Ni and Cu behave very similarly in terms of diffusion in Si. However, only Cu diffuses through the BOX at low temperature whereas Ni does not).(8-10) Because of the limited efficiency of gettering sites located in the substrate for contamination in the top layer, the primary focus for SOI wafers used in volume production is to very carefully control metal contamination of the wafer front side and the top layer. The importance of this point is illustrated by the following consideration: Due to the small thickness of the active layer, contamination at an area density of $1x\ 10^{10}\ cm^{-2}$ trapped above the BOX, would translate to a bulk contamination level in the $1x\ 10^{15}\ cm^{-3}$ range which is much higher than the typical critical bulk contamination on the order of $1x\ 10^{11}\ cm^{-3}$.

Metal levels can be monitored using SIMS, TXRF or, alternatively, chemical analysis techniques where the Si and oxide layer are chemically dissolved and contamination levels are subsequently determined by mass spectrometry. The strong point of SIMS is that a depth profile can provide important details about the location of concentration spikes (for example, at interfaces) but it is not feasible to measure more than a few locations to probe variations across a wafer. Chemical digestion techniques, on the other hand, can be used to evaluate the entire wafer surface but only wafer average concentrations are obtained. A SIMS profile going from the surface through the top Si layer, BOX and into the substrate is shown in Fig. 11, demonstrating essentially flat concentration profiles. For current material, integrated levels of $<5x\ 10^{10}\ cm^{-2}$ are consistently achieved for critical metals (Al, Cr, Cu, Fe, Ni) and gate oxide integrity

(GOI) degradation is not observed at these levels. Further improvement to 1×10^{10} cm^{-2} is targeted for the near future.

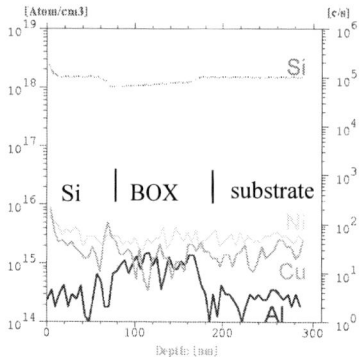

Figure 11. SIMS depth profile from surface - SOI layer- BOX – bulk Si for a total depth of approx. 300 nm. Shown are the concentrations for Ni, Cu and Al.

Slip

Slip dislocations are generated in crystalline Si as the result of plastic deformation caused by thermal and gravitational stresses at high temperatures. Slip free high temperature processing of 300 mm wafers has been an important challenge for wafer manufacturers and IC device fabs with the startup of high volume manufacturing. The challenge has been of particular relevance to manufacturing of 300 mm SOI because the processes are based on high temperature steps. For bonded SOI, temperatures at 1100 °C and above are used and a further increase has been driven by the implementation of CMP-free surface smoothing techniques discussed previously. The SIMOX process includes even higher temperatures, in the 1300°C range. Significant engineering work was required to eliminate slip step by step and tremendous progress has been made over the last several years. The goal of slip-free 300 mm SOI has essentially been achieved for current vintage material, however, the process margins still tend to be somewhat narrow so that quality monitoring of the finished SOI wafer is important.

While laser scattering inspection tools can be used to inspect SOI wafers for slip, we found that a slip inspection tool, using a high intensity collimated light source together with an imaging system, is more sensitive. Fig. 12 shows a wafer map indicating slip near the wafer edge and a micrograph of the slip line itself is shown in Fig. 13. The slip inspection system can detect not only slip in the top layer but also slip, other defects and surface anomalies in the substrate below the BOX. Verification that slip lines identified by the slip inspection tool are located below the BOX is accomplished by laser light scattering using UV light where only features in the top layer surface are detected or, alternatively, the top Si and BOX layers can be removed for direct inspection of the top surface of the substrate wafer. An example of slip in the substrate located radially toward the wafer center is shown in Fig. 14.

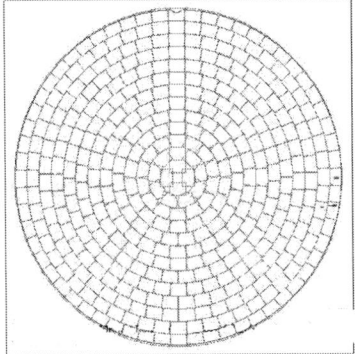

Fig. 12. Wafer map from a slip inspection tool showing slip lines at the edge of the wafer.

Figure 13. Micrograph of a slip line at the wafer edge indicated in Fig. 12.

Electrical Properties

The properties of the top Si and BOX layer are critical to the performance of CMOS devices built on SOI and therefore are important SOI wafer quality parameters. The relevant data can be obtained from device or test structures during device processing but, from the viewpoint of quality assurance, it is important to have this information available at the wafer level prior to device processing. A widely used method for this purpose not requiring lithographic processing is the point contact or pseudo (Ψ) - MOSFET technique. (1,3,4) The FET consists of two metal electrodes applied to the front surface, acting as source and drain, while the BOX functions as the gate oxide and the substrate as the gate. The method produces a well behaved MOSFET which can be analyzed in both linear and saturation regions to extract the usual transistor parameters (threshold voltage, transconductance, etc.). The FET parameters, in turn, yield the electrical characteristics of the top Si layer: (low field) electron mobility (μ_e), (low field) hole mobility (μ_h), doping level (N_A) as well as the density of fixed charges in the BOX, Q_{BOX}, and density of states at the interface between the top Si layer and the BOX, D_{it}.(4,11)

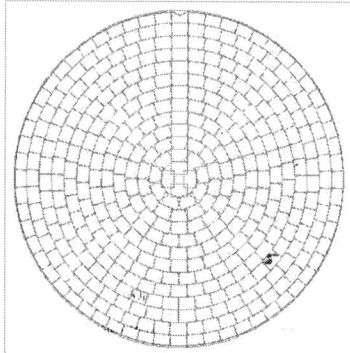

Figure 14. Wafer map from a slip inspection tool showing slip lines located in the active area of the wafer. Further examination confirms that the slip lines are located in the substrate beneath the BOX.

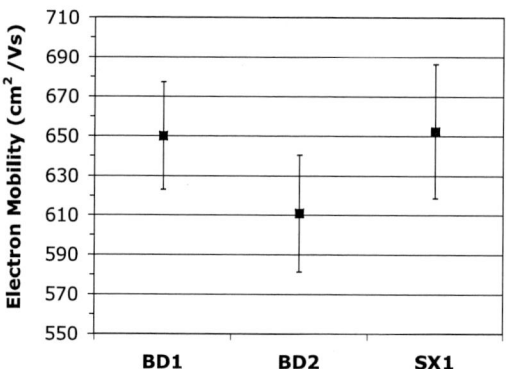

Figure 15. Low field electron mobility by wafer type. BD1 and BD2 are bonded and SX1 are SIMOX wafers. Shown are average values with the error bar indicating the standard deviation. The data were measured by HgFET on 300 mm SOI wafers at approx. 60 nm Si thickness sampled from production lots over a period of 12 months.

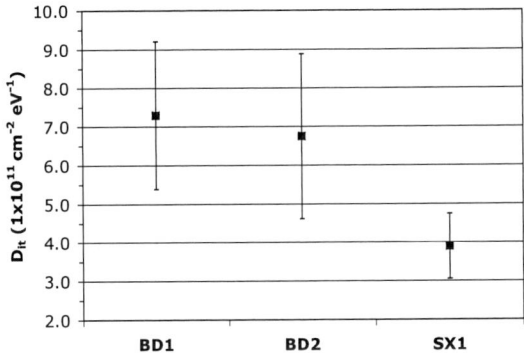

Figure 16. Interface state density D_{it} for the wafers described in Fig. 15.

To characterize SOI on a routine basis in a production environment, the HgFET technique, (11) an implementation of the Ψ-MOSFET using Hg probes for the frontside contacts, has been used for many years for both 200 and 300 mm SOI. From the electrical parameters listed above, we focus on electron mobility, BOX charge and interface state density as the main electrical quality characteristics to monitor. These three parameters are important indicators of the quality of the SOI layer – BOX structure since they can affect performance, yield and reliability of SOI devices. (1) High mobility is taken as an indication of greater top Si-BOX interface perfection with lower defects and less waviness. Electron mobility of the unprocessed SOI wafer is considered a good indicator of transistor performance. Device yield can be impacted by BOX charge via shifts in the threshold voltage. Interface states (at both interfaces of the top Si layer) can affect the threshold voltage as well via changes in their charge state induced by the transistor operation and interactions with the floating body charge.

The desired range for μ_e is greater than 550 cm^2 (Vs)$^{-1}$, less than 5x 10^{11} cm^{-2} for Q_{BOX} and less than 1x 10^{12} cm^{-2} eV^{-1} for D_{it}, as measured by HgFET. The electron mobilities (see Fig. 15) of three SOI wafer types used for high volume production, two bonded and one SIMOX, are well within the desired range, with only relatively small differences between the wafer types. Bonded wafer type BD1 and SIMOX (SX1) have slightly higher mobility than BD2. For D_{it} (see Fig. 16) and Q_{BOX} (see Fig. 17), SIMOX shows lower values than both bonded types which are at nearly identical levels. Based on SOI wafer process optimization guided by feedback from HgFET characterization, values for the two parameters have been reduced by approx. 10 % during the last 12 months bonded wafers. The key electrical parameters are now quite stable and, while lower ranges for Q_{BOX} and D_{it} are preferable, current levels are sufficiently low to avoid device issues. Overall, the electrical data reviewed in this paper are very similar to the results reported a few years ago.(1) While μ_e and Q_{BOX} are essentially unchanged, we have observed some improvements in D_{it}.

Using HgFET characterization to monitor the quality of the SOI wafer supply, we have established detailed baseline data which allow for the detection of excursions as well as gradual drifts in the electrical parameters. This information is valuable for protecting the device fabrication line and for providing feedback to wafer suppliers.

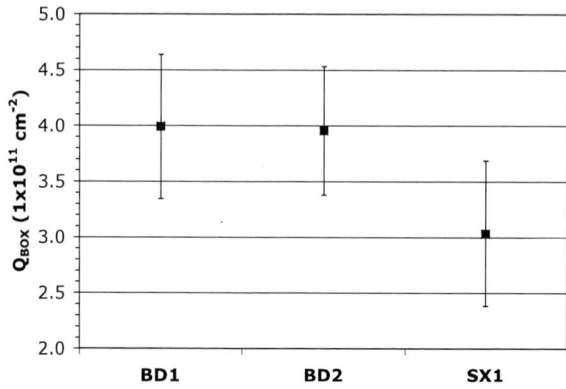

Figure 17. BOX charge Q_{BOX} for the wafers described in Fig. 15.

CONCLUSIONS

The quality of 300 mm SOI wafers has improved significantly over the last several years through extensive materials development work. For the end of 2005 time frame, both bonded and SIMOX type SOI wafers have reached the quality level required for high volume manufacturing of leading edge (90 and 65 nm) device generations.

Si layer uniformity, LPD and flatness capabilities are sufficient for current device generations. Metal contamination levels, slip performance and electrical characteristics have been improved and stabilized while additional work is being targeted to optimize capabilities. The roughness parameters of SOI prepared by CMP-free surface treatment processes are not yet equivalent to a polished surface so that additional work is needed in this area. For Si hole defects, significant further reductions in the density are clearly needed to drive device yield enhancements.

Acknowledgments

We gratefully acknowledge valuable discussions and support from D. Schepis, R. Kleinhenz, F. Schmidt, D. Sadana, P. Ronsheim, R. Murphy and J. McMurray. The authors thank G. Gardopee (ADE Corp.) for assistance with generating local Si uniformity data. We would like to thank R. Bendernagel and D. Murray for their encouragement and support.

References

1. H. Hovel, M. Almonte, P. Tsai, J.D. Lee, S.Maurer, R. Kleinhenz, D. Schepis, R. Murphy, P. Ronsheim, A. Domenicucci, J. Bettinger and D. Sadana, *Sol. State Electr.* **48**, 1065 (2004).
2. Collaborative Quality Management System, R. Krause, C.Waldenmaier and M. Schmidt, *APC/AEC Conference* 2004 (P101).

3. G. Pfeiffer and S.S. Iyer, *"Silicon Wafer Bonding Technology for VLSI and MEMS Applications"*, edited by S.S. Iyer and A.J. Auberton Herve, INSPEC, p. 83 -92 (2001).
4. G. K. Celler and S. Christoloveanu, *J. Appl. Phys.* 58, 4995 (2003).
5. http://public.itrs.net/
6. C. Maleville, C. Moulin, E. Neyret, C. Cauble, L. Cheung and R. Moirin, *Proc. of the 2002 IEEE SOI Conference, Williamsburg ,VA* (Volume 02CH37347), p. 194 (2002).
7. C. Moulin, D. Delprat, C. Maleville, W. McMillan, J. Payne, K. Birdwell, R. Brun and R. Moirin, *Proc. of the 2005 IEEE SOI Conference, Honolulu, HI,* p.146 (2005).
8. A. A. Istratov, W. Huber and E.R. Weber, *Solid State Phenomena* 93-96, 547 (2004).
9. A. A. Istratov, H. Vainola, W.Huber and E. R. Weber, *Semicond Sci. Technol.* 20, 568 (2005).
10. J.-I. Furihata, M. Nakano and K. Mitani, *Jpn. J. Appl. Phys.* 39, 2251 (2000).
11. H.J. Hovel, *Solid State Electron.* 47, 1311 (2003).

182

SESSION 3

PROCESS DEVELOPMENT AND MODELING

184

Process Development and Modeling

Ulrich Goesele
Max Planck Institute of Microstructure Physics
Weinberg 2
D-06120 Halle
Germany

Paul Packan
Intel Corporation
2501 NW 229th St.
Hillsboro, OR 97123

As device dimensions continue to shrink, higher and higher order process interactions are becoming increasingly important to understand. Modeling of these physical processes has proven to be critical to understanding these complicated interactions. In addition, modeling work has led to creative, new solutions to many problems by providing detailed physical understanding and predictions of new directions to evaluate. In this session, a wide variety of applications of theoretical modeling is presented.

A basic tenant of device scaling is the reduction of gate oxide thickness. For more than 30 years, scaling of gate oxide thickness has enabled the continued increase of transistor density and performance. Unfortunately, tradition thickness scaling of silicon dioxide is appearing to have reached its end. As gate oxides approach thicknesses of a few monolayers, quantum mechanical tunneling leads to unacceptably high gate leakage current. This leakage current creates both power and circuit functionality issues. In response, researchers have been increasingly looking to the use of high-K materials to solve this problem. However, due to surface mobility and processing issues, it appears that high performing transistors still need a silicon dioxide interface between silicon and high-K dielectric. The first two papers of this session look at the fundamental issues and properties of ultra-thin silicon dioxide layers from a modeling perspective.

The next five papers of this session deal with the extremely important area of yield and reliability. As silicon integrated circuits such as microprocessors approach the 1 billion transistor regime, material quality and defects play an even larger role. Even extremely small micro defects previously unimportant due to their small, localized size are now becoming critical to product yield. Both the fundamental microstructures of silicon substrates and gettering of metal contaminants must be understand and managed to enable a high yielding, profitable process. These papers deal with multiple aspects related to this area and provide fundamental understanding of the issues and potential solutions through advanced process modeling.

The final two papers in the session deal with fundamental material properties. In the first, basic properties of the silicon lattice and diffusion properties are investigated. This type of basic understanding underpins more advanced calculations and is essential to building a hierarchical understanding of process modeling. The final paper looks at the critical area of strain engineering through the introduction of new materials. Due to the difficulty in scaling gate oxide thicknesses described above, other new performance scaling methodologies need to be developed. Strain scaling has proved to be such a method. By straining the silicon channel through multiple techniques, energy-band splitting and band warping can be used to increase transistor mobility and thus enhance transistor performance. One of the most effective methods for increasing the strain in PMOS transistors is the use of SiGe. Due to the lattice mismatch between Si and SiGe materials, large amounts of stress can be introduced into the lattice and can dramatically change the performance of the devices. The last paper in this session deals with modeling the formation of these SiGe layers.

This session begins with an invited paper by Massoud that takes a comprehensive look at the growth kinetics and electrical properties of ultra thin silicon-dioxide layers. The paper reviews a wide range of issues including the onset of initial growth, effects of temperature, substrate orientation, dopant type and stress. An interesting treatment of differences in contact potential due to substrate orientation and processing history is presented based on partial charge transfer dipoles at the oxide/silicon interface. In the final section, the impact and implications of gate tunneling are presented.

In the second paper, Korkin et al. present theoretical work focusing on the silicon/silicon dioxide interface. Oxygen vacancies in silicon dioxide are examined by ab initio computations and are found to be more stable near the interface, thus identifying a chemical driving force for oxygen reduction at the interface. This has important implications for the thin silicon dioxide interfacial layers present in today's high-K gate stacks, including an increase in dielectric constant larger than for stochiometric silicon dioxide. This can be explained by oxygen deficiency during the growth process caused by the oxygen-vacancy interactions.

Kulkarni presents a theoretical treatment of micro defects in Czochraslski silicon crystals in the third and fourth papers of this session. This is an important subject for both yield and performance of deep sub-micron electronic devices. Kulkarni looks at the complex dynamics influenced by multiple different reactions including intrinsic point defects, oxygen and nitrogen. By approximating the defects as spherical clusters, the formation and growth of these micro defects can be calculated. The model agrees well with existing experimental data and various predictions of the model are discussed. The simplicity of the model allows prediction of complex two-dimensional micro defect distributions in the crystals for both steady and unsteady states.

The fifth paper of the session describes modeling work on the interaction of vacancies and oxygen in the oxide precipitation process and the effects of RTA treatments. Kissinger et al. show oxygen precipitation after RTA anneals are controlled by the initial concentration of interstitial oxygen and the grown-in vacancies. It is shown that the formation of tensile strained nVO_2 clusters is the preferred process for coherent nucleation of the oxide precipitates and provides a step forward in understanding and modeling oxide precipitation in CZ silicon.

In the sixth paper, Sueoka et al. presents calculations for Cu gettering by dopant and dopant-vacancy complexes. The study evaluates the effectiveness of B, Sb, As, P, C and O in the gettering process and showed that only B could be an effective gettering center. In accordance with experimental observations, these theoretical results indicate that heavily B doped p/p+ epitaxial silicon wafers will exhibit good gettering efficiency for Cu contamination. For n/n+

epitaxial silicon wafers, it was found that the dopants Sb, P, and As themselves will not facilitate good gettering, but that their complexes with vacancies, V-Sb, V-P or V-As should be effective as gettering centers. It remains to be seen whether an increase in such complexes could be obtained by some special processing to allow them to turn n+ layers into effective gettering regions. The effects of strained silicon were also evaluated with regard to the Cu atom stability in the lattice.

The next paper continues the theme of analyzing the gettering process by looking at gettering of Fe. This work by Nakamura and Tomioka describes the behavior of the internal gettering process of Fe by oxygen precipitates. The model proposed accurately describes the gettering behavior of Fe at both low and higher temperatures by considering nucleation of iron silicide at the oxygen precipitate sites. The model is able to explain both the low and high temperature dependencies of the gettering behavior on density and radius of the oxygen precipitates and thus extends earlier theoretical treatments which are valid mostly in the range of lower temperatures.

The eighth paper of this session by Matsumoto et al. use highly pure ^{30}Si epitaxial layers to measure the diffusivity of silicon self-interstitials in heavily doped B silicon. The Si self-diffusivity is determined as a function of temperature and B concentration. The Si self-diffusivity was found to increase with B concentration with larger enhancements at lower temperatures. Based on the model for Fermi level effect diffusion, energy levels of the donor Si self-interstitials and acceptor vacancies were determined from analysis of the experimental data.

SiGe has become an integral part of device performance of the past few years. The ability of the SiGe to generate strain in Si, as well as improved mobility and energy-band offset, have been used to substantially improve PMOS device performance. The final paper of this session by Yang and Tao looks at the modeling of the growth for SiGe. A kinetic model based on collision theory of chemical reactions, statistical physics and competitive adsorption is proposed for SiGe growth from SiH_4 and GeH_4 by chemical vapor deposition. Analytic equations are derived to quantitatively describe the growth rate as a function of deposition temperature, silane flow rate and germane flow rate.

188

ECS Transactions, 2 (2) 189-203 (2006)
10.1149/1.2195659, copyright The Electrochemical Society

Growth Kinetics and Electrical Properties of Ultrathin Silicon-Dioxide Layers

Hisham Z. Massoud

Semiconductor Research Laboratory,
Department of Electrical and Computer Engineering,
Edmund T. Pratt, Jr., School of Engineering, Duke University,
2 Science Drive, Durham, NC 27708–0291, USA.

In this paper, the growth kinetics and electrical properties of ultrathin silicon-dioxide layers are reviewed. Topics discussed here include the onset of oxide growth, the effects of temperature, substrate orientation, dopant type and concentration, and stress on oxide growth kinetics. The use of *in situ* real-time ellipsometry to control the rapid-thermal oxidation of silicon will be presented. The electrical properties of the ultrathin silicon-dioxide layers will be discussed, especially the contact potential difference in MOS devices and its dependence on the processing history and substrate orientation which will be explained in terms of the role of partial-charge-transfer dipoles at the oxide/silicon interface. Finally, gate tunneling considerations in MOSFET design, modeling, characterization, and circuit performance will be briefly presented.

Introduction

Growth Kinetics of Ultrathin Silicon-Dioxide Layers

The growth kinetics of ultrathin SiO_2 layers grown in dry oxygen were characterized in a comprehensive study that included many experimental parameters [1–10], including several substrate orientations, lightly and heavily doped substrates, the oxide growth temperature, and the oxygen partial pressure in the growth ambient. In this section, the experimental details of the study are given, the onset of oxide growth is characterized and modeled, and the oxide growth kinetics are analyzed. Finally, the use of *in situ* real-time ellipsometry in the control of the rapid-thermal oxidation process is presented.

Experimental Details

Single-crystal, (100)-, (111)-, and (110)-oriented silicon wafers uniformly doped with phosphorus or boron with dopant concentrations of 1.0×10^{16}, 1.8×10^{20}, or 3.2×10^{20} cm^{-3} were cleaned using a standard hydrogen-peroxide-based cleaning solutions followed by a brief native oxide etch in a dilute solution of HF and a de-ionized water rinse [1]. The wafers were immediately loaded in an automated *in situ* high-temperature, real-time, ellipsometer thickness analyzer at IBM Thomas J. Watson Research Laboratory that allowed the monitoring of the oxide growth process at high temperatures. The wafers were loaded with argon flowing to measure the native-oxide

---

Address of the 2006 ECS Electronics and Photonics Division Award.

thickness prior to switching to dry oxygen and observe the oxide growth kinetics. The oxide layers were grown at temperatures ranging from 800 to 1000 °C. The growth ambient consisted of a mixture of dry oxygen and argon that corresponded to oxygen partial pressures of 0.01, 0.1, and 1.0 atm.

Oxide Growth Kinetics – The Incubation Phase

The *in situ* monitoring of the oxide growth kinetics on silicon in the ultrathin range using the high-temperature automated ellipsometer allowed for a closer examination of the kinetics in the early stages of the oxidation process [4,7]. The wafers were introduced into the oxidation furnace and allowed to reach the oxidation temperature while idling in argon. The native oxide thickness was measured and the ambient was then switched to dry oxygen or a mixture of dry oxygen and argon. The oxide thickness was monitored as soon as oxygen was allowed in the oxidation chamber. From the geometry of the oxidation furnace and the oxygen flow rate, it was estimated that a complete furnace volume change would take less than 0.1 s. The experiment time t_{exp} was started when oxygen was turned on. It was observed for all substrate orientations that SiO_2 growth only starts after a delay period t_{del} [4], which is larger than the time needed for a complete furnace volume change from argon to oxygen. The temperature and orientation dependences of t_{del} for (100), (111), and (110) silicon wafers are shown in Fig. 1. It was found that t_{del} ranges from seconds at 1000 °C to 4 min at 800 °C [7]. This incubation period was also observed in rapid-thermal oxidation studies and room-temperature native oxide growth studies [7]. The analysis of the experimental results is shown in Fig. 1(b) and suggests that the inverse of the delay time can be expressed as

$$\frac{1}{t_{del}} = K_{del} \exp\left(-\frac{\Delta E_{del}}{k_B T}\right),$$

where K_{del}^{-1} is a pre-exponential constant and ΔE_{del} the activation energy. Both K_{del} and ΔE_{del} are orientation-dependent [7].

The mechanisms present during the incubation phase are illustrated in Fig. 2. At the onset of the experiment, the oxygen partial pressure at the native-oxide/substrate interface is lower than that needed for SiO_2 to grow and, therefore, gaseous SiO molecules are formed at the interface. Incoming streams of neutral and ionized molecular and atomic oxygen are consumed in the formation of gaseous SiO molecules at the interface. Some of these SiO molecules remain at the interface and block the formation of SiO_2 at such spots, and some out-diffuse through the SiO_2 layer. In a model of the incubation period [4], the density of surface coverage with SiO molecules was estimated to correspond to one monolayer of coverage. In other words, incoming oxygen species during the incubation phase are consumed in forming a monolayer of SiO molecules that block the formation of SiO_2 at the native-oxide/substrate interface. Once a single monolayer is reached, the incoming stream of oxygen species increases the partial pressure of oxygen at the interface and the formation of SiO_2 begins in the growth phase. It should be noted that the substrate surface is rich with silicon vacancies generated in the formation of SiO molecules. These silicon vacancies enhance the SiO_2 formation rate in the initial phases of oxide growth by providing additional oxidation sites. This description of the incubation phase does not take into account the silicon surface roughening that results from the temperature ramp-up in

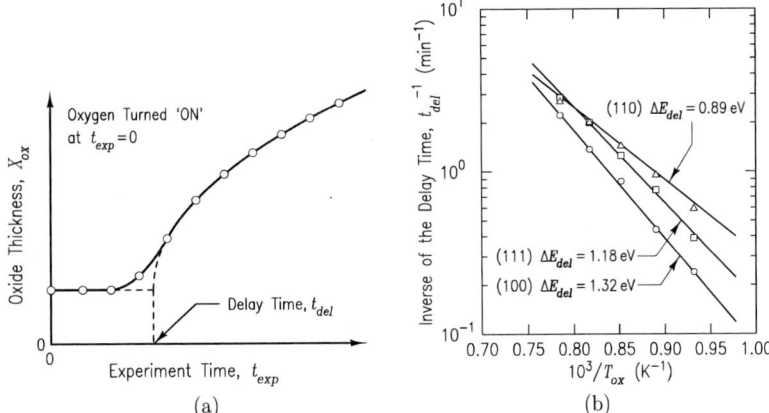

(a)   (b)

Fig. 1.   (a) Typical oxide-thickness data obtained at the onset of the oxidation
of silicon in dry oxygen. (b) Temperature and substrate orientation
dependences of the inverse of the delay time t_{del} in the 800–1000 °C
range [7].

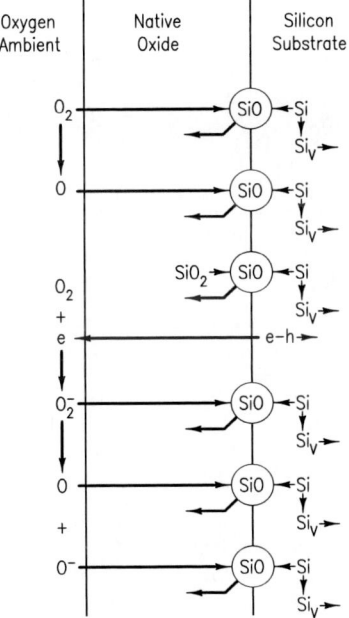

Fig. 2.   Mechanisms of oxygen dissociation at the ambient/native-oxide interface,
SiO formation at the native-oxide/substrate interface, and the generation
of silicon vacancies in the substrate [7].

argon. In addition, the SiO molecules distributed in the native oxide are converted into SiO_2 molecules at the onset of the oxide growth phase. These two additional contributions to the SiO_2 growth rate are present until the silicon vacancies are consumed and all SiO molecules are converted into SiO_2 molecules. The oxide growth phase is discussed in the next section.

Oxide Growth Kinetics – The Growth Phase

The growth phase is defined for $t_{exp} \geq t_{del}$ and the oxidation time t_{ox} is defined as the time during the growth phase beyond t_{del}, i.e., that $t_{ox} \equiv t_{exp} - t_{del}$. The dependence of the oxide growth rate in the ultrathin range on the oxide thickness was best fit with the expression [2]

$$\frac{dX_{ox}}{dt} = \frac{B}{2\,X_{ox} + A} + C_1 \exp\left(-\frac{X_{ox}}{L_1}\right) + C_2 \exp\left(-\frac{X_{ox}}{L_2}\right), \quad (1)$$

where B is the parabolic rate constant, B/A is the linear rate constant, C_1 and C_2 are constants, and L_1 and L_2 are the characteristic lengths with which the oxidation rate decays with the oxide thickness. The orientation dependence of L_1 and L_2 are shown in Table 1. The orientation dependence of C_1 and C_2 are discussed elsewhere [2,3]. The dependence of the oxide thickness on the oxidation time is obtained by numerical integration of Eq. (1)

Table 1. Orientation dependence of L_1 and L_2.

Si Orientation	L_1 (Å)	L_2 (Å)
(100)	8–12	69
(111)	11–17	78
(110)	14	60

The dependence of the oxidation rate on the oxidation time yielded an analytical relationship for the oxidation of silicon in dry oxygen given by [5]

$$X_{ox}^2 + A\,X_{ox} = B\,t_{ox} + M_1\left[1 - \exp\left(-\frac{t_{ox}}{\tau_1}\right)\right] + M_2\left[1 - \exp\left(-\frac{t_{ox}}{\tau_2}\right)\right] + M_0, \quad (2)$$

where

$$M_0 \equiv X_{nat}^2 + A\,X_{nat}, \quad (3)$$

$$M_1 \equiv K_1\,\tau_1, \quad (4)$$

$$M_2 \equiv K_2\,\tau_2, \quad (5)$$

and X_{nat} is the native-oxide thickness at the beginning of the growth phase. The temperature dependence of the constants K_1, K_2, τ_1, and τ_2 were found to fit the expressions [5]

$$K_1 = K_1^\circ \exp\left(-\frac{\Delta E_{K_1}}{k_B T_{ox}}\right), \quad (6)$$

$$K_2 = K_2^\circ \exp\left(-\frac{\Delta E_{K_2}}{k_B T_{ox}}\right), \quad (7)$$

Table 2. Orientation dependence of K_1°, K_2°, τ_1°, and τ_2°.

Orientation	(100)	(111)	(110)
K_1° (Å^2/min)	2.49×10^{13}	2.70×10^{11}	4.07×10^{10}
ΔE_{K_1} (eV)	2.18	1.74	1.54
K_2° (Å^2/min)	3.72×10^{13}	1.33×10^{11}	1.20×10^{10}
ΔE_{K_2} (eV) ·	2.28	1.76	1.56
τ_1° (min)	4.14×10^{-6}	1.72×10^{-6}	5.38×10^{-9}
ΔE_{τ_1} (eV)	1.38	1.45	2.02
τ_2° (min)	2.71×10^{-7}	1.56×10^{-7}	1.63×10^{-8}
ΔE_{τ_2} (eV)	1.88	1.90	2.12

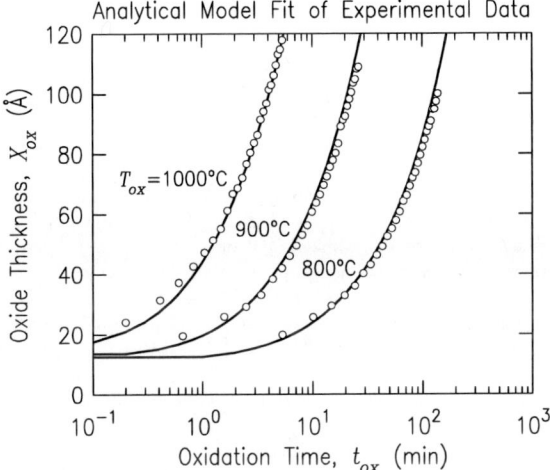

Fig. 3. Analytical relationship fit to experimental oxidation data of lightly doped (100) silicon in dry oxygen up to 100Å.

$$\tau_1 = \tau_1^\circ \, \exp\left(\frac{\Delta E_{\tau_1}}{k_B T_{ox}}\right) , \qquad (8)$$

and

$$\tau_2 = \tau_2^\circ \, \exp\left(\frac{\Delta E_{\tau_2}}{k_B T_{ox}}\right) , \qquad (9)$$

where T_{ox} is the oxidation temperature. The orientation dependence of the constants K_1°, K_2°, τ_1°, and τ_2° are listed in Table 2. The analytical relationship fit to oxidation data of lightly doped (100) silicon at 800, 900, and 1000 ℃ up to 100Å is shown in Fig. 3.

The influence of the substrate dopant concentration on the oxidation of silicon in the ultrathin regime was found to combine the generally observed increase in oxidation rate with dopant concentration with the dopant redistribution in the initial phases of oxide growth [9,10]. The oxidation of heavily phosphorus-doped (100) and (111) silicon was studied in the 800–1000 °C range in dry oxygen. In the initial phases of the oxide growth, phosphorus piles up at the Si/SiO_2 interface, thus decreasing the surface concentration of electrically active phosphorus at the interface. The oxidation rate is initially similar to that of lightly doped silicon. As the silicon dioxide layer grows, the surface concentration of electrically active phosphorus increases above its bulk value, and the oxidation rate gradually approaches that of heavily doped silicon at thicker oxides [9,10]. The dopant redistribution effects were confirmed by a two-oxidation experiment. The first oxidation time was selected such that the surface concentration of phosphorus would be depleted below its bulk value. The oxide was then etched, and the silicon sample re-oxidized. The oxide growth characteristics in the second oxidation was typical of a lightly doped silicon sample [9,10].

Oxide Growth Kinetics – Manufacturability Considerations

The challenge of small thermal budget in silicon processing for deep submicron technology generations and the continuous increase in the diameter of silicon wafers have necessitated the use of rapid-thermal processing technology. As the accurate control of the oxide thickness is the goal of the oxide-growth process, a new approach – different from the conventional combined temperature control and knowledge of the oxidation-growth kinetics – was introduced [11–20]. In this approach, single-wavelength ellipsometry was shown to be a viable *in situ* real-time technique for the simultaneous measurement of the silicon wafer temperature and the oxide thickness [11–20]. In this approach, the ellipsometric parameters ψ and Δ are first determined. Then, the substrate temperature and the oxide thickness are calculated from the knowledge of the temperature dependence of the index of refraction of Si and SiO_2 using simple polynomial expressions. It was found that for an ellipsometer operating at $\lambda = 6328$ Å, and with a resolution of 0.01° in the measurements of the ellipsometric angles ψ, Δ, and the angle of incidence ϕ, in the 0–1000 °C temperature range, the temperature worst-case error was less than ± 10 °C. For an ellipsometer operating at $\lambda = 4133$Å, the temperature worst-case error is reduced to ± 1.4 °C in the 0–700 °C range [11-18]. By changing the operating wavelength from 6328Å to 4428Å, a 30% reduction in the difference between the temperature measured by ellipsometry and that measured by a thermocouple was achieved [11-18].

Electrical Properties of Ultrathin Silicon-Dioxide Layers

The study of the electrical properties of ultrathin silicon-dioxide layers started with the experimental characterization of the oxide fixed charge density Q_f and the interface trap charge density Q_{it}. In the course of these studies, it was found that the contact potential difference φ_{MS} in MOS devices is both processing- and orientation-dependent, a result that contradicts its definition. These experimental observations were the motivation to consider the role of chemical dipoles resulting from partial-charge transfer in bonds at and near the Si/SiO_2 interface [21–29]. These studies were followed by considering the effects of carrier tunneling in ultrathin gate dielectrics on the physics, modeling, simulation, characterization, and reliability of ultrathin-oxide

MOS devices [30-44].

Electrical Properties – Partial-Charge Transfer in Interface Bonds

At the gate/gate-oxide and gate-oxide/substrate interfaces of MOS devices, atoms and molecules with different electronegativities form chemical bonds. Differences in the electronegativity of such species result in the partial charge transfer of charges in interface and near-interface bonds, and the formation of uniform dipole layers at these interfaces. A first-order model was developed to calculate the magnitude of the dipole moment at the Si/SiO_2 interface resulting from such partial charge transfer [21]. The charge transfer occurs because of the difference in electronegativity between silicon atoms and SiO_2 molecules. The partial-charge-transfer model is based on the principle of electronegativity equalization, and results obtained for (100) and (111) silicon substrates indicate that the magnitude of the interface dipole is orientation-dependent. Dipole moments at the gate/gate-oxide and gate-oxide/substrate interfaces must be included in the definition of the flat-band voltage V_{FB} of MOS structures. The metal-semiconductor contact potential difference φ_{MS} determined from capacitance-voltage measurements on (100) and (111) silicon oxidized in dry oxygen and metallized with aluminum agree with the predictions of this model.

A two-dimensional representation of the silicon substrate and the adjacent oxide layer is shown in Fig. 4. To simplify the charge-transfer calculations, the net charge transfer was calculated between silicon atoms in the substrate and whole SiO_2 molecules in the oxide. Other bonds exist at the Si/SiO_2 interface and partial charge transfer in bonds other than Si—SiO_2 bonds will be discussed later. The principle of electronegativity equalization states that when two or more atoms initially different in electronegativity combine to form a chemically stable compound, these atoms become adjusted to an equal intermediate electronegativity [21] given by the geometric mean of the electronegativities of all the atoms before molecule formation. The molecular electronegativity S_m of a compound is given by [21]

$$S_m = \left(\prod_{i=1}^{N} S_i \right)^{1/N} \tag{10}$$

where S_i is the electronegativity of the ith atom among the N atoms that form the molecule. Table 3 lists the electronegativities of the elements and compounds of interest in silicon microelectronics. All elements and compounds of interest have electronegativities larger than that of silicon, and partial charge transfer will take place by electrons being partially displaced away from the silicon atoms, thus generating a dipole where the positive charge is on the silicon side of the bond and the negative charge on the compound side, as illustrated in Fig. 4.

The charge transfer ρ_i, expressed in electronic charges per bond, which affects the ith atom in a compound is given by [21]

$$\rho_i = \frac{S_m - S_i}{\Delta S_i}, \tag{11}$$

where S_m is the molecule or compound electronegativity, S_i the electronegativity of the ith atom in the compound, and ΔS_i the change in the electronegativity of the

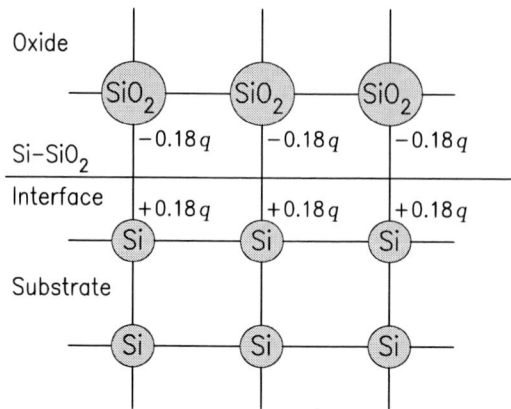

Fig. 4. Two-dimensional representation of the silicon substrate and the gate oxide used to calculate the partial charge transfer dipole in bonds between silicon atoms in the substrate and SiO_2 molecules in the oxide layer [21].

Table 3. Electronegativities of microelectronic elements and compounds.

Element	Electronegativity	Compound	Electronegativity
H	3.55	SiO	4.256
N	4.49	SiH	3.175
O	5.21	SiN	3.571
F	5.75	SiOH	3.740
Al	2.22	SiO	3.847
Si	2.84	Si_2O	3.477
Cl	4.93	Si_2O_3	4.087

atom that would result from the acquisition of a unit charge [21], which is given by

$$\Delta S_i = 2.08\sqrt{S_i}. \tag{12}$$

To calculate the partial charge transfer in the bond between a substrate silicon atom and an SiO_2 molecule in the oxide, the electronegativity of a hypothetical Si–SiO_2 molecule is first calculated, then the charge transfer ρ_i for the silicon atom in that molecule is calculated. This electronegativity $S_{Si\text{-}SiO_2}$ is given by [21]

$$S_{Si\text{-}SiO_2} = 3.477, \tag{13}$$

and the charge transfer ρ_i in the Si–SiO_2 interface bonds is

$$\rho_{Si\text{-}SiO_2} = 0.182 \text{ electron charge/bond.} \tag{14}$$

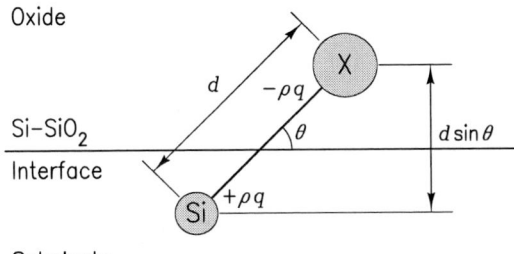

Fig. 5.　Bond between a silicon atom in the substrate and an atom or molecule X in the oxide at the Si/SiO$_2$ interface.

Charge transfer is not limited to interface bonds only, it also takes place in bonds beyond the interface both in the silicon substrate and in the silicon dioxide layer [21].

Electrical Properties – Contact Potential Difference

As shown in Fig. 5, interface bonds are, in general, inclined at an angle θ with the interface. On (100) silicon, $\theta = 35.25°$ and on (111) silicon, $\theta = 90°$. The density of bonds N_B is 6.783×10^{14} and 7.833×10^{14} bonds/cm^2 on (100) and (111) silicon, respectively [21]. The charge separation d was taken to be equal to the sum of the covalent radius of silicon, $R_{Si} = 1.11$Å, and an equivalent radius of an SiO$_2$ molecule derived from its average bulk density, $R_{SiO_2} = 1.77$Å. The net charge displacement is $d = 2.88$A. As a result of partial charge transfer in interface bonds, two charge sheets of equal charge density $q\rho N_B$, where N_B is the bond density in bonds per unit area, and opposite polarity are separated by a distance $d \sin \theta$ in a direction perpendicular to the Si/SiO$_2$ interface. The positive charge sheet will be on the silicon side and the negative charge sheet on the oxide side. The voltage φ_{DIP} across the dipole in a direction perpendicular to the interface can be expressed as

$$\varphi_{DIP} = -\frac{q \rho N_B d \sin \theta}{\epsilon}, \tag{15}$$

where ϵ is an average dielectric permittivity between that of silicon and that of silicon dioxide [21]. The applied gate bias voltage V_{GB} distribution in a metal-gate MOS capacitor can be expressed as

$$V_{GB} = V_{OX}(V_{GB}) + \varphi_{DIP} + V_{SC,B}(V_{GB}) + \varphi_{MS}, \tag{16}$$

where V_{OX} is the voltage across the oxide layer and $V_{SC,B}$ the voltage across the space-charge region in the substrate. The flat-band voltage V_{FB} can be written as

$$V_{FB} = \varphi_{MS} + \varphi_{DIP} - \frac{Q_f + Q_{it}(V_{FB})}{C_{ox}}, \tag{17}$$

where Q_f is the oxide fixed charge density and $Q_{it}(V_{FB})$ the interface trap charge density at flat-band. This relationship was the first to establish the relationship between the sample processing and chemistry, which affected φ_{DIP}, and its electrical properties, such as its flat-band voltage. It is also clear that what is usually determined from measured flat-band voltages at different oxide thicknesses is actually

$\varphi_{MS} + \varphi_{DIP}$ and not φ_{MS}. The additional dipole term explains why this quantity when determined from V_{FB} measurements exhibit an orientation and processing dependence. Experimental measurements of the contact potential difference show a substrate orientation dependence [22], where that of (100) is found consistently to be larger than that of (111) silicon by an amount up to 0.48 V on n-type samples and up to 0.30 V on p-type samples [22]. These results can be interpreted in terms of an effective φ_{MS}^{eff}, which includes the orientation-dependent contribution of the Si/SiO$_2$ interface dipole moment, rather than consisting only of φ_{MS}. It should noted that φ_{DIP} and φ_{MS} are physically separated in the MOS structure with φ_{DIP} being at the Si/SiO$_2$ interface and φ_{MS} being at the substrate/backside metal contact interface.

The general expression for φ_{MS}^{eff} is

$$\varphi_{MS}^{eff} = \varphi_{MS} - \sum_j \frac{q \, \rho_j \, N_j \, d_j \, \sin \theta_j}{\epsilon_j}, \tag{18}$$

where ρ_j is the amount of partial charge transfer in electron-charge/bond, N_j the bond density, d_j the charge separation, θ_j the angle to the Si/SiO$_2$ interface, and ϵ_j the average dielectric permittivity for type j bonds. This model predicts that the measured contact potential difference on (100) silicon could be as much as 0.475 V larger than that on (111) silicon.

In this first-order model, only bonds between silicon atoms and silicon-dioxide molecules across the interface were considered, and the model yielded that the dipole moment depends on the substrate orientation and processing conditions that the silicon wafer experiences during and after the oxide formation. The model predictions are in close agreement with experimental results obtained on (100) and (111) silicon samples. This quantitative physical model provides a framework to link the processing chemistry and the electrical properties of ultrathin SiO$_2$ layers. This model was also used to analyze heterointerface dipoles at Si–SiO$_2$, nitrided Si–N–SiO$_2$, and SiC–SiO$_2$ interfaces [25], the effects of stress on the dielectric and electrical properties of ultrathin SiO$_2$ layers [26,27], and the lateral distribution of the effective contact potential difference over the gate area of MOS devices [28,29].

Carrier Tunneling in Ultrathin Oxides

In the 1994 National Technology Roadmap for Semiconductors, trends in the oxide thickness of future generations of MOSFETs were projected to decrease to ranges in which direct tunneling of electrons and holes through ultrathin oxide layers would result in increasingly large tunneling current densities. The carrier-tunneling research studies since then were aimed at the description, modeling, and simulation of direct tunneling in ultrathin oxides on MOSFET operation, performance, characterization, design, and the ultimate tunneling-induced limits on the performance of MOS integrated circuits [30–41]. First, the MOSFET device simulator Tunnel-PISCES was introduced to account for electron and hole tunneling in ultrathin SiO$_2$ layers [30–33]. The tunneling was modeled at each node along the MOSFET channel using the independent-electron tunneling model (IETM). This implementation of carrier tunneling allowed, for the first time, to account for the effects of tunneling along the MOSFET channel for any combination of gate and drain bias voltages, and yielded the drain, gate, and substrate currents in the presence of tunneling. This implementation also yielded carrier-concentration, electrostatic-potential, and electric-field distributions along the MOSFET channel. The agreement between Tunnel-PISCES simulations and published experimental results was good, especially that only the electron effective mass in the oxide was used to fit experimental results. Tunnel-PISCES accounted initially for tunneling transitions between conduction bands (electron tunneling) and valence bands (hole tunneling). It was later extended to account for tunneling transitions between valence and conduction bands and used to extract the oxide thickness from valence-band electron tunneling characteristics [34,35]. The tunneling studies also included the effects of Ge content in poly-$Si_{1-x}Ge_x$ gate materials on the tunneling barrier in PMOSFETs [36,37], and tunneling in multilayer dielectrics, such as in SiO_2/Ta_2O_5 stacks [34,38].

The effects of carrier tunneling in future generations of MOS integrated circuits on the performance of static and dynamic MOS circuits were also studied across seven technology generations from 90 to 35 nm [39,40]. The goal of this study was to determine quantitatively the tunneling-imposed limits of MOS integrated circuits. First, gate oxides were assumed to be SiO_2 or nitrided SiO_2 layers. The goal here was to identify the SiO_2 limit technology node, and determine quantitatively in near-limit generations the degree of circuit performance degradation due to direct tunneling. It was found that MOS integrated circuits in the presence of carrier tunneling in the gate dielectric were characterized by an input current and an output current that depend on the oxide thickness and fanout. The performance of several circuits were found to degrade and ultimately fail [39,40]. For example, a minimum-geometry CMOS inverter in the 35 nm technology node ceases to operate as an inverter. The degree of degradation in near-limit technology nodes was found to be circuit-specific.

The effects of carrier tunneling on MOSFET operation and electrical characterization are also of great interest [41]. The combined effects of carrier tunneling and finite channel resistance in ultrathin-oxide MOSFETs and the two-dimensional transitions at the source-channel and drain-channel junctions result in laterally nonuniform distributions of carrier concentrations, electrostatic potential, and oxide tunneling current densities [41]. These simulated distributions were found to depend on the channel length, oxide thickness, dopant concentrations in the substrate and gate regions, and the bias conditions. Such channel-position-dependent distributions indicate that the capacitance-voltage $C(V)$ characteristics of ultrathin-oxide MOSFETs can only be studied with a distributed model [41]. From the carrier and potential distributions in the channel, it was found that there are three ranges of MOSFET channel lengths of interest. In long-channel MOSFETs, the combined tunneling and finite channel resistance result in highly nonuniform carrier and tunneling current density distributions. In short-channel MOSFETs, the source-to-channel and drain-to-channel transition regions are a substantial portion of the channel length and, again, nonuniform carrier and tunneling current density distributions are present. In mid-range, the carrier and tunneling current density distributions are nearly uniform [41]. These results highlight the need for a new methodology for the electrical characterization of ultrathin-oxide MOSFETs in the presence of gate tunneling.

Conclusions

The growth kinetics and electrical properties of ultrathin silicon-dioxide layers were reviewed. Models of the incubation and growth phases of oxidation were presented, accounting for the effects of temperature, substrate orientation, dopant type and concentration, and stress on oxide growth kinetics. The electrical properties of the ultrathin silicon-dioxide layers were also discussed, especially the contact potential difference in MOS devices and the role of partial-charge-transfer dipoles at the oxide/silicon interface. Finally, gate tunneling considerations in MOSFET design, modeling, characterization, and circuit performance were be briefly discussed.

Acknowledgments

The research results reviewed in this paper were supported by the National Science Foundation and the Semiconductor Research Corporation. The author would like to thank James D. Plummer, Eugene A. Irene, Bruce E. Deal, Charles P. Ho, John D. Shott, Reda R. Razouk, Dennis W. Hess, C. Robert Helms, Edward Poindexter, Henryk M. Przewłocki, Gerald Lucovsky, Ulrich M. Gösele, Teh Y. Tan, Jimmie J. Wortman, and John R. Hauser for their support, encouragement, and insightful discussions.

References

1. H. Z. Massoud, J. D. Plummer, and E. A. Irene, "Thermal Oxidation of Silicon in Dry Oxygen: Accurate Determination of the Kinetic Rate Constants." *J. Electrochem. Soc.*, Vol. 132, 1745 (1985).

2. H. Z. Massoud, J. D. Plummer, and E. A. Irene, "Thermal Oxidation of Silicon in Dry Oxygen: Growth-Rate Enhancement in the Thin Regime. I. Experimental Results.," *J. Electrochem. Soc.*, Vol. 132, 2685 (1985).

3. H. Z. Massoud, J. D. Plummer, and E. A. Irene, "Thermal Oxidation of Silicon in Dry Oxygen: Growth-Rate Enhancement in the Thin Regime. II. Physical Mechanisms.," *J. Electrochem. Soc.*, Vol. 132, 2693 (1985).

4. H. Z. Massoud, "The Onset of the Thermal Oxidation of Silicon from Room Temperature to 1000°C," *Microelectronic Engineering*, Vol. 28, No. 1–4, p. 109–116 (1995).

5. H. Z. Massoud and J. D. Plummer, "Analytical Relationship for the Oxidation of Silicon in Dry Oxygen in the Thin-Film Regime," *J. Appl. Phys.*, Vol. 62, 3416 (1987).

6. H. Z. Massoud, "Silicon Oxidation Kinetics in the Thin-Film Regime," in *Extended Abstracts of the International Conference on Solid-State Devices and Materials 1990*, Japan Society of Applied Physics, Tokyo, Japan, Part II, p. 1083 (1990).

7. H. Z. Massoud, "Thermal Oxidation of Silicon in the Ultrathin Regime," *Solid State Electronics*, Vol. 41, No. 7, p. 929–934 (1997).

8. E. A. Irene, H. Z. Massoud, and E. Tierney, "Silicon Oxidation Studies: Silicon Orientation Effects on Thermal Oxidation," *J. Electrochem. Soc.*, Vol. 133, 1253 (1986).

9. H. Z. Massoud, "Reverse Dopant Redistribution During the Early Stages of the Oxidation of Heavily Phosphorus-doped Silicon in Dry Oxygen," *Appl. Phys. Lett.*, Vol. 53, 497 (1988).

10. R. R. Ward, H. Z. Massoud, and R. B. Fair, "The Thermal Oxidation of Heavily

Doped Silicon in the Thin-Film Regime: Dopant Behavior and Modeling Growth Kinetics," in *Semiconductor Silicon 1990*, H. R. Huff et al., Eds., the Electrochemical Society, Pennington, NJ, Vol. 90-7, p. 405 (1990).

11. H. Z. Massoud, R. K. Sampson, K. A. Conrad, Y.-Z. Hu, and E. A. Irene, "Principles of Wafer Temperature Measurement Using *In Situ* Ellipsometry," in *Proceedings of The Third International Symposium on Ultra-Large-Scale Integration Science and Technology 1991*, J. M. Andrews and G. C. Celler, Eds., the Electrochemical Society, Pennington, NJ, Vol. PV 91-11, p. 541 (1991).

12. R. K. Sampson and H. Z. Massoud, "Simultaneous Measurement of Wafer Temperature and Native Oxide Thickness Using *In Situ* Ellipsometry," in *Proceedings of The Third International Symposium on Ultra-Large-Scale Integration Science and Technology 1991*, J. M. Andrews and G. C. Celler, Eds., the Electrochemical Society, Pennington, NJ, Vol. PV 91-11, p. 574 (1991).

13. R. K. Sampson, K. A. Conrad, E. A. Irene, and H. Z. Massoud, "Simultaneous Silicon Wafer Temperature and Oxide Film Thickness Measurement in Rapid-Thermal Processing Using Ellipsometry," *J. Electrochem. Soc.*, Vol. 140, No. 6, p. 1734–1743 (1993).

14. R. K. Sampson and H. Z. Massoud, "Resolution of Silicon Wafer Temperature Measurement by *In Situ* Ellipsometry in a Rapid Thermal Processor," *J. Electrochem. Soc.*, Vol. 140, No. 9, p. 2673–2678 (1993).

15. K. A. Conrad, R. K. Sampson, H. Z. Massoud, and E. A. Irene, "Ellipsometric Monitoring and Control of the Rapid Thermal Oxidation of Silicon," *J. Vac. Sci. Technol. B*, Vol. 11, No. 6, p. 2096–2101 (1993).

16. R. K. Sampson, K. A. Conrad, H. Z. Massoud, and E. A. Irene, "Wavelength Considerations For Improved Silicon Wafer Temperature Measurement By Ellipsometry," *J. Electrochem. Soc.*, Vol. 141, No. 2, p. 539–542 (1994).

17. R. K. Sampson, K. A. Conrad, E. A. Irene, and H. Z. Massoud, "Substrate Doping and Microroughness Effects in RTP Temperature Measurement by *In Situ* Ellipsometry," *J. Electrochem. Soc.*, Vol. 141, No. 3, p. 737–741 (1994).

18. K. A. Conrad, R. K. Sampson, H. Z. Massoud, and E. A. Irene, "Design and Construction of a Rapid Thermal Processing System for *in situ* Optical Measurements," *Review of Scientific Instruments*, Vol. 67, No. 11, p. 3954–3957 (1996).

19. R. Deaton and H. Z. Massoud, "The Effect of Thermally Induced Stresses on the Rapid-Thermal Oxidation of Silicon," *J. Appl. Phys.*, Vol. 70, 3588 (1991).

20. R. Deaton and H. Z. Massoud, "Manufacturability of Rapid-Thermal Oxidation of Silicon: Oxide Thickness, Oxide Thickness Variation, and System Dependence," *IEEE Trans. Semiconductor Manufacturing*, Vol. 5, No. 4, p. 347–358 (1992).

21. H. Z. Massoud, "Charge-Transfer Dipole Moments at the Si–SiO$_2$ Interface," *J. Appl. Phys.*, Vol. 63, 2000 (1988).

22. H. Z. Massoud, "Charge-Transfer Dipoles at the Si/SiO$_2$ Interface and the Metal-Semiconductor Workfunction Difference in MOS Devices," in *SiO$_2$ and Its Interfaces*, S. T. Pantelides and G. Lucovsky, Eds , Mat. Res. Soc. Symp. Proc., Vol. 105, p. 265 (1988).

23. H. Z. Massoud and J. D. Plummer, "A Physical Model for the Observed Dependence of the Metal-Semiconductor Work Function Difference on Substrate Orientation," p. 251 in *Physics and Chemistry of SiO₂ and the Si/SiO₂ Interface*, B. E. Deal and C. R. Helms, Eds., Plenum Press, NY (1988).

24. H. Z. Massoud, "A Simple Model of the Chemical Bonds at the Si–SiO₂ Interface and Its Influence on the Electronics Properties of MOS Devices," in *Fundamental Aspects of Ultrathin Dielectrics on Si-Based Devices*, p. 89–102, Edited by E. Garfunkel, E. P. Gusev, and A. Ya. Vul', Kluwer Academic Publishers, Dordrecht, The Netherlands (1998).

25. G. Lucovsky, H. Yang, and H. Z. Massoud, "Heterointerface Dipoles: Applications to (a) Si–SiO₂, (b) Nitrided Si–N–SiO₂, and (c) SiC–SiO₂ Interfaces," *J. Vac. Sci. Technol. B.*, Vol. 16, No. 4, p. 2191–2198 (1998).

26. H. M. Przewlocki and H. Z. Massoud, "The Effects of Stress Annealing in Nitrogen on the Effective Contact Potential Difference (ECPD), Charges, and Traps at the Si/SiO₂ Interface of MOS Devices," *J. Applied Physics*, 92, 2198 (2002).

27. H. Z. Massoud and H. M. Przewlocki, "The Effects of Stress Annealing in Nitrogen on the Index of Refraction of Silicon Dioxide Layers in MOS Devices," *J. Applied Physics*, 92, 2202 (2002).

28. H. M. Przewlocki, A. Kudla, D. Brezinska, L. Borowicz, Z. Sawicki, and H. Z. Massoud, "The Lateral Distribution of the Effective Contact Potential Difference Over the Gate Area of MOS Structures," Electron Technology – Internet Journal, 35, 1–6 (2003).

29. H. M. Przewlocki , A. Kudla , D. Brzezinska, and H. Z. Massoud, "Distribution of the Contact-Potential Difference Local Values Over the Gate Area of MOS Structures, *Microelectronic Engineering*, Vol. 72, pp. 165–173 (2004).

30. J. P. Shiely and H. Z. Massoud, "Simulation of the Drain-Current Characteristics of MOSFETs with Ultrathin Oxides in the Presence of Direct Tunneling," *Microelectronic Engineering*, Vol. 48, p. 101-104 (1999).

31. H. Z. Massoud, "MOSFET Device Simulation in the Presence of Electron and Hole Tunneling," in *Third NASA Workshop on Device Modeling*, M. Meyyappan, Ed., NASA Ames Research Center, Moffett Field, CA (1999).

32. H. Z. Massoud, "ULSI Device Simulation of MOSFETs with Ultrathin Tunneling Dielectrics," in *Proceedings of the First International Workshop on Dielectric Thin Films for Future ULSI Devices: Science and Technology*, Toyo University, Tokyo, Japan (1999).

33. H. Z. Massoud, J. P. Shiely, and A. Shanware, "Self-Consistent MOSFET Tunneling Simulations - Trends in Gate and Substrate Currents and the Drain Current Turnaround Effect with Oxide Scaling," in *Ultrathin SiO₂ and High-K Materials for ULSI Gate Dielectrics*, H. R. Huff, C. A. Richter, M. L. Green, G. Lucovsky, and T. Hattori, Eds., Materials Research Society Symposium Proceedings, Vol. 567, p. 227, San Francisco, CA (1999).

34. A. Shanware, H. Z. Massoud, E. Vogel, K. Henson, J. R. Hauser, and J. J. Wortman, "Modeling the Trends in Valence-Band Electron Tunneling in NMOSFETs with Ultrathin SiO₂ and SiO₂/Ta₂O₅ Dielectrics with Oxide Scaling," *Microelec-*

tronic Engineering, Vol. 48, p. 295–298 (1999).

35. A. Shanware, J. P. Shiely, H. Z. Massoud, E. Vogel, K. Henson, A. Srivastava, C. Osburn, J. R. Hauser, and J. J. Wortman, "Extraction of the Gate Oxide Thickness of N- and P-Channel MOSFETs Below 20Å from the Substrate Current Resulting from Valence-Band Electron Tunneling," in *Technical Digest of the 1999 IEEE International Electron Devices Meeting*, p. 815, IEEE, Piscataway, NJ (1999).

36. A. Shanware, H. Z. Massoud, A. Acker, V. Z.-Q. Li, M. R. Mirabedini, K. Henson, J. R. Hauser, and J. J. Wortman, "The Effects of Ge Content in Poly-Si$_{1-x}$Ge$_x$ Gate Materials on the Tunneling Barrier in PMOS Devices," *Microelectronic Engineering*, Vol. 48, p. 39–42 (1999).

37. A. Shanware, H. Z. Massoud, A. Acker, V. Z.-Q. Li, M. R. Mirabedini, K. Henson, J. R. Hauser, and J. J. Wortman, "Modeling the Dependence of the Gate Current in Ultrathin Dielectrics of PMOS Devices with Poly-SiGe Gate Materials on Ge Content," in *Ultrathin SiO$_2$ and High-K Materials for ULSI Gate Dielectrics*, H. R. Huff, C. A. Richter, M. L. Green, G. Lucovsky, and T. Hattori, Eds., Materials Research Society Symposium Proceedings, Vol. 567, p. 127, San Francisco, CA (1999).

38. H. Z. Massoud, "Tunneling in SiO$_2$/Ta$_2$O$_5$ Stack Dielectrics," in *Proceedings of the Third International Symposium on Control of Semiconductor Interfaces*, Nagano, Japan (1999).

39. H. Z. Massoud, "Effects of Gate Tunneling in Near-Limit CMOS Circuits," Solid-State Circuits Technical Committee Workshop on the *The Implications of Near-Limit CMOS on Circuits and Applications*, International Solid-State Circuits Conference, 2003 ISSCC, San Francisco, CA, February 2003. (Invited).

40. P. Kolar, J. Jopling, and H. Z. Massoud, "Effects of Gate Tunneling on Static and Dynamic Circuits," Techcon 2003, A Semiconductor Research Corporation Conference, Austin, Texas, August 2003.

41. M. Y.-C. Shen, J. Jopling, and H. Z. Massoud, "On the Effects of Carrier Tunneling on the Capacitance-Voltage Characteristics of Ultrathin-Oxide MOSFETs," *ECS Transactions*, Vol. 1, No. 1, pp. 283–294 (2005).

204

Oxygen Vacancies at the Si(001)/SiO$_2$ Interface

J. C. Greer[a], T. M. Hendersen[a], G. Bersuker[b], R. J. Bartlett[d], and A. Korkin[a],

[a] Tyndall National Institute, Lee Maltings, Prospect Row, Cork, Ireland
[b] SEMATECH, 2706 Montopolis Drive, Austin, Texas 78741, USA
[c] University of Florida, Gainesville, Florida 32611, USA
[d] Nano and Giga Solutions, Gilbert, Arizona 85296, USA

High-k oxides deposited on silicon substrates form interfacial silicon dioxide layers. The interface layer is of importance in microelectronics, due both to charge transport at the silicon/silicon dioxide interface and to the oxide layer influencing the equivalent oxide thickness (EOT). Electrical characterization of low temperature thermal oxides grown by atomic layer deposition of hafnia dioxide indicates a dielectric constant at the interfacial region larger than for stochiometric SiO$_2$. A proposed mechanism for the observed increase is the interface becomes oxygen deficient during the growth process. Oxygen vacancies in SiO$_2$ are examined by ab initio computations and are found to be more stable in the vicinity of the interface, thus identifying a chemical driving force for oxygen reduction at the interface.

High-k dielectrics are targeted for the 45 nanometer silicon technology node to meet transistor scaling requirements. Hf-based oxides demonstrate promising characteristics as high-k gate dielectrics, but integration into silicon fabrication processes continues to face obstacles, in particular, with respect to equivalent oxide thickness (EOT) (1) and degradation of carrier mobility at the oxide/semiconductor interface (2). An amorphous interfacial oxide layer forms in the physical thickness range of 0.6-1.1 nm at the interface and was feared to limit the gate stack scaling to an EOT range below 1 nm. It has been found the k value of the interfacial layer is higher than for stoichiometric SiO$_2$ (1, 3). It has been suggested that the higher k value is associated with oxygen deficiency in the interfacial layer. The energetics of oxygen vacancy formation in a thin SiO$_2$ layer on Si(001) is investigated by ab initio computations to investigate possible mechanisms for oxygen removal near the interface.

The Vienna ab initio Simulation Package (VASP) (4) is used for the density functional theory (DFT) calculations with periodic simulation cells. The gradient corrected PW91 approximation (5) to the exchange-correlation functional is used along with Vanderbilt ultra-soft pseudopotentials (6). An energy cut-off of 395.7 eV has been applied in the plane-wave calculations and 9.86 Å vacuum layer is included in the direction normal to the interface. A model due to Pasquarello, Hybertsen and Car (7) for Si(001)/SiO$_2$ is taken as the starting geometry for the calculations. The model is generated by bonding a seven layer slab of β-tridymite to ten layers of Si(001) and terminating all non-saturated atoms at the bottom silicon and top SiO$_2$ layers by hydrogen atoms. To account for our selection of exchange-correlation potential and other parameters associated with the DFT calculations, a silicon supercell with the same number of silicon layers and atoms per layer as for the interface model is relaxed using 2x2x1 k-points to determine the equilibrium cell parameters for the silicon substrate. For subsequent calculations allowing ionic relaxation, cell parameters parallel to the interface and bottom 4 silicon atom layers (and associated H-termination) are held fixed at their bulk values. Total energy convergence with the number of k -points was investigated with 2x2x1, 4x4x1 and 8x8x1 grids resulting in energy convergence toless than 20 meV/cell. Energies and density of states (DOS) are reported from calculations using 8x8x1 k -points; a comparable level of sampling has been shown to be adequate for calculation of the local DOS within similar simulation cells (8). To separate effects of chemical bonding from lattice constraints, molecular fragment calculations describing the local environment of the oxygen vacancies have been computed usingsecond order many body perturbation theory or Møller-Plesset (MP2) theory as implemented in the TURBOMOLE system of programs (9) using correlation consistent polarized valence double zeta (cc-pVDZ) basis sets (10). MP2 theory is chosen to avoid differences due to choosing an approximation to the exchange-correlation functional within DFT and to provide a size-extensive estimate of the vacancy formation energies. Molecular fragments have been built by considering a Si-O-Si unit with terminating hydroxyl (-OH) and silyl (-SiH₃) groups. The silicon atoms in the Si-O-Si units are terminated to construct formal charge states equivalent to the local environments of the oxygen vacancy sites within the periodic simulation cells. Prior to the removal of the oxygen atom, the fragments may be described as: Si^{+1}-O-Si^{+3}, Si^{+2}-O-Si^{+3}, Si^{+3}-O-Si^{+4}, Si^{+4}-O-Si^{+4}. Referring to Fig. 1 where the oxygen vacancy sites at the interface are labeled, these molecular fragments correspond to the back bond (A), surface bridge (B), interface bridge (C), and bulk (D) sites, respectively.

First, results from the periodic calculations are considered. Neutral oxygen vacancies V_O are introduced by removal of a single oxygen from the simulation cell and relative energies for the formation of the different vacancies are computed as the difference in the total cell energies; results for the relative formation energies are listed in Table 1. Upon removal of the oxygen atoms, the Si-Si distances at the O singlet vacancy sites relax to 2.31, 2.46, 2.34 and 2.33 Å for defects A, B, C, D, respectively. This range of values is comparable to findings for Si-Si bond formation at V_O sites in α-quartz (11), and consistent with treatment of the vacancies as neutral sites (12). The defect labeled A, referred to as a back bonding oxygen, is found to have the lowest formation energy. This could possibly be ascribed to strain relaxation upon removal of the back bond, but it is observed there is consistently lower formation energy for removal of oxygen when bonded to silicon atoms in low formal oxidation states. Oxygen atoms in these environments are predominately found in the vicinity of the Si/SiO₂ interface, indicating it is energetically favorable to create these defects near the interface region. The formation energy for V_O in α-quartz has been investigated by similar DFT calculations with a vacancy formation energy estimated to range between 5.16 to 8.10 eV, dependent on the reference chemical potential chosen for free oxygen (13). Hence the lower energies associated with formation of the vacancies near the interface, up to a 1.7 eV difference for interface sites relative to bulk sites, is significant relative to the formation energies of V_O in bulk silica.

In Fig. 2, bonding within the oxide and Si layers is compared to the bonding geometries for the simulation cell prior to the removal of an oxygen atom. The oxide is capable of relaxing upon introduction of the defect with bond angles and lengths not directly associated to the vacancy site showing some distortion after the introduction of the defect. However, the difference in the average Si-O bond lengths between the cells ranges by only a few thousandths of an Ångström. Within Fig. 2, substrate bonding is characterized by the tetrahedral distortion about each bulk silicon atom. The distortion is defined as the average of the magnitude of the vector difference for the 4 silicon-silicon bonds formed at each bulk Si atom within the simulation cells and the ideal bulk atomic positions. Clearly seen in fig2c) is the distortion away from the ideal silicon crystal as the interface is approached within the vacancy free model. In general, displacement of the substrate atoms relative to their positions in the simulation cell without vacancies (large open circles) is not large, with possibly the exception of the surface bridge vacancy situated directly at the interface. The energy DOS near the silicon band gap shows little change for the defect sites B, C, D relative to the vacancy free interface, whereas there is a slight lowering of the conduction band edge after introduction of a defect at site A; see Fig. 3. It is clear from the comparisons to the simulation cell in the absence of a vacancy, that both the oxide and silicon substrate are able to relax to reduce strain about the defect sites.

To help disentangle effects of lattice and oxide strain from chemical bonding, MP2 energies for the molecular fragments representing the local environment of the oxygen vacancy are listed in Table 2. The formation energies in the table are calculated as the difference in energy of the total molecular fragment in a singlet spin state and the fragment with the central oxygen removed, with the two neighboring silicon atoms bonding as a singlet, and the total energy of an isolated oxygen atom in the triplet ground state. Within the molecular calculations, the total range of V_O formation energies is narrowed to lie within approximately 1 eV. But again the conclusion is that the lower the oxidation states of the Si atoms nearest the oxygen vacancy site, the less energy is required to form the vacancy and again this energy difference can be an appreciable percentage of the formation energy within the bulk oxide.

From the calculations, it can be inferred that two effects will help assist oxygen removal from the silicon/silicon oxide interface. The first occurs when there are lower oxidation states for silicon atoms neighboring oxygen vacancies, and the second is strain relaxation accompanying oxygen removal. The calculations reported here indicate that both of these mechanisms can be large for oxygen vacancy formation near the Si/SiO_2 interface. Coupled to the observation that V_O formation in silica is more favorable than for HfO_2 (13), it may be concluded that O is driven away from the interface towards the high-k layer introducing a reduced stochiometry in the SiO_2 region, thus causing an increase in the interfacial dielectric constant (14).

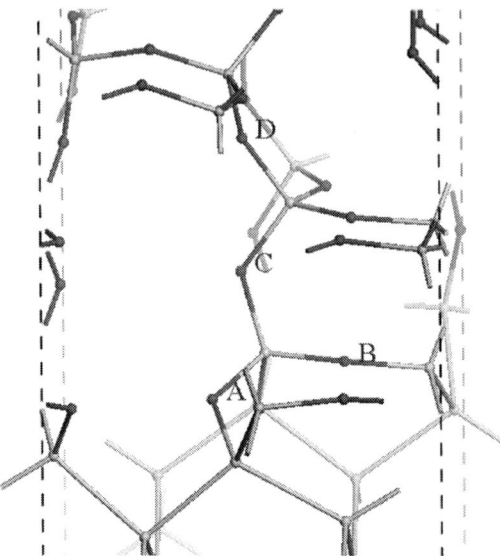

Fig. 1. Vacancy models generated by removing the oxygen atoms labeled A,B,C,D. The sites are referred to as: A- back bonded, B- surface bridge, C- interface bridge, D- Bulk. Darker sites: oxygen, lighter sites: silicon.

ECS Transactions, 2 (2) 205-211 (2006)

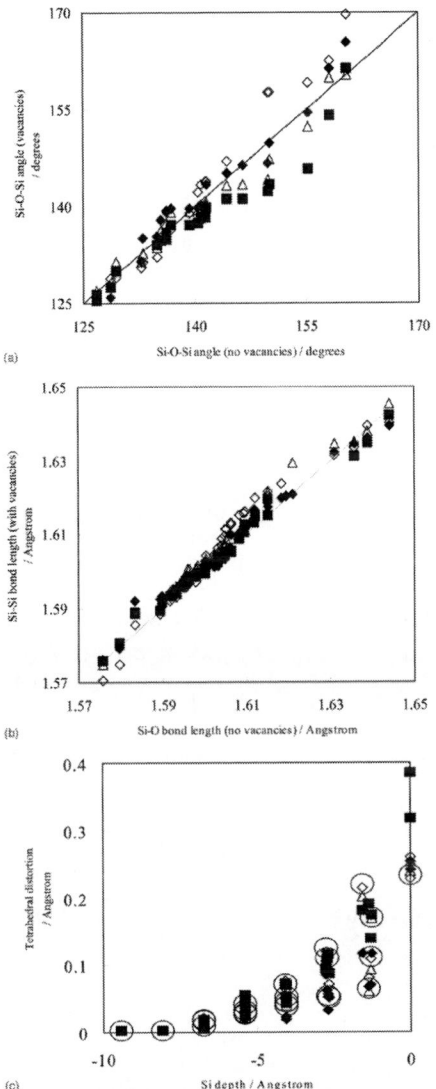

Fig. 2. Comparison of the geometries for the vacancy models against the defect free interface for ♦ back bond (A), ■ surface bridge (B), △ interface bridge (C), ◊ bulk (D): a) Si-O-Si angles. b) Si-O bond lengths. c) Distortion relative to the Si bulk equilibrium lattice. Large open circles are for the vacancy free interface.

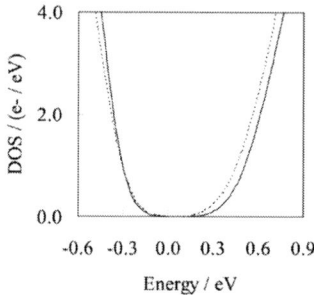

Fig. 3. Electronic density of states in the vicinity of the silicon band gap for the Si(001)/SiO$_2$ interface. Solid line - defect free structure. Dashed line - Oxygen vacancy at the back bonded (A) site.

TABLE I. Relative formation energies of oxygen vacancies near the Si(001)/SiO$_2$ interface from the periodic cell models. Defect energies given relative to dissociation to molecular and atomic oxygen

Defect	Defect Energy (eV); ½μ(O$_2$)/μ(O)	Relative Formation Energy (eV)
Back Bond (A)	4.18 / 7.21	0.00
Surface Bridge (B)	5.55 / 8.58	1.37
Interface Bridge (C)	5.68 / 8.71	1.50
Bulk (D)	5.88 / 8.91	1.70

TABLE II. Defect formation energy from molecular fragment models. Defect energies given relative to dissociation to molecular and atomic oxygen

Defect	Defect Energy (eV); ½μ(O$_2$)/μ(O)		Relative Formation Energy (eV)	
	SCF	MP2	SCF	MP2
Back Bond (A)	3.75 / 4.41	3.82 / 6.25	0.00	0.00
Surface Bridge (B)	4.22 / 4.88	4.44 / 6.87	0.47	0.62
Interface Bridge (C)	4.54 / 5.20	4.81 / 7.24	0.79	0.99
Bulk (D)	4.57 / 5.23	4.88 / 7.31	0.82	1.06

Acknowledgments

This work has been funded by Science Foundation Ireland and the National Science Foundation project ITR 032553. We are grateful to Alfredo Pasquarello for making his interface models available.

References

1. G. Bersuker *et al*, *Jap. J. Appl. Phys.* 43, 7899 (2004).
2. X.-Y. Liu, D. Jovanovic, and R. Stumpf, *Appl. Phys. Lett.* **86**, 082104 (2005).
3. G. Bersuker *et al*, *207th Electrochemical Society Meeting*, Quebec City, Canada, May 15-20 (2005).
4. G. Kresse and J. Furthmuller, *Comput. Mater. Sci.* **6**, 15 (1996); G. Kresse and J. Furthmuller, *Phys. Rev. B*, **54**, 11169 (1996).
5. J.P. Perdew *et al*, Phys. Rev. B **46**, 6671 (1992).

6. D. Vanderbilt, *Phys. Rev. B* **85**, 7892 (1990); G. Kresse and J. Hafner, *J. Phys.: Condens. Matter* **6**, 8245 (1994).
7. A. Pasquarello, M.S. Hybertsen, and R. Car, *Appl. Surf. Sci.* **104/105**, 317 (1996); A. Pasquarello, M.S. Hybertsen, and R. Car, *Appl. Phys. Lett.* **68**, 625 (1996).
8. J.B. Neaton, D. Muller, and P. Ashcroft, *Phys. Rev. Lett.* **85**, 1298 (2000).
9. S. Brode, H. Horn, M. Ehrig, D. Moldrup, J. .E. Rice, and R. Ahlrichs, *J. Comp. Chem.* **14**, 1142 (1993); F. Weigend and M.Haser, *Theor. Chem. Acc.* **97**, 331 (1997); F. Weigend, M. Haser, H. Patzelt and R. Ahlrichs, *Chem. Phys. Letters* **294**, 143 (1998).
10. T.H Dunning, Jr., J. Chem. Phys. {\bf 90} 1007 (1989);D.E. Woon and T.H. Dunning, Jr, J. Chem. Phys. {\bf 93} 1358 (1993).
11. V.B. Sulimov *et al*, *Phys. Rev. B*, **66**, 024108 (2002) and references cited therein.
12. C.J. Nicklaw *et al*, *IEEE Trans Nucl. Science* **49**, 2667 (2002).
13. W.L. Scopel, A.J.R. da Silva, W. Orellana, and A. Fazzio, *Appl. Phys. Lett.* **84**, 1492 (2004).
14. F. Giustino, A. Bongiorno, and A. Pasquarello, *J. Phys.: Condens. Matter,* **17**, S2065 (2005).

212

ECS Transactions, 2 (2) 213-228 (2006)
10.1149/1.2195661, copyright The Electrochemical Society

DEFECT DYNAMICS IN THE PRESENCE OF OXYGEN IN GROWING
CZOCHRALSKI SILICON CRYSTALS

Milind S. Kulkarni

MEMC Electronic Materials, St. Peters, MO 63376 USA

ABSTRACT

Modern Czochralski (CZ) silicon crystals contain various large defects
known as microdefects that affect the yield and the performance of
microelectronic devices. These microdefects are primarily the
agglomerates of the intrinsic point defects of silicon, vacancies and self-
interstitials, and of oxygen (silicon dioxide). The distribution of
microdefects in a CZ crystal is determined by the complex dynamics
influenced by various reactions involving the intrinsic point defects and
oxygen, and their transport. Two-dimensional oxygen influenced transient
defect dynamics in growing CZ crystals is quantified and solved. The
Frenkel reaction and the reactions between vacancies and oxygen are
considered. The formation of all microdefects is described by the classical
nucleation theory. Microdefects are assumed to be spherical clusters that
grow by a diffusion-limited kinetics. The predictions of the model agree
well with experimental data. Various predictions of the model and
experimental results are discussed.

INTRODUCTION

Silicon substrates produced from the silicon crystals grown by the modern
Czochralski process, or the CZ process, are popularly used in the fabrication of a majority
of microelectronic devices. The performance and the yield of these modern devices are
affected by many large defects formed in growing CZ crystals. These defects are
popularly known as microdefects. Microdefects are the agglomerates of the intrinsic point
defects of silicon, vacancies and self-interstitials, and of oxygen, primarily silicon
dioxide, termed oxygen clusters in this paper. The agglomerates of the intrinsic point
defects commonly exist in silicon crystals grown by both the Czochralski process and the
Float zone (FZ) process. Oxygen clusters, however, form only in CZ crystals, because CZ
crystals, during their growth, incorporate oxygen in appreciable concentration from
crucibles containing the silicon melt.

A series of studies reported before mid 1980s established that self-interstitial
agglomerates exist in a crystal either as the dislocation loops known as A defects or the
globular clusters known as B defects [1-10]; another series of studies showed that
vacancy agglomerates exist as octahedral voids termed D defects [11-14]. Oxygen
clusters in a growing CZ crystal are typically very small. These clusters facilitate the
formation of stacking faults in the crystal subjected to selective heat treatments that
generate self-interstitials. Hence, oxygen clusters are typically identified through these
stacking faults known as the oxidation induced stacking faults or OSFs [15].

213

The dynamics of the formation of various microdefects in CZ crystals is affected by many reactions involving the intrinsic point defects and oxygen, and their transport. The collective dynamics of all reactions including the agglomeration events and the transport of all participating species in a growing CZ crystal is termed the CZ defect dynamics. This dynamics in the absence of oxygen has been studied in detail [16-34]. These studies, however, do not directly quantify the formation of oxygen clusters in CZ crystals. The reported research on the direct quantification of the CZ defect dynamics in the presence of oxygen, in particular, and oxygen cluster formation in monocrystalline silicon, in general, involves various approximations and assumptions [35-41]. A complete two-dimensional quantification of the microdefect distributions in CZ crystals growing under transient conditions has not been accomplished yet. This paper addresses this need and attempts to provide insights into the CZ defect dynamics in the presence of oxygen.

REACTIONS IN GROWING CZ CRYSTALS

The essential aspect of understanding the CZ defect dynamics is the quantification of kinetics of all relevant reactions in a growing crystal. The Frenkel reaction involving the intrinsic point defects and silicon, the reactions involving vacancies and oxygen, and the agglomeration reactions forming all microdefects influence the CZ defect dynamics.

Reactions Involving No Agglomeration

The Frenkel reaction and the reactions involving vacancies and interstitial oxygen (simply, oxygen) do not directly produce microdefects. The Frenkel reaction involves the mutual annihilation of a vacancy, v, and a self-interstitial, i, by their recombination to produce a silicon lattice atom, Si, and the backward production of a pair of a vacancy and a self-interstitial from a silicon lattice atom:

$$i + v \rightleftarrows Si \tag{1}$$

Oxygen participates in a series of reversible reactions with vacancies and complexes of vacancies and oxygen in a growing CZ crystal. The following reactions involving vacancies and oxygen are of primary importance [41]:

$$v + O \rightleftarrows vO \tag{2}$$
$$vO + O \rightleftarrows vO_2 \tag{3}$$

vO and vO_2 are the vacancy-oxygen complexes. Each forward or reverse reaction listed above is considered to be an elementary reaction. The net rate of the formation of a reacting species is given by the summation of the rates of the formation of the species by each elementary reaction. At equilibrium, the net rate of production of any species is zero.

The Nucleation of the Intrinsic Point Defects

Octahedral voids or D defects are formed by the agglomeration of vacancies. Globular B defects are formed by the agglomeration of self-interstitials. A defects presumably form by the transformation of B defects. These microdefects are modeled as clusters of the intrinsic point defects. The nucleation of an intrinsic point defect species takes place through a series of elementary reactions of the following type:

$$P_{m_x} + x \rightleftarrows P_{(m+1)_x} \qquad \{x = i, v\} \tag{4}$$

where P is a cluster of the nucleating monomers. A cluster containing m intrinsic point defects of type x is represented by P_{m_x}. In an isolated element of silicon, at a fixed temperature (T) and composition (C), the free energy change associated with the formation of a cluster containing m intrinsic point defects from a solution of the intrinsic point defects (F) is given by

$$F_x(m_x) = -m_x k_b T \ln \frac{C_x}{C_{x,e}} + \lambda_x m^{\frac{2}{3}} \qquad \{x = i, v\} \tag{5}$$

where C is the concentration of any species, λ the surface energy of the cluster, and k_b the Boltzmann constant. The subscript x denotes the nucleating species as well as the type of the cluster, depending on the variable, and the subscript e denotes the equilibrium conditions. The first term on the right hand side of equation (5) represents the volume (bulk) free energy change associated with the intrinsic point defect supersaturation, and the second term is the free energy required for the formation of the new cluster surface. The number of the intrinsic point defects in the so-called 'critical cluster' is obtained by maximizing the free energy change F with respect to m. The classical nucleation theory gives the net rate of the formation of the stable spherical clusters per unit volume, defined as the nucleation rate, as a function of various properties of the critical clusters and the nucleating monomers [42-44]:

$$J_x = \left[4\pi R_x(m_x^*) D_x C_x \right] \left[\left(12\pi F(m_x^*) k_b T \right)^{-\frac{1}{2}} \left(k_b T \ln \frac{C_x}{C_{x,e}} \right) \right] \left[\rho_{site} e^{\left(-F_x(m_x^*) / k_b T \right)} \right] \qquad \{x = i, v\} \tag{6}$$

where J is the nucleation rate, R is the radius of a cluster, D is the diffusivity of any species, and ρ_{site} is the site density for nucleation. The superscript '*' indicates the critical clusters. Thus, J_x is the nucleation rate of clusters containing species x, $R_x(m^*)$ is the radius of a critical cluster of x, and $F_x(m_x^*)$ is the total free energy change associated with the formation of a critical cluster. The first term in the square brackets on the right hand side of equation (6) is the diffusion limited attachment frequency of x to a critical cluster, the second term is the Zheldovich factor, and the third term is the equilibrium concentration of the critical clusters.

An approximation of octahedral D defects and globular B defects as the spherical clusters of the intrinsic point defects is reasonably accurate. As A defects are presumed to form from B defects, the formation kinetics of B defects sufficiently describes the formation of A defects as well.

The Formation of Oxygen Clusters

Oxygen clusters are modeled as spherical agglomerates of oxygen (silicon dioxide). The specific volume of an oxygen cluster is greater than that of silicon. Thus, the formation of an oxygen cluster is associated with the generation of stress. In the presence of vacancies, however, the clusters relieve stress by the consumption of vacancies. The oxygen cluster formation proceeds through a series of reactions involving oxygen, vacancies, and complexes of vacancies and oxygen. It is, however, a good approximation to assume that the reaction rates are limited by the oxygen attachments, as the vacancy attachments are much faster. Hence, this series of reactions is written as:

$$O + P_{m_O} + \gamma v + \frac{1}{2} Si \; \rightleftarrows \; P_{(m+1)_O} \tag{7}$$

where γ is the number of vacancies absorbed per oxygen atom participating in the reaction. It must be noted that an oxygen cluster containing m oxygen atoms also contains $\frac{1}{2} m$ silicon atoms.

The volume (bulk) free energy change associated with the formation of an oxygen cluster containing m oxygen atoms in an isolated element of silicon at a fixed temperature and composition is given by the contributions from the oxygen supersaturation and the vacancy supersaturation. Thus, the total free energy change associated with the formation of an oxygen cluster containing m oxygen atoms is

$$F_O(m_O) = \left[-m_O k_b T \ln \frac{C_O}{C_{O,e}} - \gamma m_O k_b T \ln \frac{C_v}{C_{v,e}} \right] + \left[\lambda_O m_O^{\frac{2}{3}} \right] \tag{8}$$

The first term on the right hand side of equation (8) is the volume (bulk) free energy change and the second term is the oxygen cluster surface energy associated with the formation of the cluster surface. The subscript O denotes both oxygen and oxygen clusters depending on the variable.

The formation kinetics of oxygen clusters is quite complex. An oxygen cluster undergoes morphological changes as it grows. This paper does not address the details of these morphological changes. A broad macroscopic understanding of the oxygen cluster distribution is obtained by assuming these clusters to be spherical. The net rate of the formation of the stable oxygen clusters is obtained using the classical nucleation theory:

$$J_O = \left[4\pi R_O(m_O^*) D_O C_O \right] \left[\left(12\pi F_O(m_O^*) k_b T \right)^{-\frac{1}{2}} \left(k_b T \ln \frac{C_O}{C_{O,e}} + \gamma k_b T \ln \frac{C_v}{C_{v,e}} \right) \right] \left[\rho_{site} e^{\left(-F_O(m_O^*) \middle/ k_b T \right)} \right] \tag{9}$$

The attachment frequency of oxygen atoms to a critical cluster is assumed to be limited by the diffusion of oxygen. The discussed kinetics can now be applied in the development of the equations governing the CZ defect dynamics.

THE MODEL

The model quantifying the CZ defect dynamics must account for the balances of all species, the cluster formation, and the cluster growth. All microdefects are approximated as spherical clusters. D defects are termed v-clusters, A and B defects are termed i-clusters, and the agglomerates of oxygen (silicon dioxide) are termed O-clusters. At any given location of a growing CZ crystal, at a given time, one or more than one population of clusters formed at various other locations during the elapsed time period can exist. The clusters are assumed to be immobile; thus, they are only convectively carried from one location to the next by the physical movement of the growing crystal. In addition, there is a spatial distribution of these populations. A rigorous treatment of the spatial distribution of these cluster populations is computationally expensive. Kulkarni and Voronkov (2005) developed a lumped model that represents a population of clusters at any given location by an equivalent population of identical clusters [32]. In this paper, considering the complexity of the CZ defect dynamics, this lumped model is applied to quantify the cluster distributions.

The Governing Equations

The balance of self-interstitials includes their transport, their consumption by Frenkel reaction and i-clusters:

$$\frac{\partial C_i}{\partial t} + V \frac{\partial C_i}{\partial z} = \nabla \cdot (D_i \nabla C_i) + \left[k_{i-v} \left(C_{i,e} C_{v,e} - C_i C_v \right) \right] - q_i^i \tag{10}$$

The term in the square brackets in Equation (10) is the net rate of the formation (negative consumption rate) of self-interstitials per unit volume by the Frenkel reaction. The rate constant for an elementary forward or an elementary reverse reaction discussed in the previous section is denoted by k. The subscripts of k indicate the reactants involved in a forward or a reverse reaction; k_x denotes the rate constant for the elementary reaction involving only x and k_{x-y} indicates the elementary reaction involving x and y, where x and y represent the reacting species. q_x^y is the volumetric consumption rate of species y, denoted by the superscript, by the clusters containing species x, denoted by the subscript. The volumetric consumption rate of x by x-clusters is then represented by q_x^x. V is the crystal pull-rate, t is time, and z is the direction in which the crystal is pulled.

Vacancies are consumed by both v-clusters and O-clusters. In addition, vacancies participate in reactions with self-interstitials, oxygen, and VO. Hence, the vacancy balance is written as follows:

$$\frac{\partial C_v}{\partial t} + V\frac{\partial C_v}{\partial z} = \nabla \cdot (D_v \nabla C_v) + \left[k_{i-v}(C_{i,e}C_{v,e} - C_i C_v) - k_{v-O}C_v C_O + k_{vO}C_{vO} \right] - q_v^v - q_O^v$$

(11)

where the term in the square brackets is the net rate of vacancy production by reactions (1) to (3). Species VO is considered to be immobile. It is not directly consumed by clusters. Thus, the VO species balance must account only for the convection and reactions (2) and (3).

$$\frac{\partial C_{vO}}{\partial t} + V\frac{\partial C_{vO}}{\partial z} = \left[k_{v-O}C_v C_O - k_{vO}C_{vO} - k_{vO-O}C_{vO}C_O + k_{vO_2}C_{vO_2} \right]$$

(12)

Species VO_2 is also considered to be immobile and it is also not directly consumed by clusters. It participates only in reaction (3):

$$\frac{\partial C_{vO_2}}{\partial t} + V\frac{\partial C_{vO_2}}{\partial z} = \left[k_{vO-O}C_{vO}C_O - k_{vO_2}C_{vO_2} \right]$$

(13)

Oxygen is in abundance:

$$C_O = \Phi_O(r,z,t)$$

(14)

where the function Φ_O describing the oxygen concentration field in a growing CZ crystal is predetermined by the process conditions. The subscript O denotes oxygen. It is evident from equations (10) to (13) that the balance of the *excess total vacancy concentration*, defined as the difference between the sum of the concentrations of vacancies, vO species, and vO_2 species and the concentration of self-interstitials, $C_v + C_{vO} + C_{vO_2} - C_i$, is not explicitly affected by non-agglomeration reactions (1), (2), and (3); This balance is written as

$$\frac{\partial(C_v + C_{vO} + C_{vO_2} - C_i)}{\partial t} + V\frac{\partial(C_v + C_{vO} + C_{vO_2} - C_i)}{\partial z} = \nabla \cdot (D_v \nabla C_v) - \nabla \cdot (D_i \nabla C_i) - q_v^v - q_O^v + q_i^i$$

(15)

Assuming reaction equilibrium for reactions (1), (2), and (3), the species balances (10) to (13) are defined by equation (15) and the reaction equilibria defined below [16,41]:

$$C_i C_v = C_{i,e}C_{v,e}$$

(16)

$$\frac{C_{vO}}{C_v} = \sqrt{\frac{C_{vO_2,e}}{C_{v,e}}}$$

(17)

$$\frac{C_{vO_2}}{C_v} = \frac{C_{vO_2,e}}{C_{v,e}}$$

(18)

The growth of O-clusters is assumed to be limited by the consumption rate of oxygen by the clusters, when vacancies are in abundance. When the vacancy concentration is low, however, the O-cluster growth is assumed to be limited by the vacancy consumption rate. The O-cluster growth by the ejection of interstitials is not addressed in this model, as the primary focus of the model is the quantification of the O-cluster formation in the presence of excess vacancies. The diffusion-limited volumetric consumption rates of vacancies, self-interstitials, and oxygen by various clusters are defined following the methodology developed by Kulkarni and Voronkov (2005) [32]:

$$q_x^x = 4\pi D_x \left(C_x - C_{x,e}\right)\left(U_x N_x\right)^{\frac{1}{2}} \quad \{x = i,v\} \tag{19}$$

$$q_O^v = \begin{cases} \eta q_O^O & \left|D_v\left(C_v - C_{v,e}\right)\right| \geq \eta D_O\left(C_O - C_{O,e}\right) \\ 4\pi D_v\left(C_v - C_{v,e}\right)\left(U_O N_O\right)^{\frac{1}{2}} & \left|D_v\left(C_v - C_{v,e}\right)\right| < \eta D_O\left(C_O - C_{O,e}\right) \end{cases} \tag{20}$$

The evolution of the auxiliary variable U, which is proportional to the surface area of the cluster population, is described by the cluster growth equation:

$$\frac{\partial U_x}{\partial t} + V\frac{\partial U_x}{\partial z} = \frac{2D_x N_x}{\psi_x^x}\left(C_x - C_{x,e}\right) \quad \{x = i,v\} \tag{21}$$

$$\frac{\partial U_O}{\partial t} + V\frac{\partial U_O}{\partial z} = \begin{cases} \dfrac{2D_O N_O}{\psi_O^O}\left(C_O - C_{O,e}\right) & \left|D_v\left(C_v - C_{v,e}\right)\right| \geq \eta D_O\left(C_O - C_{O,e}\right) \\ \dfrac{2D_v N_O}{\eta\psi_O^O}\left(C_v - C_{v,e}\right) & \left|D_v\left(C_v - C_{v,e}\right)\right| < \eta D_O\left(C_O - C_{O,e}\right) \end{cases} \tag{22}$$

where ψ_x^x is the density of species x in a x-cluster. The total cluster density is directly obtained by the classical nucleation theory:

$$\frac{\partial N_x}{\partial t} + V\frac{\partial N_x}{\partial z} = J_x \quad \{x = i,v,O\} \tag{23}$$

The representative radius of a cluster population at any location is given as:

$$\Re_x = \left(\frac{U_x}{N_x}\right)^{\frac{1}{2}} \quad \{x = i,v,O\} \tag{24}$$

The domain of computation is transient, because a CZ crystal is continuously pulled. The equation describing this domain transience must be solved with the discussed equations.

All species are assumed to exist at equilibrium on all crystal surfaces including the melt/crystal interface. As the final size of a cluster is far greater than its critical size, the initial size of the clusters upon their formation is assumed to be zero. The initial length of a growing crystal is assumed to be finite but negligible.

The developed model describes the CZ defect dynamics in CZ crystals growing under both steady states as well as unsteady states.

PROPERTIES OF REACTING SPECIES

The model predictions strongly depend on the accuracy of the properties of all species participating in the CZ defect dynamics. The properties used in this study are listed in Table 1 :

Table 1. Key properties of various species participating in reactions in growing CZ crystals

$D_i \left(cm^2/s \right) = 4 \times 10^{-3} \times \exp\left(\dfrac{-0.3(eV)}{k_b T} \right)$	$C_{vO_2,e} \left(cm^{-3} \right) = \dfrac{C_O^2}{5 \times 10^{22}} \times \exp\left(\dfrac{-0.5(eV)}{k_b T} \right)$
$D_v \left(cm^2/s \right) = 2 \times 10^{-3} \times \exp\left(\dfrac{-0.38(eV)}{k_b T} \right)$	$\lambda_i (eV)) = 2.75$
$D_O \left(cm^2/s \right) = 1.3 \times 10^{-1} \times \exp\left(\dfrac{-2.53(eV)}{k_b T} \right)$	$\lambda_v (eV)) = 1.75$
$C_{i,e} \left(cm^{-3} \right) = 4.7625 \times 10^{27} \times \exp\left(\dfrac{-4.3492(eV)}{k_b T} \right)$	$\lambda_O (eV)) = 1.7$
$C_{v,e} \left(cm^{-3} \right) = 1.2 \times 10^{27} \times \exp\left(\dfrac{-4.12(eV)}{k_b T} \right)$	$\gamma = 0.42$
$C_O \left(cm^{-3} \right) = 9 \times 10^{22} \times \exp\left(\dfrac{-1.52(eV)}{k_b T} \right)$	

DEFECT DYNAMICS IN ONE-DIMENSIONAL CRYSTAL GROWTH

A CZ crystal growing at a fixed rate through a fixed temperature field remains at a steady state, with respect to a fixed coordinate system, far away from the regions formed at the beginning of the growth. A solution of the one-dimensional version of the developed model assuming only the axial variation of the microdefect distribution provides insights into the basics of the CZ defect dynamics in the presence oxygen. For these simulations, the crystal was assumed to grow through a temperature (T) profile described by the linear dependence of $1/T$ with respect to z:

$$\frac{1}{T} = \frac{1}{T_m} + \frac{1}{T_m^2} Gz \qquad (25)$$

where G is the magnitude of the temperature gradient at the melt/crystal interface. The subscript m denotes the conditions at the interface.

Voronkov described the conditions leading to the formation of various microdefects in growing FZ and CZ crystals in the early 1980s, in the absence of oxygen [16]. According to Voronkov's theory, an interplay between the Frenkel reaction and the transport of the intrinsic point defects of silicon determines their concentration fields in the vicinity of the melt/crystal interface. Vacancies and self-interstitials are assumed to exist at equilibrium at the interface. The temperature drop in the crystal in the vicinity of the interface drives the recombination of vacancies and self-interstitials, decreasing their concentrations. The developed concentration gradient drives the diffusion of vacancies and self-interstitials from the interface into the crystal. Vacancy concentration at the interface is higher than the concentration of self-interstitials, whereas self-interstitials diffuse faster. Thus, when the convection dominates the diffusion, vacancies remain the dominant species in the crystal; when the diffusion dominates the convection, self-interstitials are replenished at a higher rate from the interface and become the dominant species. Voronkov approximately quantified the relative effect of the convection over the diffusion by the ratio of V to G (V/G). At a higher V/G, the convection dominates; at a lower V/G, the diffusion dominates; at the critical V/G, the flux of vacancies is equal to the flux of self-interstitials. This analysis does not take into account the effect of oxygen. Oxygen introduces reactions involving vacancies and oxygen into this dynamics. In the presence of oxygen, free vacancies for the recombination with self-interstitials are supplied from the dissociation of vO and vO_2 species as well as from the interface, which is an infinite source. Hence, the presence of oxygen shifts the balance of this dynamics in favor of vacancies.

The evolution of the concentrations of vacancies, vO, and $vO2$ species as functions of the temperature in a CZ crystal growing at a very high V/G, or under vacancy-rich conditions, is shown in Figure 1. Near the interface, where the recombination rate is significant, concentrations of all three species decrease, as both free vacancies and bound vacancies (bound as vO and vO_2) participate in the recombination reaction; the participation of free vacancies in the recombination is direct, whereas the participation of bound vacancies results through the coupling of reactions (1), (2), and (3). Once the recombination rate decreases, the total vacancy concentration, $C_v + C_{vO} + C_{vO_2}$, remains essentially constant. The bound vacancy concentration, $C_{vO} + C_{vO_2}$, increases with decreasing temperature because of a shift in the reaction equilibrium. Free vacancies, however, remain dominant and nucleate at around 1100 ^{0}C. The growth of voids predominantly consumes all vacancy species, as shown in Figure 1. It must be noted, however, that the residual total vacancies concentration left at lower temperatures remains appreciable because of the binding between vacancies and oxygen.

At close to the critical yet marginally vacancy-rich conditions, the free vacancy concentration does not remain high enough to form voids at higher temperatures, in a CZ crystal. As the temperature further drops away from the interface, the concentration of bound vacancies increases. Under these conditions, free vacancies facilitate the O-cluster formation at lower temperatures. The formation and growth of O-clusters predominantly consumes both free and bound vacancies, as shown in Figure 2. The predicted total vacancy concentration at lower temperatures is approximate, because of the assumptions discussed in the previous section.

The conditions leading to the growth of crystals free of large v-clusters and i-clusters are desired in many microelectronic applications. Hence, the range of the pull-rate within which a CZ crystal free of large clusters can be grown at different oxygen concentrations is of primary interest in industrial crystal growth. This range can be determined by a series of simulations at different oxygen concentrations at different fixed pull-rates. These simulations, however, are computationally expensive. An approximate pull-rate range that allows the growth of a crystal without large v-clusters and i-clusters, at a given oxygen concentration, can be determined by simulating the growth at a continuously decreasing rate such that the microdefect distribution in the crystal continuously shifts as a function of the pull-rate. A direct but approximate correlation between the microdefect distribution and the pull-rate can thus be obtained, as shown by Kulkarni et al (2004) [31]. Figure 3 shows one such simulation defining the microdefect distribution as a function of the pull-rate, for a given oxygen concentration and temperature profile. It must be noted that the region free of large v-clusters and i-clusters contains O-clusters. O-clusters are quite large closer to the boundary between large v-clusters and O-clusters, as they are formed at higher temperatures in the presence of a relatively higher vacancy concentration. The size of O-clusters decreases as the vacancy concentration during their formation decreases. A series of such simulations at different oxygen concentrations shows how oxygen affects the range of the pull-rate within which the growth of a crystal free of large v-clusters and i-clusters is possible. Oxygen clearly expands this range because of the binding between vacancies and oxygen, as shown in Figure 4. Figure 4 also shows how V/G defining the boundary between O-clusters and i-clusters, known as the v/i boundary, shifts with the oxygen concentration in the discussed one-dimensional crystals growing through the temperature profile defined in Equation (25). As discussed before, the presence of vO and vO_2 species near the interface increases the total vacancy concentration available for the recombination with self-interstitials, thus decreasing the V/G marking the v/i boundary; in effect, the crystal becomes marginally more vacancy-rich in the presence of oxygen.

The series of one-dimensional simulations discussed in this section establishes the salient effects of oxygen on the CZ defect dynamics.

DEFECT DYNAMICS IN TWO-DIMENSIONAL CRYSTAL GROWTH

The radial variation of the temperature field in a growing CZ crystal and the radial diffusion of the intrinsic point defects, induced by the lateral surface of the crystal and the radial variation of the intrinsic point defect concentration, introduce a two-dimensional variation of the microdefect distribution in the crystal. In addition, variation of the crystal pull-rate, commonly observed in modern CZ processes, introduces an axial variation of the microdefect distribution. Hence, it is necessary to validate the developed model by a comparison of its predictions with the microdefect distribution observed in a crystal grown under an unsteady state representing a variety of possible conditions in modern CZ growth. An experimental crystal was grown by the varying rate shown in Figure 5. The crystal was cut longitudinally and the microdefect distribution was characterized by the method of copper decoration followed by etching [45,46]. The crystal was assumed to grow through a fixed temperature field predicted by the commercial software MARC, using the algorithm developed by Virzi [47]. As shown in Figure 5, v-clusters are observed in the regions of the crystal grown at higher rates, and i-clusters are observed in the regions grown at lower rates. The dense bands surrounding the region containing i-clusters indicate the presence of large O-clusters in higher densities. This observation is only qualitative, as all O-clusters are not revealed by the applied characterization technique. The featureless bands close to the region of i-clusters can contain unobservable, small microdefects.

The predicted microdefect distributions are also shown in Figure 5. As i-clusters and v-clusters do not coexist, a distribution of $\Re_v - \Re_i$ simultaneously shows the distribution of both types of clusters; a positive value of $\Re_v - \Re_i$ indicates the size of v-clusters and a negative value of $\Re_v - \Re_i$ indicates the size of i-clusters. Similarly, the distribution of the densities of i-clusters and v-clusters are quantified by the distribution of $N_v - N_i$. The size and the density distributions of O-clusters are shown separately in Figure 5. It must be noted that the lumped model predicts the representative size of the entire population of microdefects present at any given location. In the crystal regions formed under the dominance of the convection of the intrinsic point defects, the dominant clusters formed are v-clusters; in the regions formed under the dominance of the diffusion of the intrinsic point defects, the dominant clusters formed are i-clusters; near the critical but marginally vacancy-rich conditions, O-clusters are dominant. The microdefect distribution in the experimental crystal is accurately captured by the developed model.

CONCLUSIONS

Silicon crystals grown by the Czochralski (CZ) process can contain large defects known as microdefects formed by the agglomeration of the intrinsic point defects of silicon, vacancies (v) and self-interstitials (i), and by the vacancy-assisted agglomeration of oxygen (O) with silicon. The quantification of the distributions of microdefects in growing CZ crystals can be accomplished by treating reactions involving the intrinsic point defects of silicon and oxygen, along with the formation and growth of microdefects. The presence of oxygen primarily generates two bound vacancy species, vO and vO_2 [39,41]. The agglomerates of vacancies are modeled as spherical v-clusters; the agglomerates of self-interstitials are modeled as spherical i-clusters; and the agglomerates of oxygen, primarily silicon dioxide, are modeled as spherical O-clusters. The complexity of this treatment is reduced by the application of the lumped model developed by Kulkarni and Voronkov (2005) that approximates a population of clusters of different sizes at any given location in a CZ crystal as an equivalent population of identical microdefects [32].

The key element of the developed model is the vacancy-assisted formation of O-clusters. Effectively, all large O-clusters in the CZ growth are formed by absorbing vacancies, as the specific volume of O-clusters is greater than that of silicon. The growing O-clusters directly consume only free vacancies; as the free vacancy concentration decreases, however, more free vacancies are generated by the disassociation of vO and vO_2 species. Thus, both free vacancies and vacancies bound in vO and vO_2 species are consumed.

The developed model can quantify the microdefect distributions in CZ crystals growing under steady states as well as unsteady states. The type of microdefect formed in a given region in the crystal depends on the concentration of the intrinsic point defects and vO and $vO2$ species established a short distance away from the interface. In the regions marked by a high free vacancy concentration, voids or v-clusters are formed at higher temperatures by the nucleation of vacancies. The v-cluster growth consumes both free and bound vacancies. In the regions marked by a moderate free vacancy concentration, v-cluster formation is suppressed at higher temperatures; free and bound vacancies are consumed by the formation and growth of O-clusters. The binding between vacancies and oxygen allows survival of vacancies in the bound form in very low concentrations at lower temperatures even in the presence of voids and O-clusters. In the regions marked by the dominance of self-interstitials, i-clusters are formed. The concentration fields of the intrinsic point defects in the vicinity of the interface are established primarily by the interplay between the Frenkel reaction and the intrinsic point defect transport. Oxygen increases the effective vacancy concentration available for the recombination with self-interstitials by increasing the concentration of vO and vO_2 species and marginally aids the conditions leading to the survival of vacancies as the dominant intrinsic point defect species, for fixed crystal growth conditions. The increase in the pull-rate range within which crystals free of large v-clusters and i-clusters can be grown, with increasing oxygen concentration, is also predicted and explained by the model. This behavior is caused by an increase in the concentration of bound vacancies with increasing oxygen concentration, for fixed crystal growth conditions.

Qualitative microdefect distributions in CZ crystals reported in the literature can be quantified by the developed model. The simplicity of the model allows the prediction of the very complex two-dimensional microdefect distributions in CZ crystals growing under unsteady states.

REFERENCES

1. T. Abe, T. Samizo, and S. Maruyama, *Jpn. J. Appl. Phys.*, **5**, 458 (1966).
2. A. J. R. de Kock, *Appl. Phys. Letters*, **16**, 100 (1970).
3. A. J. R. de Kock, *J. Electrochem. Soc.*, **118**, 1851 (1971).
4. A. J. R. de Kock, *ACTA ELECTRONICA*, **16**, 4, 303 (1973).
5. A. J. R. de Kock, P. J. Roksnoer and P. G. T. Boonen, *J. Cryst. Growth*, **22**, 311 (1974).
6. A. J. R. de Kock, P. J. Roksnoer and P. G. T. Boonen, *J. Cryst. Growth*, **28**, 311 (1975).
7. P. M. Petroff and A. J. R. de Kock, *J. Cryst. Growth*, **30**, 117 (1975).
8. P. M. Petroff and A. J. R. de Kock, *J. Cryst. Growth*, **36**, 4 (1976).
9. H. Föll, U. Gösele and B. O. Kolbesen, *J. Cryst. Growth*, **40**, 90 (1977).
10. A. J. R. de Kock and W. M. van de Wiljert *J. Cryst. Growth*, **49**, 718 (1980).
11. P. J. Roksnoer and M. M. B. van den Boom, *J. Cryst. Growth*, **53**, 563 (1981).
12. P. J. Roksnoer, *J. Cryst. Growth*, **68**, 596 (1984).
13. M. Kato, T. Yoshida, Y. Ikeda, and Y. Kitagawara, *Jpn. J. Appl. Phys.*, **35**, 5597 (1996).
14. T. Ueki, M. Itsumi, and T. Takeda, *Appl. Phys. Lett.*, **70**, 1248 (1997).

15. M. Hasebe, Y.Takeoka, S. Shinoyama, and S. Naito, *Jpn. J. Appl. Phys.*, **28**, L 1999 (1989).
16. V. V. Voronkov, *J. Crystal Growth*, **59**, 625 (1982).
17. S. Sadamitsu, S. Umeno, Y. Koike, M. Hourai, S. Sumita, and T. Shigematsu, *Jpn. J. Appl. Phys.*, **32**, 3675 (1993).
18. R. A. Brown, D. Maroudas, and T. Sinno, *J. Cryst. Growth*, **137**, 12 (1994).
19. W. von Ammon, E. Dornberger, H. Oelkrug, H. Weidner, *J. Crystal Growth*, **151**, 273 (1995).
20. E. Dornberger and W. von Ammon, *J. Electrochem. Soc.*, **143**, 1648 (1996).
21. E. Dornberger, W. von Ammon, N. Van den Bogaert, and F. Dupret, *J. Cryst. Growth*, **166**, 452 (1996).
22. K. Nakamura, T. Saishoji, T. Kubota, T. Iida, Y. Shimanuki, T. Kotooka, and J. Tomioka, *J. Cryst. Growth*, **180**, 61 (1997).
23. R. Falster, V. V. Voronkov, J. C. Holzer, S. Markgraf, S. McQuaid, and L. Muléstagno, *Electrochemical Society Proceedings*, **98-1**, 468 (1998).
24. T. Sinno, R. A. Brown, W. von Ammon, and E. Dornberger, *J. Electrochem. Soc.*, **145**, 302 (1998).
25. T. Sinno and R. A. Brown, J. Electrochem. Soc.,**146**, 2300 (1999).
26. V. V. Voronkov and R. Falster, *J. Appl. Phys.*, **86**, 5975 (1999a).
27. E. Dornberger, W. von Ammon, J. Virbulis, B. Hanna, and T. Sinno, *J. Cryst. Growth*, **230**, 291 (2001).
28. R. A. Brown, Z. Wang, and T. Mori, *J. Cryst. Growth*, **225**, 97 (2001).
29. M. Okui, and M. Nishimoto, *J. Cryst. Growth*, **237-239**, 1651 (2002).
30. V. V. Kalaev, D. P. Lukanin, V. A. Zabelin, Yu. N. Makarov, J. Virbulis, E. Dornberger, and W. von Ammon *J. Cryst. Growth*, **250**, 203 (2003).
31. M. S. Kulkarni, V. V. Voronkov, and R. Falster, *J. Electrochem. Soc.*, **151**, G663 (2004).
32. M. S. Kulkarni and V. V. Voronkov, *J. Electrochem. Soc.*, **152**, G781 (2005).
33. M. S. Kulkarni, *Ind. Eng. Chem. Res.*, **44**, 6246 (2005).
34. M. S. Kulkarni, J. C. Holzer, and L. W. Ferry, *J. Cryst. Growth*, **284**, 353 (2005).
35. J. Esfandyari, C. Schmeiser, S. Senkader, G. Hobler, and B. Murphy, *J. Electrochem. Soc.* **143**, 1995 (1996).
36. V. V. Voronkov and R. Falster, *J. Cryst. Growth*, **194**, 76 (1998).
37. V. V. Voronkov and R. Falster, *J. Cryst. Growth*, **204**, 462 (1999b).
38. Z. Wang and R. Brown, *J. Cryst. Growth*, **231**, 442 (2001).
39. V. V. Voronkov and R. Falster, *J. Appl. Phys.* **91**, 5802 (2002).
40. K. Sueoka, M. Akatsuka, M. Okui, and H. Katahama, *J. Electrochem. Soc.*, **150**, G469 (2003).
41. V. V. Voronkov and R. Falster, *J. Electrochem. Soc.*, **149**, G167 (2002).
42. D. Turnbull and J. C. Fisher, *J. Chem. Phys.*, **17**, 71 (1949).
43. A. S. Michaels, *Nucleation Phenomena*, American Chemical Society, 1966.
44. D. Kashchiev, *Nucleation, Basic Theory with Applications*, Butterworth- Heinemann, 2000.
45. M. S. Kulkarni, J. Libbert, S. Keltner, and L. Mulestagno, *J. Electrochem. Soc.*, **149**, G153-G165 (2002).
46. M. S. Kulkarni, *Ind. Eng. Chem. Res.*, **42**, vol 6, (2003).
47. A.Virzi, *J. Crys. Growth*, **112**, 699 (1991).

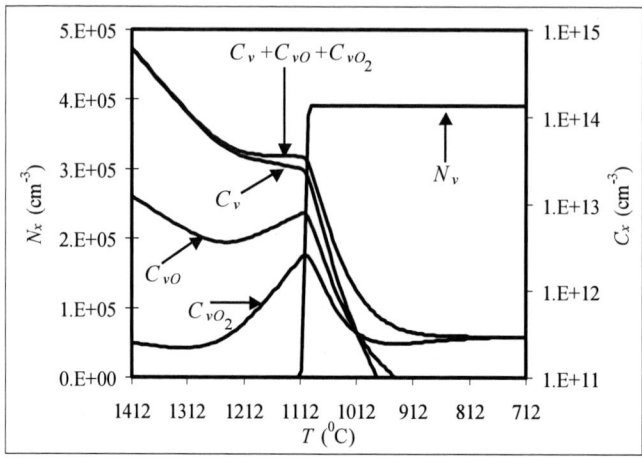

Figure 1. Evolution of the concentrations of various reacting species and the density of v-clusters in a CZ crystal growing at a high rate (G = 3.5 K/mm, V = 0.7 mm/min, C_O = 6.25×10^{17} cm^{-3} or 12.5 ppma)

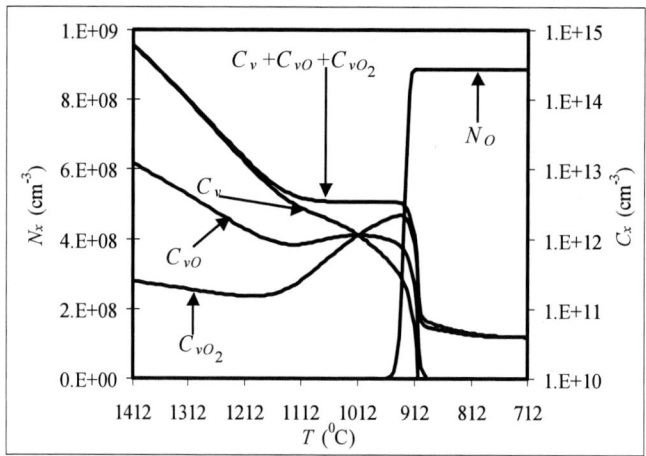

Figure 2. Evolution of the concentrations of various reacting species and the density of O-clusters in a CZ crystal growing close to the critical conditions (G = 3.5 K/mm, V = 0.48 mm/min, C_O = 6.25×10^{17} cm^{-3} or 12.5 ppma)

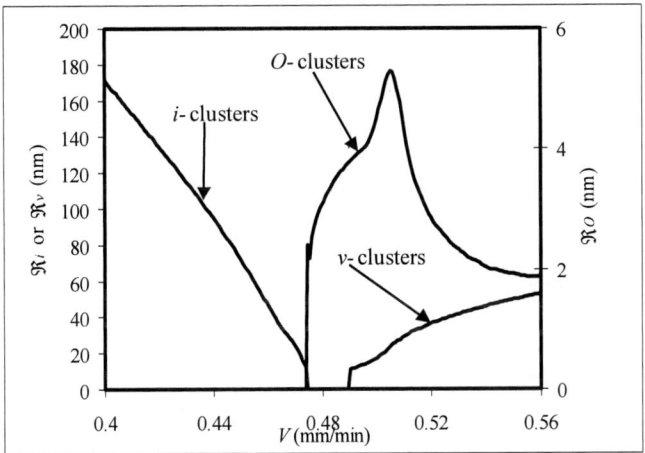

Figure 3. The simulated sizes of various clusters as the functions of the CZ crystal pull-rate.

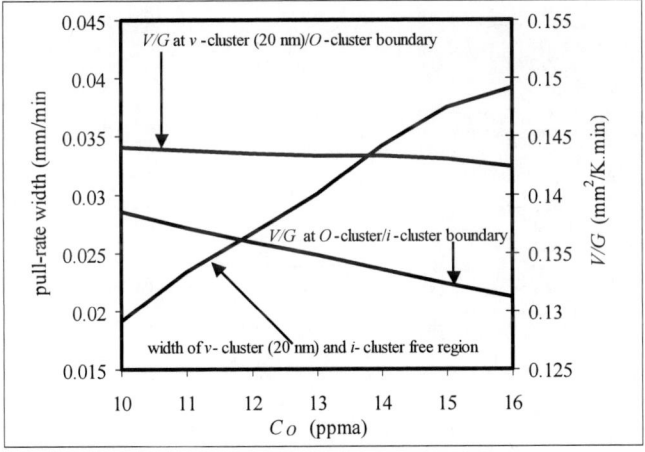

Figure 4. The sensitivity of the microdefect distribution near the critical conditions to the oxygen concentration. Note: The v-cluster and i-cluster free region includes v-clusters smaller than 20 nm (radius).

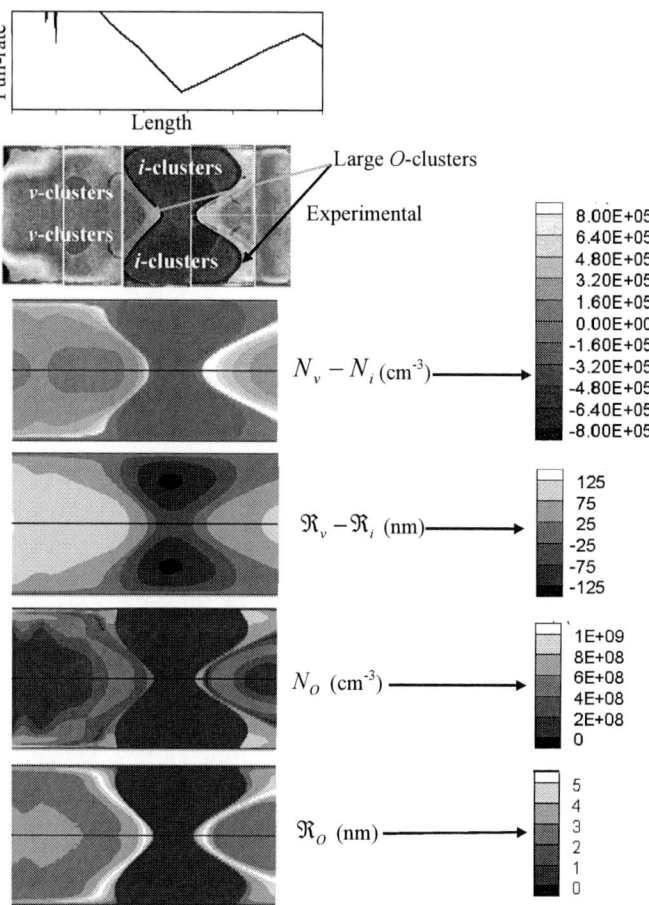

Figure 5. A comparison of the predictions of the developed model with the experimental observations. The dense bands along the periphery of the region containing i-clusters indicate intense oxygen precipitation.

ECS Transactions, 2 (2) 229-245 (2006)
10.1149/1.2195662, copyright The Electrochemical Society

DEFECT DYNAMICS IN THE PRESENCE OF NITROGEN AND OXYGEN IN
GROWING CZOCHRALSKI SILICON CRYSTALS

Milind S. Kulkarni

MEMC Electronic Materials, St. Peters, MO 63376 USA

ABSTRACT

Many structural imperfections form in silicon crystals during their
Czochralski growth. The distribution of these microdefects can be strongly
influenced and controlled by the addition of impurities such as nitrogen to
the crystal. A model describing the Czochralski defect dynamics in the
presence of nitrogen and oxygen is proposed and solved. The reactions
between vacancies and self-interstitials, nitrogen monomers and dimers,
nitrogen and vacancies, and the reactions involving vacancies, oxygen,
and complexes of vacancies and oxygen are incorporated, along with the
formation of various microdefects–the agglomerates of vacancies, self-
interstitials, and of oxygen (silicon dioxide). All microdefects are
approximated as spherical clusters. The formation of all clusters is
described by the classical nucleation theory. The clusters, once formed,
grow by a diffusion-limited kinetics. The microdefect distributions in
Czochralski crystals growing under steady state as well as unsteady state
are discussed.

INTRODUCTION

Silicon crystals grown by the Czochralski (CZ) process typically contain many
structural imperfections known as microdefects. These microdefects form during the
growth of the crystals. The agglomerates of vacancies known as voids or D defects, the
agglomerates of self-interstitials known as A defects and B defects, and the agglomerates
of oxygen, primarily silicon dioxide, termed oxygen clusters in our paper are the three
common microdefects observed in CZ crystals. The structures of the agglomerates of
vacancies and self-interstitials in Czochralski and Float Zone (FZ) crystals and the
structure of oxygen clusters in Czochralski crystals have been well established by a series
of studies in the last five decades [1-42]. The theory of initial point defect incorporation
in growing FZ and CZ crystals proposed by Voronkov (1982) provided a breakthrough in
understanding the FZ and CZ defect dynamics [16]. A series of studies following this
breakthrough accomplished the quantitative prediction of the microdefect distributions in
the both presence and absence of oxygen in CZ crystals [18,22,24-42].

As microdefects in silicon substrates produced from CZ crystals can adversely affect
the performance of devices built on them, development of crystal growth processes to
reduce the size of the microdefects in growing crystals is of industrial significance.
Various studies have shown that the microdefect distribution in CZ crystals can be
influenced in the presence of nitrogen, in general, and the size of voids can be reduced, in
particular [43-46]. Various mechanisms to describe the effect of nitrogen on CZ defect

229

dynamics have been proposed [43,45,46]. Voronkov and Falster (2005) quantified the effect of nitrogen on void distribution in an isolated element of a growing CZ crystal [46]. An accurate model for the prediction of the effect of nitrogen on CZ defect dynamics, however, must account for the diffusion of reacting species in a growing crystal. It must also incorporate reactions involving vacancies, self-interstitials, oxygen, complexes of vacancies with oxygen and nitrogen, along with the formation of various microdefects in CZ crystals growing under both steady states as well as unsteady states. This paper proposes and solves such a model to quantify both steady state and unsteady state CZ defect dynamics in the presence of nitrogen and oxygen.

MODEL

An accurate quantification of CZ defect dynamics requires a treatment of a series of elementary reactions describing the evolution of the populations of various clusters, which is computationally impractical. Therefore, reasonable approximations are required for the quantification of microdefect distribution in a CZ crystal. A lumped model proposed by Kulkarni and Voronkov (2005) that approximates a population of microdefects of various sizes present at a given location in the crystal as an equivalent population of identical microdefects is adopted for the treatment of the growth of microdefects [32]. All microdefects are approximated as spherical clusters. The stable clusters are assumed to form according to the classical nucleation theory and grow by a diffusion-limited kinetics. The cluster formation and growth kinetics is treated simultaneously with the reactions involving vacancies, self-interstitials, oxygen, nitrogen, and complexes of oxygen and nitrogen.

Reactions in Growing CZ Crystals

The primary reaction influencing the CZ defect dynamics is the Frenkel reaction, which involves the mutual annihilation of a vacancy and a self-interstitial to produce a silicon lattice atom, and the generation of a pair of a vacancy and a self-interstitial from a lattice silicon atom:

$$i + v \rightleftarrows Si \tag{1}$$

where i is a self-interstitial, v is a vacancy, and Si a silicon lattice atom. This reaction is assumed to be in equilibrium. Assuming an abundance of silicon, the equilibrium of Frenkel reaction is described by

$$C_i C_v = C_{i,e} C_{v,e} \tag{2}$$

where the concentration of a species is denoted by C and the type of the species is denoted by the subscript; the subscript e stands for the equilibrium conditions. Oxygen participates in reactions with vacancies and the complexes of vacancies and oxygen [41]:

$$v + O \rightleftarrows vO \tag{3}$$
$$vO + O \rightleftarrows vO_2 \tag{4}$$

The species O is assumed to exist in the interstitial form and vO and vO_2 can be viewed to exist in the substitutional form in silicon. Thus, the two vacancy-oxygen complexes, vO and vO_2, are considered to influence the CZ defect dynamics, while other complexes are assumed to be present in negligible concentrations. Assuming the same free energy change for reactions (3) and (4), the reaction equilibria for a given concentration of oxygen are described as

$$\frac{C_{vO}}{C_v} = \sqrt{\frac{C_{vO_2,e}}{C_{v,e}}} \tag{5}$$

$$\frac{C_{vO_2}}{C_v} = \frac{C_{vO_2,e}}{C_{v,e}} \tag{6}$$

Nitrogen exists both a monomer and a dimer, and interacts with vacancies [46]:

$$N + N \rightleftarrows N_2 \tag{7}$$

$$v + N \rightleftarrows vN \tag{8}$$

The species N and N_2 are assumed to exist in the interstitial form in silicon and vN can be viewed to exist in the substitutional form. The equilibria for reactions (7) and (8) are given by

$$\frac{C_N^2}{C_{N_2}} = K_7 \tag{9}$$

$$\frac{C_v C_N}{C_{vN}} = K_8 \tag{10}$$

where K is the equilibrium constant; the subscripts 7 and 8 specify the reaction type.

The agglomeration of the intrinsic point defects, vacancies and self-interstitials, is assumed to take place by a series of reactions of the following type:

$$P_{m_x} + x \rightleftarrows P_{(m+1)_x} \qquad \{x = i, v\} \tag{11}$$

where P is a cluster of monomers. A cluster containing m monomers of type x, or an x-cluster, is denoted by P_{m_x}. The total free energy change associated with the formation of a cluster containing m monomers from a solution of monomers, at a given temperature (T), and composition is

$$F_x(m_x) = -m_x k_b T \ln \frac{C_x}{C_{x,e}} + \lambda_x m_x^{2/3} \qquad \{x = i, v\} \tag{12}$$

where $F_x(m_x)$ is the described total free energy change, k_b is the Boltzmann constant, and λ_x is the surface energy coefficient for a cluster of x. The first term on the right hand side of equation (12) is the volume free energy change and the second term is the free energy change required to create the new cluster surface. The free-energy change associated with the formation of a so-called critical cluster and the number of monomers in it are obtained by maximizing F with respect to m. The classical nucleation theory provides the rate of formation of the so-called stable clusters per unit volume, known as the nucleation rate, as a function of various properties of the critical clusters and nucleating monomers [47-49]:

$$J_x = \left[4\pi R_x(m_x^*)D_x C_x\right]\left[\left(12\pi F(m_x^*)k_bT\right)^{-\frac{1}{2}}\left(k_bT\ln\frac{C_x}{C_{x,e}}\right)\right]\left[\rho_{site}e^{\left(-F_x(m_x^*)/k_bT\right)}\right] \qquad \{x=i,v\}(13)$$

where J is the nucleation rate, R is the cluster radius, D the diffusivity of monomers, and ρ_{site} the site density for nucleation. The subscript x denotes monomers x as well as clusters of x depending on the variable; the superscript '*' denotes the critical clusters. The first term in the square brackets on the right hand side of equations (13) denotes the diffusion limited attachment frequency of monomers to a critical cluster, the second term is the Zheldovich factor, and the third term is the equilibrium concentration of the critical clusters.

The agglomeration of oxygen produces clusters of silicon dioxide termed O-clusters. The specific volume of an O-cluster is greater than that of silicon. Therefore, the formation of an O-cluster generates stress. In the presence of vacancies, the stress relief occurs by the consumption of vacancies by the clusters. The formation of O-clusters proceeds through a series of reactions involving oxygen, vacancies and complexes of vacancies and oxygen. The kinetics of O-cluster formation, however, can be assumed to be limited by the oxygen attachment. Thus, the following reaction is assumed to describe the O-cluster formation with reasonable accuracy:

$$O + P_{m_O} + \gamma v + \frac{1}{2}Si \rightleftarrows P_{(m+1)_O} \qquad (14)$$

where γ is the number of vacancies absorbed per oxygen atom by the cluster. The volume free energy change associated with the formation of an O-cluster includes contributions from both the oxygen and the vacancy supersaturation. Thus, the total free energy change associated with the formation of an O-cluster containing m oxygen atoms, from a solution at a given temperature and composition is

$$F_O(m_O) = \left[-m_O k_b T \ln\frac{C_O}{C_{O,e}} - \gamma m_O k_b T \ln\frac{C_v}{C_{v,e}}\right] + \left[\lambda_O m_O^{2/3}\right] \qquad (15)$$

The first term in the square brackets on the right hand side of equation (15) is the volume free energy change, and the second term is the energy required for the formation of a cluster. The contribution of the strain to the free energy change is small, because of the vacancy assistance. The classical nucleation theory provides the stable O-cluster formation rate as

$$J_O = \left[4\pi R_O \left(m_O^*\right) D_O C_O\right] \left[\left(12\pi F_O \left(m_O^*\right) k_b T\right)^{-\frac{1}{2}}\left(k_b T \ln\frac{C_O}{C_{O,e}} + \gamma k_b T \ln\frac{C_v}{C_{v,e}}\right)\right]\left[\rho_{site} e^{\left(-F_O\left(m_O^*\right)/k_b T\right)}\right]$$

(16)

As discussed before, the first term in the square brackets on the right hand side of equation (16) is the diffusion limited attachment frequency of oxygen atoms to a critical cluster, the second term is the Zheldovich factor, and the third term is the equilibrium concentration of the critical clusters.

The discussed reaction equilibria and nucleation kinetics are incorporated in the governing equations describing the CZ defect dynamics.

The Governing Equations

The governing equations describe the balance of all species participating in the CZ defect dynamics. The total vacancy concentration defined as the sum of the concentration of free vacancies, v, and of the cumulative concentration of bound vacancies, vO, vO_2, and vN, is not affected by reactions (3), (4), (7), and (8); both free vacancies and self-interstitials are consumed by the Frenkel reaction at an equal rate and are generated at an equal rate. Hence, the rate of change of the total excess vacancy concentration, defined as the difference between the total vacancy concentration and the concentration of self-interstitials, is affected only by the transport of individual species, and their consumption by the clusters:

$$\frac{\partial\left(C_v + C_{vO} + C_{vO_2} + C_{vN} - C_i\right)}{\partial t} + V\frac{\partial\left(C_v + C_{vO} + C_{vO_2} + C_{vN} - C_i\right)}{\partial z} = $$
$$\nabla\cdot\left(D_v\nabla C_v\right) - \nabla\cdot\left(D_i\nabla C_i\right) - q_v^v - q_O^v + q_i^i$$

(17)

All species except for vacancies and self-interstitials are assumed to be immobile. The crystal is pulled in the z direction at rate V, which varies as a function of time t. The volumetric consumption rate of species y by x-clusters is denoted by q_x^y. Equation (17) can also be derived from individual balances for all species. The assumption of equilibrium of reactions (1), (3), (4), (7), and (8) does not necessitate the solution of the individual rate equations. Equation (17), along with the equations describing this equilibria, equations (2), (5), (6), (9), and (10), define the balances of i, v, vO, vO_2, and vN.

In the presence of excess vacancies, the O-cluster growth is assumed to be determined by the diffusion limited consumption of oxygen by the clusters. At lower vacancy concentrations, the O-cluster growth is assumed to be determined by the diffusion limited consumption of vacancies. The cluster growth by the ejection of self-interstitials is not incorporated in the model. This assumption is approximate but reasonable, because the focus of the model is on capturing the O-cluster distribution in the presence of excess vacancies. The volumetric consumption rates of different species by different clusters is given by the methodology developed by Kulkarni and Voronkov [32]:

$$q_x^x = 4\pi D_x \left(C_x - C_{x,e}\right)\left(U_x n_x\right)^{\frac{1}{2}} \quad \{x = i, v\}$$

(18)

$$q_O^v = \begin{cases} \gamma q_O^O & \left|D_v\left(C_v - C_{v,e}\right) \geq \gamma D_O\left(C_O - C_{O,e}\right)\right. \\ 4\pi D_v\left(C_v - C_{v,e}\right)\left(U_O n_O\right)^{\frac{1}{2}} & \left|D_v\left(C_v - C_{v,e}\right) < \gamma D_O\left(C_O - C_{O,e}\right)\right. \end{cases}$$

(19)

where n is the cluster density and U is the auxiliary variable proportional to the surface area of the clusters in a cluster population at any location. The equation describing the evolution of U was derived by Kulkarni and Voronkov (2005) using the cluster growth equations [32]:

$$\frac{\partial U_x}{\partial t} + V\frac{\partial U_x}{\partial z} = \frac{2 D_x n_x}{\psi_x^x}\left(C_x - C_{x,e}\right) \quad \{x = i, v\}$$

(20)

$$\frac{\partial U_O}{\partial t} + V\frac{\partial U_O}{\partial z} = \begin{cases} \dfrac{2 D_O n_O}{\psi_O^O}\left(C_O - C_{O,e}\right) & \left|D_v\left(C_v - C_{v,e}\right) \geq \gamma D_O\left(C_O - C_{O,e}\right)\right. \\ \dfrac{2 D_v n_O}{\gamma \psi_O^O}\left(C_v - C_{v,e}\right) & \left|D_v\left(C_v - C_{v,e}\right) < \gamma D_O\left(C_O - C_{O,e}\right)\right. \end{cases}$$

(21)

where the density of species x in an x-cluster is denoted by ψ_x^x. The classical nucleation theory provides the cluster balance:

$$\frac{\partial n_x}{\partial t} + V\frac{\partial n_x}{\partial z} = J_x \quad \{x = i, v, O\}$$

(22)

The representative size of a cluster population is given by

$$\mathfrak{R}_x = \left(\frac{U_x}{n_x}\right)^{\frac{1}{2}} \quad \{x = i, v, O\}$$

(23)

The domain of the calculation is transient, because a CZ crystal is continuously pulled. The described equations must be solved with the equations defining the domain transience, the temperature field in the crystal, and the segregation of nitrogen between the melt and the crystal at the melt/crystal interface. An approximate segregation coefficient describing the relationship between the total nitrogen concentration in the crystal and that in the melt is used, in accordance with the popular practice [43]. It must be noted that this segregation coefficient is just an engineering parameter fitted to describe the experimental observations. Reaction equilibrium (reactions (7) and (8)) is assumed at the interface. The heat transport dynamics is assumed to be faster than the defect dynamics. Hence, quasi-steady state assumptions for the calculation of the temperature field suffice, especially for small sections in the crystal. The temperature field is first calculated by the commercial software MARC, using the algorithms developed by Virzi, and then corrected using the experimentally measured interface shape [32,34,50].

The accuracy of the developed model depends on the accuracy of the properties of all reacting species and the parameters defining the kinetics and the thermodynamics of the discussed reactions. The parameters proposed by Kulkarni et al. (2006) for the simulation of the CZ defect dynamics in the presence of oxygen are used in this study [42]. In addition, the equilibrium constants K_7 and K_8 are defined as [46]

$$K_7 = 4.5821699 \times 10^{24} \times e^{-\frac{2.9}{k_b T}}$$ (24)

$$K_8 = 1.041555 \times 10^{26} \times e^{-\frac{3.4}{k_b T}}$$ (25)

The developed model can predict the microdefect distributions in the presence of oxygen and nitrogen in CZ crystals growing under steady state as well as unsteady state.

CZ DEFECT DYNAMICS IN ONE-DIMENSIONAL CRYSTAL GROWTH

The important effects of nitrogen on the CZ defect dynamics are better demonstrated by the simulation of the growth of one-dimensional crystals showing no radial variation in the microdefect distributions. Further simplification can be achieved by simulating a steady state crystal growth. A crystal growing at a fixed rate through a fixed temperature field reaches a steady state far away from the regions formed in the beginning of the growth. For a simulation of crystal growth at steady state, the crystal is assumed to grow through a fixed representative temperature profile.

The microdefect distributions observed in CZ and FZ crystals in the absence of oxygen and nitrogen and ignoring the effects of the radial diffusion of the intrinsic point defects can be explained very well by Voronkov's theory [16]. According to this theory, an interplay between the recombination of the intrinsic point defects driven by the decrease in the temperature of a growing crystal near the melt/crystal interface and the transport of the intrinsic point defects from the interface by the convection and the diffusion determines the final microdefect distribution in the crystal. The intrinsic point defects are assumed to exist at equilibrium at the interface. Hence, the vacancy concentration is higher than the self-interstitial concentration at the interface. Self-interstitials, however, diffuse faster at higher temperatures. Hence, when the convection dominates the diffusion, the flux of vacancies from the interface into a fixed volume element near the interface is greater than the flux of self-interstitials, resulting in the dominance of vacancies away from the interface, while self-interstitials are annihilated to very low concentrations by the recombination; when the diffusion dominates the convection, self-interstitials are replenished at a higher rate than vacancies from the interface to compensate for their consumption by the recombination, resulting in the survival of self-interstitials as the dominant species. The dominant species nucleates at lower temperatures to form microdefects. When the total flux of vacancies is comparable to the total flux of self-interstitials, both species annihilate each other to lower concentrations, forming no microdefects. Voronkov (1982) quantified the relative effects of the convection and the diffusion by the ratio of the crystal pull-rate (V) to the magnitude of the axial temperature gradient in the crystal at the interface (G) or V/G [16]. The convection dominates at a higher V/G, the diffusion dominates at a lower V/G, and the flux of vacancies is equal to the flux of self-interstitials at the critical V/G.

Nitrogen affects the reaction dynamics near the interface by interacting with vacancies according to reaction (8). The vN species produced by this interaction provides free vacancies by dissociation for the annihilation of self-interstitials. As free vacancies are also available from the interface, which is an infinite source, the interaction of nitrogen with vacancies shifts the balance of this dynamics in favor of vacancies. The crystal becomes richer in vacancies in the presence of nitrogen. Oxygen also shifts the this balance in favor of vacancies by introducing vO and vO_2 species, which provide free vacancies by their dissociation, but its effect is weaker than that of nitrogen [41, 46].

The evolution of the concentrations of key reacting species, under the dominance of convection, as functions of the temperature along the length of a growing CZ crystal is shown in Figure 1. Near the interface, all species containing vacancies participate in the recombination with self-interstitials; the participation of free vacancies is direct, and the participation of vN, vO and vO_2 species is indirect through their dissociation. After the self-interstitial concentration drops significantly, the total vacancy concentration, $C_v + C_{vO} + C_{vO_2} + C_{vN}$, remains essentially constant until v-clusters are formed in appreciable numbers at lower temperatures. Free vacancies directly participate in the formation and growth of v-clusters; the bound vacancies, vO, vO_2, and vN species, participate indirectly by their dissociation. The strong binding of vacancies with nitrogen reduces the free vacancy concentration, before appreciable nucleation, compared with that in the absence of nitrogen, decreasing the vacancy nucleation temperature. The free vacancy consumption rate by v-clusters is low in the presence of lower vacancy concentration, facilitating their formation in appreciably higher densities. As the supply of total vacancies is fixed, the formation of v-clusters in higher densities reduces their size. The effect of nitrogen on the CZ defect dynamics is studied effectively by a comparison of the evolution of various reacting species in a growing CZ crystal in the presence of nitrogen (Figure 1) with that in the absence of nitrogen (Figure 2), under otherwise same conditions.

A distinct effect of nitrogen is the facilitation of the formation of O-clusters in the presence of voids in appreciable densities. The strong binding between vacancies and nitrogen maintains appreciably high concentration of bound and free vacancy species at lower temperatures where vacancy facilitated O-cluster formation and growth mark the CZ defect dynamics. After their formation, O-clusters become the dominant vacancy sinks compared to v-clusters, as shown in Figure 1. In the absence of nitrogen, however, vN species is also absent and both weakly bound (vO and vO_2) and free vacancies are consumed at higher temperatures primarily by v-clusters, suppressing the formation of O-clusters (Figure 2). In a nitrogen doped crystal exhibiting the moderate dominance of vacancies near the interface, the free vacancy concentration does not remain high enough to form v-clusters at higher temperatures; at lower temperatures, however, O-clusters are formed in the absence of v-clusters (Figure 3). It must be noted that the total vacancy concentration even in the presence of clusters remains negligible but finite in a CZ crystal because of the binding of vacancies with oxygen at lower temperatures.

The pull-rate marking the boundary between O-clusters and i-clusters, known as the v/i boundary, represents the critical operating condition. Therefore, the quantification this pull-rate, for otherwise fixed growth conditions, is important. For a given temperature field in a crystal-puller, this pull-rate can be quantified by a series of crystal growth simulations at various fixed pull-rates. These computations are numerous and can be computationally expensive. An approximate variation of the cluster distribution in a crystal as a function of its pull-rate can be captured by predicting this distribution in a crystal pulled by either continuously increasing or decreasing rate as shown in Figure 4. One such simulation provides the approximate correlation between the cluster distribution in a crystal and its pull-rate.

The crystal growth conditions marking the v/i boundary can also be quantified in terms of the popular but approximate growth parameter, V/G. The V/G marking this boundary, known as the critical V/G, continuously decreases with increasing nitrogen concentration (Figure 5). The total vacancy concentration near the interface increases with increasing nitrogen concentration because of the strong bonding between vacancies and nitrogen, resulting in a decrease in the critical V/G. This shift is particularly significant at higher nitrogen concentration.

The conditions that allow the growth of CZ crystals in the absence of large clusters are also of interest in the crystal growth industry. Therefore, the quantification of the range of the crystal pull-rate or V/G within which CZ crystals can grow free of large v-clusters and i-clusters is important. The large cluster free range is bracketed by the v/i boundary and the boundary between O-clusters and large v-clusters approximately 20 nm in radius, termed O/v boundary in this paper. It must be noted that the crystals free of such large clusters contain O-clusters, as discussed before. The size of O-clusters increases with increasing concentration of the free vacancies available during their formation, as shown in Figure 4. The V/G marking the O/v boundary increases with the nitrogen concentration first, reaches a maximum, and then decreases; the increase in this V/G at lower nitrogen concentration, in spite of the decrease in the V/G marking the v/i boundary, is caused by the strong bonding between v and N that reduces the free vacancy concentration near the vacancy nucleation temperature range. At moderately higher nitrogen concentrations, however, the shift in the V/G marking the v/i boundary is too strong because of a significant increase in the total vacancy concentration near the interface, forcing a decrease in the V/G defining the O/v boundary. The pull-rate and the V/G range allowing the large cluster-free crystal growth still increases until the nitrogen concentration is significantly high. In the presence of nitrogen in very high concentrations, the total vacancy concentration in the crystal near the interface dramatically increases such that the free vacancy concentration around the vacancy nucleation temperature increases. Hence, the pull-rate range (as well as V/G range) within which crystals free of large clusters can be grown also marginally decreases at very high nitrogen concentration, as shown in Figure 5. The predicted increase in the size of v-clusters with increasing nitrogen concentration does not agree with reported data [51]. Therefore, the model predictions are far less accurate at very high nitrogen doping levels.

The salient features of the CZ defect dynamics and the strengths and weaknesses of the solved model are thus adequately described by the discussed simulations of one-dimensional crystal growth.

CZ DEFECT DYNAMICS IN TWO-DIMENSIONAL CRYSTAL GROWTH

The observed microdefect distributions in a CZ crystal vary along the axis and the radius of the crystal. This radial variation is caused by the radial variation of the temperature field and the radial diffusion of the intrinsic point defects induced by the temperature field and the lateral surface of a growing crystal. Axial variations are primarily induced by the change in the pull-rate and, in the case of nitrogen doped crystals, by the axial variation in the nitrogen concentration. Quantification of these 2-dimensional microdefect distributions requires a solution of the discussed model for 2-dimensional crystal growth.

The microdefect distribution in a crystal, at the end of its growth, pulled at a fixed rate through a fixed temperature field is shown in Figure 6. It must be noted that the cooling rates of the crystal are very low. The oxygen concentration in the crystal is assumed to be 16 ppma. The nitrogen concentration in the cylindrical section of the crystal varies from 1.02×10^{14} cm^{-3} to 5.87×10^{14} cm^{-3}. The crystal growth at a fixed rate equal to 0.4 mm/min is assumed. The crystal grows under the dominance of convection. The central region of the crystal contains larger v-clusters and the annular region is dominated by O-clusters. This variation is caused by the radial variation in the temperature field; the relative effect of the convection with respect to the diffusion becomes weaker along the radial position of the crystal because of the radial variation of the temperature field. The vacancy-rich region, defined as the region marked by the presence of v-clusters and O-clusters at lower temperatures or the presence of vacancies in higher concentrations at higher temperatures, expands in the crystal as the nitrogen concentration increases. As the simulation is performed for a crystal at a high nitrogen doping level, larger voids appear at the end of crystal growth at higher nitrogen concentration, showing the limitations of the model. The region free of v-clusters near the interface mark the vacancy nucleation front; the region free of O-clusters near the interface marks the O-cluster formation front.

The capability of the developed model can be demonstrated by predicting the microdefect distribution in a crystal pulled at a varying rate shown in Figure 7. The oxygen and nitrogen concentrations in the crystal were set at the same level as those set for the simulation of the crystal discussed in Figure 6. Hence, the effect of varying pull-rate can be observed by a comparison between the cluster distributions shown in Figure 6 and those shown in Figure 7. The varying pull-rate generates a distribution of all discussed clusters in this paper. As i-clusters and v-clusters do not coexist, the simultaneous distributions of these clusters can be quantified by the distributions of $\Re_v - \Re_i$ and $n_v - n_i$. The positive numbers indicate the distribution of v-clusters and the negative numbers indicate the distribution of i-clusters. Considering the difference in the densities, however, the density distributions of i-clusters and v-clusters are shown separately in Figure 7. At a higher pull-rate, v-clusters and O-clusters are formed; at a lower pull-rate i-clusters are formed; at a pull-rate marking the moderate dominance of vacancies closer to the v/i boundary, only large O-clusters are formed.

The discussed model captures the salient features of the typical 2-dimensional microdefect distributions in the presence of nitrogen observed in CZ crystals. At higher nitrogen concentration, the model predictions are less accurate.

CONCLUSIONS

Silicon single crystals grown by the Czochralski (CZ) process typically contain microdefects formed by the agglomeration of the intrinsic point defects of silicon, vacancies and self-interstitials, and of oxygen, primarily silicon dioxide. The microdefect distribution is strongly influenced by impurities such as nitrogen. A model is adopted to quantify the distribution of these microdefects in CZ crystals and solved.

A reasonably good approximation of the dynamics of the microdefect formation in growing CZ crystals requires the quantification of Frenkel reaction involving the recombination of vacancies and self-interstitials, the interactions between oxygen and vacancies, between nitrogen and vacancies, along with the agglomeration events. Vacancies and self-interstitials annihilated by the recombination near the interface are partly supplied by different reacting species and their transport from the interface. Nitrogen binding with vacancies is much stronger than that of oxygen. The complexes of vacancies with nitrogen in particular, and oxygen to some extent, provide free vacancies by dissociation for the annihilation of self-interstitials near the melt/crystal interface. These free vacancies are available in addition to the vacancies from the interface, which is an infinite source. This dynamics increases the total concentration of vacancies in all forms, defined as the total vacancy concentration, and establishes relatively vacancy-rich conditions compared to those in the absence of nitrogen. In addition, nitrogen reduces the free vacancy concentration at lower temperatures reducing their nucleation temperature. The formation of vacancy agglomerates in the presence of lower vacancy concentration results in an increase in their density at the expense of their size. The strong binding between vacancies and nitrogen provides appreciable supply of vacancies for the facilitation of the formation and growth of the agglomerates of oxygen with silicon at lower temperatures. Hence, vacancy agglomerates and oxygen agglomerates coexist in a wide range of crystal growth conditions in the presence of nitrogen.

The adapted model approximates all agglomerates as spherical clusters. Classical nucleation theory captures the formation of all clusters with a reasonable accuracy. The formed clusters grow by a diffusion-limited kinetics. Vacancy clusters and self-interstitial clusters are formed by the homogeneous nucleation of vacancies and self-interstitials, respectively. Oxygen clusters, because of their higher specific volume, are formed by the facilitation of vacancies. The growth of oxygen clusters is limited by the diffusion of oxygen when vacancies are in abundance and by the diffusion of vacancies when vacancies are scarce. The population of clusters at a given location in a growing crystal is approximated by an equivalent population of identical microdefects.

The model quantifies the conditions leading to the formation of different microdefects in CZ crystal growth. The effects of varying pull-rate and the nitrogen concentration are also captured. The model quantifies the microdefect distributions in CZ crystals growing under both steady states as well as unsteady states. The model predictions agree well with the reported microdefect distributions in the presence of moderate and high nitrogen concentrations. In the presence of very high nitrogen concentrations, however, the model predictions are less accurate.

REFERENCES

1. T. Abe, T. Samizo, and S. Maruyama, *Jpn. J. Appl. Phys.*, **5**, 458 (1966).
2. A. J. R. de Kock, *Appl. Phys. Letters*, **16**, 100 (1970).
3. A. J. R. de Kock, *J. Electrochem. Soc.*, **118**, 1851 (1971).
4. A. J. R. de Kock, *ACTA ELECTRONICA*, **16**, 4, 303 (1973).
5. A. J. R. de Kock, P. J. Roksnoer and P. G. T. Boonen, *J. Cryst. Growth*, **22**, 311 (1974).
6. A. J. R. de Kock, P. J. Roksnoer and P. G. T. Boonen, *J. Cryst. Growth*, **28**, 311 (1975).
7. P. M. Petroff and A. J. R. de Kock, *J. Cryst. Growth*, **30**, 117 (1975).
8. P. M. Petroff and A. J. R. de Kock, *J. Cryst. Growth*, **36**, 4 (1976).
9. H. Föll, U. Gösele and B. O. Kolbesen, *J. Cryst. Growth*, **40**, 90 (1977).
10. A. J. R. de Kock and W. M. van de Wiljert *J. Cryst. Growth*, **49**, 718 (1980).
11. P. J. Roksnoer and M. M. B. van den Boom, *J. Cryst. Growth*, **53**, 563 (1981).
12. P. J. Roksnoer, *J. Cryst. Growth*, **68**, 596 (1984).
13. M. Kato, T. Yoshida, Y. Ikeda, and Y. Kitagawara, *Jpn. J. Appl. Phys.*, **35**, 5597 (1996).
14. T. Ueki, M. Itsumi, and T. Takeda, *Appl. Phys. Lett.*, **70**, 1248 (1997).
15. M. Hasebe, Y.Takeoka, S. Shinoyama, and S. Naito, *Jpn. J. Appl. Phys.*, **28**, L 1999 (1989).
16. V. V. Voronkov, *J. Crystal Growth*, **59**, 625 (1982).
17. S. Sadamitsu, S. Umeno, Y. Koike, M. Hourai, S. Sumita, and T. Shigematsu, *Jpn. J. Appl. Phys.*, **32**, 3675 (1993).
18. R. A. Brown, D. Maroudas, and T. Sinno, *J. Cryst. Growth*, **137**, 12 (1994).
19. W. von Ammon, E. Dornberger, H. Oelkrug, H. Weidner, *J. Crystal Growth*, **151**, 273 (1995).
20. E. Dornberger and W. von Ammon, *J. Electrochem. Soc.*, **143**, 1648 (1996).
21. E. Dornberger, W. von Ammon, N. Van den Bogaert, and F. Dupret, *J. Cryst. Growth*, **166**, 452 (1996).
22. K. Nakamura, T. Saishoji, T. Kubota, T. Iida, Y. Shimanuki, T. Kotooka, and J. Tomioka, *J. Cryst. Growth*, **180**, 61 (1997).
23. R. Falster, V. V. Voronkov, J. C. Holzer, S. Markgraf, S. McQuaid, and L. Muléstagno, *Electrochemical Society Proceedings*, **98-1**, 468 (1998).
24. T. Sinno, R. A. Brown, W. von Ammon, and E. Dornberger, *J. Electrochem. Soc.*, **145**, 302 (1998).
25. T. Sinno and R. A. Brown, J. Electrochem. Soc.,**146**, 2300 (1999).
26. V. V. Voronkov and R. Falster, *J. Appl. Phys.*, **86**, 5975 (1999a).
27. E. Dornberger, W. von Ammon, J. Virbulis, B. Hanna, and T. Sinno, *J. Cryst. Growth*, **230**, 291 (2001).
28. R. A. Brown, Z. Wang, and T. Mori, *J. Cryst. Growth*, **225**, 97 (2001).
29. M. Okui, and M. Nishimoto, *J. Cryst. Growth*, **237-239**, 1651 (2002).
30. V. V. Kalaev, D. P. Lukanin, V. A. Zabelin, Yu. N. Makarov, J. Virbulis, E. Dornberger, and W. von Ammon *J. Cryst. Growth*, **250**, 203 (2003).
31. M. S. Kulkarni, V. V. Voronkov, and R. Falster, *J. Electrochem. Soc.*, **151**, G663 (2004).
32. M. S. Kulkarni and V. V. Voronkov, *J. Electrochem. Soc.*, **152**, G781 (2005).
33. M. S. Kulkarni, *Ind. Eng. Chem. Res.*, **44**, 6246 (2005).
34. M. S. Kulkarni, J. C. Holzer, and L. W. Ferry, *J. Cryst. Growth*, **284**, 353 (2005).

35. J. Esfandyari, C. Schmeiser, S. Senkader, G. Hobler, and B. Murphy, *J. Electrochem. Soc.* **143**, 1995 (1996).
36. V. V. Voronkov and R. Falster, *J. Cryst. Growth*, **194**, 76 (1998).
37. V. V. Voronkov and R. Falster, *J. Cryst. Growth*, **204**, 462 (1999b).
38. Z. Wang and R. Brown, *J. Cryst. Growth*, **231**, 442 (2001).
39. V. V. Voronkov, R. Falster, *J. Appl. Phys.* **91**, 5802 (2002).
40. K. Sueoka, M. Akatsuka, M. Okui, and H. Katahama, *J. Electrochem. Soc.*, **150**, G469 (2003).
41. V. V. Voronkov and R. Falster, *J. Electrochem. Soc.*, **149**, G167 (2002).
42. M. S. Kulkarni, V. V. Voronkov, and R. Falster, in *Silicon Materials Science and Technology X*, H. Huff, Editors, This PV, The Electrochemical Society Proceedings Series, Pennington, NJ (2006).
43. M. Iida, W. Kusaki, M. Tamatsuka, E. Iino, M. Kimura, S. Muraoka, in *Defects in Silicon III*, T. Abe, W.M. Bullis, S. Kobayashi, W. Lin, and P. Wagner, Editors, PV 99-1, p. 499, The Electrochemical Society Proceedings Series, Pennington, NJ (1999).
44. K. Nakai, Y. Inoue, H. Yokota, A. Ikari, J. Takahashi, K.Kitahara, Y. Ohta, W. Ohashi, *J. Appl. Phys.* **80**, 4301 (2001).
45. W. von Ammon, R. Holzl, J. Virbulis, E. Dornberger, R. Schmolke, and D. Graf, *J. Crystal Growth*, **226**, 19 (2001).
46. V. V. Voronkov and R. Falster, *J. Cryst. Growth*, **273**, 412 (2005).
47. D. Turnbull and J. C. Fisher, *J. Chem. Phys.*, **17**, 71 (1949).
48. A. S. Michaels, *Nucleation Phenomena*, American Chemical Society, 1966.
49. D. Kashchiev, *Nucleation, Basic Theory with Applications*, Butterworth- Heinemann, 2000.
50. A.Virzi, *J. Crys. Growth*, **112**, 699 (1991).
51. J. Takahashi, K. Nakai, K. Kawakami, Y. Inoue, H. Yokota, A. Toshikawa, A. Ikari, and W. Ohashi, *Jpn. J. Appl. Phys.* **42**, 363 (2003).

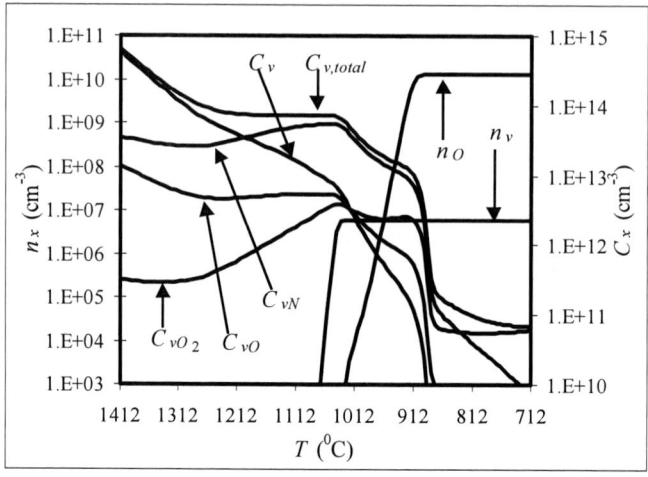

Figure 1. Evolution of the concentrations of various reacting species in a nitrogen doped CZ crystal growing at a high rate (G = 3.9 K/mm, V = 0.8 mm/min, $C_{N,total}$ = 5 × 10^{14} cm^{-3}, C_O = 7 × 10^{17} or 14 ppma)

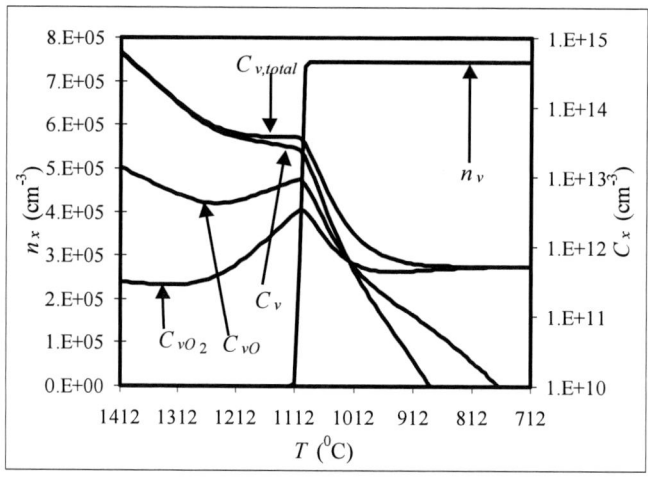

Figure 2. Evolution of the concentrations of various reacting species in a CZ crystal growing at a high rate (G = 3.9 K/mm, V = 0.8 mm/min, $C_{N,total}$ = 0 cm^{-3}, C_O = 7 × 10^{17} or 14 ppma)

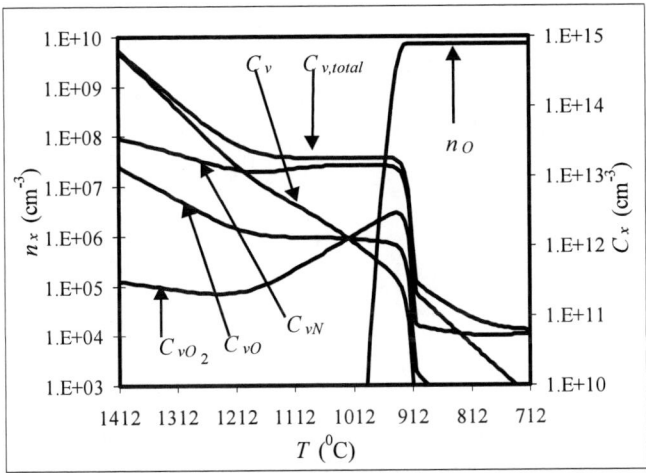

Figure 3. Evolution of the concentrations of various reacting species in a nitrogen doped CZ crystal growing close to the critical condition ($G = 3.9$ K/mm, $V = 0.45$ mm/min, $C_{N,total} = 5 \times 10^{14}$ cm^{-3}, $C_O = 7 \times 10^{17}$ or 14 ppma)

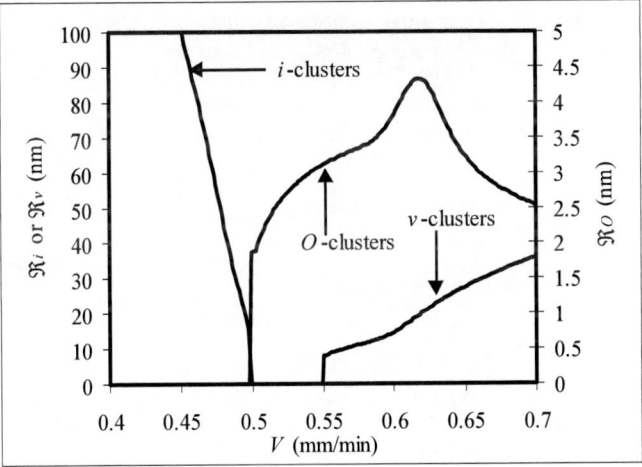

Figure 4. The simulated sizes of various clusters as the functions of the CZ crystal pull-rate ($G = 3.9$ K/mm, $C_{N,total} = 1 \times 10^{14}$ cm^{-3}, $C_O = 7 \times 10^{17}$ or 14 ppma)

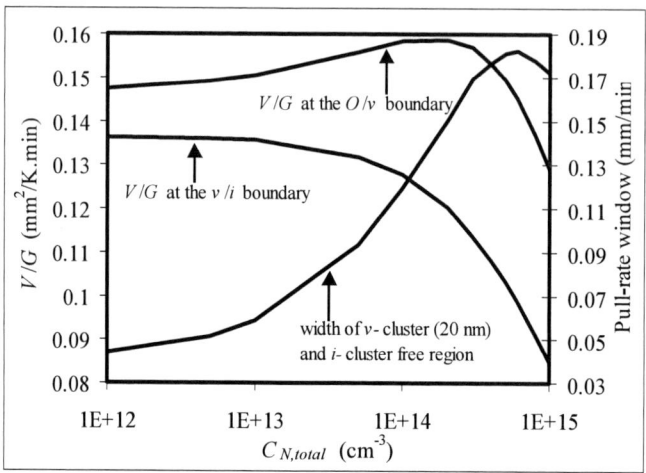

Figure 5. The sensitivity of the cluster distributions near the critical conditions to the nitrogen concentration. Note: The v-cluster and i-cluster free region includes v-clusters smaller than 20 nm ($G = 3.9$ K/mm, $C_O = 7 \times 10^{17}$ or 14 ppma)

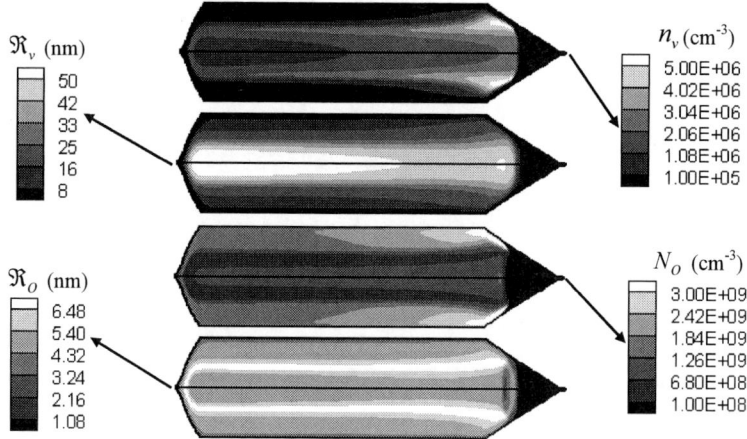

Figure 6. The model predicted cluster distributions in a CZ crystal at the end of its growth Note: $C_{N,total}$ varies from 1.02×10^{14} cm^{-3} to 5.84×10^{14} cm^{-3} in the cylindrical body of the crystal; $C_O = 16$ ppma. The crystal cooling rates are very low.

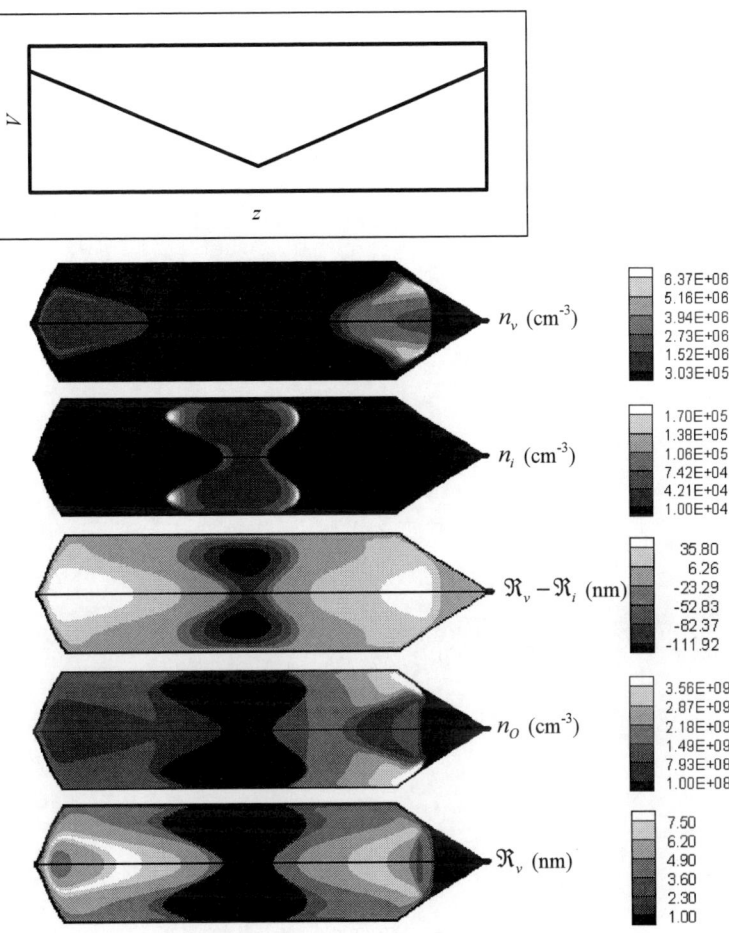

Figure 7. The predicted cluster distributions in a simulated crystal at the end of its growth under transient conditions. Note: $C_{N,total}$ varies from 1.02×10^{14} cm^{-3} to 5.84×10^{14} cm^{-3} in the cylindrical body of the crystal; $C_O = 16$ ppma. The crystal cooling rates are very low. The legend for each plot is on its right hand side.

ECS Transactions, 2 (2) 247-260 (2006)
10.1149/1.2195663, copyright The Electrochemical Society

Analytical Modeling of the Interaction of Vacancies and Oxygen for Oxide
Precipitation in RTA Treated Silicon Wafers

G. Kissinger[a,c], J. Dabrowski[a], A. Sattler[b], C. Seuring[b], T. Müller[b], H. Richter[a,c],
and W. von Ammon[b]

[a] IHP, Im Technologiepark 25, 15236 Frankfurt (Oder), Germany
[b] Siltronic AG, Johannes-Hess-Strasse 24, 84489 Burghausen, Germany
[c] IHP/BTU Joint Lab, Universitätsplatz 3-4, 03044 Cottbus, Germany

We have investigated the impact of RTA induced vacancy supersaturation on oxide precipitation based on as much as possible experimental and theoretical values. Oxygen precipitation after RTA
processing was found to be controlled by the initial concentration
of interstitial oxygen in a sixth power dependency and frozen vacancies just in a cubic dependency. The formation of tensile
strained $n\mathrm{VO_2}$ clusters seems to be the favored process for coherent nucleation of oxide precipitates. The reduction of interstitial
oxygen can be accurately modeled for the temperature range from
1150 °C to 1250 °C using Ham's theory for precipitate growth and
an empirical relation based on nucleation of oxide precipitates by
agglomeration of $\mathrm{VO_2}$ complexes. During RTA treatments at temperatures ≥ 1300 °C vacancies seem to be consumed by other processes. Below RTA temperatures of 1150 °C, oxide precipitation is
dominated by shrunken as-grown precipitate nuclei because as-
grown nuclei can be dissolved only at RTA temperatures
≥ 1150 °C.

1. Introduction

The silicon oxide precipitation in Czochralski (CZ) silicon is strongly related to
intrinsic point defects (1, 2). Supersaturation of vacancies is known to enhance oxide precipitation while it is retarded by interstitial supersaturation. This is due to the fact that the
compressive strain which results from the volume difference between silicon and the pre-
cipitating silicon oxide can be reduced either by the absorption of vacancies or the emission of interstitials.

RTA is a method that is used to generate vacancy profiles in silicon wafers which
serve for creation of defect denuded zones near the surface and high densities of bulk microdefects (BMDs) (3). It is believed that the nucleation rate is controlled by the concentration of vacancies and is thus independent of the oxygen content (3). The vacancy supersaturation due to RTA is assumed to be stored in the form of $\mathrm{VO_2}$ vacancy oxygen
complexes (4). The agglomeration of the $\mathrm{VO_2}$ species is regarded as possible path for nucleation also which would need shorter incubation periods than usual oxygen agglomeration at lower vacancy concentration (3, 4).

In Ref. 5, more experimental results about the impact of RTA on oxygen precipitation
were presented than in Ref. 3. A correlation between the excess vacancy concentration
and the density of bulk microdefects (BMDs) has shown that the BMD densities scatter
over a wide range for the same vacancy concentration.

247

The aim of this work was to investigate the impact of RTA induced vacancy super-saturation on oxide precipitation based on a broad set of experimental values obtained from variations of RTA temperature, cooling rate, vacancy concentration, interstitial oxygen concentration, and crystal pulling conditions in order to gain a most comprehensive overview on the various parameter dependencies of the oxygen precipitation. Based on these correlations, an analytical model was developed which can describe the reduction of interstitial oxygen during a process consisting of an RTA treatment and a so-called BMD test comprising a two-step anneal at 780 °C for 3 h and at 1000 °C for 16 h. In addition, an atomistic model was developed that describes a coherent nucleation process for oxide precipitates in agreement with the results of the analytical model.

2. Experimental

In order to obtain a correlation which is as much general as possible, wafers of various initial concentrations of interstitial oxygen C_0 ranging from 4.5×10^{17} cm^{-3} to 8×10^{17} cm^{-3}, wafer diameters of 150 mm, 200 mm, and 300 mm, obtained from crystals grown in various vacancy-rich growth modes and growth modes resulting in extremely tiny agglomerates of intrinsic point defects, were used in the experiments. All wafers were moderately B-doped with resistivities between 2 and 20 Ωcm. The initial concentration of substitutional carbon was always below 10^{16} cm^{-3}.

These wafers were first heat treated in a rapid thermal anneal (RTA) at various temperatures (1000, 1050, 1100, 1150, 1175, 1200, 1225, 1230, 1250, 1300, 1325, or 1350 °C) for 10 or 30 s in argon or mixtures of nitrogen and oxygen. The cooling rates used were 5, 15, 30, 50, 60, 70, 75, 90, or 100 K/s. After the RTA treatment all wafers received the same two step anneal consisting of 780 °C for 3 h followed by 1000 °C for 16 h in oxygen. During the RTA step vacancy profiles are created with a high concentration of vacancies in the bulk and a vacancy denuded zone near the surface. The concentration in the bulk depends first of all on temperature and duration of the RTA treatment whereas the vacancy denuded zone can be influenced by the cooling rate. The thermal treatment at 780 °C comprises a so-called nucleation step which serves to stabilize the precipitate nuclei. Then these nuclei are grown to a detectable size by annealing at 1000 °C for 16 h.

The oxygen concentration was measured by Fourier Transform Infrared Spectroscopy (FTIR) using a conversion factor of 2.45×10^{17} cm^{-2} (new ASTM, DIN 50438/1). The average density of bulk microdefects (BMD) was determined after etching the (110) cleavage plane perpendicular to the surface using Wright etchant (6) or Secco etchant (7). At least 3 measurement points per wafer depending on the diameter were analyzed before and after all heat treatments. In total, 187 data sets containing initial oxygen concentration C_0, oxygen concentration after the heat treatments $C(t)$, and BMD density N_{BMD} were included into the correlation analysis.

3. Simulated Input Data

In addition to the measured data sets, simulated values were generated which are based on commonly accepted theories. In this way, important intrinsic point defects and their complexes with oxygen became part of the correlation analysis also. In order to obtain the average concentrations of vacancies C_V, VO complexes C_{VO}, and VO$_2$ complexes C_{VO2} in the wafers, the vacancy and interstitial profiles were numerically simulated for each RTA process used. The numerical simulation was based on point defect diffusion and the Frenkel-pair mechanism for generation and recombination of vacancies and interstitials. The following basic equations were used.

$$\frac{\partial}{\partial t} C_I = D_I \frac{\partial^2}{\partial x^2} C_I - k_{IV} \cdot (C_I \cdot C_V - C_{Ieq} \cdot C_{Veq}) \qquad [1]$$

$$\frac{\partial}{\partial t} C_V = D_V \frac{\partial^2}{\partial x^2} C_V - k_{IV} \cdot (C_I \cdot C_V - C_{Ieq} \cdot C_{Veq}) \qquad [2]$$

Here, D_I and D_V are the diffusion coefficients of interstitials and vacancies, respectively, as obtained from Ref. 8. C_I is the interstitial concentration, t is the time, and x stands for the depth from the wafer surface. The equilibrium concentrations of interstitials C_{Ieq} and vacancies C_{Veq} were obtained from Ref. 8. The second term of Eqs. 1 and 2 describes the Frenkel-pair mechanism with k_{IV} being the reaction constant which is obtained by Eq. 3 according to Ref. 9

$$k_{IV} = 4 \cdot \pi \cdot r_C / (\Omega \cdot C_{at}) \cdot (D_I + D_V) \cdot \exp\left[(-\Delta G_{IV} / (k \cdot T)\right] \qquad [3]$$

Here, T and k are the temperature and Boltzman's constant, respectively. The atomic density of silicon C_{at}, amounts to 5×10^{22} cm^{-3}, the capture radius r_C is 1×10^{-7} cm, and the volume of the silicon unit cell Ω equals to one eighth of the lattice constant of silicon. The free energy barrier for recombination is a controversial parameter which varies between 0 and 1.5 eV (9). In our calculations of the intrinsic point defect profiles, the energy barrier was set to zero.

The simulation of the point defect profiles of the RTA treatments was frozen at 1000°C where VO2 becomes the dominating vacancy containing species and diffusion of intrinsic point defects decreases markedly (4). Based on the average concentration of vacancies in the wafer the concentrations of the VO and VO2 complexes were determined using the relations given in Eqs. 4 and 5 obtained from Refs. 4 and 10.

$$C_{VO2} = (C_V / C_{Veq}) \cdot (C_0^2 / C_{at}) \cdot \exp\left[- 0.39\text{eV} / (k \cdot T)\right] \qquad [4]$$

$$C_{VO} = (C_V \cdot C_{VO2})^{(1/2)} \qquad [5]$$

For our correlation analysis, only average values of vacancies and complexes are needed because the oxygen concentrations measured by FTIR are also obtained in transmission from the whole thickness of the wafers. However, the calculation of the profiles is required because the cooling rate after RTA strongly impacts the average concentration via the out-diffusion of interstitials and vacancies during cooling. In summary, this simulation adds the parameters C_V, C_{VO}, and C_{VO2} to the data sets.

Ham's theory (11) for diffusion limited growth of particles is commonly accepted for describing the growth of oxide precipitates. It is based on the initial assumptions that the concentration of interstitial oxygen C_0 is uniform, the concentration of precipitates N remains constant during the whole period of heat treatment, and that the precipitates are of spherical shape with the radius r which is small compared to the distance between adjacent particles. The particles are distributed uniformly. Each of the particles is surrounded by a spherical region of equal volume, the equivalent sphere of a cubic cell, of radius r_e. The concentration of oxygen at the interface between matrix and precipitate is equal to the solubility of oxygen in silicon C_S at the annealing temperature. The particles consist of silicon oxide in which the oxygen concentration is C_P. Eqs. 6 and 7 are the basic equations in Ham's theory which hold until about 50% of the supersaturated oxygen is precipitated.

$$\frac{C(t) - C_S}{C_0 - C_S} = \exp\left(-\beta \cdot t^{3/2}\right) \qquad [6]$$

$$\beta = (2 \cdot D_{OX} / r_e^2)^{3/2} \cdot [(C_0 - C_S) / (C_P - C_S)]^{1/2} \qquad [7]$$

$$(4/3) \cdot \pi \cdot r_e^3 \cdot N = 1 \qquad [8]$$

Here, D_{OX} is the diffusivity of oxygen. C_P being 2.90×10^{22} cm^{-3} according to Ref. 12 is several orders of magnitude larger than C_S. $C(t)$ is the concentration of interstitial oxygen which is measured after the thermal treatment. Then, Eqs. 6 and 7 can be summarized into Eq. 9.

$$\frac{C(t) - C_S}{C_0 - C_S} = \exp\left\{-(2 \cdot D_{OX} \cdot t / r_e^2)^{3/2} \cdot [(C_0 - C_S) / C_P]^{1/2}\right\} \qquad [9]$$

Both, the temperature dependent diffusivity D_{OX} and solubility C_S of interstitial oxygen were obtained from Ref. 13. Thus, r_e remains the only unknown variable. This offers the possibility to calculate the density of growing oxide precipitates N_{calc} from the measured difference of interstitial oxygen before and after the heat treatments according to Eq. 10 which was obtained from Eqs. 8 and 9. The advantage of such a theoretically obtained value would be that it is independent of detection limits, which can affect measurement values of BMD density (14).

$$N_{calc} = 3 \cdot \left\{-\ln\left[(C(t) - C_S) / (C_0 - C_S)\right]\right\} / \left\{4 \cdot \pi \cdot (2 \cdot D_{OX} \cdot t)^{3/2} \cdot [(C_0 - C_S) / C_P]^{1/2}\right\}$$

$$[10]$$

In Fig. 1, the calculated oxide precipitate density N_{calc} is compared to the BMD density N_{BMD} measured by cleave and etch. It can be assumed that nearly all BMDs are oxide precipitates. Part of the measured values correlate quite well with the calculated values but some measured densities are much lower than the calculated densities. The latter case was only observed on a few wafers with lower C_0. It is assumed that due to the lower growth rate in silicon with low oxygen concentration some of the precipitates remained below the detection limit. In summary, applying Ham's theory an additional value N_{calc} is added to the data sets for the correlation analysis.

4. Results and Discussion

At first, the precipitated oxygen was correlated to the frozen-in vacancy concentration after RTA which was obtained from the simulated profiles in order to check if the correlation between oxygen precipitation and vacancy concentration holds also for the large variety of experimental parameters. The results in Fig. 2 demonstrate that the percentage of supersaturated interstitial oxygen that is precipitated varies over a wide range. Generally, higher vacancy concentrations can result in higher oxide precipitation but in many cases they do not. For high vacancy concentrations, the variation ranges from 15-70% oxygen precipitated which implies that the vacancy concentration alone does not control the oxygen precipitation.

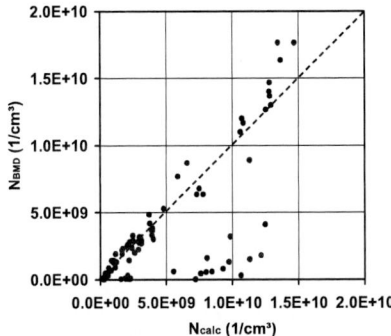

FIGURE 1. Comparison between BMD density measured by cleave and etch after the full thermal cycle consisting of RTA + 780 °C 3 h + 1000 °C 16 h and oxide precipitate density calculated according to Ham's theory.

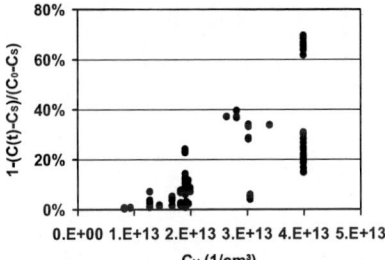

FIGURE 2. Precipitated oxygen determined after the full thermal cycle consisting of RTA + 780 °C 3 h + 1000 °C 16 h shown for RTA temperatures between 1150 °C and 1250 °C as a function of the vacancy concentration C_v frozen during the RTA treatments.

It was further checked if the oxygen concentration impacts the oxygen precipitation. This is shown in Fig. 3. The result looks quite similar to the correlation with the frozen vacancy concentration. Low oxygen concentration results in low precipitation and high oxygen concentration exhibits a variation from 1-70% oxygen precipitated. Thus, precipitation is not controlled by oxygen alone.

A detailed investigation of the variations at high vacancy concentration and high oxygen concentration has elucidated that a high vacancy concentration can only enhance oxide precipitation if the oxygen concentration is also high.

A correlation of oxygen precipitation with the concentration of VO complexes was not found either and values are scattering in the same manner as observed for correlations of oxygen and vacancy concentrations with precipitated oxygen.

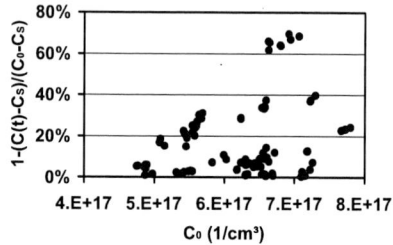

FIGURE 3. Precipitated oxygen determined after the full thermal cycle consisting of RTA + 780 °C 3 h + 1000 °C 16 h shown for RTA temperatures between 1150 °C and 1250 °C as a function of the initial concentration of interstitial oxygen C_0.

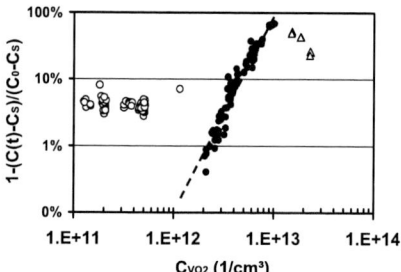

FIGURE 4. Precipitated oxygen determined after the full thermal cycle consisting of RTA + 780 °C 3 h + 1000 °C 16 h shown as a function of the concentration of VO_2 complexes for RTA treatments at 1000, 1050, and 1100 °C (empty circles), 1150, 1175, 1200, 1225, 1230, and 1250 °C (full circles), and 1300, 1325, and 1350 °C (empty triangles).

The best correlation of the precipitated supersaturated oxygen was obtained with the concentration of VO_2 complexes as shown in Fig. 4. However, this holds only for RTA temperatures from 1150-1250°C. Here the precipitated oxygen is proportional to the third power of the concentration of VO_2 complexes. The exact correlation obtained is

$$[(C(t) - C_S) / (C_0 - C_S)] = 1 - (k_R \cdot C_{VO2}^m)$$ [11]

with m = 2.99, $k_R = 1.21 \times 10^{-39}$ cm^{3m}, and a squared correlation coefficient according to Pearson of $R^2 = 0.943$. With respect to Eq. 4, it can be said that the precipitated oxygen is proportional to the sixth power of oxygen concentration and the cube of the frozen vacancy concentration. Thus speaking in terms of interstitial oxygen and vacancies, both species control the oxygen precipitation by VO_2 formation and the impact of interstitial oxygen is much stronger than the impact of vacancies. VO_2 formation is a possibility of vacancy conservation (3). Free vacancies would be at equilibrium concentration in the whole wafer already after annealing for about 30 min at 780°C. The total concentration of VO_2 complexes, which strongly impact oxide precipitation, is five orders of magnitude lower than the total concentration of interstitial oxygen. Thus, the question arises what the enhancing mechanism of VO_2 could be. A self-evident mechanism would be that the VO_2 complexes serve as nucleation centers for precipitation. Agglomeration of VO_2 would provide a form of coherent nucleation. Consequently, $Si_2V_2O_4$ would be the smallest coherent oxide precipitate nucleus which is formed by two VO_2 complexes but four VO_2 complexes are needed in order to form the smallest SiO_2 where one Si atom is bound to oxygen atoms only. As long as the agglomeration proceeds by VO_2 attachment only, the nuclei could remain coherent. The simultaneous attachment of a VO complex and an interstitial oxygen would be of the same effect.

In order to obtain the correlation between the density of oxide precipitate nuclei and C_{VO2}, the precipitate nuclei density calculated using Ham's theory N_{calc} was plotted as a function of C_{VO2} as seen in Fig. 5. The following correlation was obtained:

$$N_{calc} = a \cdot C_{VO2}{}^3 \qquad [12]$$

with the constant factor $a = 3.81 \times 10^{-29}$ cm^6 and a squared correlation coefficient according to Pearson of $R^2 = 0.963$.

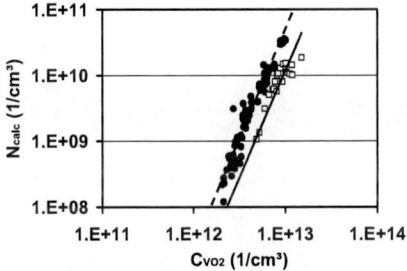

FIGURE 5 Oxide precipitate density calculated according to Ham's theory after the full thermal cycle consisting of RTA + 780 °C 3 h + 1000 °C 16 h shown for RTA temperatures between 1150 °C and 1250 °C as a function of the concentration of VO_2 complexes (full circles, dashed line). For comparison, oxide precipitate density versus vacancy concentration measured by Pt diffusion as obtained from Refs. 3 and 10 is included into the graph (empty squares, full line). It is assumed here that the vacancy concentration measured by Pt diffusion is actually the concentration of VO_2 complexes.

As shown in Fig. 5, the results of our correlation analysis are in a very good agreement with the relation between precipitate density (after RTA + 800 °C 4 h + 1000 °C 16 h) and vacancy concentration measured by platinum diffusion after RTA as published in Refs. 3 and 10. We assume here as proposed in Ref. 3 that the vacancy concentration measured by Pt diffusion is actually the concentration of VO_2 complexes because the vacancies are stored in VO_2 complexes below 1000 °C. The Pt diffusion was retarded in CZ silicon compared to FZ silicon (3). The good agreement between the results compared in Fig. 5 demonstrates again that the nucleation of oxide precipitate nuclei is not independent of the oxygen concentration because oxygen is a crucial component in the storage mechanism of vacancies and in the formation of VO_2 nucleation sites. The power of 3 indicates that the formation of $3VO_2$ clusters is critical for the nucleation of oxide precipitates under the thermal treatments applied subsequent to the RTA treatments.

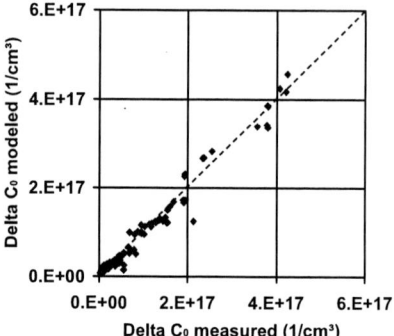

FIGURE 6. Modeled decrease of interstitial oxygen versus the measured decrease of interstitial oxygen both related to the full thermal cycle consisting of RTA + 780 °C 3 h + 1000 °C 16 h (RTA at 1150-1250 °C).

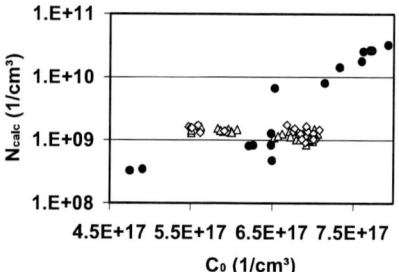

FIGURE 7. Oxide precipitate density calculated according to Ham's theory after the full thermal cycle consisting of RTA + 780 °C 3 h + 1000 °C 16 h shown as a function of the initial concentration of interstitial oxygen C_0 for RTA temperatures of 1000 °C (empty diamonds), 1100 °C (empty triangles) and for a thermal process without RTA (full circles).

With Eq. 12 and the equations of Ham's theory Eqs. 6-8, an analytical model is obtained which allows determining the reduction of interstitial oxygen after a process consisting of RTA and BMD test. The input parameters for modeling the reduction of oxygen are the RTA conditions and C_0. The RTA conditions are needed to numerically calculate the average concentration of VO_2 complexes which is needed to calculate the precipitate nuclei density for use in Ham's equations. This model is valid for RTA temperatures between 1150 °C and 1250 °C. Fig. 6 provides an accuracy check comparing the modeled ΔC_0 with the measured ΔC_0. The squared correlation coefficient according to Pearson amounts to 0.972 confirming a high accuracy of the model. Eq. 11 can also be used for modeling purposes.

Coming back to Fig. 4, all the correlations discussed above hold for RTA temperatures between 1150 °C and 1250 °C only. For higher RTA temperatures ($\geq$ 1300 °C), the precipitated oxygen decreases with increasing concentration of VO_2 complexes. In the framework of this model, this can be explained only by missing vacancies for VO_2 formation. A part of the vacancies seems to be consumed more and more by other processes as the RTA temperature exceeds 1300 °C.

For RTA temperatures lower than 1150 °C, the oxygen precipitation becomes independent of the concentration of frozen VO_2 complexes as shown in Fig. 4. It is assumed that a temperature of at least 1150 °C is needed in order to dissolve the grown-in oxide precipitate nuclei because they still seem to dominate oxygen precipitation after RTA below 1150 °C. The nucleation due to the quite low concentration of frozen VO_2 at RTA temperatures $\leq$ 1150 °C is just weak and is by far not exceeding the density of grown-in nuclei. RTAs at low temperatures obviously just shrink the as-grown nuclei what would lead to more homogeneous precipitate densities for a given oxygen concentration because size differences of the grown-in nuclei are reduced. This homogenization becomes obvious in Fig. 7 which compares the oxide precipitate density of low temperature RTA treated wafers with wafers which have not received an RTA prior to the BMD test. While the precipitate density of the wafers without an RTA is scattering and dependent on the interstitial oxygen, the precipitate density of the wafers with an RTA treatment at 1000 °C or 1050 °C remains at a constant level of about 10^9 cm^{-3} for a wide range of initial concentration of interstitial oxygen.

Already during crystal pulling, the same processes lead to VO_2 formation and nucleation of course. In the bulk crystal, out-diffusion of intrinsic point defects plays a minor role and is important only in the edge region of the crystal. Therefore, at the much lower cooling rate during crystal pulling supersaturated vacancies can be frozen as VO_2 complexes and nuclei can be formed already during cooling of the crystal. The nuclei density would be influenced first of all by the pulling conditions and the resulting supersaturation of intrinsic point defects. Because the grown-in nuclei are dissolved at RTA temperatures $\geq$ 1150 °C, it would be possible to reinstall a precipitate nuclei density at a lower or higher level than the grown-in nuclei density. This can be managed by adjusting the RTA conditions for VO_2 formation.

5. Atomistic Nucleation Model

In order to gain further insight into the details of the process of nucleation and growth of SiO_2 precipitates under vacancy supersaturation conditions, we performed ab initio calculations for total energies and atomic structures of point defects (O_i, VO_2, cf. Fig. 8a), point defect clusters (nVO_2 for n=2, 3, 4, cf. Fig. 8b), and bulk models of SiO_2 precipitates (amorphous SiO_2 and "seed"-SiO_2 in a structure coherent with VO_2 clusters, as explained below). The calculations were done with the ab initio pseudopotential plane wave code fhi96md (16), i.e., under periodic boundary conditions. We applied the Local Density Approximation (LDA) for the exchange and correlation energy (16, 17) and nonlocal pseudopotentials in the Trouller-Martins scheme (18, 19) with 40 Ry cut-off for plane waves. The point defects and point defect clusters were contained in tetragonal supercells of two sizes: 3×3×8, consisting of eight Si(001) layers (72 Si atoms) and 4×4×8, consisting of eight Si(001) layers (128 Si atoms). The Brillouin zone was sampled at the (¼,¼,¼) point of the supercell. We verified that this sampling results in total energy differences converged to at least 30 meV, which is accurate enough for our purposes. More concern is due to the size of the supercell. For example, the difference of the chemical potential of oxygen, $\mu(O)$, between a 2VO$_2$ cluster in a 3×3×8 cell and in a 4×4×8 eV amounts to 90 meV. Nevertheless, the main physical features of the system can be extracted from these results.

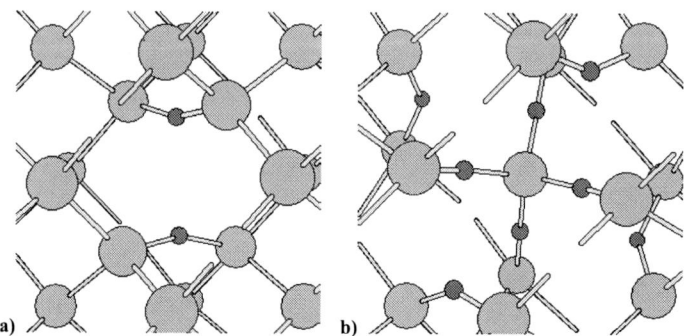

FIGURE 8. Ball-and-stick models of a) isolated VO$_2$ defect and b) 4VO$_2$ cluster. Oxygen atoms are dark. Note that the 4VO$_2$ cluster is shown from a different perspective than the VO$_2$ defect.

The "seed"-SiO$_2$ (seed-SiO$_2$) bulk model is obtained by a periodic repetition of the 4VO$_2$ cluster consisting of four VO$_2$ defects. In its ideal geometry, such a crystal is coherent with Si bulk (Fig. 9a-b): it can be obtained by removing each second (001) layer from the Si lattice and connecting with oxygen bridges the dangling bonds of the resulting Si vacancies. The size of the primitive cell of the ideal seed-SiO$_2$ corresponds then to the 1×1×4 Si(001) cell. Such geometry is the natural extension of nVO$_2$ clusters to the infinite size limit.

The first observation, evident already from the examination of interatomic distances in the ideal structure of seed-SiO$_2$, is that there is a considerable misfit between the lattices of seed-SiO$_2$ and of Si. As a matter of fact, when the symmetry constrains within the primitive cell of seed-SiO$_2$ are lifted (but the cell itself remains coherent with Si), the Si and O atoms reorganize spontaneously, whereby all Si atoms convert to triply oxidized silanone configuration (Fig. 9c). Each Si atom is thus connected with two Si neighbours through regular O bridges, and its remaining two orbitals are saturated by a non-bridging O atom forming a silanone (Si=O) bond.

We relaxed the volume of the seed-SiO$_2$ crystal, keeping for simplicity the c/a ratio of the tetragonal cell fixed. As seen in Fig. 10, the volume of seed-SiO$_2$ shrinks and reaches equilibrium at $\Delta c = -0.5$ Å. This would correspond to linear compression $\Delta a/a$ of the Si lattice by about 10 %. At the equilibrium lattice constant, the seed-SiO$_2$ crystal conserves the ideal topology of the VO$_2$ cluster (Fig. 9d). The transition from silanone to silica is reflected in the energy diagram in Fig. 10 by the appearance of a kink in the right-hand side of the energy curve around $\Delta c = -0.2$ Å.

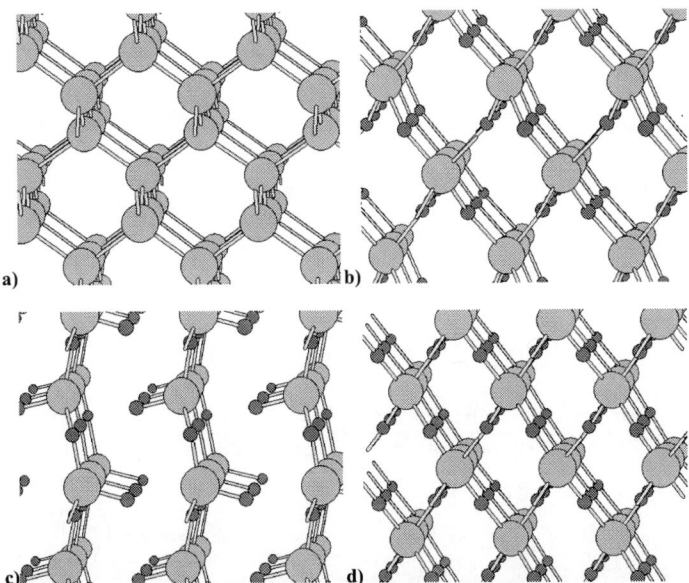

FIGURE 9. Ball-and-stick models of **a)** perfect Si bulk, **b)** ideal seed-SiO_2 crystal, **c)** relaxed seed-SiO_2 crystal having lattice coherent with that of Si and containing silanone (Si=O) bonds, and **d)** seed-SiO_2 crystal relaxed under the constraint of constant c/a ratio and containing only Si-O-Si bridges. Oxygen atoms are dark. All models are shown from the same perspective.

There are three fundamental facts which are crucial to the nucleation process:

1. VO_2 defects tend to form nVO_2 clusters (cf. Fig. 10). This is because the Si atom in the centre of each $4VO_2$ cluster can fully relax its bond angles with all its neighbours; this is not possible in an isolated VO_2.
2. Due to a significant (10 %) lattice mismatch, nVO_2 clusters cannot grow indefinitely; the specific density of Si in a cluster coherent with the host lattice is too small to support the formation of a continuous silica bond network.
3. Although the chemical potential $\mu(O)$ of oxygen is noticeably (by approx. 0.2 eV, cf. Fig. 10) lower in O_i than in VO_2 and even lower than in $4VO_2$, the conversion of VO_2 into $2O_i$ is hindered by the kinetic barrier caused by the high energy (exceeding 4 eV) of the Si vacancy produced when VO_2 dissociates. Therefore, oxygen atoms trapped in VO_2 defects may lower their energy only when nVO_2 clusters are formed.

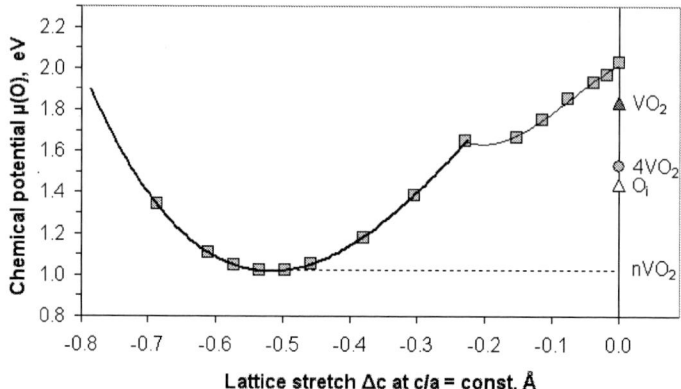

FIGURE 10. Relaxation of seed-SiO₂ crystallite. The ratio c/a was kept constant and equal to that of the tetragonal 1×1×4 silicon cell with the c axis along (001) and the a axis along (110). The energy co-ordinate in the diagram corresponds to the chemical potential μ(O) of oxygen atom with respect to that in amorphous SiO₂ in thermodynamic equilibrium with Si bulk. The values of μ(O) for VO₂, 4VO₂, and Oᵢ defects in the Si bulk are indicated.

These observations allow us to formulate a simple atomistic model of the process of nucleation and growth of SiO_2 precipitates in vacancy supersaturated material. Precipitate nuclei are formed as the first step. They consist of tiny nVO_2 clusters and have locally the geometry similar to that of seed-SiO_2. The nuclei formation volume ΔV_0 (with respect to perfect Si) is negative, that is, the Si host and the seed-SiO_2 precipitate nuclei are under tensile strain. The growth of the nuclei by attachment of further VO_2 is limited by the 10 % lattice mismatch between seed-SiO_2 and silicon. Therefore, at a certain point the precipitation can continue only through oxidation of the silicon around the nuclei. Interstitial oxygen atoms, O_i, are now attracted by the tensile strain field of the nucleus. They oxidize the Si-Si bonds in the immediate vicinity of the nVO_2 cluster, so that a cristobalite-like mantle is grown. Indeed, the cristobalite structure may be visualised as the Si structure with all Si-Si bonds substituted by oxygen bridges. The compressive strain produced by the clustering of O_i into the cristobalite mantle is annihilated by the tensile strain due to the presence of the seed-SiO_2 cluster core. In the course of this process, the formation volume ΔV_0 of the precipitate changes from negative towards zero. As further O_i are attached, ΔV_0 may become positive.

The precipitate grown in this way is thus built of a seed-SiO_2 core with a cristobalite mantle. The process consumes VO_2 and interstitial oxygen. When the crystallite is large enough and the annealing temperature high enough, the structure of the precipitate may change to amorphous.

6. Conclusions

We have investigated the impact of RTA induced vacancy supersaturation on oxide precipitation based on as much as possible experimental and theoretical values obtained from variations of RTA temperature, vacancy concentration, and interstitial oxygen concentration in order to gain connections being as general as possible. In summary, the following conclusions can be drawn:

1. Oxygen precipitation after RTA processing was found to be controlled by the initial concentration of interstitial oxygen in a sixth power dependency and frozen vacancies in just a cubic dependency.
2. The favored process for nucleation of oxide precipitates after RTA treatments is the agglomeration of VO_2 complexes. This type of coherent nucleation seems energetically favored because it would not require the generation and emission of silicon self-interstitials for strain release. The density of oxide precipitate nuclei was found to be proportional to the cube of VO_2 complexes formed during cooling after RTA.
3. The reduction of interstitial oxygen during a process consisting of an RTA treatment and a so-called BMD test comprising a two-step anneal at 780 °C for 3 h and at 1000 °C for 16 h can be accurately modeled for the temperature range from 1150 °C to 1250 °C using Ham's theory for precipitate growth and an empirical connection based on nucleation of oxide precipitates by agglomeration of VO_2 complexes.
4. During RTA treatments at temperatures ≥ 1300 °C vacancies seem to be consumed by other processes and VO_2 based nucleation is decreasing with increasing RTA temperature.
5. Below RTA temperatures of 1150 °C, oxide precipitation seems to be dominated by shrunken as-grown precipitate nuclei. This homogenizes precipitate density and makes it independent of interstitial oxygen over a wide range but it does not enhance oxide precipitation. Obviously, as-grown nuclei are dissolved only at RTA temperatures ≥ 1150 °C.

With respect to these conclusions, we developed an atomistic model that describes the coherent nucleation process of oxide precipitates via VO_2 clustering in a vacancy-rich environment based on ab initio calculations for total energies and atomic structures of point defects and point defect clusters. The model can be summarized as follows:

1. VO_2 point defect complexes tend to form nVO_2 clusters because in contrast to isolated VO_2 the Si atoms surrounded by $4VO_2$ can fully relax its bond angles with its neighbors. These tiny nVO_2 clusters have locally the geometry of seed-SiO_2 which is obtained by the periodic repetition of the $4VO_2$ cluster. They are coherent with the lattice of bulk silicon.
2. The molecular volume of seed-SiO_2 is lower than the molecular volume of silicon. In contrast to any other SiO_2 phase considered for oxide precipitate nucleation, these coherent nuclei are tensile strained with a lattice mismatch of 10 %.
3. Unlimited coherent growth of nuclei by further attachment of VO_2 is not possible because of the increasing strain energy due to the misfit and because the concentration of VO_2 complexes is orders of magnitude smaller than the concentration of interstitial oxygen O_i. However, interstitial oxygen is attracted by the tensile strain field of the nucleus and oxidizes the Si-Si bonds in the immediate vicinity of the nVO_2 cluster thus forming a cristobalite-like mantle. In the initial phase, the compressive strain in the cristobalite mantle can be annihilated by the tensile strain in the seed-SiO_2 core. During further growth and at high enough temperature, the structure of the precipitate may change to amorphous.

The results presented in this contribution provide a step forward to further understand and model oxide precipitation in CZ silicon. Until now, modeling is related to the VO_2 concentration frozen after RTA. Next steps would need to be targeted on understanding the kinetics of VO_2 decay after out-diffusion of supersaturated vacancies which is assumed to influence nucleation kinetics. Modeling VO_2 formation during crystal pulling would help to understand formation of as-grown nuclei which are important for many thermal processes.

Acknowledgments

Ab initio calculations have been performed on IBM Regatta p690+ cluster in the von Neumann Institute for Computing, Jülich, Germany (project hfo06).

The authors would like to thank Dr. U. Lambert from Siltronic AG for his valuable comments.

References

1. J. Vanhellemont and C. Claeys, *J. Appl. Phys.,* **62**, 3960 (1987).
2. J. Vanhellemont, *Appl. Phys. Lett.,* **68**, 3413 (1996).
3. R. Falster, V. V. Voronkov, and F. Quast, *phys. Stat. Sol. (b),* **222**, 219 (2000).
4. V. V. Voronkov and R. Falster, *J. Electrochem. Soc.,* **149**, G167 (2002).
5. M. Akatsuka, M. Okui, N. Morimoto, and K. Sueoka, *Jpn. J. Appl. Phys.,* **40**, 3055 (2001).
6. M. Wright Jenkins, *J. Electrochem. Soc.,* **124**, 757 (1977).
7. F. Secco d'Aragona, *J. Electrochem. Soc.,* **199**, 948 (1972).
8. T. Sinno, R. Brown, W. von Ammon, and E. Dornberger, *Appl. Phys. Lett.,* **70**, 2250 (1997).
9. R. A. Brown, D. Maroudas, and T. Sinno, *J. Cryst. Growth,* **137**, 12 (1994).
10. V. V. Voronkov and R. Falster, *J. Appl. Phys.,* **91**, 5802 (2002).
11. F. S. Ham, *J. Phys. Chem. Solids,* **6**, 335 (1958).
12. J. Vanhellemont, *J. Appl. Phys.,* **78**, 4297 (1995).
13. J. C. Mikkelsen Jr., *Mater. Res. Soc. Symp. Proc.,* **59**, 19 (1986).
14. G. Kissinger, U. Lambert, M. Weber, F. Bittersberger, T. Müller, H. Richter, and W. von Ammon, Proc. DRIP XI, September 15-19, 2005, Beijing, China, to appear in phys. stat. sol. (c).
15. M. Bockstedte, A. Kley, J. Neugebauer, and M. Scheffler, *Comp. Phys. Comm.,* **107**,187 (1997).
16. D.M Ceperley and B.J. Alder, *Phys. Rev. Lett,.* **45**, 567 (1980).
17. J. P. Perdew and A. Zunger, *Phys. Rev. B,* **23**, 5048 (1981).
18. D. R. Haman, Phys. Rev. B, **40**, 2980 (1989).
19. G. B. Bachelet, D. R. Hamann, and M. A. Schlüter, *Phys. Rev. B*, **26**, 4199 (1982).

ECS Transactions, 2 (2) 261-273 (2006)
10.1149/1.2195664, copyright The Electrochemical Society

FIRST PRINCIPLES CALCULATION FOR CU GETTERING BY DOPANT OR DOPANT-VACANCY COMPLEX IN SILICON CRYSTAL

Koji Sueoka, Shigehiro Ohara, Seiji Shiba and Seishiro Fukutani

Department of System Engineering, Okayama Prefectural University,
111 Kuboki, Soja, Okayama 719-1197, Japan

For finding the effective gettering center of cupper (Cu) atom in silicon (Si) crystal, an interaction between interstitial Cu atom and dopant (B, Sb, As, P), C or O atom was studied with first principles calculation. It was found that only B could be an effective gettering center. This result indicates that heavily B doped p/p+ epitaxial wafers will show a sufficient gettering efficiency for Cu contamination. In order to design the gettering center of Cu in n/n+ epitaxial wafers, the interaction between vanancy (V)-Sb, V-As or V-P complexes and Cu was investigated. It was found that these complexes could be effective gettering centers. The mechanism of Cu gettering by oxide precipitates was also studied with further calculations. It was found that the stabilities of Cu atom in β-crystobalite or in strained Si were not superior to that in strain-free Si. The trapping at the interface or the interaction with emitted interstitial Si (I) from oxide precipitates should be the possible mechanisms for Cu gettering.

INTRODUCTION

Among the all transition metals, Cu has the highest solubility (1) and diffusivity (2) in Si. Cu diffusion is observed even at room temperature (3,4). Recently, Cu gettering is of significant concern (3-10) because of the application of Cu wiring technology. It has been experimentally confirmed that Cu is gettered by B atoms in the substrate of p/p+ epitaxial wafer (5,7,9,10). On the other hand, n type dopant, such as Sb, As or P atom does not show a sufficient gettering efficiency for Cu (5). These results are explained with the coulomb attraction between Cu^+ and B^-, and the coulomb repulsion between Cu^+ and Sb^+, As^+ or P^+ in Si (5). The possibility of covalent bonding is not fully investigated. In order to explain the gettering efficiency quantitatively, the binding energy of Cu and dopant should be estimated. Further, the electron density distribution around Cu-dopant complex should be calculated to explain the interaction between Cu and dopant.

First principles calculation is the most effective tool to estimate the binding energy and to explain the interaction between Cu and dopant. Theoretical studies with first principles on this problem have started in these years. Recently, Matsukawa *et al* (9) obtained the binding energy of Cu-B complex, which value is 0.57eV. Michikita *et al* (11) reported that not ionic bonding but covalent bonding contributes to the formation of stable Cu-B complex. In this paper, we have performed the first principles calculation to find the effective gettering center of Cu. The binding energy of Cu and dopant (B, Sb, As, P), C or O atom was calculated. The interaction between Cu and dopant, C or O atom is also explained with the calculated electron density distribution. Further, the interaction

261

between V-Sb, V-As or V-P complexes and Cu was investigated to design the gettering center of Cu in n/n+ epitaxial wafer.

The gettering of Cu by distributed oxide precipitates (Internal Gettering (IG)) in lightly doped p- polished or p/p- epitaxial wafers is technologically important tool due to the application of double-side polishing. It is known that punched-out dislocations and/or dislocations around stacking faults formed by oxide precipitates become effective gettering center of Cu contaminations than oxide precipitates (12,13). The oxide precipitates themselves become IG sites for Cu (6,14) when the precipitates are not accompanied with these secondary defects. Sueoka et al (8,15) showed that the critical precipitate size, above which the precipitates generate punched-out dislocations, is about 140nm. In most of the current and future device processes with rapid thermal annealing (RTA), the size of oxide precipitates is less than this critical size. Therefore, IG by oxide precipitates themselves becomes most important gettering technique. Recently, IG criterion of oxide precipitate size and density for Cu gettering was given (5). This criterion is close to that for Ni gettering (5,6,15). Further, a practical simulator for IG of Cu by oxide precipitates was developed (8). The gettering site of Cu atom in precipitate/matrix Si system was not clear, but some of the proposed gettering sites in literature, such as (i) inside oxide precipitate (16), (ii) in the region of strained Si around precipitate (17), (iii) at the interface (8), and (iv) emitted interstitial Si (I) from oxide precipitates (18) should be possible. We believe that the theoretical study will contribute to clear the gettering mechanism of IG for Cu.

In this paper, we have theoretically studied the gettering mechanism of Cu by oxide precipitates with first principles calculation. The stability of Cu atom in β-crystobalite or in strained Si was compared to that in strain-free Si. Further, the interaction between Cu atom with dangling bonds at Si (100) surface, which is the simple model replace for the interface structure, was analyzed. The binding energy of Cu and I was also calculated.

SIMULATION DETAILS

The calculations were based on the local density approximation (19,20) with ultra-soft pseudopotential method (21) and the plane waves as a basis set for efficient structure optimization. The cutoff energy of the plane waves is 310eV. For the exchange-correlation energy in the generalized gradient approximation (GGA), the expression proposed by Perdew et al (22) was used. The CASTEP code was used to solve the Kohn-Sham equation self-consistently with three dimensional periodic boundary condition (23). The Density Mixing Method (24) and BFGS Geometry Optimization Method (25) were used to optimize the electric structure and atomic configurations, respectively. k point sampling is performed at 2x2x2 special points of Monkhorst-Pack grid (26). Only the neutral state of the systems was considered in this work.

We used a super cell of Si 64 atoms, which is large enough to neglect interactions between the neighbouring complexes in each super cell. Figure 1 shows the configuration of Si 64 atoms in the super cell used in this study. The lattice constant $L =$ 10.862Å. Figures 2(a) and 2(b) show the typical interstitial sites of Si crystal, Tetrahedral (T) site, Hexagonal (H) site, Bond center (Bc) site and <110> Dumbbell (D) site. Most stable interstitial site of Cu atom was determined with optimization of atomic configuration. Next, dopant (B, Sb, As, P) or C atom was located at substitutional site by replacing the Si atom at the center of the super cell, and Cu atom was located at some interstitial sites around dopant or C atom. Most stable configurations of Cu-dopant complex or Cu-C complex were obtained with optimization of atomic configuration. By using the calculated total energies, binding energy E_b of Cu and dopant, and E_b of Cu and C were obtained. Similar calculations were performed to interstitial O atom, *i.e.*, Cu-O complex. Binding energy E_b of Fe and B, and E_b of Ni and B were also obtained and compared with E_b of Cu and B.

In order to design the gettering center for Cu in n/n+ epitaxial wafers, the binding energies E_b of vanancy (V)-Sb, V-As or V-P complexes and Cu were calculated. The structures of V-Sb, V-As or V-P complexes were obtained with optimization of atomic configuration after the elimination of one Si atom next to n type dopant.

The mechanism of Cu gettering by oxide precipitates without secondary defects was also studied with further calculations. It is known that the typical precipitate morphology is platelet lying on the {100} plane of matrix Si (27). The platelet length is from several ten to hundred nanometers (27) depending on some parameters, such as annealing temperature, annealing time, oxygen concentration so on. One-axial Si [100] compressive strain of about 0.4% in Si crystal perpendicular to the platelet surface was detected with convergent beam electron diffraction (28). On the other hand, some different results were reported on precipitate structure (amorphous or crystobalite (29)) and precipitate phase (SiO_2 or close to SiO (30)). The gettering site of Cu atom in precipitate/matrix Si system was not clear, but several models for gettering site, such as (*i*) inside oxide precipitate, (*ii*) in the region of strained Si around precipitate, (*iii*) at the interface, and (*iv*) emitted I from oxide precipitates as shown in fig.3, have been proposed. Since it is difficult to form the accurate model of precipitate/matrix Si system due to the uncertainty of the structure and the phase of precipitate, we used the following simple models. For gettering site (*i*), we used the super cell of β-crystobalite structure including Si 32 atoms and O 64 atoms as shown in fig.4. The lattice constant $L_x = 7.322$Å, $L_y = L_z = 14.638$Å. For gettering site (*ii*), we used the super cell of Si 64 atoms under one-axial compressive stress of 0.5GPa in Si [100] direction. The compressive strain in Si [100] direction is 0.35%, while the tensile strain in Si [010] and [001] direction is 0.09%. These strain values almost equal to the experimentally determined values by Yonemura *et al* (28). For gettering site (*iii*), we used the Si (100) surface replaced for the unknown structure of the interface. The model is a Si 108 atoms super cell including a vacuum slab in 9.96Å thickness above Si (100) surface. The surface has the area of 11.38Å x 15.18Å, and includes 12 dangling bonds. For gettering site (*iv*), we considered I at T, H, Bc or D site. The total energy changes ΔE from Cu in strain-free Si to be in β-crystobalite, to be in strained Si, to be at interface (replaced by Si (100) surface) or to be a complex of Cu-I were calculated to determine the possible mechanism of IG for Cu by oxide precipitates.

RESULTS AND DISCUSSION

Most stable interstitial site for Cu and binding energy of Cu and B

From the calculated total energies, the most stable site for Cu was determined to be T site. On the other hand, the most stable site for Fe was T site, and that for Ni was H site. Further, the binding energy E_b of interstitial metal X and substitutional B atom was calculated with eq.[1].

$$E_b = \{E_{tot}(Si_{64}X_1) + E_{tot}(Si_{63}B_1)\} - \{E_{tot}(Si_{63}X_1B_1) + E_{tot}(Si_{64})\}. \qquad [1]$$

Here, X indicates Cu, Fe or Ni atom. For example, $E_{tot}(Si_{63}X_1B_1)$ indicates the total energy of the super cell including one X-B complex. The first term of the right side in eq.[1] indicates the sum of the total energies when X and B atom locate in each super cell independently. The second term of the right side in eq.[1] indicates the sum of the total energies when X-B complex is formed in one of the two super cells. $E_b > 0$ indicates that stable complex can be formed.

Table 1 summarizes the calculated binding energy E_b and distance between X and B. It is found that E_b for Cu is the largest while E_b for Ni is the smallest among these three metals. This result indicates that Cu is most effectively gettered, while Ni is diffuicult to be gettered by B atoms in p/p+ epitaxial wafers. This tendency well corresponds to the recent experimental results (5,7). The obtained E_b for Cu is 1.15eV, which is relatively larger than the value of 0.57eV reported by Matsukawa et al (9). This difference may be due to the use of different calculation method in each study. Figure 5 shows the stable configuration and valence electron density of interstitial Cu and substitutional B complex. It is obvious that (*i*) Cu becomes closer to B atom than to Si atoms, and (*ii*) the valence electron density is high at bond center of Cu-B due to the combination of Cu3d-B2p orbits. That is, covalent bonding is formed between Cu and B atom. We have further calculated the number of valence electrons possessed by each atom with Mulliken population analysis (31), after the projection of the plane waves to the atomic orbits formed from ultra-soft pseudopotentials. It was found that 10.28 valence electrons are possessed by Cu, while 3.76 electrons by B atom. That is, Cu is charged plus while B is charged minus. Therefore, both covalent and ionic bonding between Cu and B results in large binding energy.

Binding energy of Cu and n type dopant, C or O atom

Total energies of Cu-Sb, Cu-As, Cu-P, Cu-C and Cu-O complexes were calculated. By changing B to Sb, As, P, C or O and X = Cu in eq.[1], binding energies E_b of Cu and n type dopant, C or O were obtained. Table 2 summarizes the calculated binding energy E_b and distance between Cu and these elements. It is found that Sb and P do not form stable complexes with Cu. Although E_b of Cu and As is positive, the value is very small. These results indicate that n type dopant, such as Sb, As or P atom, cannot become an effective gettering center. Figure 6 shows the stable configuration and the valence electron density of interstitial Cu and substitutional Sb complex. It is obvious that (*i*) Cu is apart from Sb atom, and (*ii*) valence electron density between Cu and Sb is low. Mulliken population analysis showed that 10.63 valence electrons are possessed by Cu, while 4.32 electrons by Sb atom. That is, both Cu snd Sb are charged plus. Similar results were obtained for Cu-As and Cu-P complexes. These are the reasons why E_b of Cu and n type dopant is negative or very small positive values.

The calculated results for C and O atom are also summarized in table 2. It is found that $E_b < 0$ for C atom; C does not form stable complex with Cu. This result well agrees with the calculated result by Michikita et al (11). For p/p- epitaxial wafers, a technique using heavily C doped substrate is recently proposed to enhance oxygen precipitation (32). However, this technique is not effective for Cu gettering in n/n+ epitaxial wafers. The calculated result for interstitial O atom showed that O does not form stable complex with Cu. This result can be naturally accepted because the lightly doped CZ-Si wafer including O atoms with the concentration of 10^{18} atoms/cm^3 shows poor gettering efficiency for Cu.

Cu gettering technique by the complex of V and n type dopant in n/n+ epitaxial wafer

In order to find an effective gettering center for Cu in n/n+ *epi* wafers, we have considered the gettering center of vacancy V and n type dopant (Sb, As, P) complexes. First, the stable configurations of V-Sb, V-As or V-P complexes were obtained. Then, the binding energy E_b of V and n type dopant D was calculated with eq.[2].

$$E_b = \{E_{tot}(Si_{63}V_1) + E_{tot}(Si_{63}D_1)\} - \{E_{tot}(Si_{62}V_1D_1) + E_{tot}(Si_{64})\}. \qquad [2]$$

Table 3 summarizes the calculated E_b of V and n type dopants. It was found that stable V-Sb, V-As or V-P complexes can form as already estimated theoretically (33,34). We write these complexes as V-D.

Next, we have investigated the interaction between Cu and V-D complex. In the calculation, Cu is located at some interstitial sites around V-D complex. By the optimization of atomic configuration, it is found that interstitial Cu interacts with V, then forms stable complexes of Cu_s-Sb, Cu_s-As and Cu_s-P. Here, Cu_s indicates substitutional Cu atom. The binding energy E_b of Cu and V-D complex was calculated with eq.[3].

$$E_b = \{E_{tot}(Si_{62}V_1D_1) + E_{tot}(Si_{64}Cu_1)\} - \{E_{tot}(Si_{62}Cu_{s1}D_1) + E_{tot}(Si_{64})\}. \qquad [3]$$

Table 4 summarizes the calculated E_b of Cu and V-D complex. The distance between substitutional Cu_s and n type dopant D in stable complexes is also shown. It was found that the values of E_b are about 3eV. This result indicates that V-D complexes can be effective gettering centers for Cu.

Figure7 shows the stable configuration and valence electron density of Cu_s-Sb complex. It is obvious that covalent bonding is formed between Cu_s and Sb, and Cu_s and Si atom. Similar results were obtained for Cu_s-As and Cu_s-P complexes. On the other hand, Mulliken population analysis showed that (*i*) 10.86 valence electrons are possessed by Cu_s while 4.34 electrons by Sb in Cu_s-Sb complex, (*ii*) 10.78 valence electrons are possessed by Cu_s while 4.66 electrons by As in Cu_s-As complex, and (*iii*) 10.67 valence electrons are possessed by Cu_s while 5.27 electrons by P in Cu_s-P complex. That is, coulomb repulsion works between Cu_s and Sb or As, while coulomb attraction works between Cu_s and P. On the basis of above analysis, it is concluded that interstitial Cu interacts with V of V-D (D = Sb, As or P) complex, then stable Cu_s-D complex forms. The mechanism of the bonding between Cu_s and D is the covalent bonding. Further, ionic bonding works additionally when D = P.

The present result proposes a new Cu gettering technique in n/n+ epitaxial wafers with installing V-Sb, V-As or V-P complexes. Heat-treatment of n/n+ epitaxial wafers in a nitrogen atmosphere and/or with RTA (35) under optimized condition is one of the possible techniques to install V-Sb, V-As or V-P complexes.

Mechanism of Cu gettering by oxide precipitates

Stability of Cu atom in β-crystobalite. From the calculated total energies, the most stable site for Cu in β-crystobalite was determined to be T site. By calculating the total energy change ΔE with eq.[4], the stability of Cu in β-crystobalite was compared to that of Cu in strain-free Si.

$$\Delta E = \{E_{tot}(Cu_1 \text{ in β-crystobalite } (Si_{32}O_{64}Cu_1)) + E_{tot}(Si_{64})\} - \{E_{tot}(\text{β-crystobalite } (Si_{32}O_{64})) + E_{tot}(Si_{64}Cu_1)\}. \quad [4]$$

Here, $\Delta E < 0$ indicates that Cu in β-crystobalite is more stable than Cu in strain-free Si. The calculated $\Delta E = 2.60eV$ as summarized in table 5. This result indicates that Cu should not be gettered inside the oxide precipitate, if the precipitate has a β-crystobalite structure. Figure 8 shows the stable configuration and valence electron density of interstitial Cu in β-crystobalite. It is obvious that the valence electron density between Cu and O or Si is low, *i.e.*, covalent bonding is not formed around the interstitial Cu in β-crystobalite.

Stability of Cu atom in strained Si. By calculating the total energy change ΔE with eq.[5], the stability of Cu in strained Si formed around the oxide precipitate (0.35% compressive strain in Si [100] direction, while 0.09% tensile strain in Si [010] and [001] direction) was compared to that of Cu in strain-free Si.

$$\Delta E = \{E_{tot}(Cu_1 \text{ in strained Si } (Si_{64}Cu_1)) + E_{tot}(Si_{64})\} - \{E_{tot}(\text{strained Si } (Si_{64})) + E_{tot}(Si_{64}Cu_1)\}. \quad [5]$$

Here, $\Delta E < 0$ indicates that Cu in strained Si is more stable than Cu in strain-free Si. The calculated $\Delta E = 0.00eV$ as summarized in table 5. This result indicates that Cu should not be gettered in strained Si formed around oxide precipitate.

Interaction with dangling bonds at Si (100) surface. Figure 9(a) shows the obtained structure of Si (100) surface in the present work. The dangling bonds at the surface form dimers of p(2x1) symmetry (36). Figure 9(b) shows the stable configuration of Cu atom at Si (100) surface. It is found that the nearest two Si atoms at the top layer become closer to Cu in comparison to relaxed Si (100) surface in fig.9(a). By calculating the total energy change ΔE with eq.[6], the interaction of Cu with dangling bonds at Si (100) surface was estimated.

$$\Delta E = \{E_{tot}(Cu_1 \text{ at Si } (100) \text{ surface } (Si_{108}Cu_1)) + E_{tot}(Si_{64})\} - \{E_{tot}(Si (100) \text{ surface } (Si_{108})) + E_{tot}(Si_{64}Cu_1)\}. \quad [6]$$

Here, $\Delta E < 0$ indicates that Cu at Si (100) surface is more stable than Cu in strain-free Si. The calculated $\Delta E = -0.36eV$ as summarized in table 5. This result indicates that Cu should be gettered at the interface of precipitate and matrix, if the interface structure is close to Si (100) surface. The effective charges for Cu and Si atoms around Cu calculated by Mulliken population analysis are shown in fig.9(c). This result indicates that Cu is charged plus while nearest Si atoms at second and third layers are charged minus. Figure 9(d) shows the stable configuration and valence electron density of Cu in Si (011) plane near surface. It is obvious that covalent bonding is formed between Cu and two nearest Si atoms. Therefore, both covalent and ionic bonding between Cu and Si atoms around Cu results in the gettering of Cu at Si (100) surface.

Interaction with emitted interstitial Si. From the calculated total energies, the most stable site for I was determined to be D site. Figure 10(a) shows the stable configuration of Cu-I complex in a super cell. I is shown by one of the two black balls. Cu-I complex lies on Si {111} plane, and Cu is located at T site closest to I. The total energy change ΔE through the reaction of Cu + I → Cu-I was calculated with eq.[7].

$$\Delta E = \{E_{tot}(Si_{64}Cu_1I_1) + E_{tot}(Si_{64})\} - \{E_{tot}(Si_{64}I_1) + E_{tot}(Si_{64}Cu_1)\}. \qquad [7]$$

Here, $\Delta E < 0$ indicates that stable Cu-I complex can be formed. The calculated $\Delta E = -0.90eV$ as summarized in table 5. This result indicates that Cu can be gettered by interstitial Si I. Figure 10(b) shows the stable configuration and valence electron density of Cu-I complex. The maximum black contour of the density is set to 0.8 electron /$Å3$. It is obvious that covalent bonding is formed between Cu and two Si atoms (one Si atom is I). Mulliken population analysis showed that 10.46 valence electrons are possessed by Cu while 4.04 electrons by I (two Si atoms contributing to I at D site have same value). This indicates that Cu is charged plus while I is almost neutral. Therefore, the covalent bonding between Cu and I results in the formation of stable Cu-I complex.

In summary, the stabilities of Cu atom in β-crystobalite or in strained Si were not superior to that in strain-free Si. The trapping at the interface or the interaction with emitted I from oxide precipitates should be the possible mechanisms for Cu gettering.

CONCLUSION

In order to find the effective gettering center of Cu atom in Si crystal, we have studied the interaction between interstitial Cu atom and dopant (B, Sb, As, P), C or O atom with first principles calculation. It was found that only B could be an effective gettering center. This result indicates that heavily B doped p/p+ epitaxial wafers will show a sufficient gettering efficiency for Cu contamination. To design the gettering center of Cu in n/n+ epitaxial wafers, we have further calculated the binding energy between V-Sb, V-As or V-P complexes and Cu. It was found that these complexes could be effective gettering centers. We have also studied the mechanism of Cu gettering by oxide precipitates theoretically with using simple models of precipitate/matrix Si system. It was found that the stabilities of Cu atom in β-crystobalite or in strained Si were not superior to that in strain-free Si. The trapping at the interface or the interaction with emitted I from oxide precipitates should be the possible mechanisms for Cu gettering.

REFERENCES

1. E. R. Weber, *Appl. Phys. Lett.*, **A30**, 1 (1983).
2. A. A. Istratov, C. Flink, H. Hieslmair and E. R. Weber, *Phys. Rev. Lett.*, **81**, 1243 (1998).
3. M. B. Shabani, T. Yoshimi and H. Abe, *J. Electrochem. Soc.*, **143**, 2025 (1996).

4. A. A. Istratov, H. Feick, S. A. McHugo, W. Seifert, H. Hieslmair, T. Heiser and E. R. Weber, *Phys. Rev. Lett.*, **85**, 4900 (2000).
5. R. Hoelzl, M. Blietz, L. Fabry and R. Schmolke, in *Semiconductor Silicon/2002*, H. R. Huff, L. Fabry and S. Kishino, Editors, **PV 2002-2**, p.608, The Electrochemical Society Proceedings Series, Pennington, NJ (2002).
6. M. Seacrist, M. Stinson, J. Libbert, R. Standley and J. Binns, in *Semiconductor Silicon/2002*, H. R. Huff, L. Fabry and S. Kishino, Editors, **PV 2002-2**, p.638, The Electrochemical Society Proceedings Series, Pennington, NJ (2002).
7. M. B. Shabani, Y. Shiina and Y. Shimanuki, in *Proceedings of the Forum on the Science and Technology of Silicon Materials/2003*, H. Yamada-Kaneta and K. Sumino, Editors, p.97 (2003).
8. K. Sueoka, *J. Electrochem. Soc.*, **152**, G731 (2005).
9. K. Matsukawa, N. Hattori, S. Maegawa, K. Shirai and H. Katayama-Yoshida, in *Gettering and Defect Engineering in Semiconductor Technology/2005*, B. Pichaud, A. Claverie, D. Alquier, H. Richter and M. Kittler, Editors, Solid State Phenomena **PV 108-109**, p.115 (2005).
10. M. B. Shabani, K. Hirano, Y. Shiina, T. Kihara and T. Shingyoji, *ibid*, p.385 (2005).
11. T. Michikita, K. Shirai and H. Katayama-Yoshida, in *Proceedings of Silicon Technology*, **59**, p.47, The Japan Society of Applied Physics (2004) (Japanese).
12. H. Ewe, D. Gilles, S. K. Hahn, M. Seibt and W. Schröter, in *Semiconductor Silicon/1994*, H. R. Huff, W. Bergholz and K. Sumino, Editors, **PV 94-10**, p.796, The Electrochemical Society Proceedings Series, Pennington, NJ (1994).
13. B. Shen, T. Sekiguchi and K. Sumino, in *Defects in Semiconductors/18*, M. Suezawa and H. Katayama-Yoshida, Editors, Materials Science Forum **PV 196-201**, p.1207 (1995).
14. R. Falster, G. R. Fisher and G. Ferrero, *Appl. Phys. Lett.*, **59**, 809 (1991).
15. K. Sueoka, S. Sadamitsu, Y. Koike, T. Kihara and H. Katahama, *J. Electrochem. Soc.*, **147**, 3074 (2000).
16. S. Sadamitsu, M. Hourai and K. Sueoka, in *Proceedings of the Forum on the Science and Technology of Silicon Materials/2001*, H. Yamada-Kaneta and K. Sumino, Editors, p.273 (2001).
17. R. Falster, V. V.Voronkov, V. Y. Resnik and M. G. Milvidskii, in *High Purity Silicon VIII*, C. L. Claeys, M. Watanabe, R. Falster and P. Stallhofer, Editors, **PV 2004-05**, p.188, The Electrochemical Society Proceedings Series, Pennington, NJ (2004).
18. A. Ourmazd, *Appl. Phys. Lett.*, **45**, 781 (1984).
19. P. Hohenberg and W. Kohn, *Phys. Rev.*, **136**, B864 (1964).
20. W. Kohn and L. Sham, *Phys. Rev.*, **140**, A1133 (1965).
21. D. Vanderbilt, *Phys. Rev.*, **B41**, 7892 (1990).
22. J. Perdew, K. Burke, M. Ernzerhof, *Phys. Rev. Lett.*, **77**, 3865 (1996).
23. The *CASTEP* code is available from Accelrys Software Inc.
24. G. Kresse and J. Furthmuller, *Phys. Rev.*, **B54**, 11169 (1996).
25. T. Fischer and J. Almlof, *J. Phys. Chem.*, **96**, 9768 (1992).
26. H. Monkhorst and J. Pack, *Phys. Rev.*, **B13**, 5188 (1976).
27. For example, K. Sueoka, N. Ikeda, T. Yamamoto and S. Kobayashi, *J. Appl. Phys.*, **74**, 5437 (1993).
28. M. Yonemura, K. Sueoka and K. Kamei, *Jpn. J. Appl. Phys.*, **38**, 3440 (1999).
29. F. Shimura, H. Tsuya and T. Kawamura, *Appl. Phys. Lett.*, **37**, 483 (1980).
30. J. Vanhellemont, *J. Appl. Phys.*, **78**, 4297 (1995).
31. R. Mulliken, *J. Chem. Phys.*, **23**, 1833 (1955).

32. K. Sueoka, M. Akatsuka, M. Yonemura, T. Ono, E. Asayama, Y. Koike and S. Sadamitsu, in *Gettering and Defect Engineering in Semiconductor Technology/2001*, V. Raineri, F. Priolo, M. Kittler and H. Richter, Editors, Solid State Phenomena **PV 82-84**, p.49 (2001).
33. R. Car, P. J. Kelly, A. Oshiyama and S. T. Pantelides, *Phys. Rev. Lett.*, **54**, 360 (1985).
34. D. C. Mueller, E. Alonso and W. Fichtner, *Phys. Rev.*, **B68**, 045208 (2003).
35. R. Falster, in *Proceedings of the Forum on the Science and Technology of Silicon Materials/2001*, H. Yamada-Kaneta and K. Sumino, Editors, p.241 (2001).
36. A. Ramstad, G. Brocks and P. J. Kelly, *Phys. Rev.*, **B51**, 14504 (1995).

Table 1 Calculated binding energy E_b and distance between interstitial metal and substitutional B atom. $E_b > 0$ indicates that stable complex can be formed.

Metal	Site of metal	E_b (eV)	Distance (Å)
Cu	T	1.15	2.07
Fe	T	0.73	2.12
Ni	H	0.49	2.12

Table 2 Calculated binding energy E_b and distance between interstitial Cu and substitutional Sb, As, P, C or interstitial O atom. $E_b > 0$ indicates that stable complex can be formed.

Dopant, Impurity	Site of Cu	E_b (eV)	Distance (Å)
Sb	T	-1.28	2.85
As	T	0.06	2.61
P	T	-0.14	2.47
C	T	-0.13	2.29
O	T	-0.15	3.01

Table 3 Calculated binding energy E_b between V and substitutional Sb, As or P atom. $E_b > 0$ indicates that stable complex can be formed.

Dopant	E_b (eV)
Sb	1.48
As	1.30
P	1.25

Table 4 Calculated binding energy E_b between interstitial Cu and V–dopant complex. The distance of Cu and dopant in the stable complex is also shown. $E_b > 0$ indicates that stable complex can be formed.

Dopant	E_b (eV)	Distance of Cu and Dopant (Å)
Sb	2.85	2.45
As	3.18	2.33
P	3.32	2.27

Table5 Calculated total energy changes ΔE from Cu in strain-free Si to be in β-crystobalite, to be in strained Si around precipitate, to be at interface (replaced by Si (100) surface) or to be Cu-I complex. $\Delta E < 0$ indicates that Cu can be gettered at the gettering site.

Gettering site	ΔE (eV)
in β-crystobalite	2.60
in strained Si around precipitate	0.00
at interface	-0.36
emitted I at D site	-0.90

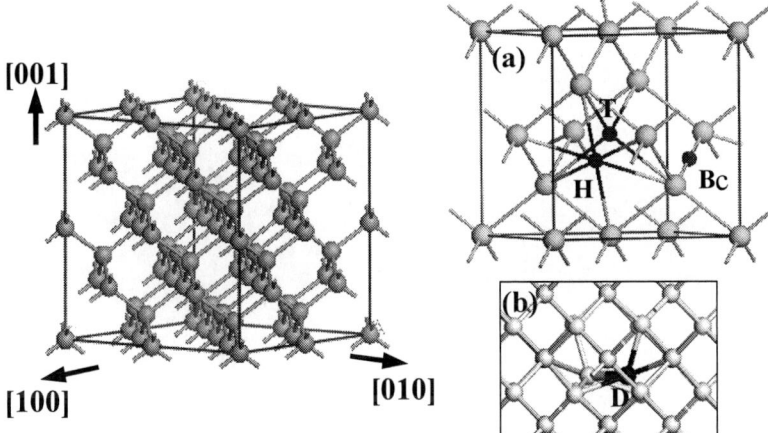

Fig.1 Configuration of Si 64 atoms in the super cell used in this study.

Fig.2 Typical interstitial sites of Si crystal, (a) T, H, Bc and (b) D site.

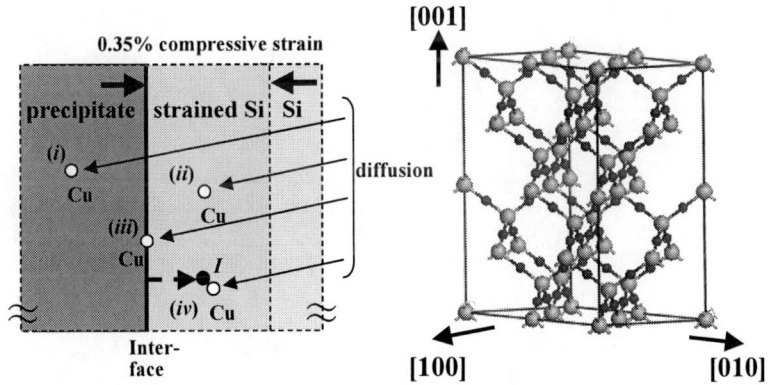

Fig.3 Schematic illustration of the proposed gettering sites for Cu by oxide precipitate.

Fig.4 Configuration of Si 32 atoms and O 64 atoms in the super cell of β-crystobalite used in this study.

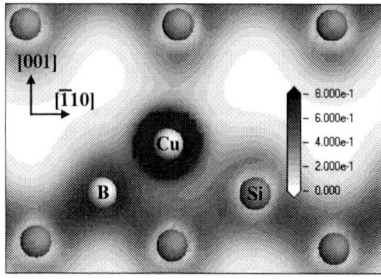 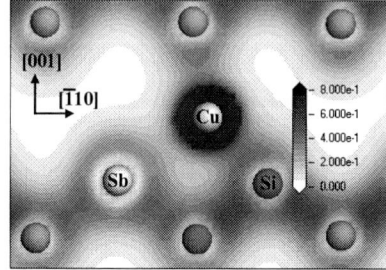

Fig.5 Stable configuration and valence electron density of interstitial Cu and substitutional B complex in a Si (110) plane. The maximum contour of the density is set to 0.8 electron /Å³.

Fig.6 Stable configuration and valence electron density of interstitial Cu and substitutional Sb complex in a Si (110) plane. The maximum contour of the density is set to 0.8 electron /Å³.

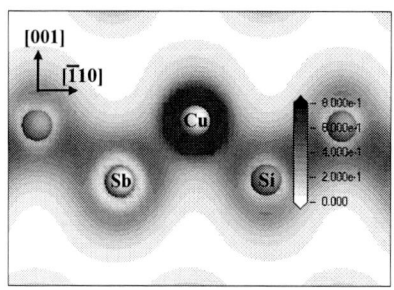 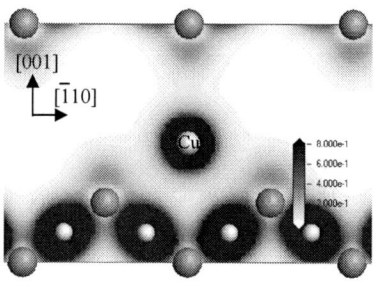

Fig.7 Stable configuration and valence electron density of substitutional Cu and substitutional Sb complex in a Si (110) plane. The maximum contour of the density is set to 0.8 electron /Å³.

Fig.8 Stable configuration and valence electron density of interstitial Cu in a (110) plane of β-crystobalite. Smaller white balls indicate O atoms, while larger gray balls indicate Si atoms. The maximum contour of the density is set to 0.8 electron /Å³.

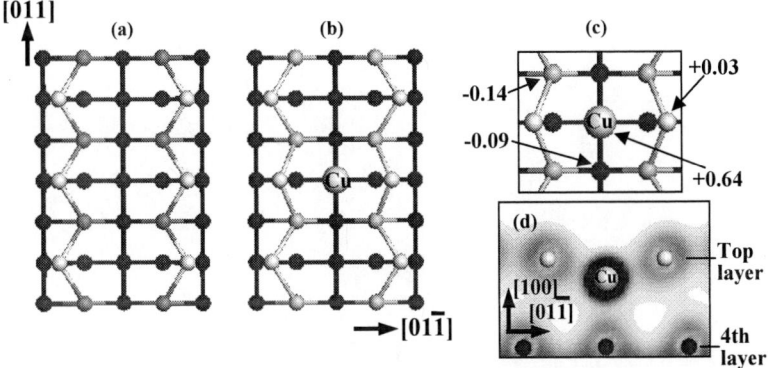

Fig.9 (a) Structure of Si (100) surface, (b) stable configuration of Cu atom at Si (100) surface, (c) effective charges of Cu and Si atoms around Cu, and (d) stable configuration and valence electron density of Cu in Si (011) plane. White balls indicate Si atoms at the top layer, gray balls indicate Si atoms at the second layer and black balls indicate Si atoms at and lower than third layer. The maximum black contour of the density is set to 0.8 electron /Å³.

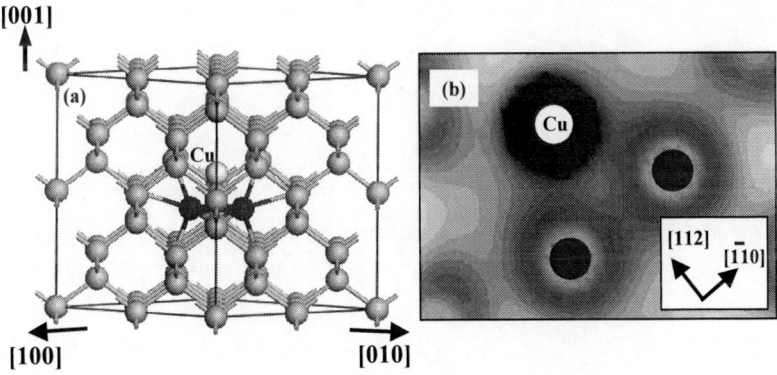

Fig.10 (a) Stable configuration of Cu-I complex in a super cell. Cu is located at T site closest to I at D site. One of the two Si atoms shown with black balls indicates I at D site. (b) Stable configuration and valence electron density of Cu-I complex in a Si (11$\bar{1}$) plane. White ball indicates Cu atom, while one of the two black balls indicates I at D site.

274

ANALYSIS OF INTERNAL GETTERING OF IRON BASED ON THE NUCLEATION MODEL OF IRON PRECIPITATION

Kozo Nakamura, Junsuke Tomioka

Komatsu Electronic Metals Co., Ltd., Technology Division

1324-2, Masuragahara, Ohmura, Nagasaki, 856-8555   JAPAN

ABSTRACT

We propose a new simulation model to describe the behavior of the internal gettering of Fe by oxygen precipitates. The internal gettering of Fe has typically been described by the diffusion-limited model proposed by Gilles et al. While the diffusion-limited model is effective in describing the gettering behavior at low temperature ($< 400^{\circ}$C), it fails to do so at practical gettering temperature of 600°C and above. The model proposed here accurately describes the internal gettering behavior of Fe at both low and practical temperatures by considering the nucleation of iron silicide on the oxygen precipitates. Our model also explains how the gettering behavior is influenced by the product of NRo at low temperature and by the product of NRo^3 at high temperature, where N and Ro are the density and radius of the oxygen precipitates, respectively.

1 INTRODUCTION

The integrated circuit (IC) industry has used the internal gettering effects of oxygen precipitates in silicon wafers to improve device yields. Because oxygen precipitates act as the heterogeneous precipitation sites for the supersaturated impurity metals such as Fe, Ni, Cu and Co, these impurity metals precipitate on the oxygen precipitates. As a result, the oxygen precipitates confer a gettering effect. As Fe is the most typical metal contaminat during device process, various studies to date have focused on the effect of oxygen precipitates in gettering of Fe [1-5]. Few of these studies, however, have attempted to quantify gettering efficiency [1,5].

Gilles et al. [1] showed that the gettering rate depended on the product of NRo at low temperature (280 $^{\circ}$C), where N and Ro are the density and radius of oxygen precipitates, respectively. The results from their experiments corresponded to the behavior of the gettering model in which iron silicides precipitate on the surfaces of oxygen precipitates via a diffusion-limited process. The diffusion-limited model can only provide valid results at low temperature, however.

The gettering rates reported experimentally by Aoki et al. [2,3], Ogushi et al. [4] and Hieslmair et al. [5] in the temperature range from 600 °C to 900 °C were lower than the rates estimated by the diffusion-limited model with the difference in the orders-of-magnitude. In the experiments by Ogushi et al. [4], the gettering rate depended not on the product of NRo, but on the total volume of oxygen precipitates (NRo^3). In the research by Hieslmair et al. [5], the gettering rate demonstrated experimentally diverged from the rate predicted by the diffusion-limited model under the conditions with a high oxygen precipitate density and high temperatures (> 400 °C).

The nucleation of iron silicide on the surface of oxygen precipitates plays an important role in the gettering process. The diffusion-limited model typically used to describe the gettering of Fe neglects the effect of the rate of this process. We carried out the present study to develop and investigate a new method for analyzing the behavior of the internal gettering of Fe by oxygen precipitates, by considering the nucleation of iron silicide on oxygen precipitates. Our model accurately describes the internal gettering behaviors of Fe at both low and high temperatures. The results obtained from this model also explain the discrepancies in the data reported in earlier studies.

2 INTERNAL GETTERING MODEL

2-1 Diffusion-limited model

We will begin by explaining the diffusion-limited gettering model proposed by Gilles et al. [1]. In this model, the supersaturated Fe atoms precipitate on the surface of oxygen precipitates as iron silicide ($FeSi_2$). The precipitation rate of iron silicide on the surface of oxygen precipitates is so rapid that the precipitation rate of Fe is only limited by the diffusion rate of Fe in the silicon crystal.

Gilles et al. assumed that the gettering sites are spheres with a radius of Ro. The concentration on the surface of the gettering sites is the equilibrium concentration of Fe. Thus, the diffusion flux of Fe to one gettering site is described by Eq. 1:

$$J = 4\pi RoD \ (C(t) - C^{eq}) \qquad 1)$$

where D is a diffusion coefficient of Fe, C(t) denotes the concentration of Fe at time t, and C^{eq} is the equilibrium concentration of Fe. Eq. 2 shows the concentration change of Fe with the density of gettering sites N:

$$\frac{\partial \, C(t)}{\partial t} = - 4\pi R_o D \Big(C(t) - C^{eq} \Big) N \qquad 2)$$

When the initial concentration is C_{ini} and the concentration after infinite time is C^{eq}, Eq.2 can be solved by Eq. 3 and 4.

$$\frac{C(t) - C^{eq}}{C_{ini} - C^{eq}} = \exp\left(-\frac{t}{\tau}\right) \qquad 3)$$

$$\frac{1}{\tau} = 4\pi R_o DN \qquad 4)$$

where τ is the relaxation time of the gettering process and $1/\tau$ denotes the gettering rate. Gilles et al. demonstrated that the decay curve of Fe concentration could be expressed as relaxation equation by Eq. 3. Further, they derived the product of NRo at Eq. 4 by fitting of decay curve to Eq. 3. Here, NRo correspond to the product of the density and radius of gettering site of Fe. On the other hand, the decay rate of the oxygen concentration induced by the additional heat treatment of 1050 °C depends on the product of NRo of oxygen precipitates. Accordingly, we can know the product of NRo of oxygen precipitates by the additional heat treatment. When they derived the same NRo by two methods, they therefore concluded that the gettering sites are the oxygen precipitates and that the gettering of Fe is a diffusion-limited process. Hieslmair et al. [5] later confirmed the validity of Gilles' model in describing gettering process at low temperatures.

The precipitation rate of iron silicide in the diffusion-limited model is limited by the diffusion rate of Fe in the silicon crystal. This presents a problem, however, as it compromises the results of the model when the flux of the precipitation induced by the nucleation and growth of iron silicide is smaller than the diffusion flux. From this point of view, we can understand that the diffusion-limited model well explains the gettering process at low temperatures at which the diffusion rate is very low.

2-2 This model

In this section, we explain the design of our model of the internal gettering of Fe based on the nucleation of iron silicide on the surfaces of the oxygen precipitates.

Eq.5 describes the energy change of a spherical silicide formation with a radius of Rs.

$$\Delta G = -4\pi/(3\Omega)\, Rs^3\, k_B T \ln C(t)/C^{eq} + 4\pi Rs^2\, \sigma \qquad 5)$$

where Ω is a volume of iron silicide per Fe atom, k_B is the Boltzmann constant, T is absolute temperature, and σ is the surface energy at the interface of iron silicide and silicon. Eq.6 gives the nucleation rate on the surfaces of the oxygen precipitates.

$Is = \{4 \pi Rs^*D\ C(t)\}\ Z(T)\ \rho\ \exp(-\Delta G^*/k_B T)$ $[1/\text{sec cm}^2]$ 6)

Rs^* is a critical radius of iron silicide derived by $\partial \Delta G/\partial r = 0$. ρ is the site density of the surface of oxygen precipitates (atoms/cm^2). $\rho\ \exp(-\Delta G^*/k_B T)$ is the equilibrium concentration of the critical nucleus, and $Z(T)$ is the Zeldovich factor.

The density of the oxygen precipitates with iron silicide formed from time t to t + Δt is

$\Delta Ns(t) = Is(t)\ 4\pi R^2\ (N - Ns)\ \Delta t$ (/cm^3)   7)

where Ns is a density of the oxygen precipitates with iron silicide. For simplification, we set the assumptions:

1) The oxygen precipitates immediately become diffusion-limited gettering sites when the iron silicide nucleates on their surfaces. In other words, we skipped over the stage when the nucleus of the iron silicide grows on the oxygen precipitate and envelops it (a stage we consider to be negligibly short).
2) The oxygen precipitates immediately cease to act as sites of iron silicide nucleation when the iron silicide nucleates on the oxygen precipitates. They act solely as the diffusion-limited gettering sites without the nucleation process.
3) The oxygen precipitates without iron silicide do not contribute the gettering.

That is, we introduced the nucleation barrier into the diffusion-limited gettering model. The oxygen precipitate density with iron silicide is calculated by Eq. 8.

$$N_S(t) = \int_{t=0}^{t=t} I_S(t)\ 4\pi R_o^2 \left(N - N_S(t)\right) dt \qquad 8)$$

We can derive the concentration change of Fe at time t by substituting Ns for N in Eq. 2. More specifically, we numerically calculate the density of the effective gettering sites by Eq. 8 and then find the concentration change of Fe by Eq. 2 at each time point in the gettering process. These steps make it possible to calculate the retained Fe concentration in silicon crystal.

We determined the surface energy σ at the interface of iron silicide, an unknown parameter, by fitting it together with $Z(T)$ using the experimental results of Hieslmair et al. [5].
We will introduce results obtained by Hieslmair et al. [5] in experiments to quantitatively investigate Fe gettering in silicon containing oxygen precipitates of different densities under a wide range of temperatures. Hieslmair's group derived the relaxation time τ by

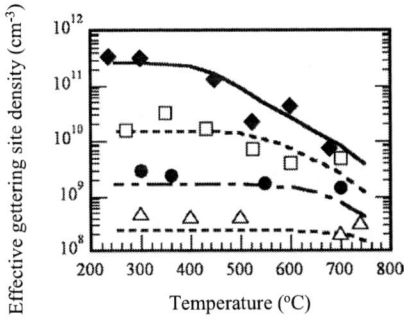

Fig. 1 Relationship between the temperature and the density of effective gettering sites with various oxygen precipitate states.
The lines in this figure denote the calculation results by our model.

Exp.	Cal.	Density (cm^{-3})	Radius (nm)
△	-----	2.3×10^8,	83.8
●	— · · —	1.6×10^9,	54.6
□	- - - -	1.5×10^{10},	41.5
◆	———	2.5×10^{11},	22.5

fitting the decay curve of Fe concentration to Eq. 3, then, estimated the density of effective gettering sites by Eq. 4 using the derived τ.

$$N = 1/(4\pi Ro\, D\tau) \qquad 9)$$

Fig. 1 shows their experimental result. The oxygen precipitates in the silicon wafers used had the following densities and radii: 2.3×10^8 cm^{-3}, 83.8 nm, 1.6×10^9 cm^{-3}, 54.6 nm, 1.5×10^{10} cm^{-3}, 41.5 nm and 2.5×10^{11} cm^{-3}, 22.5 nm. The radii as spheres were estimated by the precipitation content of the oxygen and density. As we see from the figure, the densities of the effective gettering sites agree with the density of the oxygen precipitates when the temperatures or densities are low. In other word, the prediction of the diffusion-limited model agrees with the experimental result under these conditions, but differs from the experimental result when the temperatures and densities of oxygen precipitates are high.

Next, we determine the surface energy σ and Z(T) using the data which diverged most sharply from the prediction as the fitting target (i.e., data plotted at an oxygen precipitate density of 2.5×10^{11} cm^{-3}).

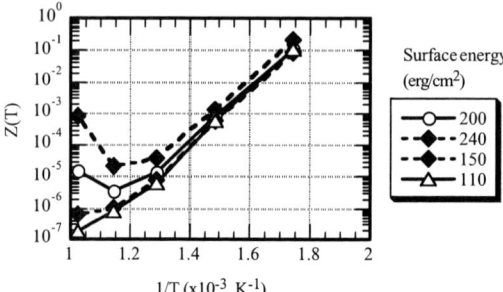

Fig. 2 Relationship between the temperature and Zeldovich factor, a variable representing the density of effective gettering sites, in the experimental results obtained with an oxygen precipitate density of 2.5×10^{11} cm^{-3} in Fig. 1 under the assumption of various surface energies.

Fig. 2 shows Z(T), a variable representing the density of effective gettering sites, in the experimental results obtained with an oxygen precipitate density of 2.5×10^{11} cm^{-3}. Z(T) was determined at each temperature under various surface energies. Z(T) revises the nucleation rate of iron silicide, hence we can get Z(T) representing the density of effective gettering sites in the experimental results by a trial-and error method. When the nucleation rate accelerates, all of the oxygen precipitates become effective gettering sites. When the nucleation rate slows, on the other hand, a few of the gettering sites formed earlier begin to decrease the Fe concentration and thereby prevent the nucleation from proceeding. This, in turn, reduces the density of the effective gettering sites. For this reason, we can specify Z(T) providing a result corresponding to the sites density of the experiments.

Fig.2 also informs us that Z(T) is an Arrhenius type function for the surface energy of 110 erg/cm^2. This is only natural. Given this condition, we selected the surface energy of 110 erg/cm^2 and the corresponding Z(T), that is, Z(T) = 3.43×10^{-16} exp(1.63 eV/k_BT). Although Z(T) is too small as a Zeldovich factor, we can presume that it corrects all of the error by simplifying the model. We therefore treated Z(T) as a fitting factor.

Using the σ and Z(T) determined, we calculated the relaxation time of the Fe concentration for the other oxygen precipitate states, that is: 2.3×10^8 cm^{-3}, 83.8 nm, 1.6×10^9 cm^{-3}, 54.6 nm and 1.5×10^{10} cm^{-3}, 41.5 nm. The densities of the effective gettering sites were derived by Eq. 9 using the calculated τ. The lines in Fig. 1 show the calculation results. The agreement between the experimental plots and the calculation results at the oxygen precipitate density of 2.5×10^{11} cm^{-3} is entirely to be expected. Notable, however, are the agreements between the calculation results and the

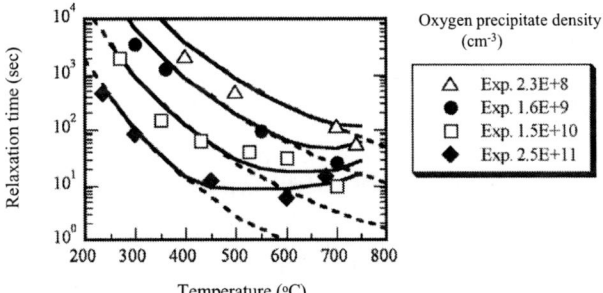

Fig. 3 Relationship between the temperature and relaxation time of gettering under various oxygen precipitate states.
Plots : experimental result [5]
Solid lines denote the calculation results by this model. Dotted lines denote the calculation results by the diffusion-limited model.

experimental plots for the other oxygen precipitate states. That is, we can estimate the gettering behavior under all conditions using the σ and Z(T) determined.

Fig.3 shows the relation between the temperature and the relaxation time. We prepared this figure by changing the vertical axis from the density of effective gettering site (the vertical axis used Fig. 1) to the relaxation time of Fe concentration. Solid and dotted lines denote the calculation results by this model and by the diffusion-limited model, respectively. The calculation results by the two models under low-temperature conditions are in agreement. The following physical parameters were used in this calculation:

$D = 1.3 \times 10^{-3} \exp(-0.68 \text{ eV}/k_B T) \text{ cm}^2/\text{s}$ [7], $C^{eq} = 4.3 \times 10^{22} \exp(-2.1 \text{ eV}/k_B T) \text{ cm}^{-3}$ [2], $\Omega : 3.9 \times 10^{-23} \text{ cm}^3$ [8], $\rho : 1.35 \times 10^{15} \text{ cm}^{-2}$

Sueoka [8] reported that the internal gettering of nickel took place at a rate several orders of magnitude lower than that predicted by the diffusion-limited model. Sueoka supposed that the interface reaction on the surfaces of the oxygen precipitates limited the gettering reaction (that is, the reaction-limited model). When we consider the delay of the internal gettering of Fe using the reaction-limited model as an alternative, the gettering rate at high temperature only becomes slower than that predicted by the diffusion-limited model if we assume that the reaction constant of the interface reaction is temperature-independent. This seems to correspond to the experimental results of Hieslmair. In contrast, however, the gettering rate calculated by the reaction-limited model is delayed both at a low and high oxygen precipitate density at high temperature. The model fails to explain why the gettering rate is only delayed at a high oxygen precipitate density. Therefore, we cannot apply the reaction-limited model for the internal gettering of Fe.

3 CALCULATION RESULT

3-1 Dependence on the oxygen precipitate density

The most characteristic feature in the internal gettering of Fe is the influence of the oxygen precipitate density in changing the delay of the gettering rate relative to the rate predicted by the diffusion-limited model at high temperatures. In this section we attempt to explain this process by our simulation model. Figs. 4 a), b) and c) denote the changes of Fe concentration and the oxygen precipitate density with and without iron silicide at oxygen precipitate densities of 10^8, 10^9 and 10^{10} cm^{-3}, respectively. The temperature is 700 °C and the initial concentration of Fe is 3×10^{13} atoms/cm^3.

In the case of a low oxygen precipitate density (see Fig. 4a), the iron silicides immediately nucleate on all of the oxygen precipitates, whereupon the oxygen precipitates act as effective gettering sites for a very short time. In the case of high oxygen precipitate density (see Fig. 4c), on the other hand, a small fraction of the oxygen precipitates becomes effective gettering sites with iron silicide. This can be explained by the sharp drop of the Fe concentration at the high oxygen precipitate density and the halting of the iron silicide nucleation before the iron silicides nucleate on all of the oxygen precipitates. Thus, only a small fraction of the oxygen precipitates act effective gettering sites. This remarkable feature only appears at high temperature.

The agreement between the results calculated by our model and by the diffusion-limited model at the low temperature range in Fig. 3 can be explain by the temperature-dependence of the nucleation rate. This temperature-dependence is determined by the changing of the degree of super-saturation and the diffusion coefficient by temperature. The activation energies of the diffusion coefficient and solubility of Fe are 0.68 and 2.1 eV, respectively. This system can therefore be assumed to be strongly controlled by the degree of super-saturation, and the nucleation rate increases at lower temperature. This is why all of the oxygen precipitates become effective gettering sites at lower temperatures.

Our model thus explains why the gettering rate at high temperature and high oxygen precipitate density are only delayed relative to the rate predicted by the diffusion-limited model.

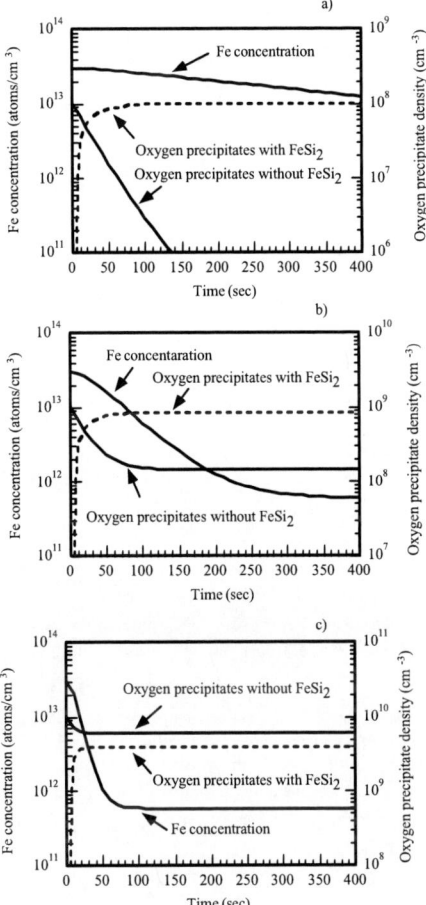

Fig. 4 Changes of Fe concentration and oxygen precipitate density with and without iron silicide at oxygen precipitate densities of a) 10^8, b) 10^9 and c) 10^{10} cm^{-3}, respectively. The temperature is 700°C and the initial concentration of Fe is 3×10^{13} atoms/cm^3.

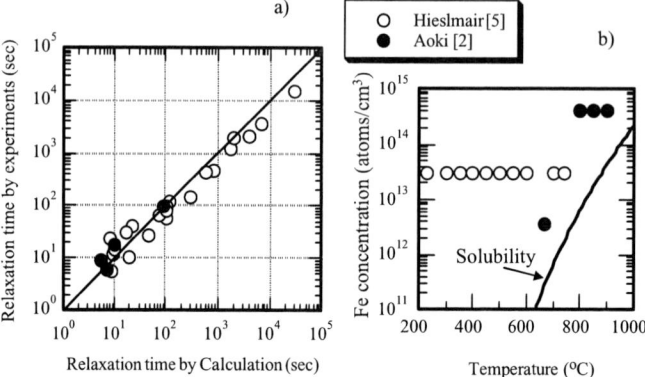

Fig. 5 Comparison of the relaxation time between the experiment and calculation by our model a).
b) the temperature and Fe concentration in the experiments.

3-2 Comparison with the experiment by Aoki

Aoki et al. [2,3] conducted one of the few studies to clarify both the densities and radii of the oxygen precipitates in the internal gettering of Fe. They described the density and the precipitation content of the oxygen in the wafer used. We estimated the radii of oxygen precipitates as spheres $(2.5 \times 10^{10}$ cm^{-3}, 34.2 nm). Fig. 5 b) shows the temperature of gettering and the Fe concentration under Aoki's experimental conditions, together with corresponding data from the experiment by Hieslmair et al. [5]. Fig. 5 a) compares the relaxation time in the experiment by Aoki and Hieslmair with that in the calculation by our model. Given the high temperature and high oxygen precipitate density in the experimental condition of Aoki et al., we know that the diffusion-limited model cannot explain the relaxation times in their results. We see from Fig. 5 a), however, that our calculation results agree well with their results.

3-3 Dependence of the gettering rate on the total volume of oxygen precipitates (NRo^3)

Ogushi et al. [4] reported that the gettering rate of Fe depends on the total volume of oxygen precipitates. In this section we discuss how N and Ro influence the gettering rate by investigating the relationship among the gettering relaxation time and density and radius of oxygen precipitates in our calculation model. Fig. 6 a) and b) denote the relationship among the gettering relaxation time (expresses as the contour lines) and the

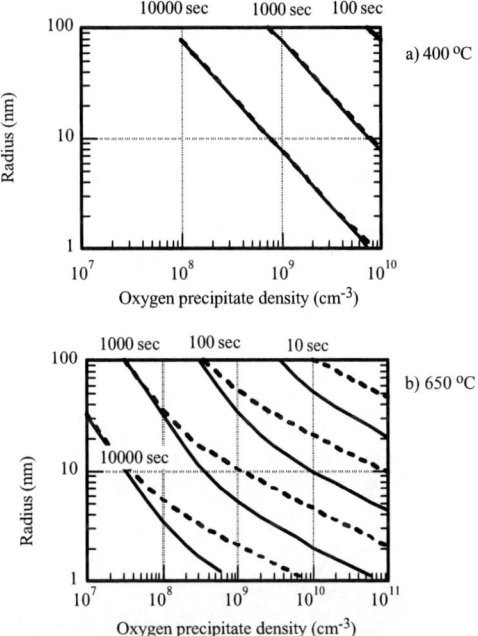

Fig. 6 Relationship among the relaxation time of gettering (expressed as a contour line) and the density and radius of oxygen precipitates at a) 400 °C and b) 650 °C, respectively.
Solid and dotted contour lines denote the Fe contamination of 10^{14} atoms/cm^3 and 10^{13} atoms/cm^3, respectively.
The solid and dotted lines overlapped each other in Fig. 6 a).

density and radius of the oxygen precipitates at 400 °C and 650 °C, respectively. The solid and dotted lines denote Fe contaminations of 10^{14} atoms/cm^3 and 10^{13} atoms/cm^3 in the initial concentrations, respectively. We see, in Fig. 6 a), that the solid and dotted lines overlap. This tells us that the relaxation time does not depend on the initial Fe concentration at 400 °C. The contour lines are straight-lines and the product of NRo on each contour line is constant in Fig. 6 a). In this case, the relaxation time depends on the product of NRo. This describes the behavior predicted by the diffusion-limited model.

In Fig. 6 b), however, we see that the slopes of the contour lines change at high oxygen precipitate density. At 650 °C, the relaxation time depends on the product of NRo at a low oxygen precipitate density and on NRo^3 (the total volume of oxygen precipitates) at high oxygen precipitate density. And, at high oxygen precipitate density, the relaxation time also depends on the initial Fe concentration. In other words, the gettering rate in the condition controlled by nucleation depends on NRo^3 and the initial concentration.

Given that the experiments of Ogushi were conduced at a high temperature (600 °C) and at a high oxygen precipitate density (10^5 and 10^6 cm^{-2}), we know that the condition was controlled by nucleation. Accordingly, the gettering rate of Fe must have been depended on the total volume of oxygen precipitates.

We now turn to the question of why the relaxation time depends on NRo^3 under the condition controlled by nucleation. The density of gettering sites (Ns) depends on the nucleation rate under this condition. The nucleation rate is therefore proportional to the total surface area of the nucleation site ($4\pi Ro^2$ N), which means that Ns $\propto 4\pi Ro^2$ N.

$$\frac{1}{\tau} = 4\pi R_0 DNs \propto 4\pi R_0 D\left(4\pi R_0^2 N\right) \propto N R_0^3 \qquad 10)$$

Thus, the relaxation time depends on NRo^3 under the condition controlled by nucleation. From Fig. 6 b), we find that the relaxation time varies as the initial Fe concentration changes. Because our calculation results well correspond to the experimental results obtained using a wide range of the Fe concentration, as shown in Fig 5, we conclude that the concentration-dependence of our model is valid.

4 SUMMARY

We propose a new simulation model to describe the behaviors of the internal gettering of Fe by oxygen precipitates, with considering the nucleation of iron silicide on the oxygen precipitates. This proposed model accurately describes the internal gettering behaviors of Fe at both low and high temperatures and also explain how the gettering behavior is influenced by the product of NRo at low temperature and by the product of NRo3 at high temperature, where N and Ro are density and radius of oxygen precipitates, respectively.

REFERENCES

[1] D.Gilles and E.R.Weber, Physical Review Letters, 64, 196 (1990)

[2] M.Aoki, A.Hara and A.Ohsawa, J.Appl.Phys. 72, 895 (1992)

[3] M.Aoki, A.Hara and A.Ohsawa, Jpn.J.Appl.Phys. 30, 3580 (1991)

[4] S.Ogushi, S.Sadamitsu, K.Marsden, et al., Jpn.J.Appl.Phys. 36, 6601 (1997)

[5] H.Hieslmair, A.A.Istratov, S.A.McHugo, C.Flink and E.R.Weber, J. Electrochem. Soc. 145, 4259 (1998)

[6] V.V. Voronkov and R.Falster, J Crystal Growth 194, 76 (1998)

[7] E.R. Weber, Appl.Phys. A30, 1 (1983)

[8] H.Hieslmair, A.A.Istratov, T.Heiser, and E.R.Weber, J.App.Phys. 84, 713 (1998)

[9] K.Sueoka, J.Electrochem.Soc. 152, G731 (2005)

ECS Transactions, 2 (2) 287-297 (2006)
10.1149/1.2195666, copyright The Electrochemical Society

Silicon Self-diffusion in Heavily B-doped Si Using Highly Pure [30]Si Epitaxial Layer

S. Matsumoto[a], S. R. Aid[a], S. Seto[a], K. Toyonaga[a], Y. Nakabayashi[a], M. Sakuraba[b], Y. Shimamune[b], Y. Hashiba[b], J. Murota[b], K. Wada[c], and T. Abe[d]

[a] Department of Electronics and Electrical Engineering, Keio University
Hiyoshi, Yokohama 223-8522, Japan
[b] Research Institute of Electrical Communication, Tohoku University
Sendai 980-8577, Japan
[c] Department of Materials, University of Tokyo, Hongo, Tokyo 113-8656, Japan
[d] Shin-Etsu Handoutai, Isobe, Gunma 379-01, Japan

Si self-diffusivity in heavily B-doped Si at 867-1067°C has been determined directly using highly pure isotope [30]Si as a diffusion source. Samples consist of [30]Si epi-layer/B-doped natural Si epi-layer/natural Si substrates. Si self-diffusivity increases with the increase of B concentration and its enhancement degree is larger at lower diffusion temperatures. Based on the model of the Fermi level effect, energy levels of donor Si self-interstitial and acceptor vacancy are determined from the analysis of the experimental data.

Introduction

Toward the fabrication of smaller transistors, formation of very shallow and highly doped source and drain junctions becomes an every more important challenge. For such junction formation, it is indispensable to develop a predictive dopant diffusion model with physically meaning parameters in Si. Since dopant atoms are mediated via point defects, i.e., vacancies (V) and Si self-interstitials (I), information about point defect properties in Si is essential.

It has been known that vacancies in Si have different charge states, i.e., neutral (V^x), singly-charged acceptor-type (V^-), doubly-charged acceptor-type ($V^=$) and singly-charged donor-type (V^+) from electron paramagnetic resonance (EPR) experiments at cryogenic temperatures (1,2). Therefore, the equilibrium concentration of respective charge-state vacancies depends on the position of the Fermi level in Si (3), i.e., the concentration of dopant atoms. On the other hand, there is a lack of reliable information about the charge states and the energy levels in the band-gap of Si self-interstitials. It is, however, suggested that they also have acceptor and donor levels (4) in the band-gap.

Until now, the Fermi level effect on point defect concentration has been investigated intensively in different ways in order to reveal information about charge states of point defects. Giles extracted the position of the singly-charged acceptor level, E_I^-, and donor level E_I^+, of Si self-interstitals from analysis of oxidation-enhanced diffusion experiments under extrinsic conditions (5,6). Roth et al. also determined the E_I^-, and E_I^+ (7). These data are much different each other and its main reason is considered that they were estimated indirectly from the experiments of the high concentration dopant diffusion. Then more direct estimation of the energy levels of point defects is expected.

Since self-diffusion in Si is mediated via point defects directly, the study of Si self-diffusion in extrinsic Si is very suitable to unveil the Fermi level effects on point

287

defect concentrations. The growth of epitaxial layer with highly pure stable isotopes of ^{28}Si or ^{30}Si enabled us to investigate the above Fermi level effect. The preliminary works have been performed using isotopically enriched Si epitaxial layer structures under restricted annealing temperatures and doping levels (8,9), but systematic study has not been carried out yet.

In this work, Si self-diffusion under extrinsic conditions at different temperatures has been investigated by monitoring the diffusion of isotope ^{30}Si in Si doped with various boron (B) concentrations. Using these data, the energy levels of point defects in Si have been determined.

Experimental

B-doped or non-doped Si epitaxial-layers were grown on CZ-Si substrates with a thickness of about 500 nm with low-pressure CVD (LCVD) at 700℃ using normal SiH$_4$ and B$_2$H$_6$. B concentrations in the epitaxial layer were varied from 2×10^{18}cm^{-3} to 1×10^{21}cm^{-3} by controlling B$_2$H$_6$/ SiH$_4$ flow rate. After depositing the B-doped epitaxial layers, carrier concentration was measured with four-point probe method and it was confirmed that almost all dopants in the epitaxial layer were electrically activated except very heavily doped sample (1×10^{21}cm^{-3}). Next, isotopically pure ^{30}Si epitaxial layers with a thickness of 30~40 nm were grown on B-doped Si epitaxial layers with LCVD at 700℃ using enriched ^{30}SiH$_4$ purchased from Kurchatov Institute, Russia. The epitaxial growth by LCVD was performed at Tohoku University.

These samples were annealed at 867℃ for 264 h, 967℃ for 10 h and 1067℃ for 1h in Ar (99.999%) ambient. The concentrations of ^{30}Si isotope and B in the sample before and after annealing were analyzed with secondary ion mass spectrometry (SIMS) with the ATOMIKA SIMS instrument with a 3 keV O^{2+} primary beam. Diffusivity of ^{30}Si in B-doped Si epitaxial layers was determined using a numerical fitting process by solving Fick's equation. In this process, the as-grown ^{30}Si profiles were used as an initial condition. The best fit was determined by minimizing the root-mean-square error. The diffusivity of ^{30}Si determined in such a procedure is referred as the self-diffusivity of Si in the present work. The experimental error is estimated within ±20%. The largest contributor to the error is the uncertainty in crater depth measurement of SIMS.

Results and Discussion

Dependence of Si Self-Diffusivity on B concentration

Figure 1 shows the SIMS profiles of Si isotopes (^{28}Si, ^{29}Si, ^{30}Si) in the ^{30}Si epitaxial and B-doped Si epitaxial layers of the as-grown samples. It is found that highly pure ^{30}Si layer with a thickness of 30~40 nm was grown on the B-doped Si epitaxial layers. In B-doped Si epitaxial layers, natural abundance of the respective Si isotopes is obtained. In order to avoid the complex of the figure, B concentration profiles are not included in Fig. 1.

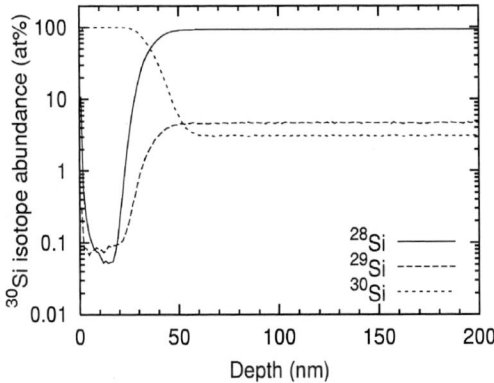

Fig. 1 Respective Si isotope SIMS profiles of as-grown sample.

Figure 2 shows a typical example of B concentration profiles of the as-grown and after-annealing (967℃,10 h) samples with SIMS analysis. As shown in the figure, B diffuses from the uniform B-doped epi-layer after annealing and B concentration near the surface changes from $2 \times 10^{19} cm^{-3}$ to $1 \times 10^{19} cm^{-3}$ in this case.

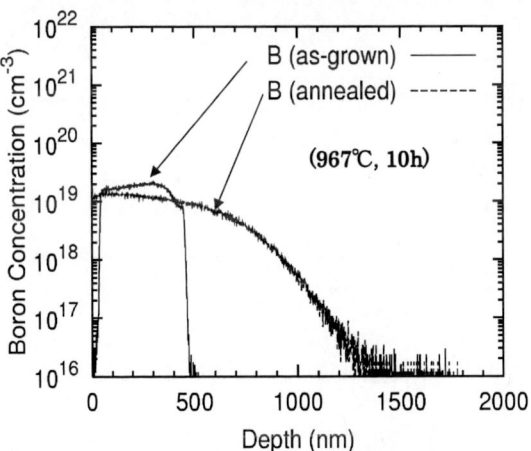

Fig. 2 B concentration SIMS profiles before and after annealing

For other samples, B concentration near the surface is also reduced after annealing. It is noted that in the discussion of B concentration dependence of Si self-diffusivity, B concentration near the surface after annealing are used instead of the initial B concentration in the epitaxial layer before annealing.

Although respective isotopes (^{28}Si, ^{29}Si, ^{30}Si) diffuse each other, the movement of ^{30}Si diffusing into semi-infinite Si substrates is monitored as Si self-diffusion in this work. Curves in Fig. 3 show the SIMS data of ^{30}Si concentration annealed at 967℃ for 10 hr in Si with B concentrations of $2x10^{18}$, $1x10^{19}$ and $3x10^{19}$. These B concentrations are the values after-annealing obtained from SIMS analysis as described above. In the figure, the as-grown profile shown with the solid curve is also included, which was used as the initial condition in the numerical simulation. Since the intrinsic carrier concentration at 967 ℃ is $5.2x10^{18}$cm^{-3}, the sample with B concentration of $2x10^{18}$cm^{-3} is regarded as the intrinsic Si. The Si self-diffusivity obtained under this intrinsic condition at 967℃ agrees well with the intrinsic value previously reported by our group (8,11) as well as Bracht et al (10). On the other hand, in extrinsic conditions, diffusion of ^{30}Si is enhanced with the increase in B concentration. The simulated curves obtained from the numerical calculation fitted well with the experimental SIMS data, indicating a very accuracy of the data of Si self-diffusivity. The Si self-diffusivity for different annealing temperatures (867℃ and 1067℃) is also determined for each B concentration.

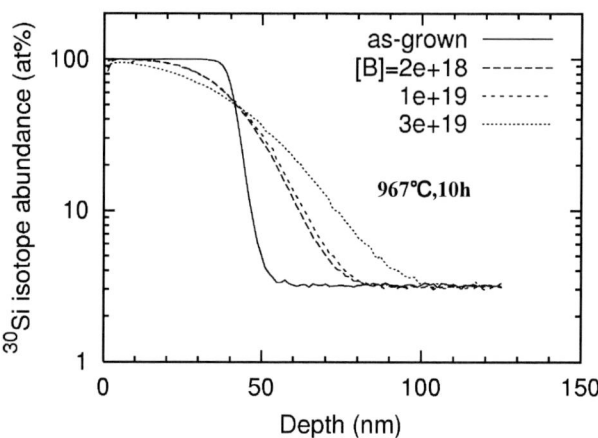

Fig. 3 ^{30}Si SIMS profiles in various B-doped Si epitaxial layers.

The extracted Si self-diffusivity is shown as a function of B concentrations for 867, 967 and 1067℃ in Fig. 4. The intrinsic carrier concentrations, n$_i$, at 867, 967 and 1067℃ are $2.8x10^{18}$cm^{-3}, $5.2x10^{18}$cm^{-3}, and $9.3x10^{18}$cm^{-3}, respectively. Since B in epitaxial layers is electrically activated except very high B concentration, B concentration is regarded as the carrier concentration. It is found from Fig. 4 that Si self-diffusivity is constant below the intrinsic carrier concentration for respective annealing temperatures, and it increases almost linearly with the carrier concentration above n_i except above $1 \times 10^{20} cm^{-3}$. Above $1 \times 10^{20} cm^{-3}$, Si self-diffusivity seems to be saturated. This is due to the fact that at very heavy B concentration above $1 \times 10^{20} cm^{-3}$, the deactivation of B occurred and the carrier concentration was restricted to about $1 \times 10^{20} cm^{-3}$, which was confirmed with the spreading resistance measurements.

The linear dependence of Si self-diffusivity on the carrier concentration is due to the increase in the concentration of point defects. For the mechanism of the increase in point defect concentration, both physical (strain effect) and electronic effects (Fermi level effect) have been considered. Although the strain due to the difference in covalent radius between Si and B may affect the point defect concentration, it is not clear to include its effect quantitatively at present and then the Fermi level effect is only considered in this discussion.

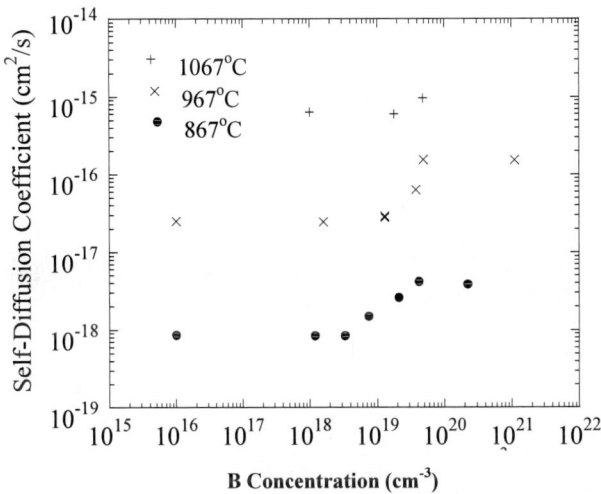

Fig. 4 Si self-diffusivity as a function of B concentration

Determination of the position of acceptor vacancy and donor Si self-interstitial energy levels

Using the data of concentration dependence of Si self-diffusivity in Fig 4, energy levels of point defects in the band gap will be extracted as follows.

Si self-diffusivity, D^{SD}, is given by a sum of vacancy and Si self-interstitial terms with different charge states as follows.

$$D^{SD} = \phi_V C_V^r D_V^r + \phi_I C_I^r D_I^r \qquad [1]$$

where C_V^r and D_V^r are the concentration and diffusivity of vacancies with charge state r ($=0$, ±1 and ±2), respectively. Similarly, C_I^r and D_I^r are the concentration and diffusivity of Si self-interstitials. ϕ_V and ϕ_I are the correlation factors for the vacancy and interstitialcy diffusion mechanisms, where $\phi_V=0.5$ (12) and $\phi_I=0.73$ (13).

The Fermi level, E_F, dependence on C_V^r or C_I^r is given originally by the treatment of Shockley and Last (3). For example, concentrations of singly-charged acceptor-type vacancy (C_V^-) and singly-charged donor-type vacancy (C_V^+) are expressed by

$$C_V^- = C_V^0 \exp[(E_F-E_V^-)/kT] \qquad [2]$$

$$C_V^+ = C_V^0 \exp[(E_V^+-E_F)]/kT] \qquad [3]$$

where C_V^0 is the concentration of neutral vacancy, and E_V^- and E_V^+ are the energy levels of V^- and V^+, respectively. For singly-charged acceptor-type Si self-interstitial (C_I^-) and singly-charged donor-type Si self-interstitial (C_I^+), the similar expressions are given by replacing V with I in Eq. [2] and Eq. [3], respectively. For doubly-charged acceptor-type and donor-type point defects, the expressions are given in the reference (3).

In a general treatment of the Fermi level effect, contribution of double and more charge point defects ($r \geq 2$) might to be included. However, in this work, only singly-charged vacancy and Si self-interstitial are considered. It is considered reasonable from the fact that Si self-diffusivity increases almost linearly with B concentration in the present experimental conditions, suggesting the main contribution of singly-charged point defects.

Regarding singly-charged point defects, there are four species, that is, acceptor-type vacancy, V^-, donor-type vacancy, V^+, acceptor-type Si self-interstitial, I^-, and donor-type Si self-interstitial, I^+. The concentration of these point defects depends on the position of the Fermi level, E_F as expressed by Eq. [2] and Eq. [3]. In discussing the type of dominate point defects in extrinsic Si, Nakabayashi et al. (14) investigated the Si self-diffusion in heavily As- ([As]= 3×10^{19}cm^{-3}) and B-([B]=2×10^{19}cm^{-3}) doped Si under oxidizing ambient, that is, the introduction of Si self-interstitials from the Si surface. They showed that Si self-diffusivity was reduced in heavily As-doped Si, while it was enhanced in heavily B-doped Si. Their result clearly indicated that excess vacancies were generated in heavily As-doped Si, while excess Si self-interstitials were generated in heavily B-doped Si. Taking account of Eq. [2] and Eq. [3], V^- is found to be dominant in n-type Si and I^+ is dominant in p-type Si. Therefore, as singly-charged acceptor-type point defects, only V^- is considered and similarly I^+ is considered as singly-charged donor-type point defects in the present work.

In respect to diffusivity of point defects with different charge states, point defects with each charge state could have a different free energy of migration that also depends on E_F. However, since the dependence of diffusivity on E_F is not clear at present, we make the following assumption that each diffusivity with singly-charged point defects is same as that of neutral state.

$$D_{V^-}= D_{V0}, \qquad D_{I^+}= D_{I0}, \qquad [4]$$

Assuming that only I^+ and V^- contribute to Si self-diffusion as stated above, the ratio of Si self-diffusivity under extrinsic conditions ($D_{ext}{}^{SD}$) to that under intrinsic conditions ($D_{int}{}^{SD}$) is given by

$$\frac{D_{ext\,SD}}{D_{int\,SD}} = \frac{D_{vo} + D_{vo}\exp\left(E_i - E_{V-}/kT\right)\left(n/n_i\right) + D_{Io} + D_{Io}\exp\left(E_{I+} - E_i/kT\right)\left(p/n_i\right)}{D_{vo} + D_{vo}\exp\left(E_i - E_{V-}/kT\right) + D_{Io} + D_{Io}\exp\left(E_{I+} - E_i/kT\right)}$$

[5]

where the following relationships are used.

$$\frac{n}{n_i} = \exp[\frac{E_F - E_i}{kT}] \quad \text{and} \quad \frac{p}{n_i} = \exp[\frac{E_i - E_F}{kT}]$$

[6]

By defining $\gamma^- = \exp[(E_i - E_V{}^-)/kT]$, $\gamma^+ = \exp[(E_I{}^+ - E_i)/kT]$ and $\alpha = D_{V0}/D_{I0}$, we have

$$\frac{D_{ext\,SD}}{D_{int\,SD}} = \frac{1 + \gamma - (n/n_i)}{(1+\gamma_+) + 1/\alpha(1+\gamma_-)} + \frac{1 + \gamma + (p/n_i)}{(1+\gamma_-) + \alpha(1+\gamma_+)}$$

[7]

The experimental data of the diffusivity ratio can be fitted with the calculated results of Eq. [7] with parameters of γ^-, γ^+, and α. In the fitting procedure, the data obtained by Nakabayashi et al. (8), where Si self-diffusion was performed under extrinsic n-type ([As]= 3×10^{19}cm^{-3}) Si in inert ambient, were also used in the above calculation as well as the present results obtained from extrinsic B-doped p-type Si.

Figure 5 shows an example of diffusivity enhancement at 967°C as a function of p/n_i from the above fitting procedure. The point below 1 in p/n_i of the horizontal axis in Fig 5 is the data of Nakabayashi et al. (8) as described above. From these results, the levels of singly-charged acceptor-type vacancy, $E_V{}^-$ and singly-charged donor-type Si self-interstitial, $E_I{}^+$ are obtained at temperatures of 867, 967 and 1067°C, respectively.

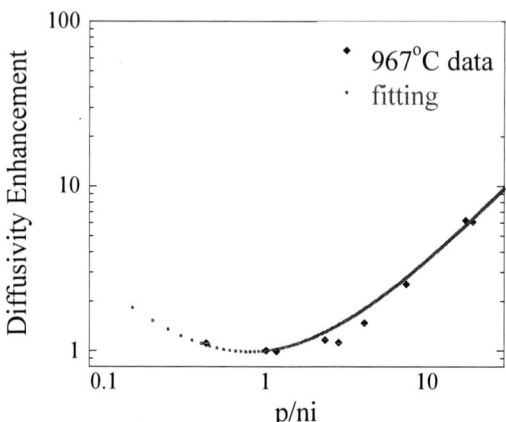

Fig. 5 Diffusivity enhancement as a function of p/n$_i$ at 967℃.
Plots are experimental data and a solid curve is the fitting curve.

Figure 6 (a), (b) and (c) shows the position of vacancy acceptor level (E_V^-) and Si self-interstitial donor level (E_I^+) in Si at 867, 967 and 1067℃, respectively. Values of energy band gap of Si are regarded with the temperature dependence of the band gap. The following results are extracted at the temperature range from 867-1067℃.

$$E_V^- = E_C - 0.37 \sim 0.45 \text{eV} \qquad [8]$$

$$E_I^+ = E_V + 0.37 \sim 0.42 \text{eV} \qquad [9]$$

It has been known from electron paramagnetic resonance (EPR) experiments at cryogenic temperatures (1,2) that singly-charged vacancy acceptor has a relatively deep energy level in the band-gap ($\sim$0.44eV). Although the experimental temperatures between the present work and EPR work are so different each other, very similar values are obtained in both works. Regarding Si self-interstitial donor level, Giles (5,6) reported from the experiment of extrinsic background doping in dopant diffusion that it has a value of 0.39eV above the valence band edge (E_V). The present results are good agreement with Giles's work and also the previously proposed value by Frank et al. (4). Thus both singly charged vacancy acceptor and singly charged Si self-interstitial donor have deep energy levels in the band gap.

Ural et al. (9) tried to estimate the position of energy levels of point defects based on the Fermi level effect model, but a very unrealistic assumption, that is, the position of energy levels of vacancy and Si self-interstitial is same, was made in their analysis.

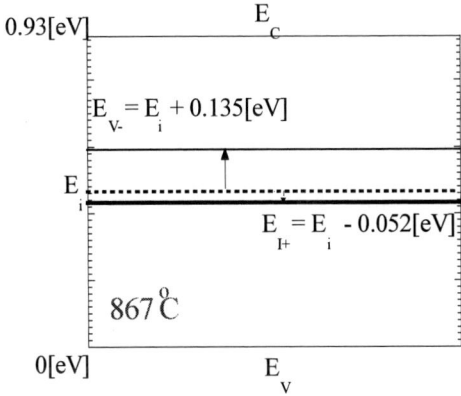

Fig. 6(a) Vacancy acceptor level (E_V^-) and Si self-interstitial
donor level (E_I^+) in Si at 867℃.

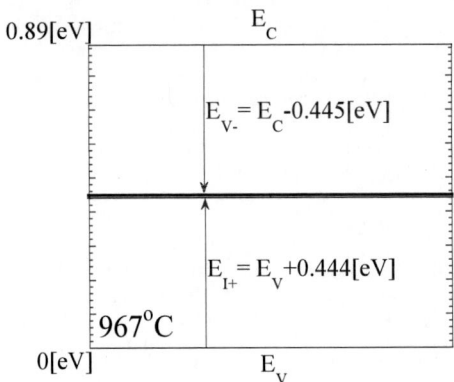

Fig. 6(b) Vacancy acceptor level (E_V^-) and Si self-interstitial
donor level (E_I^+) in Si at 967℃.

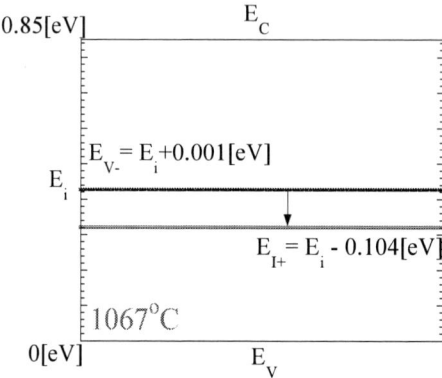

Fig. 6(c) Vacancy acceptor level ($E_V{}^-$) and Si self-interstitial donor level ($E_I{}^+$) in Si at 1067°C.

Conclusion

Si self-diffusivity in heavily B-doped Si at 867-1067°C has been determined directly using highly pure isotope ^{30}Si as a diffusion source. Samples consist of ^{30}Si epi-layer/boron-doped natural Si epi-layer/natural Si substrates. Si self-diffusivity increases with the increase of B concentration and its enhancement degree is larger at lower diffusion temperatures. We determined the position of energy levels of singly-charged vacancy acceptor and singly-charged Si self-interstitial donor in Si at 867-1067°C based on the model of Fermi level effect on point defect concentration.

Acknowledgments

We would like to thank M. Suzuki, A. Takano and Y. Sato of NTT-AT for measuring SIMS profiles. This work was partly supported by a Grant-in-Aid for Scientific Research (A) (10305030) from the Ministry of Education, Culture, Sports, Science and Technology, and by the Foundation for the Promotion of Material Science and Technology of Japan.

References

1. G. D. Watkins and J. W. Corbett, *Phys. Rev.*, **134**, 1359(1964).
2. G. D. Watkins, in *Lattice Defects in Semiconductors 1974,* **23**, F. A. Huntley, Ed. (Inst. Phys. Cnf. Series). London, 1975
3. W. Shockley and J. T. Last, *Phys. Rev.*, **107**, 392(1957).
4. W. Frank, U. Gosele, H. Mehrer and A. Seeger, in *Diffusion in Crystalline Solids*, edited by G. E. Murch and A.S. Nowick(Academic, new York, 1984)
5. M. D. Giles, *IEEE Trans. On CAD*, **8**, 460(1989).
6. M. D. Giles, *Appl. Phys. Lett.*, **58**, 2399(1991).
7. D. J. Roth and J. D. Plummer, *J. Electrochem. Soc.*, **141**, 1074(1994).
8. Y. Nakabayashi, H. I. Osman, T. Segawa, K. Saito, S. Matsumoto, J. Murota, K. Wada, T. Abe, *Jpn. J. Appl. Phys.* **42**, 3304(2000).
9. A. Ural, P. B. Griffin and J. D. Plummer, Aool. *Phys. Lett.*, **79**, 4328(2001).
10. H. Bracht, N. A. Stlwijk, and H. Mehrer, *Phys. Rev.* **B52**, 16542(1998).
11. S. R. Aid, T. Sakaguchi, K. Toyonaga, Y. Nakabayashi, S. Matsumoto, M. Sakuraba, Y. Shimamune, Y. Hashiba, J. Murota, K. Wada, T. Abe, *Materials Science and Engineering B* **114-115**, 330(2004).
12. K. Compaan and Y. Haven, *Trans. Faraday Soc.* **52**, 786(1956).
13. K. Compaan and Y. Haven, *Trans. Faraday Soc.* **54**, 1498(1958).
14. Y. Nakabayashi, H. I. Osman, K. Yokota, K. Toyonaga, S. Matsumoto, J. Murota, K. Wada, T. Abe, *Materials Science in Semiconductor Processing*, **6**, 15(2003).

ECS Transactions, 2 (2) 299-309 (2006)
10.1149/1.2195667, copyright The Electrochemical Society

Modeling Growth Behavior for $Si_{1-x}Ge_x$ from SiH_4 and GeH_4 by CVD

X. L. Yang* and M. Tao

Department of Electrical Engineering, University of Texas at Arlington,
Arlington, Texas 76019, USA
* xxy2570@exchange.uta.edu

A kinetic model based on the collision theory of chemical reactions, statistical physics, and the concept of competitive adsorption is proposed for $Si_{1-x}Ge_x$ growth from SiH_4 and GeH_4 by chemical vapor deposition. It takes into account both homogeneous and heterogeneous reactions, which involve the precursors (SiH_4 and GeH_4) and the homogeneous decomposition product of germane, germylene (GeH_2), and three types surface sites, silicon sites, hydrogen-terminated silicon sites, and germanium sites. The growth of $Si_{1-x}Ge_x$ can be divided into two regimes: a heterogeneous decomposition dominated regime and a homogeneous decomposition dominated regime. Analytical equations are derived to quantitatively describe growth rate as a function of deposition conditions, including deposition temperature, silane flow rate, and germane flow rate, for the heterogeneous regime. Homogeneous decomposition of germane into germylene causes precursor depletion, and an empirical linear relation is employed to describe the growth behavior in the homogeneous regime. The model agrees well with the experimental data.

Introduction

A recent report by Thompson et al[1] drew more attention to $Si_{1-x}Ge_x$ alloy for its important application in strained-Si complementary MOSFET for mobility enhancement.[2-4] They indicated an increase in hole mobility by more than 50% in p-MOSFET and an increase in electron mobility by about 20% for n-MOSFET. Various chemical vapor deposition (CVD) techniques (such as ultrahigh vacuum CVD,[5,6] low pressure CVD,[7,8] very low pressure CVD,[9,10] atmospheric pressure CVD,[11,12] rapid thermal CVD,[13,14] and plasma-enhanced CVD[15]) have been used to deposit $Si_{1-x}Ge_x$. Several precursors have been investigated for $Si_{1-x}Ge_x$ deposition as the Ge source, such as germane (GeH_4)[5-15] and digermane (Ge_2H_6),[6,16] and as the Si source, such as silane (SiH_4),[5,6,8-10] disilane (Si_2H_6),[6,16] and dichlorosilane (SiH_2Cl_2).[12,14,17] $Si_{1-x}Ge_x$ growth by CVD is a much more complicated process than the growth of pure Si by CVD. Different behaviors in growth rate have been observed with respect to the effect of Ge content incorporated in the film. With GeH_4 and SiH_2Cl_2 as the precursors, a monotonic increase in growth rate has been reported with increasing Ge content by Kamins et al[12] in atmospheric pressure CVD at deposition temperature of 625°C, by Garone et al[14] in rapid thermal CVD between 625°C and 700°C, and by Hoyt et al[17] in rapid thermal CVD at 640°C. With SiH_4 and GeH_4 as the precursors, Jang et al[18] demonstrated that the effect of Ge content on growth rate is more complicated in very low pressure CVD. A monotonic growth rate increase was observed with Ge content at 570°C, but a monotonic growth rate decrease with Ge content at 700°C. Between 600°C and 675°C, the growth rate exhibited first an increase and then a decrease with

299

Ge content. Using ultrahigh vacuum CVD, Meyerson et al[5] reported a growth rate increase with Ge at 550°C, Robbins et al[8] reported a growth rate decrease with Ge at 700°C, and Racanelli et al[19] reported that the growth rate first increased and then decreased with Ge between 577°C and 665°C.

Several kinetic models have been proposed to explain the complicated growth behavior in $Si_{1-x}Ge_x$ CVD from SiH_4 and GeH_4. Robbins et al[8] attributed the growth rate peak with Ge content to a competition between desorption of H from the surface and adsorption of H from the gas phase. Russell et al[20] proposed a model by treating SiH_4 and GeH_4 deposition as two parallel processes, each of which is a competition between adsorption of precursor and desorption of H. Maik et al[21] attempted to explain the growth behavior on the basis of Langmuir-Hinshelwood's adsorption theory. Lee et al[22] presented a model based on competitive adsorption of GeH_4 and SiH_4 on vacant surface sites, coupled with GeH_4-assisted H desorption. Jang et al[18] explained the growth behavior with two mechanisms: (1) enhanced H desorption by Ge at low Ge content and/or temperature and (2) reduced reaction probability of GeH_4 and SiH_4 due to Ge at high Ge content and/or temperature. Although these models attempted to quantitatively explain the growth behavior of $Si_{1-x}Ge_x$ by various CVD processes, they are best described as semi-quantitative or qualitative models and a truly quantitative or comprehensive model has yet to appear. In this paper, we present a quantitative kinetic model for $Si_{1-x}Ge_x$ growth by CVD with GeH_4 and SiH_4 as the precursors. The model is based on the collision theory of chemical reactions, statistical physics, and the concept of competitive adsorption.[23,24] It takes into account both heterogeneous and homogenous reactions, which involve the precursors (GeH_4 and SiH_4) and the homogenous decomposition product of GeH_4, germylene (GeH_2), and three types of surface sites, H-free Ge sites, H-free Si sites, and H-terminated Si sites. Homogeneous decomposition of SiH_4 into silylene (SiH_2) and H-terminated Ge sites are assumed negligible. An analysis of both heterogeneous and homogeneous reactions leads to the conclusion that $Si_{1-x}Ge_x$ growth by CVD can be divided into two regimes: a heterogeneous decomposition dominated regime and a homogeneous decomposition dominated regime. Analytical and empirical equations are derived to describe the growth behavior of $Si_{1-x}Ge_x$ as a function of deposition conditions, including deposition temperature, SiH_4 flow rate, and GeH_4 flow rate.

Kinetics

Based on the collision theory of chemical reactions and statistical physics, the activated flux of a precursor, i.e. the number of precursor species that decomposes upon collision with the substrate, can be approximated by[25,26]

$$J = \frac{P}{(2\pi mkT)^{1/2}} (\frac{E_a}{kT} + 1) \exp(-\frac{E_a}{kT})$$ [1]

where P, m, and E_a are the precursor's partial pressure, molecular mass, and activation energy for its heterogeneous decomposition reaction on the substrate, respectively.

Heterogeneous decomposition.—In $Si_{1-x}Ge_x$ CVD, there are four types of surface sites on the substrate: H-terminated Si sites (H-Si), H-terminated Ge sites (H-Ge), H-free Si sites (-Si), and H-free Ge sites (-Ge). With two precursors involved (SiH_4 and GeH_4), there are a total of eight heterogeneous reactions on the substrate, which are listed in Table I and depicted in Fig. 1. In our model, four of the eight reactions are considered: GeH_4 flux on Ge sites ($J_{GeH4/Ge}$), GeH_4 flux on Si sites ($J_{GeH4/Si}$), SiH_4 flux on Ge sites ($J_{SiH4/Ge}$), and SiH_4 flux on Si sites ($J_{SiH4/Si}$). This is

300

justified by the fact that reactions on H-terminated Si and Ge sites require much larger activation energies and thus have much smaller rates. Based on Eq. 1, we can write down the equations for these four reaction fluxes as shown in Table II. In these equations, x is the Ge content in the film. θ_{Si} and θ_{Ge} represent the ratio of H-terminated Si sites to all Si sites and the ratio of H-terminated Ge site to all Ge sites, respectively. P_{SiH4}, m_{SiH4}, P_{GeH4}, and m_{GeH4} are the SiH$_4$ partial pressure, SiH$_4$ molecular mass, GeH$_4$ partial pressure, and GeH$_4$ molecular mass, respectively. $E_{GeH4/Ge}$, $E_{GeH4/Si}$, $E_{SiH4/Ge}$, and $E_{SiH4/Si}$ are the activation energy for GeH$_4$ decomposition on H-free Ge sites, GeH$_4$ on H-free Si sites, SiH$_4$ on H-free Ge sites, and SiH$_4$ on H-free Si sites, respectively.

Table I. Heterogeneous reactions in Si$_{1-x}$Ge$_x$ CVD from GeH$_4$ and SiH$_4$

Notation	Description	Reactions
$J_{GeH4/Ge}$	GeH$_4$ flux on Ge sites	GeH$_4$(g) + 2-Ge(s) = H$_3$Ge-Ge(s) + H-Ge(s)
$J_{GeH4/H-Ge}$	GeH$_4$ flux on H-Ge sites	GeH$_4$(g) + H-Ge(s) = H$_3$Ge-Ge(s) + H$_2$(g)
$J_{GeH4/Si}$	GeH$_4$ flux on Si sites	GeH$_4$(g) + 2-Si(s) = H$_3$Ge-Si(s) + H-Si(s)
$J_{GeH4/H-Si}$	GeH$_4$ flux on H-Si sites	GeH$_4$(g) + H-Si(s) = H$_3$Ge-Si(s) + H$_2$(g)
$J_{SiH4/Ge}$	SiH$_4$ flux on Ge sites	SiH$_4$(g) + 2-Ge(s) = H$_3$Si-Ge(s) + H-Ge(s)
$J_{SiH4/H-Ge}$	SiH$_4$ flux on H-Ge sites	SiH$_4$(g) + H-Ge(s) = H$_3$Si-Ge(s) + H$_2$(g)
$J_{SiH4/Si}$	SiH$_4$ flux on Si sites	SiH$_4$(g) + 2-Si(s) = H$_3$Si-Si(s) + H-Si(s)
$J_{SiH4/H-Si}$	SiH$_4$ flux on H-Si sites	SiH$_4$(g) + H-Si(s) = H$_3$Si-Si(s) + H$_2$(g)

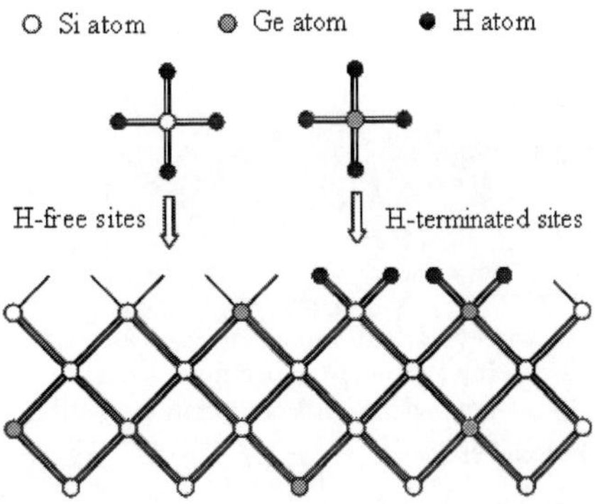

FIGURE 1. Heterogeneous reactions in Si$_{1-x}$Ge$_x$ CVD from GeH$_4$ and SiH$_4$.

Surface H coverage.–The effect of surface H coverage, θ_{Si} and θ_{Ge}, needs to be taken into account to understand the growth behavior of Si$_{1-x}$Ge$_x$ CVD. Liehr et al[27] and Greenlief et al[28] showed that in SiH$_4$ CVD, the surface H coverage of Si sites, θ_{Si}, decreases with higher temperatures and increases with larger flow rates of SiH$_4$. Surface H coverage in Si$_{1-x}$Ge$_x$ growth is more complicated due to the presence of Ge. Ning et al[29] observed H desorption at lower temperatures with increasing Ge content on the surface. Suemitsu et al[30] confirmed these results. The adsorption of SiH$_4$ and GeH$_4$ brings H to the surface and H desorption removes H from the surface. In addition, H atoms diffuse between Ge and Si sites on the surface

H-Si(s) + -Ge(s) = H-Ge(s) + -Si(s)

which enhances H desorption. Fig. 2 depicts these reactions. All of these factors make it difficult to derive an analytical equation for surface H coverage in $Si_{1-x}Ge_x$ CVD. For simplification, our model assumes that θ_{Ge} is negligibly small, i.e. $\theta_{Ge} \approx 0$. This is supported by the fact of the weaker Ge-H bond (90.8 kJ/mol enthalpy of formation for GeH_4) compared with the stronger Si-H bond (34.3 kJ/mol enthalpy of formation for SiH_4).[31] For θ_{Si}, we have found that one of the simplest mathematical relations provides a good fit to the experimental data at different GeH_4/SiH_4 ratios in the temperature range of 570–750°C

$$\theta_{Si} = \frac{1}{g(T, P_{SiH_4}) + h(T, P_{SiH_4})P_{GeH_4}}$$ [2]

where $g(T, P_{SiH_4})$ and $h(T, P_{SiH_4})$ are dependent only on temperature and SiH_4 partial pressure. Eq. 2 indicates that, at any given temperature and SiH_4 partial pressure, θ_{Si} decreases with increasing GeH_4 partial pressure. This agrees with the fact that Ge incorporation enhances H desorption from the surface.

Table II. Primary Ge and Si fluxes in $Si_{1-x}Ge_x$ CVD from GeH_4 and SiH_4

Flux	Equation
$J_{GeH4/Ge}$	$x(1-\theta_{Ge})\dfrac{P_{GeH_4}}{(2\pi m_{GeH_4}kT)^{1/2}}(\dfrac{E_{GeH_4/Ge}}{kT}+1)\exp(-\dfrac{E_{GeH_4/Ge}}{kT})$
$J_{GeH4/Si}$	$(1-x)(1-\theta_{Si})\dfrac{P_{GeH_4}}{(2\pi m_{GeH_4}kT)^{1/2}}(\dfrac{E_{GeH_4/Si}}{kT}+1)\exp(-\dfrac{E_{GeH_4/Si}}{kT})$
$J_{SiH4/Ge}$	$x(1-\theta_{Ge})\dfrac{P_{SiH_4}}{(2\pi m_{SiH_4}kT)^{1/2}}(\dfrac{E_{SiH_4/Ge}}{kT}+1)\exp(-\dfrac{E_{SiH_4/Ge}}{kT})$
$J_{SiH4/Si}$	$(1-x)(1-\theta_{Si})\dfrac{P_{SiH_4}}{(2\pi m_{SiH_4}kT)^{1/2}}(\dfrac{E_{SiH_4/Si}}{kT}+1)\exp(-\dfrac{E_{SiH_4/Si}}{kT})$

Homogeneous decomposition.—SiH_4 can homogeneously decompose into SiH_2 and H_2 at temperatures above 950°C.[32,33] However, the temperatures for $Si_{1-x}Ge_x$ growth by CVD range from 450°C to 750°C, and thus homogeneous decomposition of SiH_4 is not considered in our model. Hall[34] studied the decomposition of GeH_4 and found that heterogeneous decomposition of GeH_4 predominates at 300°C and/or small GeH_4 flow rates and homogeneous decomposition of GeH_4 predominates at 450°C and large GeH_4 flow rates. He also observed Ge deposition on the reactor wall at high temperatures and large GeH_4 flow rates due to homogeneous decomposition of GeH_4. A side effect of this Ge deposition on reactor wall is the depletion of the Ge source and thus a decrease in $Si_{1-x}Ge_x$ growth rate, especially at high temperatures and/or large GeH_4 flow rates. Since the temperature for homogeneous decomposition of GeH_4 is close to the deposition temperature of $Si_{1-x}Ge_x$ by CVD, the product of GeH_4 homogeneous decomposition, GeH_2, is taken into account in our model.

Discussion

Jang et al.[18] reported an extensive study on ultrahigh vacuum CVD of $Si_{1-x}Ge_x$ from GeH_4 and SiH_4. In one experiment, they changed the flow ratio of GeH_4/SiH_4 from 1.25% to 5.75%

and the deposition temperature from 570°C to 700°C, while keeping a constant SiH₄ flow rate of 40 sccm and a constant H₂ flow rate of 80 sccm as the carrier gas. The resultant deposition pressure varied from 15 mTorr to 17 mTorr. Another experiment was carried out at a constant deposition temperature of 620°C with the SiH₄ flow rate ranging from 20 sccm to 80 sccm (GeH₄/SiH₄ ratio from 1.25% to 5.75%) without a carrier gas. In both experiments, they found that, as the Ge content in the film increased, the growth rate first increased and then decreased. The decrease in growth rate at high deposition temperatures and/or large GeH₄ flow rates (i.e. high Ge content) can be attributed to depletion effects of GeH₄ homogeneous decomposition. Therefore, Si$_{1-x}$Ge$_x$ growth by CVD can be divided into two regimes: (1) a heterogeneous decomposition dominated regime at low temperatures and/or small GeH₄ flow rates where the growth rate increases with Ge content and (2) a homogeneous decomposition dominated regime at high temperatures and/or large GeH₄ flow rates where the growth rate decreases with Ge due to depletion effects. The two regimes are illustrated in Fig. 3.

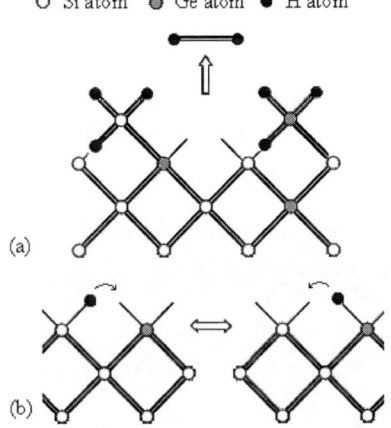

FIGURE 2. (a) H desorption from surface sites and (b) H diffusion between Ge and Si sites.

Growth Rate in Heterogeneous Regime.—In the heterogeneous regime, the overall growth rate is proportional to the sum of four reactions fluxes

$$G \propto J_{GeH_4/Ge} + J_{GeH_4/Si} + J_{SiH_4/Ge} + J_{SiH_4/Si} \qquad [3]$$

substituting the four equations in Table II into Eq. 3 and rearranging

$$G \propto xP_{GeH_4}A(T) + (1-x)(1-\theta_{Si})P_{GeH_4}B(T) + xP_{SiH_4}C(T) + (1-x)(1-\theta_{Si})P_{SiH_4}D(T) \qquad [4]$$

where $A(T)$, $B(T)$, $C(T)$, and $D(T)$ are dependent only on temperature

$$A(T) = \frac{1}{(2\pi m_{GeH_4}kT)^{1/2}} \left(\frac{E_{GeH_4/Ge}}{kT} + 1\right) \exp\left(-\frac{E_{GeH_4/Ge}}{kT}\right)$$

$$B(T) = \frac{1}{(2\pi m_{GeH_4} kT)^{1/2}} (\frac{E_{GeH_4/Si}}{kT} + 1) \exp(-\frac{E_{GeH_4/Si}}{kT})$$

$$C(T) = \frac{1}{(2\pi m_{SiH_4} kT)^{1/2}} (\frac{E_{SiH_4/Ge}}{kT} + 1) \exp(-\frac{E_{SiH_4/Ge}}{kT}) \qquad [5]$$

$$D(T) = \frac{1}{(2\pi m_{SiH_4} kT)^{1/2}} (\frac{E_{SiH_4/Si}}{kT} + 1) \exp(-\frac{E_{SiH_4/Si}}{kT})$$

Eq. 5 suggests that the growth rate is a function of Ge content (x), GeH$_4$ partial pressure (P_{GeH4}), SiH$_4$ partial pressure (P_{SiH4}), and surface H coverage (θ_{Si}) at any given temperature. In several previous studies,[8,19,20] the relation between Ge content in the film and GeH$_4$ and SiH$_4$ partial pressures is approximated by

$$x = m \frac{P_{GeH_4}}{P_{SiH_4} + P_{GeH_4}} \qquad [6]$$

where m is an enhancement factor, which is reported to be around 2.92–3.12 by Robbins et al,[8] 3.06 by Racanelli et al,[19] and 2.7–3.1 by Suemitsu et al.[30] In our model, we choose $m = 3.13$ for all the deposition temperatures. By substituting Eqs. 2 and 6 into Eq. 4, the growth rate can be calculated for the heterogeneous regime as a function of Ge content at any temperature. The values for $g(T)$ and $h(T)$ in Eq. 2 used to obtain a good fit in Fig. 3 are listed in Table III.

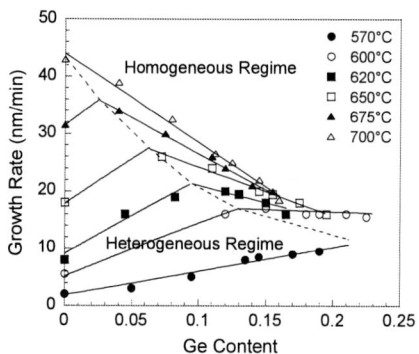

FIGURE 3. Growth rate as a function of Ge content at different temperatures. SiH$_4$ flow rate is 40 sccm. Solid lines are model predictions and experimental data from Ref. 18. The dash line is the boundary between homogeneous and heterogeneous regimes.

Growth Rate in Homogeneous Regime.—In the homogeneous regime, the growth rate decreases with Ge content due to precursor depletion. Due to the difficulty in simulating depletion effects, we have applied one of the simplest mathematical relations to describe the growth rate in the homogeneous regime

$$G = -k(T)x + b \qquad [7]$$

where $k(T)$ is assumed to be only temperature-dependent and b is dependent on both temperature and SiH$_4$ partial pressure.

Table III. Fitting parameters used in Eq. 4 and thus Eq. 6 for Fig. 3

	$G(T)$	$h(T)$ (1/sccm)
570°C	2.41	1.02
600°C	9.01	1.98
620°C	14.22	5.02
650°C	31.94	7.94
675°C	59.81	1.20

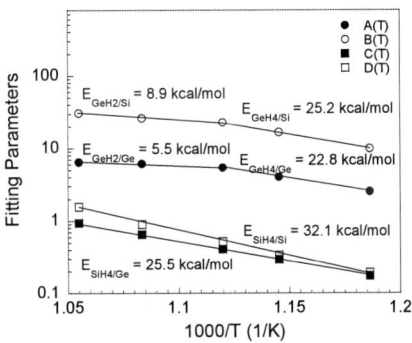

FIGURE 4. Arrhenius plot for fitting parameters used in Eq. 4. Activation energies for different heterogeneous elemental reactions on the surface are extracted.

Table IV. Activation energy for GeH_4 and GeH_2 adsorption (kcal/mol)

$E_{GeH4/Ge}$	$E_{GeH4/Si}$	$E_{GeH2/Ge}$	$E_{GeH2/Si}$
22.8	25.2	5.5	8.9

Figure 3 shows the growth rate as a function of Ge content in the film at different deposition temperatures. The model includes two parts, heterogeneous growth from Eq. 4 and homogeneous growth from Eq. 7. The Arrhenius plot for the parameters $A(T)$, $B(T)$, $C(T)$, and $D(T)$ used in our model for Eq. 4 is shown in Fig. 4, from which the activation energies for GeH_4 and SiH_4 decomposition on H-free Ge and Si sites can be estimated according to Eq. 5. The activation energies for SiH_4 adsorption on H-free Ge and Si sites are $E_{SiH4/Ge} = 25.5$ kcal/mol and $E_{SiH4/Si} = 32.1$ kcal/mol, respectively. They do not change in the temperature range of 570–675°C, suggesting that SiH_4 does not undergo homogeneous decomposition in the temperature and pressure ranges involved. The activation energy for SiH_4 adsorption on H-free Si sites, 32.1 kcal/mol, also agrees well with a previously reported value of 30 kcal/mol.[25] However, the activation energies for GeH_4 adsorption on H-free Ge and Si sites decrease with increasing temperature, as listed in Table IV. This suggests that GeH_4 undergoes homogeneous decomposition into GeH_2 and the kinks in Fig. 4 indicate that this homogenous decomposition starts at 620°C. Therefore, the activation energies extracted from the Arrhenius plot at high temperatures correspond to GeH_2 adsorption on H-free Ge and Si sites.

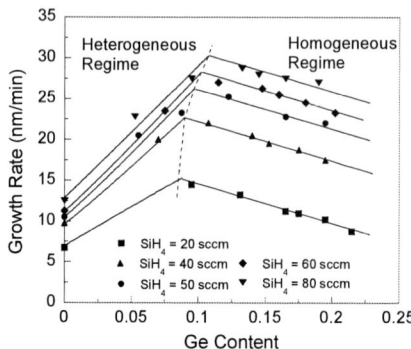

FIGURE 5. Growth rate as a function of Ge content at different SiH$_4$ flow rate. The deposition temperature is 620°C. Solid lines are model predictions and experimental data from Ref. 18. The dash line is the boundary between homogeneous and heterogeneous regimes.

Table V. Fitting parameters used in Eq. 4 and thus Eq. 6 for Fig. 5

SiH$_4$ flow rate (sccm)	$g(P_{SiH4})$	$h(P_{SiH4})$ (1/sccm)
20	3.01	1.10
40	1.21	0.21
50	1.67	0.58
60	1.58	0.69
80	1.46	0.78

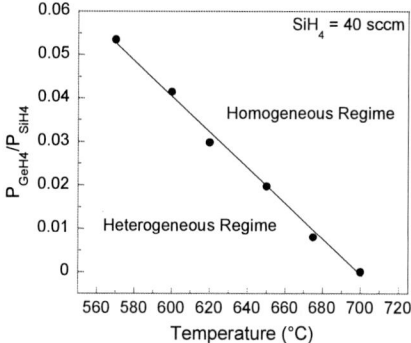

FIGURE 6. The boundary between heterogeneous and homogeneous regimes as defined by P_{GeH4}/P_{SiH4} ratio and temperature.

Figure 5 shows a comparison between the experimental data[18] and our model for growth rate at 620°C as a function of Ge content at different SiH$_4$ flow rates. The model again includes two parts, heterogeneous growth from Eq. 4 and homogeneous growth from Eq. 7. The values for

$g(P_{SiH_4})$ and $h(P_{SiH_4})$ in Eq. 2 for a good fit in Fig. 5 are listed in Table V. An excellent agreement is achieved between our model and experimental data in Figs. 3 and 5.

Conditions for Homogeneous Decomposition.—The growth rate peaks in Figs. 3 and 5 are the boundary between the heterogeneous regime and homogeneous regime. If we plot the P_{GeH_4}/P_{SiH_4} ratio at these peaks in Fig. 3 as a function of deposition temperature, a linear relation is obtained, as shown in Fig. 6. The growth is dominated by homogeneous decomposition of GeH$_4$ if the deposition conditions (P_{GeH_4}/P_{SiH_4} ratio and temperature) are above this line and it is dominated by heterogeneous decomposition of GeH$_4$ if the deposition conditions are below this line. This line also suggests that it takes lower GeH$_4$ partial pressures at higher temperatures or lower temperatures at higher GeH$_4$ pressures for homogeneous decomposition to occur.

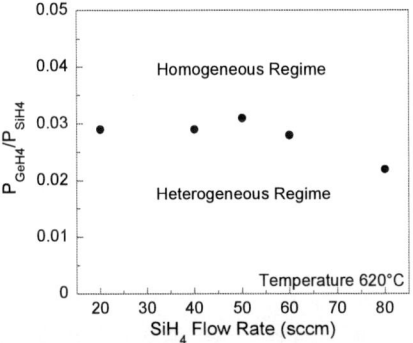

FIGURE 7. The boundary between heterogeneous and homogeneous regimes as defined by P_{GeH_4}/P_{SiH_4} ratio and SiH$_4$ flow rate.

Figure 7 shows the P_{GeH_4}/P_{SiH_4} ratio at growth rate peaks in Fig. 5 as a function of SiH$_4$ flow rate, which gives the boundary between heterogeneous and homogeneous regimes. As SiH$_4$ flow rate increases from 20 sccm to 60 sccm, the boundary line keeps constant at about 0.03, and then decreases slightly to about 0.02 with SiH$_4$ flow rate reaching 80 sccm. This suggests that GeH$_4$ homogeneous decomposition at 620°C is more or less independent of SiH$_4$ flow rate.

Conclusion

A quantitative kinetic model based on the collision theory of chemical reactions, statistical physics, and the concept of competitive adsorption is proposed for Si$_{1-x}$Ge$_x$ growth from SiH$_4$ and GeH$_4$ by CVD. It takes into account both homogeneous and heterogeneous reactions, which involve SiH$_4$, GeH$_4$, and the homogeneous decomposition product of GeH$_4$, GeH$_2$, and three types of surface sites, H-free Si sites, H-terminated Si sites, and H-free Ge sites. The growth of Si$_{1-x}$Ge$_x$ is divided into two regimes: a heterogeneous decomposition dominated regime and a homogeneous decomposition dominated regime. Analytical equations are derived to describe growth rate as a function of deposition conditions, including deposition temperature, GeH$_4$ flow rate, and SiH$_4$ flow rate, for the heterogeneous regime. Homogeneous decomposition of GeH$_4$

into GeH_2 causes precursor depletion, and an empirical linear relation is employed to describe the deposition kinetics in the homogeneous regime. At low temperatures and/or low GeH_4/SiH_4 ratios, the growth is dominated by heterogeneous decomposition and the growth rate increases with increasing Ge content. At high temperatures and/or high GeH_4/SiH_4 ratios, the growth is dominated by homogeneous decomposition of GeH_4 and the growth rate decreases with Ge content due to depletion effects. The temperature and GeH_4/SiH_4 ratio conditions for homogeneous decomposition of GeH_4 are established from experimental data. The model agrees well with the experimental data for $Si_{1-x}Ge_x$ growth from GeH_4 and SiH_4 by CVD.

Acknowledgement

This work was supported in part by the National Science Foundation (grant No. ECS-0322762) and the Petroleum Research Fund, administrated by the American Chemical Society (grant No. 38563-AC5).

References

1. S. E. Thompson, M. Armstrong, C. Auth, M. Alavi, M. Buehler, R. Chau, S. Cea, T. Ghani, G. Glass, T. Hoffman, C.-H. Jan, C. Kenyon, J. Klaus, K. Kuhn, Z. Ma, B. McIntyre, K. Mistry, A. Murthy, B. Obradovic, R. Nagisetty, P. Nguyen, S. Sivakumar, R. Shaheed, L. Shifren, B. Tufts, S. Tyagi, M. Bohr, and Y. El-Mansy, *IEEE Trans. Electron Devices*, **51**, 1790 (2004).
2. J. Welser, J. L. Hoyt, S. Woo, J. S. Park, K. L. Wang and K. P. MacWilliams, *IEDM Tech. Dig.*, 373 (1994).
3. Z.-Y. Cheng, M. T. Currie, C. W. Letiz, G. Taraschi, E. A. Fitzgerald, J. L. Hoyt, and D. A. Antoniadas, *IEEE Electron Device Lett.*, **22**, 321 (2001).
4. D. K. Nayak, K. Goto, A. Yutani, J. Murota, and Y. Shiraki, *IEEE Trans. Electron Devices*, **43**, 1709 (1996).
5. B. S. Meyerson, K. J. Uram, and F. K. Legoues, *Appl. Phys. Lett.*, **53**, 2555 (1988).
6. C. Li, S. John, E. Quinones, and S. Banerjee, *J. Vac. Sci. Technol. A*, **14**, 170 (1996).
7. S. Suzuki and T. Itoh, *J. Appl. Phys.*, **54**, 6385 (1983).
8. D. J. Robbins, J. L. Glasper, A. G. Cullis, and W. Y. Leong, *J. Appl. Phys.*, **69**, 3729 (1983).
9. S.-M. Jang and R. Reif, *Appl. Phys. Lett.*, **59**, 3162 (1991).
10. S.-M. Jang and R. Reif, *Appl. Phys. Lett.*, **60**, 707 (1992).
11. W. B. de Boer and D. J. Meyer, *Appl. Phys. Lett.*, **58**, 1286 (1991).
12. T. I. Kamins and D. J. Meyer, *Appl. Phys. Lett.*, **59**, 178 (1991).
13. C. M. Gronet, C. A. King, W. Opyd, J. F. Gibbons, S. D. Wilson, and R. Hull, *J. Appl. Phys.*, **61**, 2407 (1987).
14. P. M. Garone, J. C. Sturm, P. V. Schwartz, S. A. Schwartz, and B. J. Wilkens, *Appl. Phys. Lett.*, **56**, 1275 (1990).
15. T. Hsu, B. Anthony, R. Qian, D. Kinosky, A. Mahajan, S. Banerjee, C. Magee, and A. Tasch, *J. Electron. Mater.*, **21**, 65 (1992).
16. H. Kim, N. Taylor, T. R. Bramblett, and J. E. Greene, *J. Appl. Phys.*, **84**, 6372 (1998).
17. J. L. Hoyt, C. A. King, D. B. Noble, C. M. Gronet, J. F. Gibbons, M. P. Scott, S. S. Laderman, S. J. Rosner, K. Nauka, J. Turner, and T. I. Kamins, *Thin Solid Films*, **184**, 93

(1990).

18. S.-M. Jang, K. Liao, and R. Reif, *J. Electrochem. Soc.*, **142**, 3513 (1995).

19. M. Racanelli and D. W. Greve, *Appl. Phys. Lett.*, **56**, 2524 (1990).

20. N. M. Russell and W. G. Breiland, *J. Appl. Phys.*, **73**, 3525 (1993).

21. R. Malik, E. Gulari, S. H. Li, and P. K. Bhattacharya, *J. Appl. Phys.*, **73**, 5193 (1993).

22. I. M. R. Lee and C. G. Takoudis, *J. Electrochem. Soc.*, **143**, 1719 (1996).

23. M. Tao, *J. Appl. Phys.*, **87**, 3554 (2000).

24. B. Mehta and M. Tao, *J. Electrochem. Soc.*, **152**, G309 (2005).

25. M. Tao, *Thin Solid Films*, **223**, 201 (1993).

26. M. Tao, *Thin Solid Films*, **307**, 71 (1997).

27. M. Liehr, C. M. Greenlief, S. R. Kasi, and M. Offenberg, *Appl. Phys. Lett.*, **56**, 629 (1990).

28. C. M. Greenlief and M. Armstrong, *J. Vac. Sci. Technol.* B, **13**, 1810 (1995).

29. B. M. H. Ning and J. E. Crowell, *Appl. Phys. Lett.*, **60**, 2914 (1992).

30. M. Suemitsu, K. J. Kim, and N. Miyamoto, *J. Vac. Sci. Technol.* A, **12**, 2271 (1994).

31. *CRC handbook of Chemistry and Physics*, 85nd ed., p. 5-17 and 5-19, CRC Press, (2004-2005).

32. C. G. Newman, H. E. O'Neal, M. A. Ring, F. Leska, and N. Shipley, *Int. J. Chem. Kinet.*, **11**, 1167 (1979).

33. L. J. Giling, H. H. C. de Moor, W. P. J. H. Jacobs, and A. A. Saaman, *J. Cryst. Growth*, **78**, 303 (1986).

34. L. H. Hall, *J. Electrochem. Soc.*, **119**, 1593 (1972).

310

SESSION 4

MATERIALS AND PROCESS INTEGRATION

312

ECS Transactions, 2 (2) 313-315 (2006)
10.1149/1.2195668, copyright The Electrochemical Society

Materials and Process Integration

Shuji Ikeda
Sematch ATDF
2706 Montopolis Drive
Austin, TX 78741

Mark Rodder
Texas Instruments, Inc.
13560 North Central Expressway
Dallas, Texas 75243

Materials and process integration issues are becoming more complex as technology is scaled to the 45nm node and beyond. Key issues are whether present materials and integration approaches can continue to be scaled or whether new materials and new integration approaches are required, and if so, when. Primary focus areas for materials and process integration include a) gate stack, b) device structure, c) substrate engineering, and d) junctions and contacts.

It is presently considered that transistor scaling is slowing, primarily due to the difficulty in the scaling of an SiON gate dielectric to an equivalent oxide thickness (EOT) substantially below 1.2nm with acceptable gate current. Additionally, the gate depletion associated with a polycrystalline-Si gate electrode results in difficulty in scaling the associated effective inversion oxide thickness below 2.0nm. This difficulty in scaling the gate stack is thus slowing the scaling of transistor gate length due to increased short-channel effects (SCE) and drain-induced barrier lowering (DIBL) associated with poor gate control of channel electrostatics.

Nonetheless, to maintain desired transistor and circuit performance improvement trends, different approaches are being investigated in industry and academia to overcome these above scaling difficulties. First, new gate stacks utilizing high-k gate materials and/or metallic gate electrodes are being investigated for significant scaling of the inversion oxide thickness. At the same time, novel approaches are being investigated to scale the SiON dielectric below the present perceived 'limit'. Second, new device structures utilizing narrow vertical channels rather than thin planar channels, i.e. multiple-gate vertical structures for improved gate control, are being considered for ultimate transistor scaling. Third, increase in carrier mobility by means of process-induced strain or substrate-engineering, including substrate-induced strain or newer high-mobility channel materials, is significantly being implemented and/or being further investigated for higher driving current. Fourth, the reduction in parasitic source/drain or contact resistance remains an important area of research for improving overall transistor performance, notably as channel resistance reduces as a result of higher channel mobility. As is typical with any new approaches, there are associated concerns of scalability, material compatibility with present silicon technology, process integration limitations, and defectivity.

In this session, papers are presented discussing materials and process integration issues, including defectivity issues, associated with these scaling approaches.

The invited paper by Ishimaru et al. discusses gate stack scaling with emphasis on the gate dielectric material and integration. A novel SiON process is described allowing for further reduction of EOT to 0.7nm, well below the present perceived 'limit', and with acceptable gate leakage for high performance technology to the 22nm node with the use of metallic gate electrodes. Further gate dielectric scaling with high-k material such as HfSiON is considered for

313

low-standby power technologies which require significantly reduced gate current, and the means to scale the gate stack by variation of the high-k material is presented.

The invited paper by Maszara et al. discusses the benefits of multiple-gate devices as well as the process integration challenges associated with the evolution of conventional thin-film planar structures to double-gated or triple-gated vertical structures. The improved gate control of the channel potential by use of multiple-gates and narrow vertical channel enables electrostatic scalability beyond that achievable with ultra-thin film fully-depleted planar SOI devices, and offers another approach to gate length scaling beyond that which could be achieved by aggressive gate dielectric scaling alone. High channel mobility can also be achieved in narrow channel, multiple-gate devices since negligible channel dopant concentration is required, offering an additional approach to high mobility beyond that which could be achieved by strain techniques alone. Nonetheless, Maszara et al. describe the many challenges associated with multiple-gate devices which may delay significant introduction of this technology to the 22nm node and beyond.

The invited paper by Diaz reviews process-induced strain engineering approaches which have become a critical aspect of process integration for advanced technology nodes. A key concern is the scalability of these strain approaches which will require further development of process integration techniques and improved materials properties to enable higher strain in dense scaled layouts with small device layout dimensions.

The fourth paper by Claeys et al. also discusses strain engineering but by means of substrate induced strain. In particular, thin strained-Si layers are formed on thin strain-relaxed-SiGe-buffers, whereby the strain relaxation is achieved utilizing a thin C-doped SiGe layer. The process integration issues associated with this strain approach are described, with emphasis on junction properties affected by the strain-related defectivity.

The fifth paper by Muller et al. presents a three-dimensional model of surface diffusion describing topography changes of deep trenches utilizing low pressure hydrogen annealing. While low-temperature hydrogen annealing has been previously utilized for bottom corner rounding and reduction of surface roughness in deep trenches for trench capacitors, this paper further describes the formation of Silicon-on-Nothing structures by means of hydrogen induced surface diffusion. (It is noted that the understanding of surface diffusion by hydrogen annealing is important for multiple-gate device structures as well since hydrogen annealing can reduce the surface roughness of the narrow vertical channel films inherent in such multiple-gate device structures.)

The sixth paper by Danielson et al. describes the future materials and process integration issues associated with scaling the channel thickness of planar SOI devices to ultra-thin channel films. Present planar SOI device technology is transitioning from partially depleted to fully-depleted devices for improved performance at scaled gate lengths, with the latter requiring ultra-thin channel films. However, Danielson et al. describe a possible scaling roadblock for planar devices using ultra-thin channel films due to agglomeration of the ultra-thin films if the process integration temperature is significantly above 700°C. For example, this temperature limitation may prevent incorporation, and/or require modification of selective epitaxial deposition processes to lower the series resistance in ultra-thin film planar SOI devices. Agglomeration is not specific to ultra-thin planar silicon films on insulator but can occur for other materials as well, such as GeOI and III-V-on-insulator films as well, requiring careful consideration of material and process integration issues for future nodes. (This agglomeration effect may be another reason to migrate from planar to vertical SOI device structures which utilize thicker starting semiconductor films while still achieving an ultra-thin channel layer.)

Although prior papers in this session have considered defectivity associated with intentionally added strain or associated with ultra-thin films, the seventh paper by Hirano et al. consider yield limiting defectivity associated with the conventional Si substrate itself, for such products as flash memory. Hirano et al. discuss failures associated with substrates having low oxygen concentration and demonstrate that failures from slip propagation can be reduced by optimizing the density of oxygen precipitates.

The final paper by Zhu et al. addresses the scaling problem of parasitic contact resistance which becomes a larger fraction of the overall device resistance if the intrinsic channel resistance is reduced, e.g. by use of strain or low-doped channels to improve carrier mobility. Zhu et al. investigate the Ti/n-type Si structure and show that use of Se passivation can result in a significant reduction in contact resistivity.

316

Scaled CMOS with SiON and high-k

K. Ishimaru, M. Takayanagi, T. Watanabe, S. Inaba, M. Fujiwara, and *D. Matsushita

SoC R&D Center, Toshiba Corp. Semiconductor Company, JAPAN
*Corporate R&D Center, Toshiba Corp., JAPAN

A MOSFET dimension is continuously scaled down to improve its performance. Gate insulator thickness reaches, however, almost its physical limitation in the case of SiON film. Recently, fabrication process of SiON film for less than 1nm EOT was presented. This technology can extend applicability of SiON film beyond 65 nm node.

High-k film is a promising candidate to replace SiON film with lower leakage current. HfSiON is one of the candidates for high-k materials. Despite the intensive studies by many researchers, pMOS Vth shift caused by Fermi-level pinning is still the biggest problem when combined with poly-Si gate electrode.

This paper addresses issues for scaled CMOS with SiON and high-k. Process optimization method with device characteristics is demonstrated for 45 nm node and beyond.

Introduction

Recently, scaling of MOSFET with performance improvement becomes more difficult. Although new materials are introduced every generation, such as Ti, Co, Ni, SiGe, Cu, Low-k, and high stress SiN film, the red brick wall in ITRS roadmap will come out within a couple of years [1]. One of the obstacle components in MOSFET scaling is the gate insulator. A silicon dioxide has been used for a long time because of its excellent reliability with low interface state density regardless of its simple fabrication process. Scaling of the gate insulator thickness improves MOSFET performance, however, it increases gate leakage current and also causes boron penetration from p+ poly-Si gate electrode into the substrate as drawbacks. Nitridation of SiO_2 [2] was investigated to suppress the boron penetration and SiON film was used as a gate insulator below 0.25μm generation. Figure 1 shows the relation of equivalent oxide thickness (EOT) of gate insulator and its leakage current density, Jg for each technology platform such as HP, LOP, and LSTP. As shown in the figure, conventional SiON film cannot satisfy the leakage current requirement beyond 65nm node.

High-k gate insulator is the promising candidate to replace the SiON. So many materials were evaluated and hafnium based insulator is becoming the most promising candidate as a high-k material at present. Nevertheless the intensive study by many researchers, there still exists Vfb shift problem for pMOS with poly-Si gate by the Fermi-level pinning [3]. On the contrary, extendability of SiON film below 1nm EOT is reported [4]. This new SiON is expected to sustain SiO_2 based gate insulator until the high-k film will really come into the mass production.

This paper addresses issues for scaled CMOS with SiON and high-k (HfSiON) gate insulator. Process optimization method with device characteristics is demonstrated for 45 nm node and beyond.

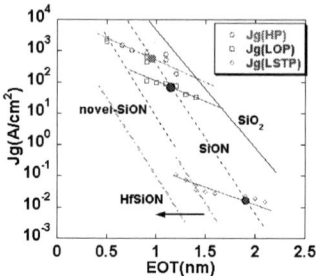

Figure 1 : EOT dependence of leakage current density, Jg for each technology platform.

MOSFET with SiON Gate Insulator

Conventional SiON

Although MOSFET operation with sub 2nm gate oxide thickness has been demonstrated [5], suppression of boron penetration without mobility degradation was the main issue for SiON gate insulator. The MOSFET performance and reliability strongly depends on the nitridation process.

NO Oxynitride

It is known that gate leakage current of NO oxynitride (fabricated by NO gas anneal after SiO_2 formation) decreases monotonically as increasing nitrogen concentration [6]. Figure 2(a) shows areal density (C_N) dependence of the gate leakage current at given physically equivalent oxide thickness (T_{OX}^{EQ}). High C_N achieves high dielectric constant, which realizes the same gate capacitance with increasing physical thickness. It should be noted that $C_N > 1E15cm^{-2}$ shows more than one order of magnitude lower gate leakage current compare to the pure oxide. Although the higher C_N is beneficial to suppress the threshold voltage variation caused by boron penetration for pMOS as shown in the Fig. 2(b), excessive nitrogen generates positive fixed charge at the dielectric interface leading to threshold voltage increase (Fig. 2(c)).

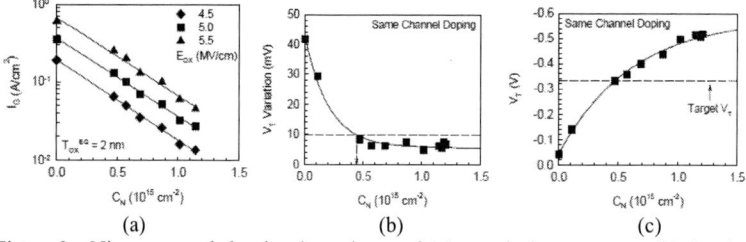

| (a) | (b) | (c) |

Figure 2 : Nitrogen areal density dependence of (a) gate leakage current, (b) threshold voltage variation caused by boron penetration for pMOS, (c) threshold voltage shift by positive charge for pMOS.

Therefore, in order to keep low gate leakage current without performance degradation, channel or other doping profile optimization is required. Figure 3 shows Ion and Ioff characteristics at V_D=1.2V for nMOS with high and low C_N oxynitride. Higher C_N device with optimized extension/halo profiles shows better performance compare to the lower C_N device.

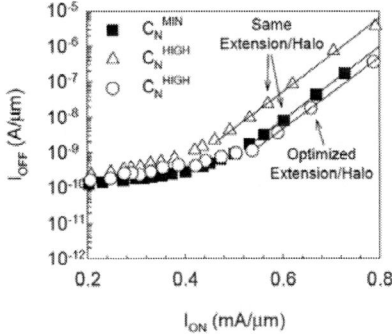

Figure 3 : nMOS Ion and Ioff characteristics with high and low C_N oxynitride.

Above study was done by relatively thick oxynitride (T_{OX}^{EQ} ~2nm). A source/drain was fabricated by conventional RTA process. To realize sub-50nm gate length MOSFETs for 65nm node, thickness of oxynitride should be scaled down to less than 1.5nm. Simple scaling of oxynitride thickness, however, causes significant increase of boron penetration and gate leakage current. Higher C_N enables lower gate leakage current with preventing boron penetration, but high nitrogen concentration at Si/SiO₂ interface degrades both mobility and reliability as drawbacks. Reduction of the process temperature, such as S/D RTA, decreases boron penetration from p+ poly-Si gate to substrate, though it degrades performance by poly depletion and parasitic resistance increase. Thanks to the advanced annealing process, spike RTA, it achieves higher temperature with less thermal budget. It realizes not only shallow junction but also less boron penetration. Figure 4 shows C_N dependence of both threshold voltage and its variation for pMOS. It should be noted that almost the same threshold voltage and variation showed in Fig. 2 are obtained even though the oxynitiride thickness is scaled down to 1nm [7].

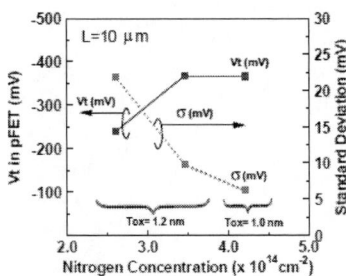

Figure 4 : Nitrogen concentration dependence of both threshold voltage and its variation applied spike RTA for S/D activation.

Nevertheless the small threshold voltage variation, nitrogen has its peak in profile at Si/SiO₂ boundary [7]. This is the limitation of NO oxynitride process. From both performance and reliability point of view, nitrogen profile should have its peak at poly-Si/SiO₂ boundary and as low as at SiO₂/Si boundary.

Plasma nitrided SiON

The concept of plasma nitrided SiON is to nitride top surface of SiO₂ by exposing nitrogen plasma ambient after gate oxidation. Figure 5(a) shows plasma immersion time dependence of nitrogen concentration in the oxynitride [8]. Two different plasma pressures, indicated as (A) and (B), were investigated. As shown in the figure, a3 and b1 have the same nitrogen concentration. Physical thickness of oxynitride was also measured by XPS as shown in Fig. 5(b). The thickness is also proportional to plasma immersion time. In order to determine the optimum nitridation condition, NBTI characteristics were also investigated. Figure 6 shows time-to-failure (TTF) of oxynitride fabricated by the process b1 and a3. Only a3 condition satisfies 10 years of TTF under the criteria of less than 10% Ion degradation at 105C. This result suggests plasma nitridation by longer process time, in other words, gentle nitridation process is beneficial to obtain high quality oxynitride film under given EOT. By applying this plasma nitridation process, higher transconductance Gm value is obtained compare to that of NO oxynitride (Fig. 7).

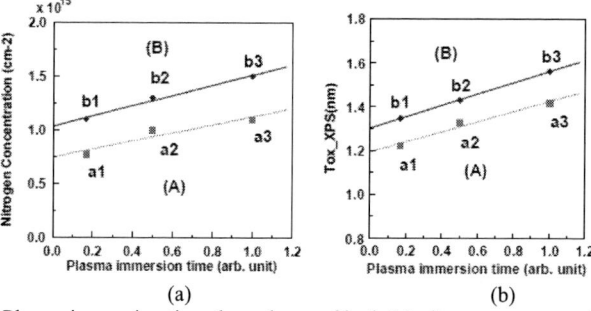

(a)   (b)

Figure 5 : Plasma immersion time dependence of both (a) nitrogen concentration and (b) oxynitride thickness under the different plasma pressure (A) and (B).

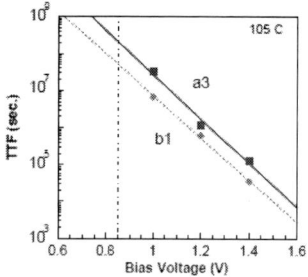

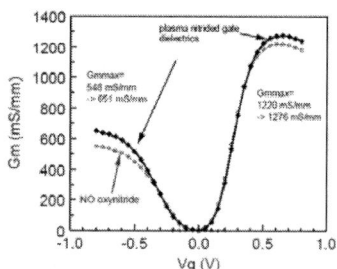

Figure 6 : TTF vs. stress bias for NBTI.

Figure 7 : Comparison of Gm for both plasma and NO nitridation process.

Novel SiON

It is difficult to achieve less than 1nm EOT either by conventional, NO or plasma nitrided SiON. In order to prevent gate leakage current increase and boron penetration, high nitrogen concentration at the top surface of SiO_2 is required. Plasma nitridation process realizes steep nitrogen profile in the SiO_2, however, nitrogen diffuses into the Si substrate when the SiO_2 thickness becomes less than 1nm. High nitrogen concentration at Si/SiON boundary causes significant degradation in both performance and reliability. Use of SiN instead of the SiON is one of the solutions, if poor Si/SiN interface quality can be improved. To overcome this problem, novel SiON fabrication process has been reported as shown in Fig. 8 [4]. Difference between conventional SiON and novel SiON is the starting material of gate insulator. Instead of oxide film, novel SiON utilizes nitride layer fabricated by direct nitridation of Si substrate. After the nitridation, post oxidation was carried out. This oxidation process oxidizes not only the surface of nitride layer but also nitride/Si interface. This process reduces interface state density generated by direct nitridation. Finally, re-nitridation process was proceeded and only the top surface was nitrided. By using this process, 0.7nm of EOT oxynitride was realized and it meets the target for hp22nm node LOP device as shown in Fig. 9. The quality of novel SiON film strongly depends on the quality of starting nitride film. Both process temperature and nitridation pressure were varied and moderate temperature with certain pressure minimizes surface roughness and lead to low defect density [9]. This process condition gives better NBTI characteristics as shown in Fig. 10.

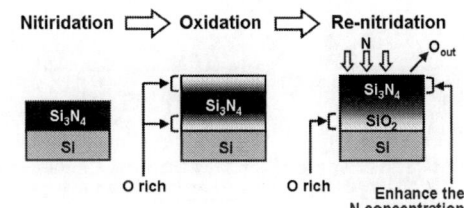

Figure 8 : Schematic diagram of novel SiON fabrication process.

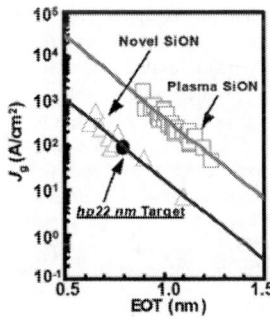

Figure 9 : Jg vs. EOT comparison between plasma and novel SiON.

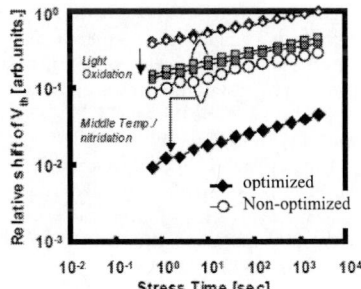

Figure 10 : NBTI characteristics with various process conditions.

Scalability of SiON

Figure 11 shows I-V characteristics of 10nm gate length MOSFET with 1nm EOT plasma SiON gate insulator. It is expected that by applying novel SiON with 0.7nm EOT, Ion-Ioff characteristics will be improved and satisfy the target of hp22nm LOP device [10]. Remaining issue is how to combine with a metal gate. Because of its thin physical thickness, it is very difficult to obtain sufficient etching selectivity of metal gate material against SiON film. FUSI is the promising candidate as a metal gate material, because conventional poly-Si etching process can be used. Figure 12 shows cross sectional TEM photograph of 10nm length Ni FUSI gate and its 1/f noise characteristics. Poly-Si gate comparable 1/f noise characteristics was obtained by FUSI gate with 1nm EOT SiON. Therefore, combination of SiON and FUSI can extend MOSFET down to hp22nm node for HP or LOP platform. For low stand-by power (LSTP) platform, further reduction of gate leakage current is required, which requires high-k gate insulator.

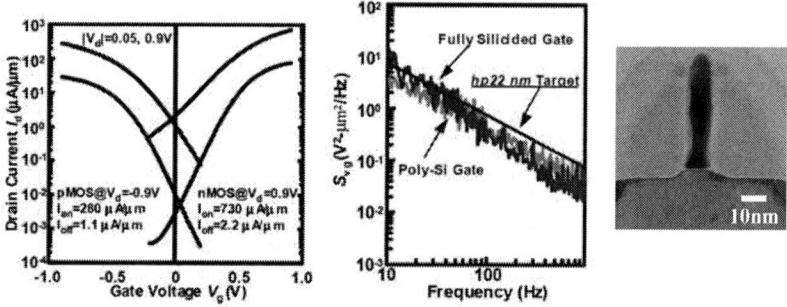

Figure 11 : I-V characteristics of 10nm gate length MOSFET with plasma SiON

Figure 12 : Cross sectional TEM of 10nm length FUSI gate and its 1/f noise characteristics.

MOSFET with HfSiON Gate Insulator

Mobile wireless market is expanding and the demand of low gate leakage current device is increasing. High-k gate insulator is inevitable component and HfSiON is one of the promising candidates. Despite the intensive studies by many researchers, pMOS Vth shift caused by Fermi-level pinning [3] is still the biggest problem when combined with poly-Si gate electrode. However, low stand-by power device uses relatively high threshold voltage and does not affect large impact on MOSFET design compare to the HP device. One of the advantages of HfSiON is the flexibility of dielectric constant design. By changing both Hf and N concentration, dielectric constant can be varied from 5 to 25 [11]. Incorporation of N realizes high thermal stability and improves breakdown characteristics [12,13]. Undesirable interfacial oxidation and boron penetration were also prevented by nitridation of HfSiO even after 1050C spike anneal process as shown in Figs. 13 and 14 [14].

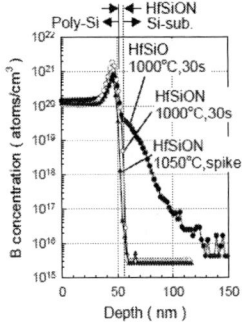

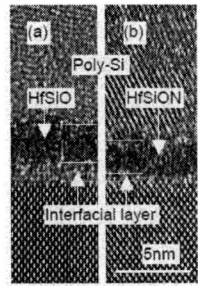

Figure 13 : Boron depth profile in gate stack measured by SIMS.

Figure 14 : Cross sectional TEM images of 3nm (a) HfSiO and (b) HfSiON after 1050C spike anneal.

Hf concentration dependence

As previously mentioned, by changing Hf contents, dielectric constant can be varied. This means there exists a couple of combination to realize given EOT as shown in Fig. 15. Therefore, it is important to know the impact of Hf concentration on device performance. In the following, Hf 30% with 2nm thickness and Hf 50% with 3nm thickness are compared. Both HfSiON has the same EOT of 1.8nm. Figure 16 shows effective electron mobility of nMOS with and without interfacial oxide layer for both Hf 30% and Hf 50% case. There are no significant difference in electron mobility between Hf 30% and Hf 50%. On the other hand, the electron mobility degrades in the case of without interfacial layer. Compare to the universal curve, 80% of the mobility was obtained at effect electric filed of 0.8MV/cm in the case of HfSiON with interfacial layer.

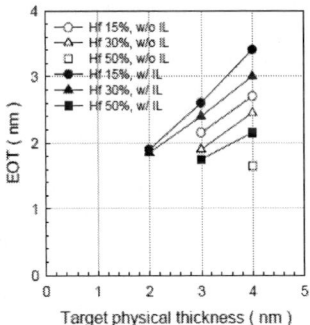

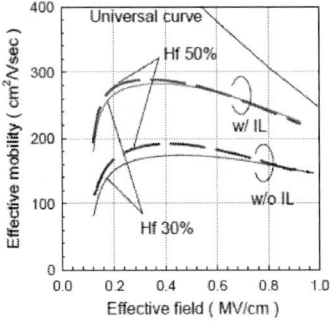

Figure 15 : Target physical thickness dependence of EOT under various Hf concentration.

Figure 16 : Effective electron mobility for nMOS with and without interfacial oxide layer both Hf 30% and 50% case.

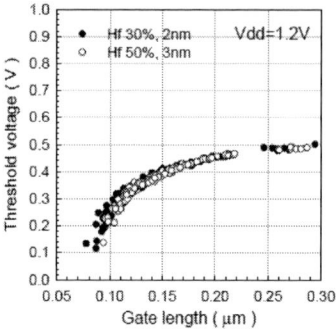

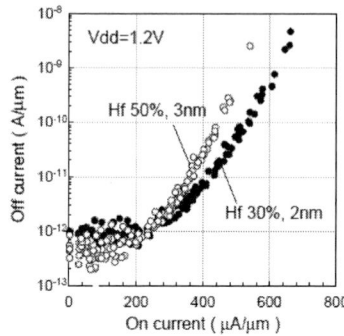

Figure 17 : Gate length dependence of threshold voltage for both Hf 30% and 50% nMOS.

Figure 18 : Ion-Ioff characteristics for both Hf 30% and 50% nMOS.

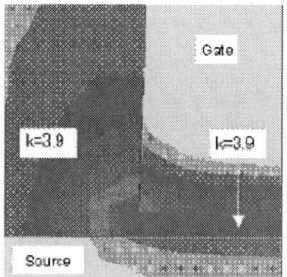

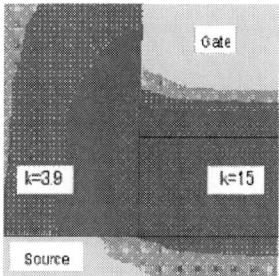

Figure 19 : Device structures used for simulation. Both devices have the same EOT.

Figure 17 shows gate length dependence of nMOS threshold voltage. There are also no significant differences between Hf 30% and 50% MOSFET. However, Ion-Ioff characteristics shows large difference between Hf 30% and 50% case and Hf 30% shows higher Ion value as shown in Fig. 18. In order to investigate this phenomenon, device simulation was carried out. Figure 19 shows device structures used in this study. Both devices have the same EOT value but have different physical thickness by varying dielectric constant. To enhance the difference both k=3.9 and k=15 was chosen in this simulation. The contour plots of source region denote electron density and those of gate insulator denote electric field. Even both devices have same EOT, physically thick insulator shows weak electric field at gate edge region, which results in less accumulation charge at source region. This causes less gate controllability and higher parasitic resistance.

The other impact of high Hf concentration to MOSFET is undesirable oxidation which causes severe reverse short channel effect [15]. Figure 20 shows cross sectional TEM photograph for Hf 30% and 50% case. In the case of Hf 30%, no oxide layer was formed between poly-Si gate and HfSiON films. On the contrary, Hf 50% device shows oxide layer between poly-Si gate and HfSiON film. Its thickness is almost comparable to that of interfacial layer between HfSiON and Si substrate. This undesirable oxide layer degrades EOT. By replacing sidewall material from SiO_2 to SiN, this oxide layer was completely suppressed.

Figure 21 shows hot carrier lifetime of L=65nm nMOS with Hf 30% and 50%. It is clearly shown that lower Hf content shows more than 10 years lifetime while Hf 50% does not satisfy the requirement. Flicker noise characteristics are also shown in Fig. 22. Hf 30% device shows almost the same characteristic as that of SiO_2, however, Hf 50% device shows more than two orders of magnitude worse characteristics.

From above studies, it is concluded that Hf concentration should be as low as possible at given EOT to realize both high performance and better reliability.

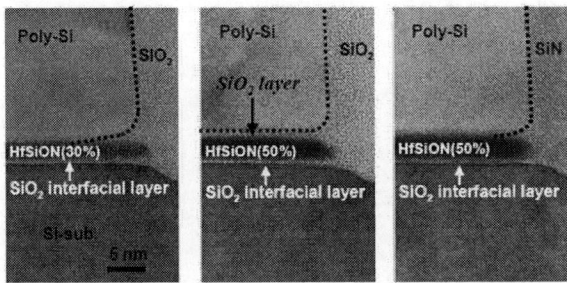

Figure 20 : Sidewall material dependence of undesired oxidation.

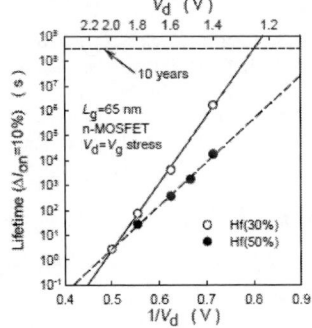
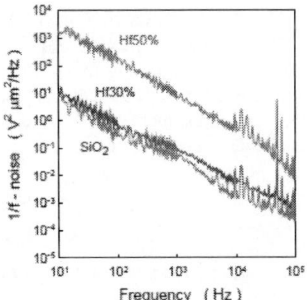

Figure 21 : Hot carrier lifetime of L=65nm nMOS with Hf 30% and 50%.

Figure 22 : Flicker noise characteristics of nMOS with Hf 30% and 50%.

Scalability of HfSiON

One of the concerns of HfSiON scalability is its thermal stability. Beyond 45nm node, it is expected to introduce milli-second anneal, MSA. Its annealing time is very short, however, peak temperature is higher than that of spike anneal and will exceed 1200C. Figure 23 shows in-plane XRD spectra after Laser Spike Anneal (LSA) process [16]. By optimizing LSA power density, crystallization of HfSiON was not observed even for Hf 50% case. Figure 24 shows EOT dependence of gate current density, Jg. It is shown that HfSiON satisfies the leakage current specification down to hp45nm node in the case of Hf 30% [17]. Beyond hp45nm, HfSiON can be extended by increasing Hf concentration. Combination of metal gate gives another room for Hf 30% because of elimination of poly depletion. Further EOT scaling, such as less than 0.5nm EOT, requires another high-k material like LaAlO$_3$ [18].

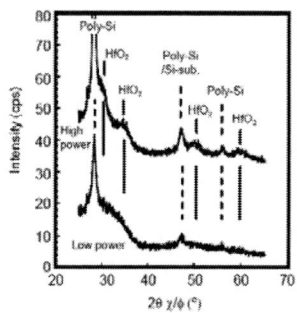

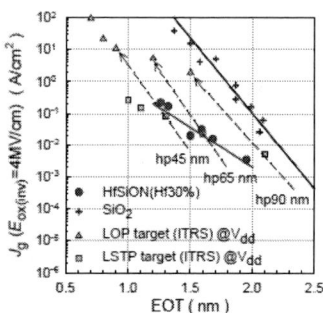

Figure 23 : In-plane XRD spectra of HfSiON after LSA.

Figure 24 : Scalability of HfSiON with ITRS target value.

CONCLUSION

MOSFET characteristics with SiON and HfSiON were discussed. For both HP and LOP technology platform, novel SiON can satisfy the gate leakage current requirement. By combining metal gate, it can be extended down to hp22nm node. For LSTP purpose, HfSiON is front up material and can be used down to hp45nm node. Beyond that node, increase of Hf concentration or combination with metal gate can sustain HfSiON. For further scaling, continuous work for material research is required.

Acknowledgments

Authors will thank H. Ishiuchi, K. Eguchi, Y. Tsunashima, N. Fukushima, A. Nishiyama, for their continuous support and encouragement.

References

1. http://public.itrs.net
2. H.S. Momose, et al., IEDM Tech. Dig., p359, 1991
3. C. Hobbs, et al., Symp. on VLSI Tech. Dig., p9, 2003
4. D. Matsushita, et al., Symp. on VLSI Tech. Dig., p172, 2004
5. H.S. Momose, et al., IEDM Tech. Dig., p593, 1994
6. M. Fujiwara, et al., IEDM Tech. Dig., p227, 2000

7. S. Inaba, et al., IEDM Tech. Dig., p423, 2001
8. S. Inaba, et al., IEDM Tech. Dig., p651, 2002
9. D. Matsushita, et al., IEDM Tech. Dig., p847, 2005
10. N. Yasutake, et al., Symp. on VLSI Tech. Dig., p84, 2004
11. M. Koike, et al., IEDM Tech. Dig., p107, 2003
12. M. Koyama, et al., IEDM Tech. Dig., p849, 2002
13. M. Koyama, et al., IEDM Tech. Dig., p38, 2003
14. T. Watanabe, et al., Symp. on VLSI Tech. Dig., p19, 2003
15. T. Watanabe, et al., IEDM Tech. Dig., p507, 2004
16. K. Adachi, et al., Symp. on VLSI Tech. Dig., p142, 2005
17. M. Takayanagi, et al., IEDM Tech. Dig., p903, 2005
18. M. Suzuki, et al., IEDM Tech. Dig., p445, 2005

Multiple Gate MOSFETs

W.P.Maszara, Z.Krivokapic, Q.Xiang, M.-R.Lin

AMD, Sunnyvale, CA 94088-3453

Planar single-gate transistors have been recently demonstrated with reasonable performance at sub-20 nm of physical gate length. However, a need for high performance transistors with channels shorter than that, as expressed by 2005 ITRS goals, requires devices with more than one gate, which facilitates better control of electrostatic charge in the channel. Double- and triple-gate transistors with their process integration complexity will likely become a device of choice for the high performance logic circuits in second decade of the 21st century. This paper will discuss various approaches to realization of those multi-gate fully depleted channel devices and their performance and process integration issues and layout design challenges for sub-20 nm gates. We also discuss issues related to designing circuits with multi-gate fin-based devices.

Multi-Gate Advantage

Shrinking dimensions of transistors with each generation impose thinner oxides and higher doping in the channel to stave off ever more challenging short channel effect. Today we are at the limit of acceptable gate leakage for high performance devices and observing serious degradation of channel mobility and increased band-to-band drain junction leakage resulting from high channel doping. To lessen the burden of those two factors, an alternative device architecture is needed, where lower gate fields and lower channel doping are used without sacrificing scaling trend and performance.

Scalable fully depleted (FD) devices can be built with undoped channel in very thin SOI film. Multiple-gate (MG) fully depleted transistors can provide even better scalability, not afforded by the single-gate devices (1,2). There is also a potential for a drive current improvement in MG devices over that of same-gate-length single gate (SG) ones. A full channel volume inversion is expected for sufficiently thin device body with two opposing gates (3). This could deliver higher mobilities due to less surface scattering. Modest performance improvements in devices with double-gates (DG) over that of an SG have been recently reported (4-6).

MG FETs are a key departure from classical single gate MOSFET transistors that have been known and utilized by industry for more than 3 decades. Multi-gate MOSFETs have been considered earlier as a promising alternative to planar single gate (SG) devices for 45/32nm technology nodes for high performance logic applications. Arrested scaling of gate SiON thickness, due to its excessive leakage and slower than anticipated progress of its high k replacement, slowed down gate length scaling. This delayed potential MG introduction probably beyond 32nm technology node, where $L_{gate} \leq 20$ nm. Well behaved MG FETs with metal gate and high k dielectric have been demonstrated (7)

Early circuit demonstrations

At this point little AC data exists to assess full performance benefit of MG devices. Several elemental circuits have been demonstrated recently with MG devices. Probably the earliest demonstration of a circuit with MG transistors was planar DG ring oscillator by Tanaka et al. (8) in 1994. The first demonstration of FinFET circuit was a 4-stage inverter by Rainey et al (9) and the earliest report of FinFET ring oscillator was published by Nowak et al. (10). Collaert et al. reported high aspect ratio Tri-gate (TG) ring oscillators (11). FinFET inverter delay was found better than SOI SG for devices with gate length L=65nm (12). Working FinFET (13) and TG (14) SRAM cells and 20Mb TG SRAM arrays have been reported (15).

Types of Multi-Gate Transistor Architecture

Numerous different device architectures have been proposed and demonstrated for double-gate MOSFETs. They can be represented by three main categories related to the directions of gate field and channel current with respect to substrate surface, schematically illustrated in Fig. 1.

- *planar*, or *standard* DG transistor has the gate field perpendicular to the substrate and its current parallel to it.
- *vertical* device has the opposite relationship: current vertical (hence, the name) and gate field parallel to the substrate surface.
- *fin* device - Fin FET has both the gate field and channel current in the plane of the wafer.

Tri-gate (TG) is a fin device with top surface of the fin also acting as a gate. Usually thick insulating cap on top of the fin in FinFET is substituted by gate dielectric in TG device. A fully depleted SG device

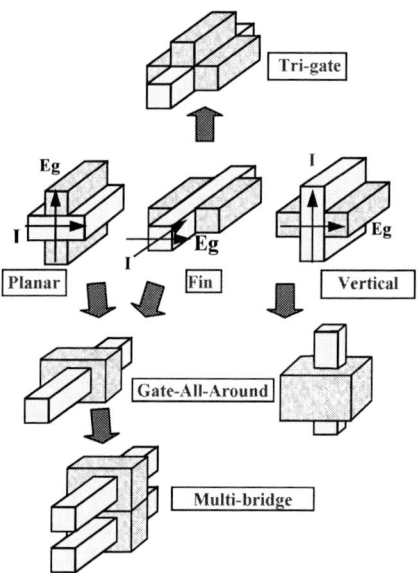

Figure 1. Types of multi-gate transistor architecture. Directions of channel current (I) and gate field (Eg) are indicated.

in mesa configuration, where very shallow isolation trenches are not filled and gate is allowed to wrap the sides of an SOI island can also be considered a TG. A true TG device, however, is a compromise between fully depleted SG device and classical, fully depleted FinFET, where all three gates are necessary to achieve full depletion operation of the transistor body (Fig.2). Since the side and the top gates help each other in achieving full depletion, critically thin dimension of body thickness (SOI film in SG and fin width in FinFET) can be relaxed, making TG more manufacturable. That very same condition, though, imposes limitations on both the height of the fin and its width, leading to rather restricted amount of effective channel width available with true TG transistors.

A pseudo-tri-gate (case A in Fig.2), while requiring thinner than optimum silicon thickness, has shown significant drive current improvements (16,17). It has been postulated that geometry of the metal gate in those transistors was favorable in stress distribution and contributed to the improvement.

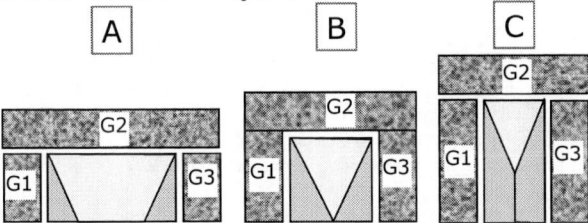

Figure 2 Optimum shape of tri-gate body is obtained when all three gates contribute to channel depletion equally (case B). Top gate dominates channel depletion in case A and side gates in case C. Case B eases up Si film constrain of A and fin width of B.

A derivative of double-gate device, gate-all-around (GAA) (18) surrounds the channel fully and can be implemented in a vertical or in-wafer-plane configuration. The latter case merges both fin and planar devices.

Fin devices appear to be the simplest to make and as such have the best chance of entering main stream product market and we will focus primarily on such devices in this paper. Specifically, we will discuss technological and design challenges multi-gate FETs face in entering the high performance CMOS product market.

A Bit of History

The most significant era of multi-gate MOSFET discoveries happened almost 20 years ago, in late 1980's. Shortly after researchers mastered etching of vertical wall trenches the availability of more than planar surface for MOSFET channel formation has been realized. Probably the first multi-gate transistor was published by Hieda et al. (19) in 1987. The authors realized that fully depleted body of this narrow tri-gate bulk Si-based transistor helps improve switching due to lessened body bias effect. A year later, the same group reported an ultimate multi-gate transistor – gate-all-around (GAA),

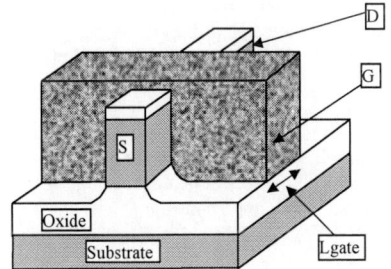

Figure 3. Fin FET on SOI wafer with etched and planarized gate (23)

which was formed around vertical pillar of silicon (20). A year after that, Hisamoto et al (21) demonstrated a predecessor of FinFET - first double gate transistor, in bulk silicon, called DELTA.

When SOI substrates became more available, a FinFET on SOI substrate followed the DELTA(22). SOI substrate offered improvement in device geometry and integration. Damascene process with spacers inside trench was used to form sub-lithographical gate lengths. An improved approach used gate structure, where gate length was defined by an etch process and allowed for formation of self-aligned source/drain and source/drain extensions (23) (Fig.3). SOI also enabled horizontal GAA transistor (18) creating a precursor to silicon nanowire devices. A tri-gate transistor built on SOI was presented in 2003 (24). Two extensions of MG transistor concept into extremely scaled regime were recently proposed: a 10nm diameter silicon nanowire with sub-10nm gates (25) and a multi-bridge GAA transistor (26), which demonstrated increased drive current capability for a given foot-print size of a transistor, due to stacked channels.

Multi-Gate Transistor Technology Challenges

Every DG transistor design and construction has to meet two unique requirements not shared with their bulk Si predecessors:
- thin body, about 1.5-2 times thinner than the length of the gate, to maintain short channel performance control, and
- gate electrodes should have the same size, and a very good alignment between them should be maintained.

The former would imply ~8-10 nm thick silicon for 15 nm gates formed with better than ±10% tolerance, or a few atomic layers. The latter requires alignment overlay better than ¼ of gate length, or in this case less than 4 nm to avoid 20-30% performance loss, as was shown in a simulated result (27). Experimental results show that 15nm misalignment (state-of-the-art lithography overlay, today) between 70nm top and 40nm bottom gate leads to almost two orders of magnitude increase in off-current (6). Since desired overlay accuracy for <20nm gates is far beyond capabilities of modern lithography tools, non-self-aligned DG architectures are unlikely candidates for highly scaled applications.

It is expected that all MG devices, by the time they are introduced into the market, will utilize metal gates and high k gate dielectrics and will have their channels undoped.

Fin Devices

FinFET and Tri-gate structure are the most conventional and the most successful among all MG devices in demonstrating high performance for scaled devices. High DC performance with good control of short channel effects has been demonstrated for NMOS and PMOS devices with L_{gate} =20 nm over 10 nm channel thickness (fin width) (23). Fin devices key challenges are:

Fin definition. Very thin fin of the transistor body is defined by etch process. Fin has to be about twice narrower than gate length, L, to keep short channel effects in check.

Gate length of today's highly scaled transistors is already defined beyond the capability of state-of-the-art lithography. Resist trimming process further decreases the width of photolithographically defined critical gate dimension by 30 to 50%. Overall, fin width in FinFET needs to be about 1/10 of minimum active layer pitch size, and about twice that much for TG. Various approaches are used to get such small dimension including trimming of active layer resist mask and oxidation-and-strip thinning of fin's sides. An elegant and more manufacturable approach utilizes spacer-defined patterning (23,28). It is done at a cost of several more process steps and two extra mask layers.

Channel mobility. Very rough surface of the channel can lead to channel mobility deterioration. Roughness is related to the etch process which transfers line edge roughness of photoresist into the channel surface. Smoothing by high temperature hydrogen annealing has been shown to significantly remedy this drawback (29)

S/D extension series resistance. Very thin fin in s/d extension area can significantly contribute to series s/d resistance. Possible remedy is shortening the extension using narrower spacers and thickening the fin by selective epi outside the spacers. This requires laser-type anneal process to minimize extension and main s/d implant diffusion into the channel. Laser anneal will also help in better activation of the extension dopants. Isotropic doping of fins with plasma immersion method could be a valuable alternative (30).

Gate patterning. Patterning of gate electrode over high steps of the fin seriously challenges lithography (depth of focus) and etch (removal of gate material stringers) and is a key limiter of available fin aspect ratios (height/width). In case of TG structures top of the fin is protected against over-etch, needed to remove stringers, only by gate oxide. Partial planarization of the gate material by polishing, prior to patterning, can alleviate this issue somewhat, but without a polish stop the removal of gate is poorly defined and gate salicidation can be impeded.

Non-self-aligned gate to s/d. Certain extended length of narrow transistor body, beyond the length of the gate, is necessary before the s/d contact area is allowed to fan-out, to assure proper placement of the gate over the fin (body) in a non-self-aligned fashion (Fig.4). An incurred penalty in this structure is the loss of symmetry of the source and drain parasitic resistances, which depends on lithographic overlay accuracy. Such variation of the s/d parasitic resistance can lead to substantial variation of drive current (26). This could be lessened by selective thickening of the fin outside the gate spacers using selective epitaxy.

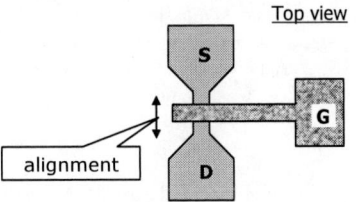

Figure 4 Non-self aligned nature of gate to s/d contact area in FinFET

Source/drain doping. Proper placement of s/d junctions along the width of the device (height of the fin) requires implantation at an angle into the sides of the fin. This leads to two challenges. First, tilted implant from two opposite sides of the wafer can serve only devices oriented in one direction on the wafer – a serious design constrain. A tilted implant, when accompanied by four wafer twists at ±45° to two principal orthogonal directions on the wafer, would accommodate devices laid-out along those two directions (31) – a standard for circuit design, but causes increased overlap of s/d to gate. Second, high implant doses used for s/d extensions will likely fully amorphize the fin. Limited amount of crystalline seed for subsequent regrowth of the amorphous material can lead to formation of highly defected silicon thus increasing s/d parasitic resistance and, possibly, junction leakage. This could be particularly of an issue in NMOS where implant specie is heavier that in PMOS. Significant over etch of the spacers is required to clear fin's sidewall for unimpeded implant access.

Adding strain. Very thin fins and their vertical orientation generally make it difficult to apply strain elements used in modern bulk and partially-depleted SOI devices (32,33) in order to boost their channel mobility. Typical such elements include embedded SiGe in PMOS s/d, strained SOI substrate or strained overlayers for both NMOS and PMOS. Adding SiGe into already very thin s/d areas may be difficult and inefficient. However, a full replacement of silicon s/d with SiGe has been attempted and resulted in 25% drive current improvement (34). Modeling studies show that small body volume devices gain more drive current from externally applied strain (35).

Layout vs. channel mobility. Layout of the transistor's fin along traditional <110> direction would place its channel on (110) plane, which improves p-channel mobility but deteriorates n-channel mobility (36,37). Alternative layout along <100> direction loses the benefit of p-channel mobility improvement.

Vertical Transistor

We have not identified a vertical type transistor that appears scalable to sub-20nm gate lengths with full benefit of fully-depleted MG device. One of more interesting implementations of a vertical MOSFET is a device called Vertical Replacement Gate (VRG) (38,39). The VRG uses sacrificial film whose thickness precisely defines transistor gate length, and two doped-oxide films below and above the sacrificial layer that facilitate self-aligned doping of s/d extensions. The body of the device is grown using selective epitaxy through a trench etched into the multi-layer structure. Since the device relies on deposited layer of PSG and BSG for doping of s/d extensions for NMOS and PMOS, respectively, integration of the two types of transistors is likely to be quite complex.

The main challenge from the scaling point of view, however, seems to be the size of transistor body which is formed in lithographically defined trench. The trench determines the body size and a 10 nm wide trench is well beyond the range of foreseeable patterning capability. Single gate length set by the thickness of the sacrificial layer for all transistors is a significant limitation for circuit design. Layout density for typical CMOS circuits is estimated to be about the same as for planar SG transistors (40).

Planar Double Gate Transistor

Most of planar double gate approaches face the challenge of some gate misalignment or gate size mismatch. Some that succeeded in both challenges are complex and come with their own challenges. Examples include devices that ingeniously utilize lateral epitaxy and sacrificial channel layer (41) or sacrificial gate electrode layer (42).

Another approach with sacrificial gate electrodes has been proposed by R.S.Shenoy and K.C.Saraswat (43). It utilizes multilayer epi structure and offers more controllable channel dimensions than (42), which relied on polishing process to define channel thickness.

For more comprehensive review of vertical and planar transistors see (44)

Gate All Around (GAA) Transistor

GAA concept presents ideal realization of channel control with a wrap-around gate, however, its practical realization appears very challenging. Early devices (18) and some more recent offerings (45,46) of GAA transistors had a key shortcoming in having the bottom part of the gate much larger and non-self-aligned with the side and top parts of the electrode.

A transitional geometry between GAA and TG is represented by "Omega" FET named after the cross-sectional shape of its gate electrode (47,48). The wrap-around gate comes short of encircling the channel but extends beyond what TG does. Excellent short channel control was demonstrated for gate length L=20nm at L/W fin ratio of ~1/3 (48).

Design Challenges

Unit performance gain of a mutli-gate transistor has to be transferable into circuits with competitive layout density. Since fin type devices can be printed lithographically at the density of one per minimum litho pitch, question arises whether those MG devices could run amounts of current comparable to SG devices per unit area of the chip.

We will briefly look at the current delivering capability of DG and TG fin devices for a range of desirable design widths of the transistor. A simple model can be constructed to compare TG and DG with SG devices.

Transistor layout efficiency

First, an effective width is assumed for single fin DG, $W_2=2*T$, and for single fin TG, $W_3=2*T+W_{fin}$; where T=silicon film thickness (=fin height) and W_{fin}= fin width. Fin aspect ratio can be defined as $R=T/W_{fin}$. These are compared to SG transistor of width equal to minimum active layer pitch, $W_1=P$. Second, we define short channel factor S, or a minimum ratio of gate length, L, to transistor body thickness, B. In DG transistor $B=W_{fin}$ and from experimental data (2,23,31) $S_2=L/W_{fin}$ =1.5-2. For this calculation we'll assume more conservative value $S_2=2$. For TG an optimum condition for short channel control appears to be when $W_{fin}{\approx}T{\approx}L$ (24), hence $S_3{\approx}1$. These conditions are associated with full depletion of transistor body by the gates. We also notice that the ratio between minimum active pitch, P, and nominal gate length of high performance logic transistors appears fairly constant over at least last four technology generations and $P{\approx}5L$ (Fig.5).

Layout efficiency of fin devices with respect to SG device can be expressed by the ratio of their available effective widths W/W1, and using the above relationships they are:

$$W_2/W_1 = 2RL/PS_2 = 2R/5S_2 \quad \text{for FinFET, and} \quad [1]$$
$$W_3/W_1 = (2R+1)L/PS_3 = (2R+1)/5S_3 \quad \text{for TG} \quad [2]$$

and both relationships show no dependence on technology node (P,L) providing P and L continue to scale at the current rate.

Height of the fin in FinFETs can be significantly more than that of TG and does not impact short channel control, but transistors with fin's high R are difficult to manufacture and this sets practical fin height limit. The difficulty lies primarily in etching the high aspect ratio fin in controllable fashion and litho patterning and etching gates placed over such large step in silicon. Related issues involve fin's mechanical stability, full removal of sidewall spacers around fin, and doping of the fin. Practical fin aspect ratio is probably around $R=T/W_{fin}{\approx}3$ and we assumed this value for our FinFET calculation. As we mentioned earlier, for TG R=1.

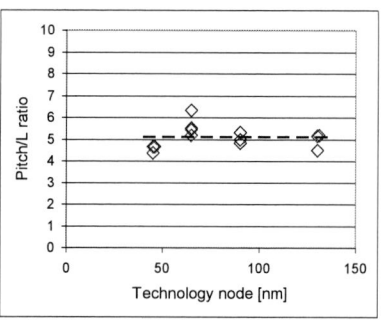

Figure 5. Ratio of minimum active layer pitch to gate length remained constant for several generations at ~5. (data compiled from 2000-2005 issues of VLSI Tech. and IEDM proceedings)

With the above relationships, formulas (1) and (2) yield the same values of relative layout efficiency: W_2/W_1=60% and W_3/W_1=60%. DG efficiency can be shown similar to that calculated by Yang and Fossum (49) if R=3 is used. They used somewhat different starting assumptions. Our calculated TG efficiency is much higher than in (49), where it was concluded that optimum body dimensions should be much smaller than indicated in (24).

Size of transistor in the direction orthogonal to W is expected to be similar for SG and fin devices and thus should not impact layout area efficiency.

Designing with digitized transistor width

As transistor designs call for widths wider than minimal pitch, the efficiency drops further down as W approaches $P+W_{FM}$, at which point design rule allows to print second fin and efficiency jumps by factor of 2 (W_{FM} is design width of the fin, a larger value than W_{fin} by the amount of fin trim applied). As W increases further, efficiency decreases again until $W=2P+W_{FM,}$ and so on. The cyclical nature of layout efficiency continues as W increases further, with decreasing amount of variation (Fig.6), and for large W its

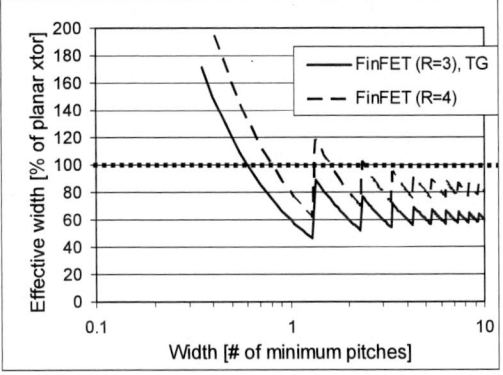

Figure 6 Layout efficiency for FinFET and TG transistors with respect to SG transistor. Fin aspect ratio of up to 4 is not sufficient to match the width of SG transistors for widths bigger than one pitch

value approaches the value it had at W=P. This implies that FinFET and TG (with the assumptions imposed above) could not compete with SG transistors in layout efficiency for designs requiring transistor widths equal or higher than critical pitch of active layer.

However for designs with very narrow transistors, less than critical pitch, e.g. in embedded SRAM cells, effective width of single fin MG transistors could be higher than SG (Fig.6), resulting in higher transistor current. It is not likely though, that actual area savings could be realized in MG SRAMs, as their layout is usually limited by S/D contact area and pitches of other layers, such as contact or metal. FinFET layout efficiency would become more competitive if higher fin aspect ratios are allowed. Efficiency of both DG and TG would increase if L scaling further slows down (i.e. P/L decreases) or devices become more under-lapped thus allowing for smaller scaling factor S.

Implicit in this model is assumption that current density per unit width of the channel is equal for all three transistor configurations. In fact the differences could be significant even within one transistor, and related to surface orientation of each channel face (37), roughness and local field distribution.

If we assume that MG do match the performance of SG per unit area for wide transistors, through enhancements in mobility or increased fin aspect ratio in case of FinFET, then migration of existing circuits layout designs would be rather straightforward for those transistor widths that are either large (say, W>~5P) or match closely effective widths of 1…5-finger fin devices. Since there are typically some narrower widths that would not match the digitized W, major circuit and layout redesign would be needed.

One might consider a hybrid approach where some of the narrower transistors be made as SG, allowing full flexibility of effective width design. The challenge here is the same as in patterning of gates over fins, discussed earlier. Silicon thickness of high performance SG (fully-depleted) would be much thinner than the fin height of MG causing concerns over depth of focus for gate lithography. An interesting approach, that might extend fin device layout efficiency, was proposed by Mathew et al (50) – an inverted 'T' shape of the channel combining SG and fin in one body of transistor. However, it faces the same step height issue as above and, in addition, relies on timed silicon etch to define the shape of the channel – a difficult condition to control in manufacturing environment.

Yet another design alternative could utilize fin devices only for very narrow transistors, e.g. in embedded SRAM, getting lower leakage current and more stability, while the rest of the circuitry could be realized with partially-depleted SOI devices using the same silicon film thickness and thus avoiding step height integration issues mentioned above (12).

Fin devices could gain major layout advantage if spacer-defined patterning is utilized (23,28), which allows for printing two fins per pitch thus doubling effective width.

Conclusions

Over two decades of multi-gate transistor explorations brought a variety of different architectures with various levels of complexity and competence. Among the investigated transistor architectures with two, three and more gates, fin devices (FinFET, Tri-gate) emerge as the most capable and most widely studied multi-gates. Although no clear AC performance advantage has been demonstrated yet, competent DC data and demonstration of some circuits exist.

Two key conditions for sub-20 nm devices: incorporation of high k gate dielectric / metal gates and ability to scale the MG transistor body below 15nm have been demonstrated. A key challenge of fin devices is its compatibility with circuit/layout designs. Single transistor width availability and limited layout efficiency for single transistor are key challenges for circuit designers. Multi-gate devices may show up in high performance logic products 6-9 years out, probably at 22nm technology node.

References

1. Y.Taur et al., IEEE Proc., 486 (1997).
2. X.Huang et al., IEDM Tech.Dig., 67 (1999) .
3. F.Ballestra et al., IEEE EDL, 410 (1987).
4. D.Esseni et al., IEDM Tech.Dig., (2001).
5. J.-T.Park et al., IEEE International SOI Conf.Proc., 115 (2001).

6. J.Widiez et al., IEEE TED, **52**, 1772 (2005)
7. N.Collaert et al, VLSI Tech.Dig., 108 (2005)
8. T.Tanaka et al., EDL, **15**, 386 (1994).
9. B.Rainey et al., Dev.Res.Conf.Proc., (2002).
10. E.Nowak et al., IEEE CiCC Conf. Proc., 339 (2003)
11. N.Collaert et al., IEEE EDL, 25, 568 (2004)
12. F.-L.Yang et al., IEDM Tech.Dig., 627 (2003)
13. E.Nowak et al., IEDM Tech.Dig., 411 (2002)
14. A. Nackaerts et al., IEDM Tech.Dig., 269 (2004)
15. J.A.Choi, et al., IEDM Tech.Dig., 647 (2004)
16. Z.Krivokapic et al., IEEE Trans.Semicond.Man. **18**, 5 (2005)
17. Z.Krivokapic et al., SSDM 760 (2003)
18. J.-P.Colinge et al., IEDM Tech.Dig. 595 (1990).
19. K.Hieda et al., IEDM Tech.Dig., 736 (1987)
20. H.Takato et al., IEDM Tech.Dig., 222 (1988)
21. D.Hisamoto et al. IEDM Tech.Dig., 833 (1989).
22. D.Hisamoto et al. IEDM Tech.Dig., 1032 (1998).
23. Y.-K.Choi et al., IEDM Tech.Dig., 421 (2001).
24. B.Doyle et al., VLSI Tech.Dig., 133 (2003)
25. F.-L.Yang et al., VLSI Tech.Dig., 196 (2004)
26. S.-Y.Lee et al., VLSI Tech.Dig., 200 (2004)
27. H.-S.Wong et al., IEDM Tech.Dig., 747 (1994).
28. A.Kaneko et al, IEDM Tech.Dig., 863 (2005)
29. J.-S.Lee, IEEE EDL, **24** 186 (2003)
30. A.Hori and B.Mizuno, IEDM Tech.Dig., 641 (1999)
31. J.Kedzierski et al., IEDM Tech.Dig., (2001).
32. T.Ghani et al., IEDM Tech.Dig., 978 (2003)
33. M.Horstman et al., IEDM Tech.Dig., 243 (2005)
34. P.Verheyen et al., VLSI Tech.Dig., 196 (2005)
35. F-L Yang et al., SSDM, 772 (2004)
36. S.Takagi et al., IEEE Trans. Electron Dev., **41**, 2363 (1994)
37. J.Kedzierski et al., IEEE TED, **50**, 952 (2003)
38. J.M.Hergenrother et al., IEDM Tech.Dig., 75 (1999).
39. S.-H.Oh et al, IEDM Tech.Dig., 65 (2000).
40. D.Monroe & J.M.Hergenrother, ISSCC Tech.Dig., 134 (2000).
41. H.-S.Wong et al., IEDM Tech.Dig., 427 (1997).
42. G.W.Neudeck et al., IEDM Tech.Dig., 169 (2000).
43. R.S.Shenoy and K.C.Saraswat, Proc.Intl.SOI Conf., 190 (2004)
44. W.P.Maszara, Mat.Res.Soc.Proc., **686**, 59 (2002)
45. S.Monfray et al., VLSI Tech.Dig., 108 (2002)
46. S.Harrison et al., IEDM Tech.Dig., 449 (2003)
47. F.-L.Yang et al., IEDM Tech. Dig., 255 (2002)
48. C.Jahan et al., VLSI Tech.Dig., 112 (2005)
49. J.-W.Yang and J.Fossum, IEEE TED, **52**, 1159 (2005)
50. L.Mathew et al., IEDM Tech.Dig., 731 (2005)

340

ECS Transactions, 2 (2) 341-348 (2006)
10.1149/1.2195671, copyright The Electrochemical Society

Mobility Enhancement and Strain Integration in Advanced CMOS

Carlos H. Diaz

TSMC, No. 8 Li-Hsin Road 6, SBIP, Hsin-Chu, Taiwan, R.O.C., Tel. 886-3-666-5890, Fax. 886-3-563-7386

E-mail: chdiaz@tsmc.com

Abstract

Mobility enhancement techniques have become pervasive in advanced CMOS technologies. Non-scalable and scalable approaches to mobility enhancement are widely used in various application segments (high-speed, low-operating power, and low-standby power). Non-scalable techniques rely on preferential channel directions having fundamentally higher carrier mobility than the conventional <110> directions of (100) substrates. On the other hand, scalable techniques rely on process built-in mechanical strain to boost mobility as a result of band-structure response impacting both carrier-effective mass and scattering rates. Scalable techniques are based on the channel strain induced by contact-etch stop layers or $Si_{(1-x)}Ge_{(x)}$ structures primarily. This paper will review both non-scalable and scalable mobility enhancement approaches from two angles, namely fundamental device physics and overall device integration schemes. The paper will also review and discuss issues related to superposition of process built-in strain for performance enhancement.

Introduction

Device current density is directly proportional to the carrier velocity. Carrier velocity is in turn proportional to carrier mobility times the electric field. Consequently mobility enhancement is a prime approach towards device performance enhancement as technology scales.

Carrier mobility improvement is also attainable by the reduction of various scattering mechanisms and/exploiting the band asymmetry on preferential directions that minimize the carrier effective mass m. Mechanical stress breaks crystal symmetry and removes the 2-fold and 6-fold degeneracy of the valence and conduction bands respectively. This leads to changes of the band scattering rates and/or the carrier effective-mass which when properly optimized improve the carrier mobility and consequently device performance.

One approach towards mobility enhancement exploits preferential directions of the crystal structure for which the carrier's effective-mass is lowest. The carrier effective mass is dependent on the crystal orientation and its generally expressed as $m \sim 1 / (d^2E / dk^2)$. Figure 1 depicts the valence band diagram for silicon along two specific crystal orientations. Conventional devices have been fabricated on (100) surface wafers with channels oriented along the <110> direction. However, as it is observed from Fig. 1, the effective mass for holes along the <100> direction is lighter than that along <110>. Consequently device performance of a pMOS device can be significantly enhanced when its channel orientation is set along the <100> direction as shown in Fig. 2. Similar results were reported earlier by Sayama et al. in [1]. This mobility enhancement scheme offers a simple and cost-effective approach towards pMOS performance improvement. However, it is not scalable. The remaining of this paper focuses on scalable mobility enhancement approaches that primarily rely on mechanical strain.

341

Strain Effects from Shallow-Trench Isolation

The electron and hole mobility response to compressive strain in the three principal device directions is qualitatively described in Fig. 3. Compressive strain along the device width direction -z is detrimental to both NFET and PFET device performance. Consequently, stress from the STI need to be reduced at any given technology node. Device isolation and active area design rules scale linearly by a factor of ~0.7 per technology generation bringing about increased levels of mechanical stress induced by abrupt material transitions. Figure 4 show that the built-in compressive stress over an active area increases as its dimensions decrease. Consequently, transistor drive current density per unit width becomes not only a function of the device width −W (an effect commonly known as narrow-width effect) but also a function of the active area dimension along the direction of the channel current flow. Shallow-trench stress lowering by co-optimization of the STI liner, gap-fill, and densification processes steps has been shown to effectively reduce device sensitivity to active area dimensions, c.f. [3]. Figures 5 and 6 demonstrate that device current density can indeed be made almost independent of the active area dimensions.

Uni-axial strain from Contact Etch Stop Layers (CESL)

As indicated in Fig. 3, uni-axial strain along the direction of MOS current flow −x can benefit either the NMOS or pMOS performance depending on the stress type. Electron mobility enhancement using tensile CESL stress was introduced by Ito et al. [4]. Figure 7 shows that nFET drive current enhancement correlates well with the process built-in stress. The associated 3% or less pFET degradation is expected and can be eliminated by selective stress relaxation techniques requiring one additional mask/implantation step [4-5]. Figure 8 indicates that electron mobility response to tensile CESL saturates at close to 15% as reported in [6]. Compressive CESL was shown to improve pFET performance as shown in Fig. 9 based on the work reported by Shimizu et al. [7], see also [8].

Strain Memorization through gate stack

NFET performance improvement by the stress memorization technique (SMT) was introduced by Chen et al. in [9]. A high tensile nitride capping layer acts as a temporary stressor to effectively modulate the channel stress. The stress effect is then enhanced and memorized by well-controlled poly amorphorization and re-crystallization procedure. This Nitride high tensile capping p-layer will be removed after annealing step, Therefore much thicker capping layer can be used to increase the stress level without any process limitation to impact the further gap filling process steps. The process sequence of SMT is illustrated in Fig. 10. After the spacer and source/drain formation, a Poly Amorphorization Implantation (PAI) of Si gate electrode by Ge is performed in NMOS. The low-temperature and high tensile (>1.5G Pa) Nitride film is deposited as Activation Capping Layer (SiN_ACL) to modulate the channel stress. As shown in Fig. 11, the STI and oxidation-induced channel compressive stress can be effectively reduced by this tensile ACL. The selective removal of SiN_ACL on PMOS was performed to alleviate the stress-induced device degradation. Gate and S/D activation were followed by ACL selective removal step. SMT has been shown to improve nFET performance by > 10% as shown in Fig. 12 [9].

Uni-axial strain from SiGe SD

Lattice-mismatched regions within the device structure can also exert uni-axial strain on the channel. Thompson et al. [10] that by forming SiGe (larger lattice constant than Si) regions in the source/drain areas sets channel under compressive strain thus effectively improving pFET mobility. This mobility enhancement approach has been shown to be highly scalable and capable of attaining > 25% pFET I_{DSAT} performance improvement over un-strained silicon at nominal gate lengths [10-11].

Superposition of uni-axial strain approaches

The uni-axial strain mobility enhancement approaches discussed above have also been shown to be additive, c.f. [12]. Performance improvement can combine both T-CESL and SMT for nFETs and e-SiGe with a compressive contact etch stop layer   for pFETs as shown in Fig. 14 based on results reported in [12].

Conclusions

Scalable strain engineering through the entire process flow has become a critical element of process optimization in advanced technologies. Device performance improvements into the 45nm technology nodes will rely primarily on the scalability of these approaches through both process integration and material properties.

References

[1] H. Sayama et al., IEEE Intl. Electron Devices Meeting, Tech Digest, 1999.
[2] C. H. Ge et al, IEEE Intl. Electron Devices Meeting, Tech Digest, 2003.
[3] M. Miyamoto et al., IEEE Trans. On Electron Devices, March 2004.
[4] S. Ito et al., IEEE Intl. Electron Devices Meeting, Tech Digest, 2000.
[5] C. Diaz et al., IEEE Intl. Electron Devices Meeting, Tech Digest 2003.
[6] K. Goto et al., IEEE Intl. Electron Devices Meeting, Tech Digest, 2004.
[7] A. Shimizu et al., IEEE Intl. Electron Devices Meeting, Technical Digest, 2001.
[8] H. Yang et al., IEEE Intl. Electron Devices Meeting, Tech Digest, 2005.
[9] C. C. Chen et al., Symposium on VLSI Technology, 2004.
[10] S. Thompson et al, pp. 61-64, IEEE Intl. Electron Devices Meeting, Tech Digest, 2002
[11] K. Mistry et al., Symposium on VLSI Technology, 2004
[12] M. Horstmann et al., IEEE Intl. Electron Devices Meeting, Tech. Digest, Dec. 2005.

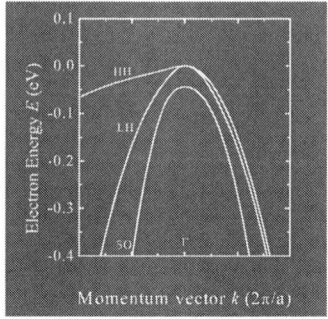

Fig. 1. Valance band for silicon substrates. The effective mass m_h along the <100> directions is lighter.

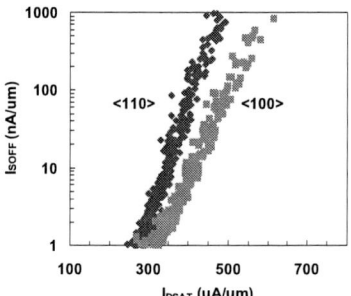

Fig. 2. pMOS I_{OFF}-I_{DSAT} dependency on channel orientation. Devices are fabricated on (100) silicon substrates.

Compressive strain increase	e mobility	h mobility
X	degrades	improves
Y	degrades	degrades
Z	improves	degrades

Fig. 3. Compressive strain from STI is detrimental to device performance [2]. Process optimization is necessary to limit compressive strain built-up from STI.

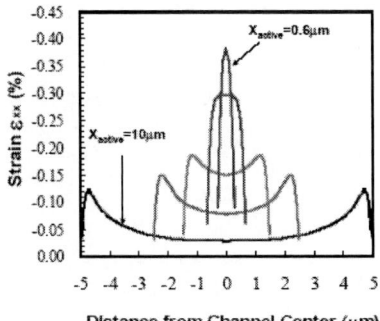

Fig. 4. Simulated strain ε_{xx} inside silicon along a line 20Å deep below Si/SiO2 interface for different active area size. Strain magnitude increases as active area size decreases.

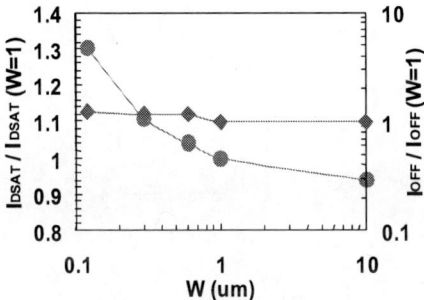

Fig. 5. Low stress STI process can mitigate device performance sensitivity to device width as shown here.

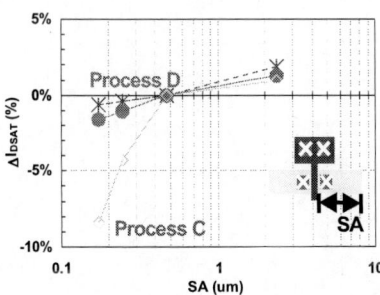

Fig. 6. Compressive STI stress reduction can also alleviate the device performance dependency to active area size.

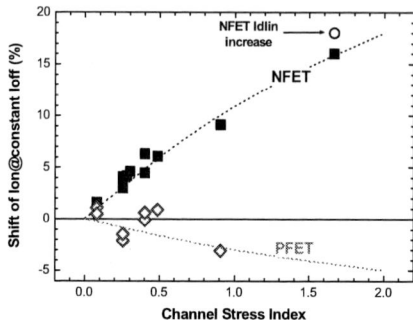

Fig. 7. n/pFET saturation and drain current shift versus channel stress index [5].

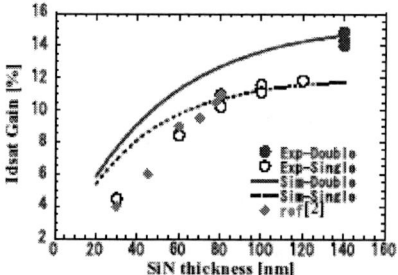

Fig. 8. Tensile contact-etch-stop layer improvement on NFET. Performance may saturate @ 15% [6].

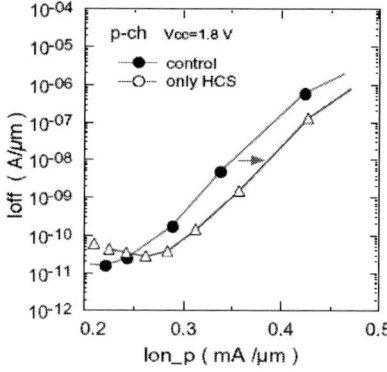

Fig. 9. Effect of a highly compressive etch stop layer (HCS) (-1.4Gpa SiN) on pFET I_{DSAT} improvement [7].

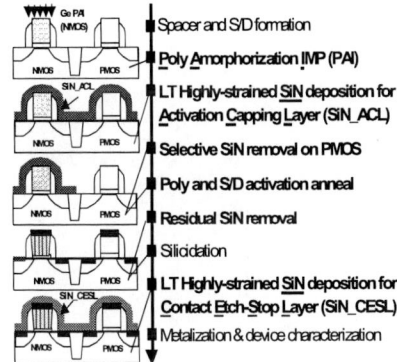

Fig. 10. Representative process sequence of the Stress Memorization Technique (SMT) [9].

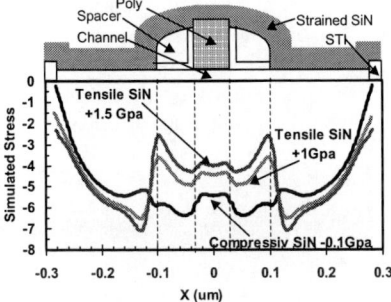

Fig. 11. Strained SiN capping layer effect on channel stress distribution (simulation) [9].

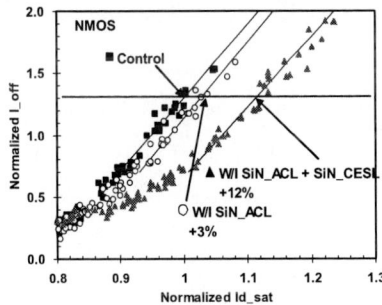

Fig. 12. Up to 12% nFET drive current improvement can be attained by inception of the SMT [9].

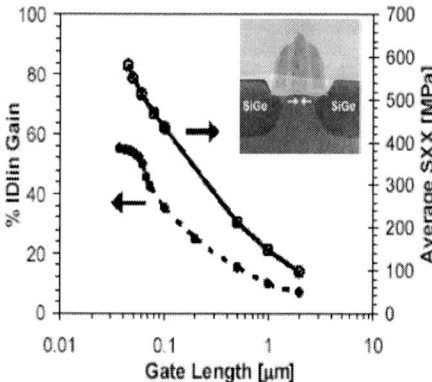

Fig. 13. Selective epitaxial SiGe at source/drain generates the uni-axial compressive strain on the channel [10]. Above 50% pMOS I_{Dlin} gain demonstrated at nominal Lg [11].

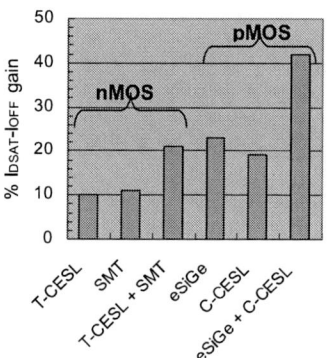

Fig. 14. Superposition of various uni-axial strain enhancement methods have been demonstrated on both n/pFETS, c.f. Horstmann et al. [12].

DEFECT ENGINEERING CONSIDERATIONS FOR STRAINED SILICON SUBSTRATES

E. Simoen[1], G. Eneman[1,2,3], C. Claeys[1,2], M. Scholz[1], R. Loo[1], P. Verheyen[1] and K. De Meyer[1,2]

[1]IMEC, Kapeldreef 75, 3001 Heverlee, Belgium
[2]ESAT-INSYS, K.U. Leuven, Kasteelpark Arenberg 10, 3001 Heverlee, Belgium
[3]Research assistant of The Fund for Scientific Research – Flanders (Belgium)

The impact of different defect-engineering parameters on the electrical performance of p-n junctions made in strained silicon (SSi) substrates is investigated. The thin SSi layers have been deposited on so-called thin Strain Relaxed SiGe Buffers (SRBs), whereby the relaxation of the strain is achieved by implementing a thin C-doped SiGe layer. Processing variables studied are the distance d_C of this layer with respect to the wafer surface, or, more importantly to the junction position and the concentration of carbon, which determines the density of threading dislocations (TDs). As references, step-graded SRB material and standard Czochralski silicon has also been used. From combined current-voltage and capacitance-voltage measurements it is concluded that within the range studied (2-$3x10^5$ to $1-2x10^7$ cm^{-2}), the TD density has the strongest impact on the reverse current density and generation lifetime. However, additional leakage current is created by defects associated with the carbon-rich layer, when it is present in the depletion region of the junction. The best performance of the thin SRB junctions is thus obtained for the lowest TD density and the largest d_C. It will, finally, be demonstrated that the electrical activity of the TDs also depends on the well doping type (n- or p), whereby the largest activity is observed in the n+/p junctions, which are technologically the more relevant ones.

INTRODUCTION

The fabrication of high-mobility Strained Silicon (SSi) layers on a $Si_{1-x}Ge_x$ virtual substrate is a nice example of defect and strain engineering in silicon materials, reaching currently a high level of maturity [1-2]. For state-of-the-art step-graded Strain Relaxed Buffers (SRBs) threading dislocation densities (TDDs) are in the low 10^5 cm^{-2} range. However, materials' cost and heat dissipation issues fuel today's developments towards thinner SRBs, employing among others point and implantation defects as nuclei for dislocation formation [3-12]. An original approach has been developed at IMEC, using a thin carbon-doped layer for strain relaxation [13]. The advantage of this technique is that it can also be implemented in a Selective Epitaxial Growth (SEG) scheme for n-channel devices [14], while the mobility of p-MOSFETs can be enhanced by the uniaxial compressive stress executed by recessed SiGe source/drain regions [15]. The penalty paid for the implementation of a C-rich layer is a higher TDD, which may affect the off-state leakage current of the devices fabricated in the resulting thin SRB SSi wafers. The aim of the present paper is to evaluate the electrical activity of the threading dislocations in SSi substrates using a combination of analytical techniques. The main part of the work relies on the analysis of the reverse current in p+/n and n+/p junctions fabricated in various SSi substrates.

In a previous study [16-17], it has been demonstrated that the electrical characteristics of junctions in thin SRB layers depend basically on three parameters: the TDD, the position of the C-rich layer (d_C from the wafer surface) with respect to the junction and the well processing, using dopant ion implantation and rapid thermal annealing. Here, a wider range of TDD and d_C is employed than before, while at the same time, commercially available SSi wafers, using a thick step-graded buffer layer are included in the experimental matrix. Besides the diode current-voltage (I-V) and capacitance-voltage (C-V) characteristics, additional analyses like EMmission MIcroscopy (EMMI) or MicroWave Absorption (MWA) recombination lifetime measurements have been performed. Detailed results of the latter technique have been reported elsewhere [18]. The interpretation of the diode characteristics is based upon a dedicated methodology developed in Refs. [19] and [20].

EXPERIMENTAL

Large-area diodes (Area=$10^5 \mu m^2$) were fabricated by standard highly-doped drain (HDD) and well implantations, placing the metallurgical junction at a depth of ~80 nm below the surface. All wafers received a standard 1000°C spike anneal to activate the dopants and nickel silicidation to lower the contact resistance. The different types of SRB's used in this experiment are shown in Table I. All SRB's have a Ge concentration of ~20% and a strained Si top layer with a thickness of ~10 nm. Samples A-D were fabricated in an ASM 2000 epsilon reactor with a 'thin-SRB' scheme, using a 5 nm C-rich layer to induce a high relaxation in the SiGe SRB of around 90% [13]. The C concentration is around 1%. TDD's were estimated using defect etching and inspection by optical microscopy. Samples A-C were grown with a similar density of threadings, but placing the C-doped layer at different positions below the surface. Sample D was grown with a deep position of the C-rich layer and with an increased density of threadings, induced by a higher C concentration. Sample E was produced, using a 'thick-SRB' approach, relying on gradually increasing the Ge concentration during growth to control the defect propagation. This process resulted in the lowest value for the TDD (Table I). Finally, diodes have also been processed in standard Czochralski (Cz) silicon substrates as references.

Diode I-V characteristics have been measured by an HP 4156 Parameter Analyzer, for temperatures between 25 and 250°C. High-frequency (1 MHz) C-V measurements were performed by an HP4284 LCR meter. From the C-V data, the depletion width W for the SSi junctions was obtained from:

$$W = \varepsilon_{SiGe}A/C \qquad (1)$$

with ε_{SiGe} the permittivity of SiGe ($12.7\varepsilon_0$ for x=20 %; ε_0 is the permittivity of vacuum) and A the junction area. The free carrier concentration profile in the depletion region has been derived from:

$$n(W) = 2/[q\ \varepsilon_{SiGe}\ A^2\ dC^{-2}/dV] \qquad (2)$$

In Eq. (2), q is the elementary charge.

EMMI measurements have been performed in a Hypervision-Visionary II photon emission microscope equipped with a highly sensitive GENIII-NIR detector [21], at large reverse bias across the junction.

Table I. Overview of the C-rich layer position and Threading Dislocation Density for the SRB-substrates used in this experiment.

	C-rich layer: depth below the surface (nm)	Threading defect density (cm^{-2})
SRB-A	100	1-2x10^7
SRB-B	200	1-2x10^7
SRB-C	270	1-2x10^7
SRB-D	270	$\sim10^9$
SRB-E	-	2-3x10^5

RESULTS AND DISCUSSION

A. Impact of the dislocation density

It has been well-established that TDs enhance the leakage current (I_R) of a junction when penetrating the depletion region [16,17,22-24]. In the first instance, one expects an increase of I_R proportional with TDD [23], which is confirmed by the results of Fig. 1 for the n+/p junctions. As shown in Fig. 2, the corresponding reverse current density J_R increases proportionally with TDD, yielding a leakage current density of ~10 pA/TD at a V_R=-1 V and T=25°C [25] for the n+/p junctions. This is in reasonable agreement with the value reported in the literature of ~2.5 pA/TD, corresponding with a depletion width of 0.5 μm and 25% Ge [23]. Here -1 V corresponds approximately to a W=80 to 90 nm. It should be remarked that the depth of the carbon layer in Fig 2 is 270 nm or ~200 nm from the metallurgical junction, which is outside the depletion region for V_R=-1 V. In this way, no direct effect of the defects related with the C-rich layer on J_R is expected [16,17].

For the p+/n junctions, there is also an increase of J_R at -1 V with TDD in Fig. 2, although it appears to be more sub-linear in this case. It again emphasizes the role of the well implantation damage and species, noted during an earlier experiment [16,17]. It is not clear for the moment why there is a different dependency on TDD for the n+/p and p+/n junctions. It could point to a different electrical activity of a threading dislocation in n- or p-type $Si_{0.8}Ge_{0.2}$, related to the electron charging of the dangling bond centers, which generates a potential barrier for electrons around the line charge.

Also the forward characteristics show a higher current at low forward bias V_F, as can be noted in Fig. 1. This is related to the lower recombination lifetime with higher TDD [18], resulting in a higher ideality factor of the junction.

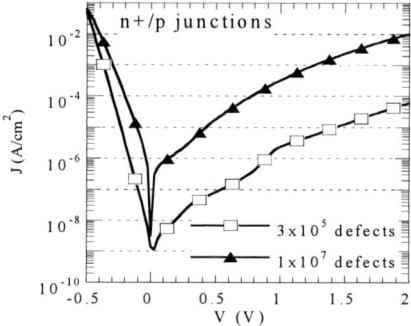

Fig. 1. Current characteristics for a thick (3×10^5 cm^{-2}) and a thin SRB (1×10^7 cm^{-2}) n+/p junction at 300 K.

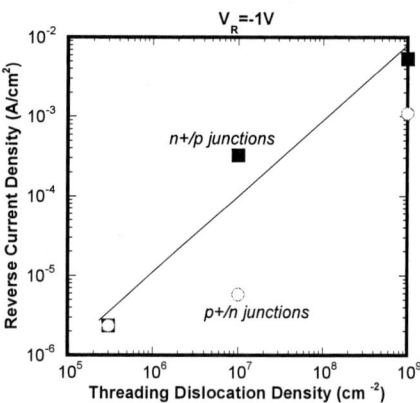

Fig. 2. Reverse current density in function of TDD for n+/p and p+/n junctions. T=300 K, V_R=-1 V and $d_C \geq$ 270 nm.

The same linear increase with TDD is reflected in the inverse of the effective generation lifetime $\tau^{-1} \sim \partial J_R / \partial W$ [19,20], extracted from the I-V curves of the n+/p junctions at small $-V_R = 25$ mV and represented in Fig. 3. Comparing with reference Cz junctions in Fig. 4, it is clear that the initial slope of the J_R versus W curve is about 1 decade higher for the thick SRB wafer, with a TDD=2×10^5 cm^{-2}. Another difference in Fig. 4 is the depth of the depletion region. This could originate from a different free carrier density n. However, this is not the case according to Fig. 5. This means that the lower W value found for $Si_{0.8}Ge_{0.2}$ results from the balance between the ~10 % higher ε_{SiGe} and the lower built-in potential [26], whereby the latter effect is clearly dominant in the case of Fig. 5.

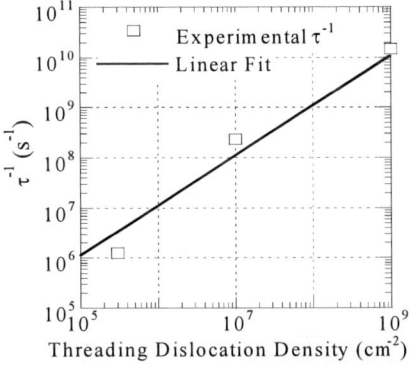

Fig. 3. Inverse generation lifetime versus threading dislocation density. The data points are calculated from current and capacitance measurements at 25 mV reverse voltage. The line shows a linear fit through the data points.

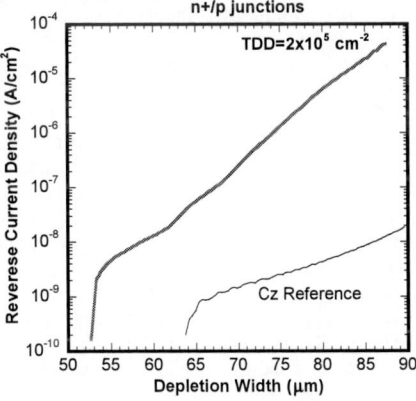

Fig. 4. Reverse current density versus depletion width for a reference Cz and a thick step-graded SRB SSi junction at 300 K.

However, as shown elsewhere, the reverse current related to the TDD does not compromise too much the power consumption of CMOS made in these SSi substrates [25]. The highest leakage current density corresponding with the highest TDD studied yields a value of 100 mA/cm^2 at a supply voltage of 1 V and a realistic operating temperature T of 100°C.

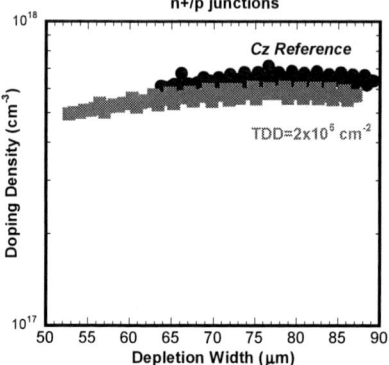

Fig. 5. Free carrier profile, derived from a 1 MHz C-V obtained on a Cz reference and a thick step-graded SRB SSi junction. A=10^5 μm^2.

B. Impact of the depth of the C-rich layer

As has already been shown, another important factor in the reverse current of the studied SSi junctions is the position of the C-rich layer with respect to the junction [16]. This is illustrated in Fig. 6 for n+/p junctions with the same TDD, for different V_R values. Especially at larger V_R, there appears to be an exponential dependence of J_R on d_C. It is clear that associated with this C-layer, there exists a high density of electrically active defects, localized around the 5 nm thick layer. This defect band will contribute to the leakage as soon as the depletion width W crosses d_C.

This is illustrated more clearly in Fig. 7, representing J_R versus W for n+/p junctions with different d_C. Similar as in a previous experiment [17], one can see that for the 100 nm depth (or ~20 nm from the junction) there is already a high leakage current density at V_R=0 V. This has been explained by considering the fact that the built-in depletion region width (W_{bi}) in Fig. 7 is clearly higher than 20 nm, so that the C-related defect band is contained within W_{bi}. This is not the case for the other two conditions, where it is observed that initially, a lower J_R is found. For the 200 nm junction in Fig. 7, the leakage current density approaches the value for the 100 nm devices at W~93 nm, indicating that the carbon-related defects start to contribute. Finally, for the 270 nm diode, J_R is always smaller, demonstrating the absence of C-related defects in the depletion layer for a reverse bias up to -2 V.

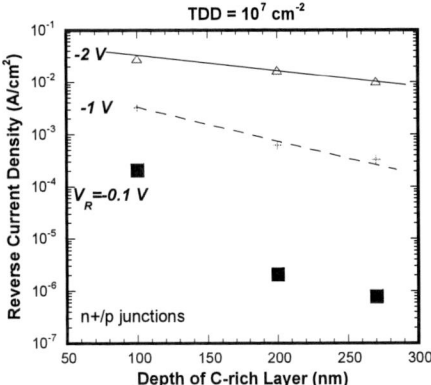

Fig. 6. Reverse current density at different V_R in function of the depth of the C-rich layer for n+/p junctions. T=300 K and TDD=10^7 cm^{-2}.

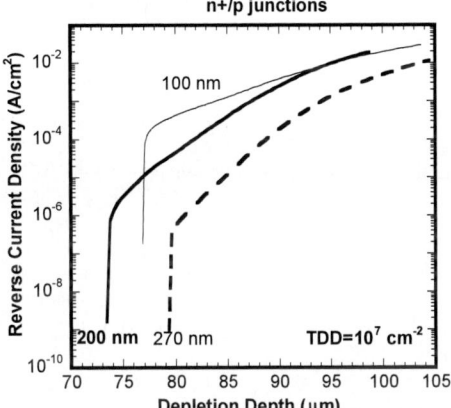

Fig. 7. Reverse current density versus depletion depth for three n+/p junctions in a thin SRB SSi substrate with TDD=10^7 cm^{-2} and corresponding with a different position of the C-rich layer.

A striking feature in Fig. 7 is the apparently larger depletion width compared with the thick SRB n+/p junctions or even with the reference device of Fig. 4. Curiously enough, this cannot be explained by a smaller free carrier concentration, as witnessed by Fig. 8. Interestingly, the nature of the profile is different for the thin SRB SSi n+/p junctions compared with the junctions without carbon doping. It is known that the presence of carbon suppresses the diffusion behaviour of boron [27], which may partly explain the observations. Another question mark relates to the exact depth of the junction, which is probably larger in the $Si_{0.8}Ge_{0.2}$ diodes compared with the standard devices. This can be derived from the fact that according to Fig. 7, the impact of the C-rich layer appears at around 80+93 nm=173 nm, which is 20 to 30 nm too low. This is also suggested by shape of the profiles in Fig. 8, which suggest that they are situated at either flank of the maximum boron well concentration for the Cz reference, on the one hand and for the thin SRB junctions, on the other. A junction shift in the range of 10 to 15 nm is derived from that. Detailed Secondary Ion Mass Spectrometry results are required to further resolve this issue.

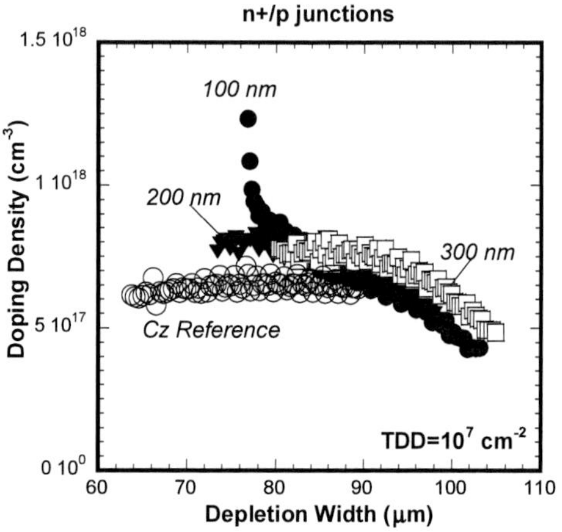

Fig. 8. Free carrier density profiles for n+/p junctions with similar TDD and A=10^5 μm^2 and corresponding with a different depth of the C-rich layer. For reference the profile for a Cz diode is also shown.

Figure 9 compares the results for the n+/p with the p+/n junctions for the same TDD (10^7 cm^{-2}) and different d_C. Overall, the p+/n junctions show a better reverse current density than the n+/p junctions; the latter are, however, the technologically relevant ones for the fabrication of n-channel MOSFETs.

C. Junction breakdown as studied by EMMI

It is known from the early days of semiconductor device physics that a junction emits visible light when operated in the breakdown mode [28-30]. This light emission will be localized in the regions of higher electric field, which in the case of a defect-free planar junction is generally the perimeter. This is illustrated in Fig. 10 for a standard Cz diode, showing local breakdown along the perimeter, where the highest electric field is present.

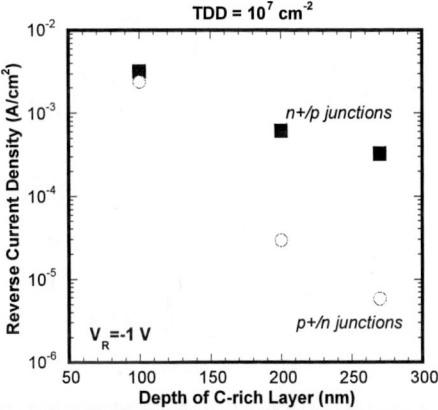

Fig. 9. Reverse current density in function of the depth of the C-rich layer for n+/p and p+/n junctions. T=300 K, V_R=-1 V and TDD=10^7 cm^{-2}.

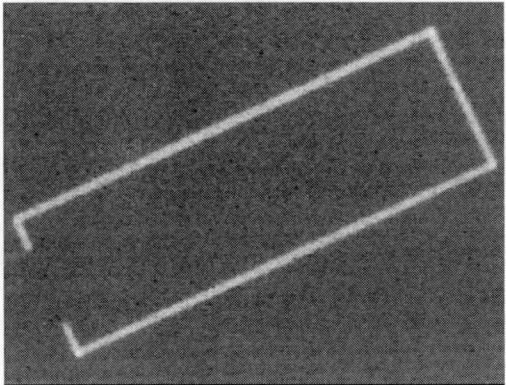

Fig. 10. EMMI of a standard Cz n+/p diode, at high reverse bias.

On the other hand, when extended defects, like dislocations are present in the junction, local avalanching will preferentially occur there [30]. This is confirmed by the results of Figs 11 to 13, corresponding with the 100, 200 and 270 nm n+/p junctions, respectively.

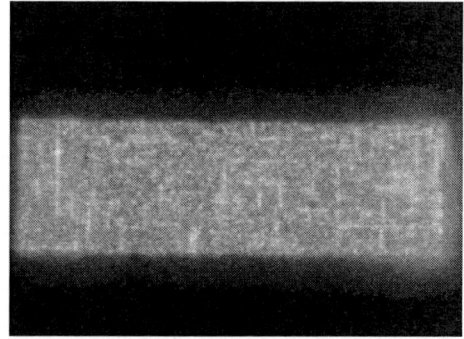

Fig. 11. EMMI of a d_C=100 nm thin diode, at high reverse bias.

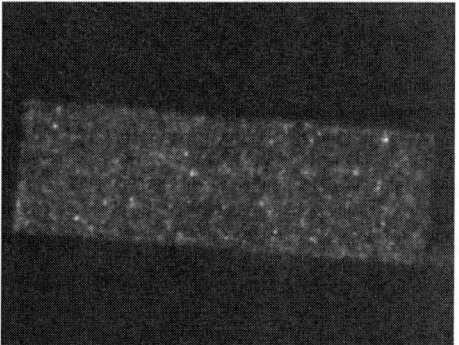

Fig. 12. EMMI of a d_C=200 nm thin diode, at high reverse bias.

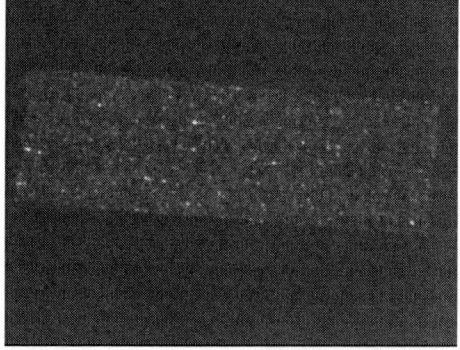

Fig. 13. EMMI of a d_C=270 nm thin diode, at high reverse bias.

There are some qualitative differences between the emission micrographs of Fig. 11, on the one hand and Figs 12 and 13, on the other. It appears that for the shallowest d_C, breakdown occurs preferentially at the misfit dislocations present at the bottom SiGe-Si interface, confirming our earlier observations [16]. This is related to the fact that the total stack thickness can be depleted by a large reverse bias. For the higher d_C values, a more uniform light pattern is found, which could correspond to breakdown at the TDs or possibly at extended defects associated with the C-rich layer. In this case, it is believed that the depletion layer during breakdown is not reaching the bottom interface, showing no evidence of breakdown by misfits. Finally, for p+/n junctions, a clear breakdown pattern associated with TDs is found in Fig. 14, irrespective of d_C. This could be again related to the space charge region surrounding a dislocation line charged by electrons in an n-type substrate [31,32], resulting in a high local electric field and an early onset of breakdown.

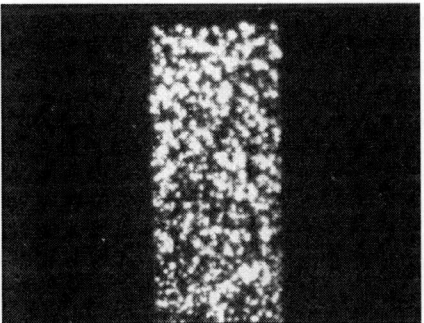

Fig. 14. Emission Microscopy (EMMI) image of a p+/n diode (Area=$10^5 \mu m^2$) during breakdown, showing microplasma breakdown at dislocation sites.

CONCLUSIONS

Concluding, it has been shown that the electrical characteristics of p-n junctions fabricated in SSi substrates depend largely on the quality of the SRB used. A strong dependence on the TDD has been observed, although within the defect density range studied here, tolerable power consumption levels related to the extra leakage current have been estimated. At the same time, for thin SRB material, one should optimize the depth of the C-rich layer (or in general, the extended defects introduced to relax the strain) with respect to the junction and make sure that the depletion width at realistic operation conditions does not overlap with this defect band.

REFERENCES

[1]   D.J. Paul, *Semicond. Sci. Technol.*, **19**, R75 (2004).

[2] M.L. Lee, E.A. Fitzgerald, M.T. Bulsara, M.T. Currie and A. Lochtefeld, *J. Appl. Phys.*, **97**, 011101-1 (2005).

[3] H. Trinkaus, B. Holländer, St. Rongen, S. Mantl, H.-J. Herzog, J. Kuchenbecker and T. Hackbarth, *Appl. Phys. Lett.*, **76**, 3552 (2000).

[4] M. Luysberg, D. Kirch, H. Trinkaus, B. Holländer, St. Lenk, S. Mantl, H.-J. Herzog, T. Hackbarth and P.F.P. Fichtner, *J. Appl. Phys.*, **92**, 4290 (2002).

[5] K. Sawano, Y. Hirose, Y. Ozawa, S. Koh, J. Yamanaka, K. Nakagawa, T. Hattori and Y. Shiraki, *Jpn. J. Appl. Phys.*, **42**, L735 (2003).

[6] Yu.B. Bolkhovityanov, A.S. Deryabin, A.K. Gutakovskii, M.A. Revenko and L.V. Sokolov, *Appl. Phys. Lett.*, **84**, 4599 (2004).

[7] J. Cai, P.M. Mooney, S.H. Christiansen, H. Chen, J.O. Chu and J.A. Ott, *J. Appl. Phys.*, **95**, 5347 (2004).

[8] P. Chen, P.K. Chu, T. Höchbauer, M. Nastasi, D. Buca, S. Mantl, N.D. Theodore, T.L. Alford, J.W. Mayer, R. Loo, M. Caymax, M. Cai and S.S. Lau, *Appl. Phys. Lett.*, **85**, 4944 (2004).

[9] K. Lyutovich, J. Werner, M. Oehme, E. Kasper and T. Perova, *Mat. Sci. in Semicond. Process.*, **8**, 149 (2005).

[10] T. Sadoh, R. Matsuura, M. Ninomiya, M. Nakamae, T. Enokida, H. Hagino and M. Miyao, *Mat. Sci. Semicond. Process.*, **8**, 167 (2005).

[11] B. Romanjuk, V. Kladko, V. Melnik, V. Popov, V. Yukhymchuk, A. Gudymenko, Ya. Olikh, G. Weidner and D. Krüger, *Mat. Sci. Semicond. Process.*, **8**, 171 (2005).

[12] J.P. Liu, L.H. Wong, D.K. Sohn, L.C. Hsia, L. Chan, C.C. Wong and H.J. Osten, *Electrochem. and Solid-St. Lett.*, **8**, G60 (2005).

[13] R. Delhougne, G. Eneman, M. Caymax, R. Loo, P. Meunier-Beillard, P. Verheyen, W. Vandervorst and K. De Meyer, *Solid-St. Electron.*, **48**, 1307 (2004).

[14] G. Eneman, P. Verheyen, R. Rooyackers, R. Delhougne, R. Loo, M. Caymax, P. Meunier-Beillard, K. De Meyer and W. Vandervorst, *Mat. Sci. in Semicond. Process.*, **8**, 337 (2005).

[15] G. Eneman, P. Verheyen, R. Rooyackers, F. Nouri, L. Washington, R. Degraeve, B. Kaczer, V. Moroz, A. De Keersgieter, R. Schreutelkamp, M. Kawaguchi, Y. Kim, A. Samoilov, L. Smith, P.P. Absil, K. De Meyer, M. Jurczak and S. Biesemans, in *2005 Symp. on VLSI Technol. Dig. of Techn. Papers*, The IEEE, p. 22 (2005).

[16] G. Eneman, E. Simoen, R. Delhougne, E. Gaubas, V. Simons, P. Roussel, P. Verheyen, A. Lauwers, R. Loo, W. Vandervorst, K. De Meyer and C. Claeys, *J. Phys.: Condens. Matter*, **17**, S2197 (2005).

[17] G. Eneman, E. Simoen, R. Delhougne, P. Verheyen, V. Simons, R. Loo, M. Caymax, C. Claeys, W. Vandervorst and K. De Meyer, *J. Electrochem. Soc.*, **153**, in press (2006).

[18] E. Gaubas, R. Tomašiūnas, G. Eneman, R. Delhougne and E. Simoen, *Semicond. Sci. Technol.*, **20**, 1052 (2005).

[19] A. Poyai, E. Simoen, C. Claeys, E. Gaubas, D. Gräf and A. Huber, *Mater. Sci. Engineer. B.*, **102**, 189 (2003).

[20] A. Poyai, E. Simoen, C. Claeys and R. Rooyackers, *J. Electrochem. Soc.*, **150**, G795 (2003).

[21] M.S. Rasras, I. De Wolf, G. Groeseneken and H.E. Maes, *J. Appl. Phys.*, **89**, 249 (2001).

[22] S. Kanjanachuchai, T.J. Thornton, J.M. Fernández and H. Ahmed, *Semicond. Sci. Technol.*, **13**, 1215 (1998).

[23] L.M. Giovane, H.-C. Luan, A.M. Agarwal and L.C. Kimerling, *Appl. Phys. Lett.*, **78**, 541 (2001).

[24] N. Sugii, M. Kondo, M. Miyamoto, Y. Hoshino, M. Hatori, W. Hirasawa, Y. Kimura, S. Kimura, Y. Kondo and I. Yoshida, in *2005 Symp. on VLSI Technol. Dig. of Techn. Papers*, The IEEE, p. 54 (2005).

[25] G. Eneman, E. Simoen, R. Delhougne, P. Verheyen, R. Loo and K. De Meyer, *Appl. Phys. Lett.*, **87**, 192112 (2005).

[26] V.P. Gopinath, H. Puchner and M. Mirabedini, *IEEE Electron Device Lett.*, **23**, 312 (2002).

[27] H. Rücker, B. Heinemann, W. Röpke, R. Kurps, D. Krüger, G. Lippert and H.J. Osten, *Appl. Phys. Lett.*, **73**, 1682 (1998).

[28] A.G. Chynoweth and K.G. McKay, *Phys. Rev.*, **102**, 369 (1956).

[29] A.G. Chynoweth and K.G. McKay, *Phys. Rev.*, **106**, 418 (1957).

[30] A.G. Chynoweth and G.L. Pearson, *J. Appl. Phys.*, **29**, 1103 (1958).

[31] W.T. Read, Jr., Phil. Mag., **45**, 775 (1954).

[32] W.T. Read. Jr., Phil. Mag., **45**, 1119 (1954).

Modelling of Morphological Changes by Surface Diffusion in Silicon Trenches

T. Müller[a], D. Dantz[c], and W. v. Ammon[a]

[a] Siltronic AG, 78434 Burghausen, P.O. 1140, Germany

J. Virbulis[b], U. Bethers[b]

[b] Center for Processes Analysis and Research Ltd., Riga, Latvia

[c] present address: Volkswagen AG, Wolfsburg, Germany

We present a three – Dimensional model of surface diffusion which describes the time evolution and topography changes of deep trenches under low pressure hydrogen anneal and compare the results with experimental observations. A phase diagram for regular arranged cylindrical trenches was derived by using a genetic optimization algorithm. Seven different structural phases of the final bubble / layer arrangement out of the trench array anneal could be predicted. The impact of perturbations in the regular trench array is discussed too and the change in the phase diagram – shrinkage of the stable region for single top layer formation – is calculated .

INTRODUCTION

In the pioneering work of Sato & Mizushima et al. [1-4] a new simple approach to establish large –scale empty spaces in Silicon (ESS) basing on microstructure transformation of silicon (MSTS) is demonstrated. Trench patterned silicon surfaces were annealed in hydrogen ambient at low pressures and result in formation of empty bubbles, pipes or plates underneath a refilled and closed silicon top layer. It is stated that the self-organizing surface migration of silicon acts in that way that the surface energy is minimized. Depending on the spacing (or pitch) and width of the initial trenches they coalesce and form in two dimensions an empty plate under the surface.

It is important to develop simulation models to study the transformation of the trench into single bubbles or empty plates in detail. Thus predictions could be made about the limitations of this Silicon- On Nothing technique e.g. the minimum allowable silicon top layer if the trench geometry is given. Developing a "phase" diagram spanned by the two fundamental geometrical parameters depth and radius for a simple cylindrical trench array would give information about stable conditions to derive a single silicon top layer. Perturbations of the lithography process can lead to not perfect closed top layers. Therefore the influence of perturbations like lateral shifts and depth variations were studied too.

EXPERIMENTAL PROCEDURE

Structuring of the experimentally demonstrated trenches was performed by lithography and reactive etching. The critical dimensions of the mask process were 0.25 μm;

measuring of the structures was performed with a JEOL 6300 SEM. The annealing was typically performed for process times of 5 - 20 min at 1100 °C in 100 % H$_2$ atmosphere at 10 Torr.

NUMERICAL MODELS

Surface diffusion model

Surface diffusion as the fundamental process for this silicon on nothing (= SON) technology can be macroscopically described by a Mullin's equation [5]:

$$\frac{\partial n}{\partial t} = B\Delta_s k ,$$
(1)

whereas k is the sum of the two principal surface curvatures (mean curvature), n is the surface position in normal direction and Δ_S is the surface Laplace operator. Coefficient B is defined as:

$$B = \frac{D_S \gamma v \Omega^2}{k_B T}$$
(2)

whereas D_S is the surface diffusion coefficient, γ is the surface tension (surface energy density), k_B is the Boltzmann's constant, Ω and v are atomic volume and atomic surface density, respectively. The surface diffusion coefficient is given by Keefe [6] as

$$D_S = 0.1\exp(-\frac{2.3\,eV}{k_B T}),$$
(3)

with pre-exponential constant of 0.1 and activation energy of 2.3 eV.

The following algorithms are used for the surface development modelling in 3D, i.e. for the triangular meshes on surfaces:

1. The algorithm for mean surface curvature (k) is adapted from Zinchenko et. al [7]. It allows determination of curvature and normal vector (**n**) at mesh nodes and it employs least square method for fitting best paraboloid surface passing through the given node with a given normal vector. Algorithm proceeds by iterating through a set of normal vectors until the convergence is achieved.
2. The algorithm for calculating the surface Laplace operator is adapted from Mayer [8]. It finds the value of the surface Laplace operator at node (i) as

$$(\Delta\kappa)_i \approx \frac{1}{\Delta S}\int_S \Delta\kappa dS = \frac{1}{\Delta S}\oint_C \frac{\partial\kappa}{\partial n}d\Gamma \approx \frac{1}{\sum_j S_j}\sum_j \frac{1}{2}\left(\frac{\kappa_{1j}-\kappa_i}{|\vec{r}_{1j}-\vec{r}_i|} + \frac{\kappa_{2j}-\kappa_i}{|\vec{r}_{2j}-\vec{r}_i|}\right)|\vec{r}_{2j}-\vec{r}_{1j}|,$$
(4)

and integration is done along the boundary of the triangular patch (C). Summation is performed over all triangles (j) of the patch. Indexes 1 and 2 correspond to two other nodes of the triangle, **r** are the co-ordinates of the nodes.
3. Moving of the mesh is performed by the first order approximation of the equation of motion:

$$\frac{d\vec{r}}{dt} = v_n \vec{n}$$
(5)

4. Mesh restructuring uses following algorithms:

4.1. Mesh edge merging is applied to edges which have a length smaller than prescribed minimum edge length.

4.2. Mesh refinement is adapted from Ruppert [9] extending it to three dimensions. Refinement algorithm proceeds by eliminating the triangles for which a ratio of the circumsphere radius to the shortest edge is greater than given prescribed value (our algorithm employs value of 1.2). Elimination is achieved by inserting node at the projection of the center of triangle circumsphere onto the surface. If such a node is in the circumsphere of the boundary edge, then the node is not inserted; instead the boundary edge is splitted.

4.3. Mesh edge swapping (node reconnection) is employed according to a criterion given by Cristini et al. [10].

4.4. Mesh splicing occurs during the narrowing of the surface neck. Algorithm for splicing detection finds the shortest possible connected path on the surface, with diameter smaller than prescribed splicing distance. Splicing occurs if for any point on this path following relation holds: ($\kappa>0$ and $v_n<0$) or ($\kappa<0$ and $v_n>0$).

4.5. Mesh joining occurs if two parts of a mesh are moving towards each other. Algorithm for joining detection checks the distance between two mesh nodes. Joining occurs if this distance is smaller than a prescribed joining length, and if nodes are moving towards each other in a prescribed sector (30°), and if the following logical relation holds true: ($\kappa>0$ and $v_n>0$) or ($\kappa<0$ and $v_n<0$).

5. As boundary conditions the symmetry conditions i.e. the triangles touching the symmetry planes are mirrored against these planes and mirrored vertices are used in calculations in p. $1 - 3$. The symmetry condition allows simulating infinite symmetric areas of trenches.

A specialized simulation software was developed using these described algorithms.

Genetic optimization algorithm

The numerical model of surface diffusion is used for parameter studies of layer thickness and uniformity. The aim is to find out the optimum configuration of the trench array which gives the minimum layer thickness and maximum layer stability against perturbations. Due to the required large number of simulations the genetic optimization method was used and it's adaptation to this problem is described here. The genetic optimization algorithms, initiated in 1960-ies (for instance by Bremermann et al [11]) and theoretically developed in 1970-ies (for instance by Rechenberg [12], Schwefel [13]), nowadays are considered among the most efficient methods in finding the global minimums of the target functions (Goldberg [14]).

The genetic algorithm was constructed as follows:

1. Let us consider a task of minimizing the function $f(P_1, P_2, \ldots, P_n)$. wheras f is the upper Si layer thickness and P_1 to P_n are the n process parameters.

2. Binary representations of the parameter sets are binary strings $\vec{R}$. The binary string is assembled taking into account the restrictions set on the particular j-th parameter (N=11 for the particular implementation)

$$R_j = 2^N \frac{Param_j - Min_j}{Max_j - Min_j}, \ j = 1,2,...,n$$

3. A finite set of the permissible process parameters forms a population. An initial current population is generated by the random formation of binary strings, and the algorithm steps 4 – 10 are repeated. The parameter sets in the population are numbered from k = 1 to N_p, and the initial ranking (numbering) of the population is random. The population size N_p used in the present report is 50.

4. (N_p – N_E) pairs of the parameter sets (parents) $\vec{R}_{parent\,1}, \vec{R}_{parent\,2}$ are chosen from the current population according to the probability

$$p = \frac{1}{Z}\exp(-\alpha k); \quad Z = \sum_{i=1}^{N_p}\exp(-\alpha k_i)$$

Here k is the number (rank) of the parameter set in the population, ensuring that the higher-ranked parameter sets have higher probability to become parents, $\alpha=0,3$. $N_E=1$ is the number of elite parameter sets, which are copied to the next population without crossover.

5. The crossover of the each parent pair is performed yielding new parameter sets (crop):

$$R_{new,i} = (1 - \lambda_i)R_{parent\,1,i} + \lambda_i R_{parent\,2,i}$$
$$\lambda_i = r_i(1 + 2\delta) - \delta$$

Here $\delta = 0,2$ is crossover extrapolation parameter, and r_i is random number [0, 1)

6. The first step of crop perturbation – mutation – is introduced as

$$R_{new,i} = R_{new,i} + \lambda_M R_{max,i}$$

Here λ_M is normally distributed random value [-0.5, 0.5] with mutation dispersion $D_m=0,1$. The mutation probability 0,2 was used in this report.

7. The second step of crop perturbation – inversion – is introduced as cyclic shift of bits by random number. The inversion probability 0,1 was used in this report.

8. The check for existence of the parameter sets of the new population in the common calculation history is performed. If any of the new parameter sets has already existed then the mutation and perturbation steps 6 – 7 for this parameter set are repeated until the parameter set becomes unique.

9. The target function (thickness of the upper Si layer) is calculated for each parameter set in the crop.

10. The crop is updated to the new population. N_E best (elite) parameter sets from the previous population are added to the new population. The new population is ranked according to the value of the target function. The algorithm proceeds to step 4.

The best value of the target function (upper Si layer thickness) and the corresponding set of parameters are maintained throughout the optimization process. Although the optimization algorithm is not deterministic one may consider that (a) the crossover step 5 ensures the minimization procedure in between the values of the most advanced parameter sets (parents), (b) mutation step 6 seeks for optimum function values in the vicinity of the parent parameter sets, and (c) inversion step 7 searches for the possible other local minimums of the target function.

The genetic algorithm was realized numerically, and implemented as a server process in the cluster of workstations. The server program performs the algorithm steps 1 - 8 and 10, and distributes the parallel calculation tasks (of the annealing process – step 9) to the workstations. The Si surface development is calculated by Surver v.2.3 software on the workstations returning the target function – the final Si upper layer thickness – to the server.

RESULTS

SON layer formation: The principle of layer formation during the trench annealing is illustrated by presenting the simulated shapes developed after a subsequent time increment. Fig. 1 shows one trench in an array of trenches in cross-section perspective (trench pitch = 1 μm, trench diameter = 0.6 μm and trench depth = 4 μm). In the first stage of annealing the main effect of surface diffusion is the formation of a neck near the wafer surface (a). The radius of the neck decreases, the trench has been closed and the underlying bubble structure is formed (b) (furtheron calling this process "splicing"). Afterwards the surface diffusion changes the bubble shape, its height decreases and the diameter increases (c). If the distance between the neighboring trenches is smaller than the diameter of the sphere the bubbles will join each other (d) and form a continuous empty layer (e).

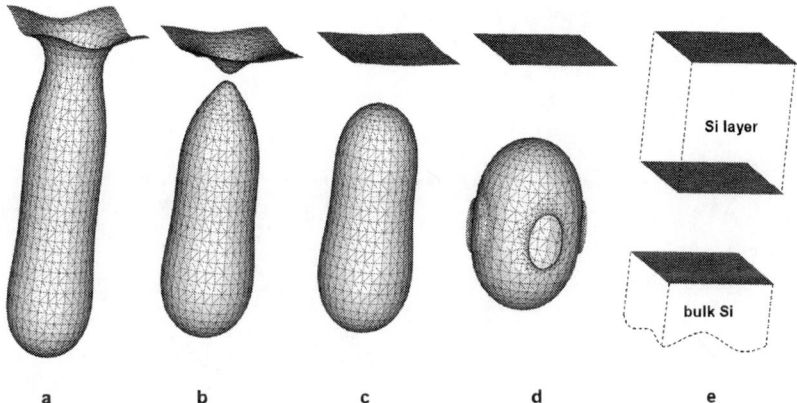

a   b   c   d   e

Figure 1. Time development of an SON layer represented by one trench out of an infinite calculated trench layer: neck formation (a), splicing of neck and bubble formation (b), bubble surface minimization (c), bubble joining with neighboring bubbles (d), continuous layer formation (e).

Comparison of bubble formation with experimental data I: from literature:
The already above mentioned experiments [2] presenting bubble formation from trench annealing were used as a starting point for our verification of the developed SON surface diffusion model.
 In Fig.2 an experimentally obtained trench geometry (rounded trench with an aspect ratio of 30) and it's annealing conditions were taken from this work as initial simulation parameter .
The silicon nitride / silicon dioxide film on the upper surface can be considered as a specific feature in this study. This film prevents the diffusion on the plane silicon surface.

Therefore the closing of the trench starts from the bottom in the annealing process. As a consequence bubbles are formed in regular distances starting from the bottom too. The simulation was able to verify this effect and same final shapes of the bubbles could be derived by our simulation approach (red lines in Fig. 2)

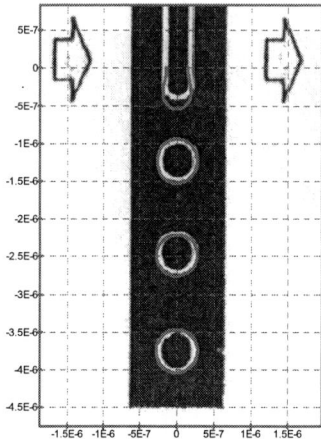

Figure 2. Comparison of simulation results (lines) with experimental bubble shapes taken from [2].

Comparison of bubble formation with experimental data II: by own experiments. For further model verifications a different trench and therefore bubble geometry was used: trench depth = 3.5 μm, trench diameter = 0.3 μm, trench pitch = 1 μm. Two different annealing times - 8 min and 20 min – were applied resulting in different bubble geometries (Fig. 3). The simulated shape (red curve) and distance between bubble and wafer surface fit well again with the measured cross sections of the experimentally obtained bubbles.

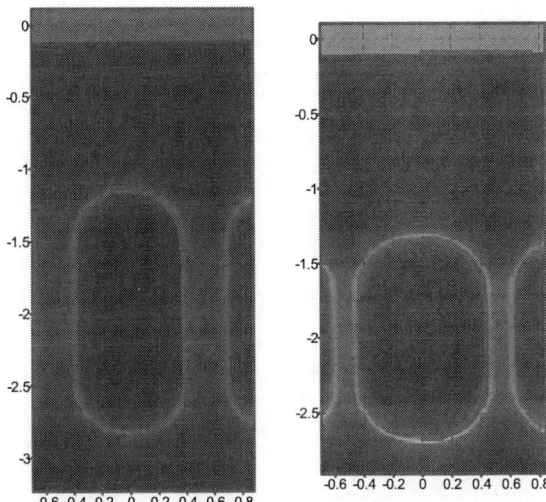

Figure 3. Comparison of 3D simulation results (red lines) with experimental SEM pictures of bubble shapes for two different annealing process times (left - 8 min, right – 20 min).

Study of layer uniformity:
The comprehensive optimization studies were performed by the developed genetic optimization method for cylindrical trenches only.The target of the optimization was defined to minimize the upper Si layer thickness Δz. Cylindrical trenches have only two (optimization) parameters – the trench radius R and the trench depth H. The used pitch of the trench array was set to 1 µm; however due to the linear process of the surface annealing a linear scaling of the geometric parameters is allowed (i.e. if the pitch and trench size are 2 times smaller the resulting layer thickness is also 2 times smaller). The time scale t of the surface diffusion changes with radius by t ~R^4 (e.g. annealing of two times smaller trenches speeds up the annealing process dramatically by a factor of 16).

The different structure types occurring after the anneal process are presented as a "phase diagram" *radius* vs. *depth*. Each point represents one simulation and seven individual structural types (Fig. 4-b)/ phase diagram zones (Fig. 4-a) can be isolated:
- (1) SON layer: the closure of the upper trench openings (splicing) occurs first, the cavities beneath the upper Si layer join each other later;
- (2) SON layer and spherical objects: this development is characteristic for the rather deep medium-radius trenches;
- (3) SON layer with trench joining first: the joining of the deeper parts of the neighboring trenches occurs before the closure of the upper trench openings and the

upper Si layer develops through the rather complicated stages of the surface diffusion process. This sequence occurs for the medium-depth trenches of close-to-maximum radius (R>40 μm);

- (4) Two SON layers: two upper Si layers (and, respectively, 2 layers of an empty space) develop for rather deep and wide trenches;

- (5) plane Si surface: the trenches disappear, and the final Si wafer surface is plane for the shallow trenches;

- (6) Bubbles: the trenches evolve into spherical enclosures which do not join each other for the thin trenches;

- (7) Two bubbles: increasing of the depth of the thin trenches leads to the evolution of two spherical enclosures which do not join the neighboring spheres.

An important result is the dependence of the thickness of the SON layer on the trench shape. The summary of seven optimization campaigns for the cylindrical trenches is given in Table 1. **The 1st to 3rd optimization campaigns** were performed for the standard genetic situation using the standard optimization algorithm. These campaigns present the high stability of the genetic algorithm. The minimum of the upper layer thickness (412 nm) was found at the maximum allowed trench radius and at a trench depth of 1.65 μm. The **4th optimization campaign** considered only the straightforward upper Si layer development when the splicing of the trench entrance occurred first followed by the forming of the joint empty space beneath it. One may consider that the straight genetic algorithm is efficient for the search of the global minimum value but in the same time may be less efficient in minimizing the target function in the vicinity of the local minimum. Therefore **the 5th optimization campaign** was initiated to test the hybrid optimization method. The scheme of this method was as follows: (1) The standard genetic optimization algorithm was used; (2) The SIMPLEX algorithm Nelder&Mead (1965) was applied for the search of the local minimum in the vicinity of the optimum value of genetic optimization process. The SIMPLEX algorithm was employed as a parallel computation to p.1.; (3) the optimum parameter set of p.2 was returned to the population of the genetic algorithm p.1. One may conclude that the convergence of the hybrid method is faster in comparison with pure genetic algorithm. At the same time there was rather insignificant improvement of the achieved minimum of the upper layer thickness (1.3 %, see Table 1). The configuration leading to a minimum layer thickness (0.63 μm) for zone 1 (splicing first) is indicated (empty circle) in the phase diagram (Fig. 4-a). This minimum is located close to the borders of zones 1, 3 and 6.

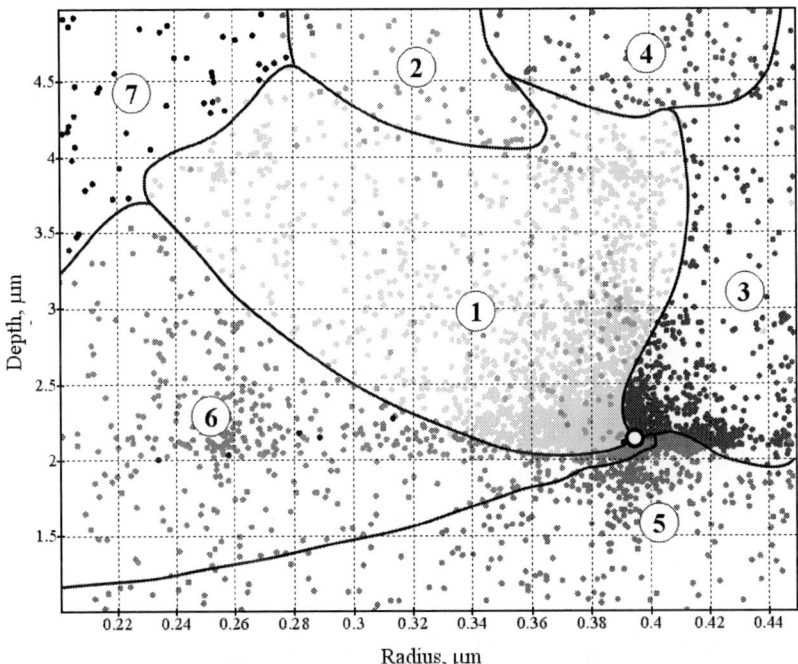

Figure 4-a. Phase diagram of surface near structures after trench anneal: SON layer (1), SON layer and spheres (2), SON layer with trench joining first (3), two SON layers (4), plane Si surface (5), bubbles (6), two bubbles (7).

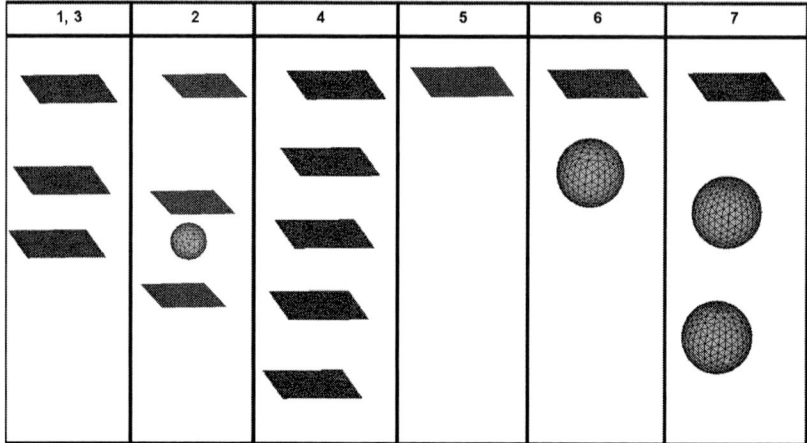

Figure 4-b. Shapes of surface near structures after trench anneal. Numbers correspond to the depiction of the zones given in Fig. 4-a.

TABLE I. Summary of the optimization campaigns for cylindrical trenches

No	Geometry range	Splicing first	Simplex	Number of runs	Min Δz	R_{opt}	H_{opt}	No. of run for Opt
1	R:0.25-0.45 H: 1- 4	-	-	713	0.466	0.450	1.656	592nd
2	R:0.25-0.45 H: 1- 4	-	-	2059	0.415	0.449	1.675	363rd
3	R:0.1-0.45 H: 1- 7	-	-	4563	0.412	0.449	1.648	3701st
4	R:0.2-0.45 H: 1- 5	X	-	1960	0.639	0.395	2.008	1650th
5	R:0.2-0.45 H: 1- 5	X	X	2192	0.631	0.396	2.023	1827th

Study of trench perturbations influence on top layer uniformity
The role of defects / perturbations of the initial trench shape and trench distribution on the upper Si layer in regard of final film perfection were analyzed too. The elementary calculation domain consisted of four quarter – trenches. One of these four trenches was perturbed changing either its radius r or depth z. The considered perturbations were modelled as increase of the trench radius by factors of 0.01, 1, 2, 5, 10 and 20% respectively. Multiple calculations of the anneal process were performed for each array perturbation level. The case with a 20% perturbation is shown in Fig. 5. Each calculation which resulted in a successful continuous layer is represented as one coloured dot in the diagram. The area of the non-perturbed „splicing first" process is surrounded by a green line whereas the perturbed "splicing first" area is indicated by a black line. Finally one can conclude that by choosing trench parameters from inside the corresponding areas the existence of continuous layers can be guaranteed for very large perturbations of the trench radius.

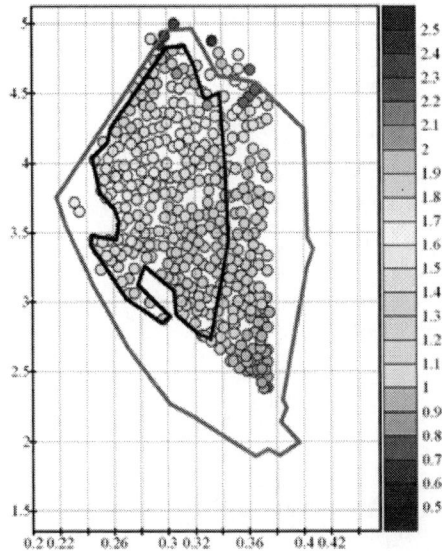

Figure 5. Layer existence area and layer thickness (in μm) as a function of the trench radius and depth (in μm). Radius perturbation rate of one trench was set to +20%. The area for the perturbed process is indicated by circles; the area of the non-perturbed „splicing first" process is indicated by the green line; the area for the perturbed "splicing first" process is indicated by the black line.

CONCLUSIONS

An efficient numerical model of the surface diffusion allows calculating complicated surface restructuring of cylindrical trench arrays during the SON anneal. Very good agreement between experimental and simulated shapes was achieved. The allowed parameter space for layer formation is found and the necessary parameter set to obtain a minimum layer thickness could be predicted. The studies of layer uniformity and perturbation shows that the formation of a continuous layer is possible even for large radius and depth perturbations.

ACKNOWLEDGMENTS

We like to thank Mr. Bearda from the IMEC Institute for performing part of the demonstrated experimental work. The authors alone are responsible for the content.

REFERENCES

[1] I. Mizushima, T. Sato, S. Taniguchi, and Y. Tsunashima Applied Phys. Lett., Vol. 77, No. 20, (2000), p. 3290

[2] Tsunashima Yoshitaka, Tsutomu Sato, Ichiro Mizushima, Proc. Electrochem. Soc. Vol. 2000-17, pp. 532-545

[3] T. Sato, I. Mizushima, J. Iba, M. Kito, Y. Takegawa, A. Sudo, Y. Tsunashima, 1998 Symposium on VLSI Tech. Dig. Tech. Paers, 1998, p. 206

[4] T. Sato, K. Mitsutake, I. Mizushima, and Y. Tsunashima, Jpn. J. Appl. Phys. Vol. 39, 5033 (2000)

[5] W. W. Mullins, J. Appl. Phys., 28, 333 (1957)

[6] M. E. Keefe, J. Phys, Chem. Solids, 55 (10), 965 (1994).

[7] A. Z. Zincenko. M. A. Rother. R.H. Davis, in J. Phys. Fluids 9 (6) (1997).

[8] U. F. Mayer, J. Appl. Math., 11 (2), pp. 61-80 (2000)

[9] J. Ruppert, J. of Algorithms 18(3), 548-585 (1995)

[10] V. Cristini, J. Blawdziewicz, M. Loewenberg, J. of Computational Physics 168, 445-463 (2001)

[11] H. J. Bremermann, J.Roghson, S. Salaff, Natural automata and useful simulations. London, MacMillan. pp. 3 – 42. (1966)

[12] I. Rechenberg, Evolutionsstrategie: Optimierung technischer Systeme nach Prinzipien der Biologischen Information. Freuburg: Fromann (1973)

[13] H. P. Schwefel, Numerical optimization of computer models. Chichester: Wiley. (1982)

[14] D.E.Goldberg, Genetic algorithms in search, optimization, and machine learning. Reading, MA: Addison-Wesley. (1989)

ECS Transactions, 2 (2) 375-389 (2006)
10.1149/1.2195674, copyright The Electrochemical Society

**Thermal Agglomeration of Ultra-Thin SOI and SSOI Films:
A Quantitative Stability Study and Physical Model to
Guide Ultra-Thin SOI Process Design**

D. T. Danielson, J. Michel, and L.C. Kimerling

Department of Materials Science and Engineering,
Massachusetts Institute of Technology, Cambridge, MA 02139

We present the results of a controlled experimental study of the
thermal agglomeration of (100)-oriented ultra-thin SOI films along
with a surface-energy-driven physical model for SOI
agglomeration to guide ultra-thin SOI process design. Through a
UHV annealing study, the SOI agglomeration edge initiation time
and edge propagation velocity were determined as a function of
SOI thickness and annealing temperature for bonded and SiMOX
SOI films. SiMOX SOI and strained SOI films were studied and
shown to have nearly identical agglomeration behavior to bonded
SOI films, clearly indicating that SOI agglomeration is not a stress-
driven phenomenon. An ultra-thin SOI defect study was also
performed in parallel with this agglomeration study to identify the
defect responsible for bulk SOI agglomeration initiation. HF
defects, dislocations, and stacking faults were ruled out and initial
evidence indicates that defects at the top Si/buried SiO_2 interface
may be responsible.

Introduction

As the semiconductor industry rapidly approaches traditional device scaling limits,
the introduction of ultra-thin body silicon-on-insulator (UBSOI) MOSFET technology is
becoming necessary to continue the long-running exponential improvements in device
performance embodied in Moore's Law. As device gate length continues to shrink, the
off-current of traditional device technologies continues to increase, resulting in
unacceptable levels of static power consumption. Fully depleted UBSOI devices solve
this problem through a reduction of the body capacitance of the transistor, which results
in nearly ideal sub-threshold behavior and correspondingly low off-currents and static
power dissipation. For this reason, according to the International Technology Roadmap
for Semiconductors, fully depleted UBSOI technology having top Si thickness < 100A
will need to be introduced by 2008 in order for Moore's Law growth to continue apace
(1).

However, a critical materials phenomenon has been encountered in recent years that
may prevent this technology from realizing its promise: the thermal agglomeration of
ultra-thin silicon-on-insulator (SOI) films. In this phenomenon, upon the application of
high temperature vacuum annealing, a sufficiently thin top Si SOI layer agglomerates
upon, or dewets, the underlying SiO_2 layer and breaks up into a discontinuous array of
islands, as illustrated in Figure 1. It has been found that the presence of capping films
(even a native oxide) prevents the agglomeration process from occurring (2). Thus,

375

technologically speaking, the critical processes susceptible to agglomeration are those that occur at high temperature and result in or require a bare SOI surface. For this reason, high temperature epitaxial film deposition steps are highly vulnerable to agglomeration. Thus far, the raised source/drain Si homoepitaxy process, in which the SOI film must be bare *and* go through a relatively high temperature surface cleaning and Si epitaxy process, has been found to be the step most susceptible to agglomeration (3).

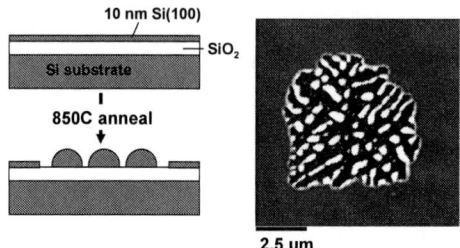

Figure 1. Illustration and AFM image of ultra-thin SOI agglomeration.

The existing literature on SOI agglomeration has resulted in a number of key observations related this phenomenon (2, 4-9). First, thinner SOI films have been observed to be more unstable to agglomeration. Second, as stated before, agglomeration has been observed to occur only in uncovered SOI films, with the ex-situ or in-situ removal of even the native oxide layer required in order for agglomeration to proceed (2). Third, agglomeration in blanket films has been observed to initiate and grow from apparently random film locations. Fourth, patterned SOI structures have been found to have lower stability than blanket films and undergo agglomeration initiation first at film edges (4, 5). Fourth, agglomeration propagation has been observed to occur in a highly characteristic void finger growth and island formation geometry with the resultant island dimensions observed to scale in proportion to the top Si thickness (6).

However, the existing work on SOI agglomeration suffers from a number of key limitations. Studies to date have been limited to isochronal annealing studies of only bonded SOI films. Furthermore, in these studies, the Si native oxide layer was not typically removed before annealing and thus the results of these studies represent a convolution of both the thermal desorption of the native oxide and the SOI agglomeration process. Therefore, these studies do not provide definitive information about the SOI agglomeration process itself. Furthermore, the consensus theory in the SOI literature to date has been that SOI agglomeration is a stress-driven phenomenon (2, 4-9). As we demonstrate in this paper, this theory does not appear to be correct.

The key goals of this work were 1) to quantitatively determine the thermal stability limits of bare ultra-thin SOI films against agglomeration, 2) to identify the SOI film defect responsible for agglomeration initiation in the film bulk, and 3) to develop a descriptive physical model for SOI agglomeration in order to inform SOI process design. UHV annealing studies were performed to determine the SOI edge agglomeration initiation time and edge agglomeration growth velocity as a function of SOI thickness and annealing temperature for bonded and SiMOX SOI films and the agglomeration behavior of strained SOI (SSOI) films was also explored. An SOI defect study was performed in parallel with this SOI agglomeration study in order to identify the defect responsible for ultra-thin SOI agglomeration nucleation in blanket film regions. In addition, a novel surface-energy-driven SOI agglomeration model is presented based upon a defect-driven film void nucleation and capillary-instability-driven void growth process. We conclude with a brief discussion of the implications of this work for ultra-thin SOI process design.

Experimental Methods

Samples and Sample Preparation

Three distinct types of state-of-the-art ultra-thin (100)-oriented SOI material were studied in this work: bonded (SmartCut™) SOI, SiMOX SOI, and strained SOI (SSOI). Bonded and SiMOX samples with top Si thicknesses of 50, 100, 150, 200, and 350A were prepared from thicker SOI layers via sacrificial thermal oxidation and HF etching. The single bonded SSOI wafer used for this study had a top Si thickness of 75A originating from the transfer of a strained Si layer epitaxially grown on a relaxed $Si_{0.80}Ge_{0.20}$ layer (σ_{SSOI}=1.4GPa). Wafers were subsequently cleaved into 40mm square coupons. Patterned top Si structures were fabricated on these samples by standard photolithographic techniques followed by selective Si wet etching using KOH (5g/L).

UHV Annealing Study

UHV SOI annealing experiments were carried out in a hot wall horizontal ultra high vacuum (UHV) furnace with a base pressure of ~1x10^{-7} torr. SOI anneal samples consisted of cleaved 10mm square coupons. The standard pre-anneal SOI cleaning procedure in this work consisted of the following steps:

(1) Compressed N_2 removal of any particles from sample cleaving
(2) 5 minute Piranha clean (1 H_2O_2 : 3 H_2SO_4) to remove organic contamination
(3) 1 minute rinse in DI H_2O
(4) 1 minute dilute HF dip (20 DI H_2O: 1 HF(49%)) to remove the SiO_2 layer
(5) Compressed N_2 removal of any residual HF drops upon removal

In order to study the SOI agglomeration process, and not a convolution of oxide desorption *and* agglomeration, bare SOI layers were prepared via a final dilute HF dip before sample annealing. Samples were shown to be HF passivated, and thus oxide-free, by their hydrophobicity. After cleaning, samples were introduced into the UHV furnace load lock within 2 minutes of the final HF dip and were introduced from the load lock into the UHV furnace within 7 minutes of the final HF dip, ensuring an oxide-free surface upon annealing. Upon sample introduction, the furnace temperature was found to decrease by approximately 10C. The finite sample temperature ramp rate was accounted for by considering the anneal start time to be when the furnace temperature rose to within 5C of the desired annealing temperature. The time between sample furnace introduction and this anneal start time was 3-5 minutes. The highest temperatures reached during anneals were < 5C above the desired annealing temperature. Anneal end times were considered to be the time at which the sample was removed from the UHV furnace into the load lock. Annealing temperatures of 700C, 750C, 800C, and 850C and annealing times of 2-60 minutes were explored.

Ultra-Thin SOI Defect Study

An ultra-thin SOI film defect etch study was also performed in this work. SOI film pinholes ("HF defects") were revealed using a concentrated HF etch (49% HF). Film dislocations and stacking faults were revealed using a standard Secco etch followed by a concentrated HF etch (49%). The density of these defects was measured using optical microscopy after defect etching.

Experimental Results

UHV SOI Annealing Study

In the SOI annealing study performed in this work, SOI agglomeration was found to initiate and grow first at patterned SOI film edges and cleaved sample edges ("Edge Initiation"). Figure 2a shows a cleaved sample edge at which agglomeration has just initiated. Upon further annealing, agglomerated regions were found to nucleate and grow from discrete and apparently random locations in the bulk of the film, as illustrated in Figure 2b ("Bulk Initiation"). These results are consistent with what has been observed in the literature (2, 4-9).

Edge Agglomeration Initiation & Growth. As patterned structures are of most significant interest for practical device applications and have been found to be the most vulnerable structures to agglomeration, this study focused on quantifying agglomeration behavior at patterned and cleaved SOI film edges. Through annealing experiments over a range of times for each SOI type, SOI thickness, and annealing temperature, and subsequent sample characterization via optical and scanning electron microscopy, the edge agglomeration initiation time was determined. These results are summarized in the Ultra-Thin SOI Agglomeration Stability Map in Figure 3.

The stability map in Figure 3 provides clear guidance for ultra-thin SOI process design. All films studied (50-350A) were observed to undergo agglomeration initiation in < 2 min at 850C, while all films studied were observed to be stable for > 60 min at 700C, with intermediate behavior observed at intermediate temperatures. Significantly, bonded and SiMOX samples were observed to have indistinguishable edge agglomeration initiation behavior, with the edge initiation time observed to be only a function of SOI thickness and annealing temperature.

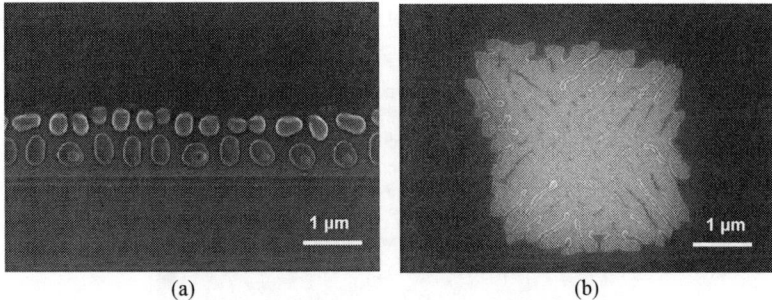

(a)   (b)

Figure 2. SEM images of annealed SOI films illustrating the two observed types of SOI agglomeration initiation. (a) Edge Initiation: 350A bonded SOI film, 850C – 30 min. (b) Bulk Initiation: 50A bonded SOI film, 750C – 15 min.

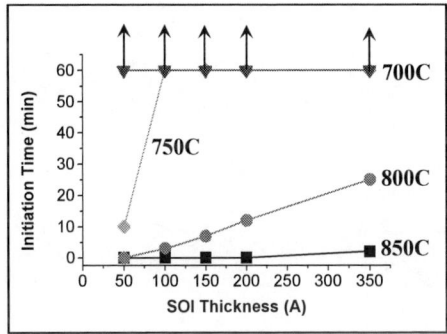

Figure 3. Ultra-Thin SOI Agglomeration Stability Map indicating edge agglomeration initiation time versus SOI thickness and annealing temperature. (Vertical arrows indicate initiation time > 60 minutes) – Bonded and SiMOX SOI behavior was virtually identical and thus this map applies to both types of material.

In addition to the edge agglomeration initiation time, the edge agglomeration distance was determined as a function of SOI thickness and annealing time and temperature for each SOI type. The edge agglomeration distance at a patterned SOI edge in an annealed sample is illustrated in Figure 4. For each sample type, SOI thickness, and annealing temperature studied, at least three different annealing times yielding measurable edge agglomeration distances were required to determine whether the edge propagation velocity was constant in time, and if so, to extract an edge agglomeration velocity. In all cases where three or more such measurements were possible, the edge agglomeration distance was found to be linear with the time elapsed since the edge initiation time, indicating a constant propagation velocity. For thin samples annealed at high temperature, the edge propagation velocity was sufficiently high that edge and bulk agglomeration areas quickly overlapped during the studies performed in this work, making propagation velocity determination impossible for thinner samples at high annealing temperatures.

Figure 5 shows the edge propagation velocities that were measured in this study. These initial data tentatively indicate that the edge propagation velocity scales with SOI thickness (t_{SOI}) with an inverse power law, $\sim t_{SOI}^{-1.5}$. Bonded and SiMOX samples were again found to have very similar agglomeration behavior, exhibiting indistinguishable agglomeration propagation velocities for the same SOI thicknesses and annealing temperatures. Since propagation velocity data was only obtained for two different temperatures, we can make no definitive statements about the temperature dependence of the propagation velocity here. This will be a subject of future study.

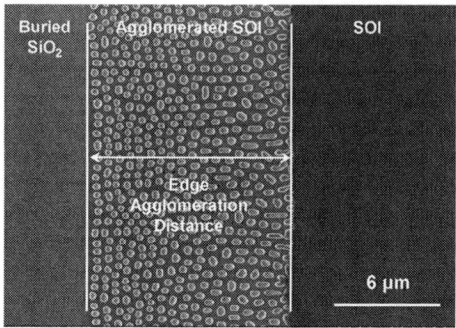

Figure 4. SEM image showing the edge agglomeration distance at a patterned film edge in a SiMOX SOI sample annealed at 850C for 30 minutes.

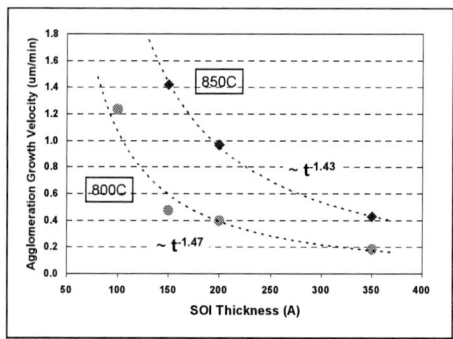

Figure 5. Plot of measured agglomeration propagation velocity versus SOI thickness and annealing temperature. Indistinguishable behavior was observed for bonded and SiMOX SOI samples.

Quite interestingly, SSOI was observed to have very similar agglomeration behavior to that observed in the nominally unstrained SOI films in this study. Although no direct comparison was possible due to the fact that none of the unstrained SOI samples had exactly the 75A top Si thickness of the SSOI sample, Figure 6 demonstrates that for fixed annealing conditions (temperature, time), the agglomeration results observed for

380

the 75A SSOI sample were clearly intermediate between those observed for 50A and 100A nominally unstrained bonded SOI samples. From the literature, nominally unstrained bonded SOI layers with thicknesses > 50A have been shown to have film stresses < 100MPa (10). The fact that SOI samples having widely different stress states but comparable thicknesses exhibit very similar agglomeration behavior strongly indicates that stress is not the predominant driving force behind SOI agglomeration.

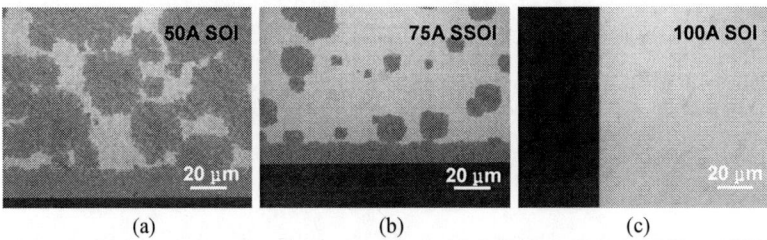

(a)   (b)   (c)

Figure 6. Optical micrographs showing agglomeration behavior of bonded SOI and SSOI films annealed at 750C for 30 min. (a) 50A SOI, (b) 75A SSOI, (c) 100A SOI.

Bulk Agglomeration Initiation and Defect Study. In order to identify the defect responsible for agglomeration initiation in the SOI film bulk (non-edge regions), the density and nature of bulk agglomeration sites was studied and compared with the results of an SOI film defect study.

In all annealed samples where bulk agglomeration was observed, bulk agglomeration nucleation was found to occur with a density in the range of $5x10^4 - 5x10^5$ cm^{-2}, illustrated in Figure 7a. No visible defects were observed at the center of these nucleated regions via optical and scanning electron microscopy. However, in AFM studies of the films observed to undergo bulk agglomeration, nominally flat regions were observed to have fine scale local surface asperities with a density of $\sim2x10^9$ cm^{-2}. As shown in Fig. 7b, local asperities with approximately the same size and density were observed both at the SiO$_2$ interface in uncovered agglomerated regions as well as in neighboring unagglomerated film regions.

The film defect content of the SOI samples studied here was evaluated via HF defect and Secco etch studies.

HF defect densities were determined to be very low in all SOI films in this study, with bonded and SiMOX samples exhibiting comparable densities in the range of 0.16-1.0 cm^{-2}. These very low densities relative to the observed bulk agglomeration initiation density ($5x10^4 - 5x10^5$ cm^{-2}) clearly indicate that HF defects are not responsible for the observed SOI bulk agglomeration nucleation.

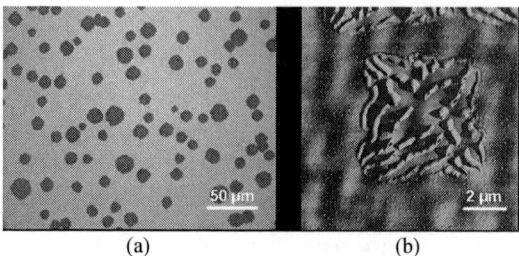

(a)   (b)

Figure 7. Illustration of bulk SOI agglomeration. (a) Optical micrograph showing the density of bulk agglomeration initiation in a 100A bonded SOI film annealed at 800C, 30 minutes: dark regions are agglomerated, bright regions are unagglomerated. (b) AFM phase contrast image showing high density asperities ($\sim 2 \times 10^9$ cm^{-2}) in unagglomerated film and at SiO$_2$ surface of agglomerated region.

Secco etch defect studies were performed on 150A, 200A, and 350A bonded and SiMOX SOI samples. Secco defect-to-film etch selectivity was insufficient to observe defects in 150A films, but repeatable results were obtained for 200A and 350A films. Secco defect density counts were comparable for bonded and SiMOX films and were found to be in the range of 700-4000 cm^{-2}. The observed densities of Secco defects were at least an order of magnitude too low to account for the observed density of bulk agglomeration initiation, indicating that the dislocations/stacking faults revealed by the Secco etch were not the defects responsible for SOI bulk agglomeration initiation in this study.

These results clearly indicate that HF defects, dislocations, and stacking faults are not the defects responsible for bulk SOI agglomeration initiation. The observed fine scale film roughening seen in annealed SOI structures indicates that a high density defect at the top Si/buried SiO$_2$ interface may be responsible for bulk agglomeration initiation. The much lower bulk agglomeration initiation density relative to the density of the observed local asperities may indicate that bulk initiation only occurs at particularly severe interfacial defects. In-situ hot stage SEM SOI agglomeration studies not presented here appear to reinforce this conclusion. However, this conclusion is only tentative at this time and further study is required to definitively identify the defect responsible for SOI bulk agglomeration initiation.

Surface Energy Driven Physical Model for SOI Agglomeration

In this section, we present a physical model for SOI agglomeration that provides a theoretical framework to guide ultra-thin SOI process design. Contrary to the literature consensus that SOI agglomeration is a stress-driven phenomenon, we present and argue for the validity of a surface-energy-driven physical model for SOI agglomeration. We begin by demonstrating that there exists a surface-energy-related driving force for SOI agglomeration and that the magnitude of this driving force far exceeds that associated with stress relaxation. We then present a physical model for SOI agglomeration based upon defect-driven void nucleation and capillary-instability-driven void growth.

Thermodynamics of SOI Agglomeration: Importance of Surface Energy

In a stress-free SOI film, the relative interfacial energies of the system dictate the equilibrium shape of the film. Assuming isotropic surface energies and a rigid SiO_2 layer, the equilibrium morphology of a thin unstrained SOI film will be that of a spherical cap that makes an equilibrium contact angle, Θ_c, with the substrate determined by a surface tension balance at the cap edge, as shown in Figure 8. This balance establishes the equilibrium contact angle for SOI to be

$$\Theta_c = \cos^{-1}\left[\ (\gamma_{SiO2\text{-}vap} - \gamma_{Si\text{-}SiO2})\ /\ \gamma_{Si\text{-}vap}\ \right], \qquad [1]$$

where $\gamma_{SiO2\text{-}vap}$ is the SiO_2-vapor interfacial energy density, $\gamma_{Si\text{-}SiO2}$ is the $Si\text{-}SiO_2$ interfacial energy density, and $\gamma_{Si\text{-}vap}$ is the Si-vapor interfacial energy density. For a (100) oriented SOI film under vacuum, $\gamma_{SiO2\text{-}Si(100)} \sim 0.3$ J/m^2 (11), $\gamma_{SiO2\text{-}vacuum} \sim 0.75$ J/m^2 (12), and $\gamma_{Si(100)\text{-}vacuum} \sim 1.5$ J/m^2 (13). Thus, ignoring surface energy anisotropy, the contact angle for the SOI system in vacuum is calculated from these surface energy values to be $\Theta_c \sim 73°$. This equilibrium contact angle value clearly indicates that the top Si film in SOI has a driving force to agglomerate due to surface energy effects alone.

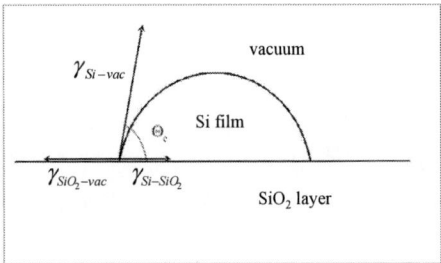

Figure 8. Equilibrium spherical cap morphology of an unstrained SOI thin film assuming isotropic surface energies and a rigid SiO_2 film.

However, simply because a surface-energy-related driving force for SOI agglomeration exists does not necessarily imply that surface energy is the dominant driving force behind SOI agglomeration. Below, we demonstrate that not only does a surface-energy-related driving force exist for agglomeration, but that the magnitude of this driving force far exceeds that associated with stress relaxation, clearly indicating that surface energy is the dominant driving force behind SOI agglomeration.

To compare the magnitudes of the surface energy and strain energy related driving forces for SOI agglomeration, we calculate the energy changes associated with the formation of each Si island via SOI agglomeration using the model system shown in Figure 9. The surface energy and strain energy changes associated with the formation of a

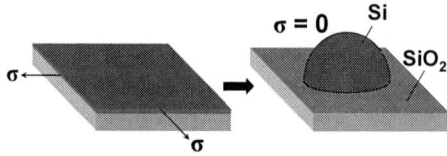

Figure 9. Model system considered in comparing surface energy and strain energy related driving forces for SOI agglomeration.

single island from an initially square biaxially stressed film region are calculated under the following simplifying assumptions: Si islands are assumed to be hemispherical ($\Theta_c=90°$), agglomeration is assumed to result in complete stress relaxation (a conservative assumption), and from the SOI agglomeration literature the Si island radius is assumed to be equal to 20 times the top Si layer thickness, t_{SOI} (6).

Under these assumptions, the surface and strain energy changes, in Joules, associated with the formation of a single Si island via SOI agglomeration are given by:

$$\Delta E_{surface} = -(\pi/3) * t_{SOI}^2 * [\ 1.36x10^{4*}\ \gamma_{Si\text{-}vac} - 1.48x10^{4*}(\ \gamma_{SiO2\text{-}vac} - \gamma_{Si\text{-}SiO2})\ ] \qquad [2]$$

$$\Delta E_{strain} = -(\pi/3) * t_{SOI}^3 * [\ (1.6x10^{4*}(1\text{-}\upsilon_{Si})*\sigma_{Si}^2\ ) / E_{Si}\ ], \qquad [3]$$

where υ_{Si} and E_{Si} are the Poisson's ratio and elastic modulus of Si, respectively, and all other parameters are as previously defined. In this calculation, the Si thickness is expressed in m, the surface energies in J/m², and the stress and modulus in Pa.

To provide an illustration of the relative magnitudes of these two driving forces, we calculate their ratio for a representative SOI sample. We use the previously assumed values for SOI surface energies and values of $\upsilon_{Si}=0.28$ and $E_{Si}=166$GPa from the literature (14). Using these values, for a 100A sample estimated to have a biaxial top Si film stress of 65MPa from the literature, we calculate the ratio of the surface energy reduction to the strain energy reduction to be ~5000. This extreme dominance of surface energy reduction over strain energy relief clearly indicates that surface energy is the dominant driving force behind SOI agglomeration.

Kinetic Mechanism of SOI Agglomeration: Void Nucleation & Growth

Void Nucleation: Film Edges & Defects. Although islanded states are preferred energetically to the flat film state in SOI from a surface energy perspective, a continuous flat SOI film is actually metastable to these lower energy islanded states. Any non-through-thickness perturbation of the film surface results in an increase in the total surface energy of the system and thus will be spontaneously restored to a flat film. It can be shown that even a through thickness film void must exceed a critical size in order for its growth to be energetically favorable. Srolovitz and Goldiner determined that this critical void radius is given by

$$r_{crit} = t_{film} / \sin \Theta_c \qquad [4]$$

where t_{film} is the film thickness (15). For the equilibrium SOI contact angle calculated in this work, their result predicts a critical void radius of $\sim 1.0 \times t_{SOI}$ for SOI.

Under this model, any film edge represents the edge of a pre-existing supercritical film void and is thus unstable to growth. This explains why SOI agglomeration is experimentally observed to initiate first at patterned film and sample edges before it initiates in blanket film regions. In blanket film regions, critical film voids must first be nucleated in order for agglomeration to proceed spontaneously. Even though the critical void size in ultra-thin SOI films does not appear to be very large ($r_{crit} \sim 50\text{-}350A$ for $t_{SOI} \sim 50\text{-}350A$), the activation energies required to form even a very small void in a very thin SOI film are prohibitively large. A first order analysis for the SOI system reveals that even in a top Si film of only 10A thickness, an activation barrier of $\sim 60eV$ exists to homogenously nucleate a critical void and that this value should scale with t_{SOI}^2. Such high activation barriers are indeed prohibitively large. Accordingly, it can be concluded that void nucleation in blanket SOI film regions must occur via a heterogeneous, defect-mediated process. As stated previously, our work has ruled out HF defects, dislocations, and stacking faults as the defects that are responsible for heterogeneous void nucleation in blanket SOI film regions and appears to indicate that asperities at the Si/buried SiO_2 interface may indeed be the critical defect in this phenomenon.

<u>Void Growth Driven by Capillary Instabilities.</u> We propose that once a critical film void is formed in an SOI film, agglomeration initiation and propagation proceeds through a capillary-instability-driven five-step kinetic void growth process first described by Jiran and Thompson (16).

The master equation for surface energy driven surface diffusion is given by

$$J_s = - (D_s \nu \Omega^2 \gamma / kT) * \nabla \kappa(s) \qquad [5]$$

where D_s is the surface diffusivity, ν the atomic surface density, Ω the atomic volume, γ the surface energy density, kT has its usual meaning, and $\kappa(s)$ is the mean surface curvature along the surface coordinate s. (The curvature convention is that convex surfaces have positive curvature, concave surfaces negative curvature). The mean surface curvature provides the driving force for surface diffusion, with mass flowing away from regions of higher curvature to regions of lower curvature to minimize surface energy. The thermally activated surface diffusivity, D_s, has a strong positive temperature dependence and provides the dominant temperature dependence in surface energy driven surface diffusion.

The SOI agglomeration initiation and growth kinetic mechanism described here is driven by two distinct capillary instabilities: the edge instability and the Rayleigh instability. In the capillary edge instability, as shown by Jiran and Thompson, any dewetting thin film edge is unstable due to its high local curvature and curvature gradient (16). This high local edge curvature drives mass away from film edges toward nearby flat film regions, causing film edges to be fundamentally unstable. Due to the increased edge curvature present in thinner films, the severity of the edge instability increases with decreasing film thickness. In the Rayleigh instability, a high aspect ratio object breaks up into a row of spherical objects having a highly characteristic size and spacing to minimize its total surface energy (17). Our kinetic agglomeration model can be understood in terms of these two instabilities.

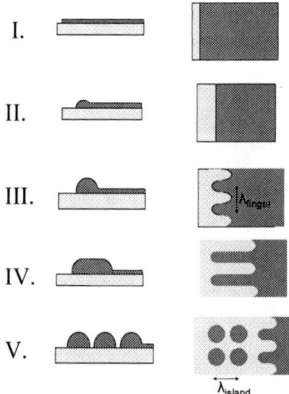

Figure 10. Schematic illustration of the 5 step SOI agglomeration initiation and growth kinetic model presented in this work.

Step I in our model is critical void formation and is illustrated in Figure 10(I). As previously described, a critical void must first pre-exist or be nucleated in an SOI film in order for agglomeration to proceed spontaneously. This step determines the location of agglomeration initiation.

Under our model, Steps II and III determine the edge agglomeration initiation behavior once a critical void exists. In Step II, void edge thickening, the edge instability drives mass to diffuse from the high curvature film edge to the neighboring flat film region, resulting in a continuous and uniform edge thickening, as illustrated in Figure 10(II). However, as the film edge thickens, the severity of the edge instability continuously decreases and thus edge thickening slows and ultimately stops. Visible edge agglomeration occurs via the break-down of this thickened edge, Step III in our model, as first proposed by Jiran and Thompson (16). Jiran and Thompson did not identify the specific break-down mechanism, but we propose that this break-down occurs via a Rayleigh instability in the thickened edge. Recent numerical modeling by Kan and Wong strongly supports this mechanism (18). Their work predicts that a thickening dewetting film edge has a Rayleigh instability whose characteristic wavelength, λ_{finger} in Figure 10(III), is proportional to the initial film thickness. Under our model, the SOI edge agglomeration initiation time is determined by the kinetics of Steps II and III. From the work of Kan and Wong, the characteristic time of SOI edge agglomeration initiation is predicted to have the following general dependencies on SOI thickness and Si surface diffusivity (18),

$$t_{\text{edge agglomeration}} \sim t_{SOI}^4 / D_s, \qquad [6]$$

exhibiting the characteristic dependencies of the Rayleigh instability. These dependencies are consistent with the experimental observation in SOI agglomeration that thinner films and higher annealing temperatures result in lower observed edge agglomeration initiation times.

Under this model, Step IV, void finger formation and growth, is responsible for agglomeration propagation. In Step IV, the locally thinner edge regions resulting from the edge break-down of Step III serve as a template for the formation of void fingers due to their high local vulnerability to the edge instability. Their high local edge curvature drives mass to flow rapidly away from these regions, resulting in the penetration and growth of equally spaced void fingers into the adjacent flat film. The tips of these fingers are able to maintain a high edge curvature due to their ability to move mass away in a number of directions. According to Jiran and Thompson, this allows the void fingers to propagate at a constant velocity dictated by the edge instability. It is this velocity that determines the agglomeration propagation velocity. Through a simplified model, Jiran and Thompson derived an expression for this constant void finger growth velocity with the following thickness and surface diffusivity dependencies (16)

$$v_{\text{void finger}} \sim D_s / t_{SOI}^3 \qquad [7]$$

These dependencies are consistent with the experimental observation that thinner films and higher annealing temperatures result in higher edge agglomeration propagation velocities. Although this simple treatment correctly predicts the general trend that thinner SOI films agglomerate faster than thinner film, the t_{SOI}^{-3} dependence of the void finger growth velocity predicted here differs from the $t_{SOI}^{-1.5}$ agglomeration propagation velocity dependency tentatively determined from our experimental work. This discrepancy will be the topic of further study.

In Step V, Si island formation, the uniformly spaced Si fingers formed between the growing void fingers of Step IV break up into discrete islands via a second Rayleigh instability, as illustrated in Figure 10(V). From the results of McCallum et al. on Rayleigh instabilities in lines of material on a substrate, it can be shown that the island spacing in Step V, λ_{island}, should be proportional to the initial Si thickness (19).

The fact that the kinetic model presented here is 1) consistent with a surface-energy driving force and 2) able to account for the key experimental observations in SOI agglomeration indicates to us that this model represents an accurate theoretical description of SOI agglomeration. First, the model is able to account for the observed spatial nature of agglomeration initiation. The local and random initiation in blanket film regions is explained to be the natural result of heterogeneous void nucleation at randomly distributed film defects. The edge agglomeration initiation that is observed at patterned film and sample edges is explained by the fact that these edges represent the edges of pre-existing unstable critical voids. Secondly, the model is able to account for the increased agglomeration instability observed for thinner SOI films and higher annealing temperatures. Third, the model presented here is able to clearly account for the highly characteristic void finger/island formation agglomeration morphology observed in SOI agglomeration. Furthermore, the observed scaling of the characteristic dimensions of agglomerated structures with the top Si thickness is easily explained by our model to be a result of the natural scaling intrinsic to Rayleigh instabilities.

Implications for Ultra-Thin SOI Process Design

The experimental results and physical model of SOI agglomeration presented here have a number of key implications for ultra-thin SOI process design.

Practically speaking, the Ultra-Thin SOI Agglomeration Stability Map obtained from this work indicates that agglomeration represents a severe process constraint at 850C for films < 350A, while at 700C even 50A films exhibit a high degree of stability. Importantly, we have demonstrated experimentally that SOI agglomeration behavior is independent of SOI type, with agglomeration behavior depending only on the SOI film thickness and annealing temperature. The fact that SSOI films have been shown to have agglomeration behavior similar to nominally unstrained SOI films provides strong experimental evidence that SOI agglomeration is not a stress-driven phenomenon. According to our SOI agglomeration model, this observed independence of agglomeration behavior on SOI type is a natural result of SOI agglomeration being driven by the very high amount of surface energy possessed by ultra-thin SOI films and not by film stress. Furthermore, our results indicate that defects at the top Si/SiO_2 interface are responsible for bulk SOI agglomeration.

These results have a number of implications for ultra-thin process design. For one, stress engineering will not allow for SOI stabilization against agglomeration. According to this work, stabilization strategies will require the suppression of surface diffusion, along with special attention being paid to patterned film edges and the top Si/SiO_2 SOI interface. According to the work presented here, promising SOI stabilization strategies include the use of processing temperatures <700C or surfactants to reduce the Si surface diffusivity, the selective use of capping films, and the development of SOI fabrication processes that eliminate the observed defects at the top Si/SiO_2 SOI interface.

A broader implication of the conclusions of this work is that other common thin film-on-insulator materials with similar interfacial energies to SOI, such as ultra-thin GeOI and III-V-on-insulator films, should share SOI's vulnerability to agglomeration. Accordingly, agglomeration will likely represent a key process problem for thin film-on-insulator materials in general in the future.

Acknowledgments

The authors would like to thank Intel Corporation and specifically Mohamad Shaheen, Peter Tolchinsky, Micheal McKeag, and Joann Qiu for invaluable support in experiment planning and execution, sample characterization, and insightful scientific discussions throughout the course of this work. The authors would also like to acknowledge Professors Carl V. Thompson and W. Craig Carter in the Department of Materials Science and Engineering at MIT for valuable insights and discussions.

References

1. International Technology Roadmap for Semiconductors, Executive Summary, p. 11 (2003).
2. B. Legrand, V. Agache, T. Melin, J.P. Nys, V. Senez, and D. Stievenard, *J. Appl. Phys.* **91**, 106 (2002).
3. C. Jahan, O. Faynot, L. Tosti, and J.M. Hartmann, *J. Cryst. Growth* **280(3-4)**, 530 (2005).
4. Y. Ishikawa, M. Kumezawa, R. Nuryadi, and M. Tabe, *Appl Surf. Sci.* **190**, 11 (2002).
5. Y. Ishikawa, Y. Imai, H. Ikeda, and M. Tabe, *Appl. Phys Lett.* **83**, 3162 (2003).
6. R. Nuryadi, Y. Ishikawa, Y. Ono, and M. Tabe, *J. Vac. Sci. Tech. B* **20(1)**, 167 (2002).
7. Y. Ono, M. Nagase, M. Tabe, and Y. Takahashi, *J. Appl. Phys.* **34**, 1728 (1995).
8. R. Nuryadi, Y. Ishikawa, and M. Tabe, *Appl. Surf. Sci.* **159-160**, 121 (2000).
9. B. Legrand, V. Agache, J. P. Nys, V. Senez, and D. Stievenard, *Appl. Phys. Lett.* **76**, 3271 (2000).
10. A. Tiberj, B. Fraisse, C. Blanc, S. Contreras, and J. Camassel, *J. Phys. Condens. Matter* **14**, 13411 (2002).
11. M Schrems, T. Brabec, M. Budil, and H. Potz, *Proceedings of the International Conference on the Science and Technology of Defects in Semiconductors*, Elsevier, New York, NY, p. 245 (1990).
12. J. Vanhellemont and C. Claeys, *J. Appl. Phys.* **62**, 3960 (1987).
13. D.J. Eaglesham, A.E. White, L.C. Feldman, N. Moriya, and D.C. Jacobson, *Phys. Rev. Lett.* **70**, 1643 (1993).
14. H. Gao and W.D. Nix, *Annu. Rev. Mat. Sci.* **29**, 173 (1999).
15. D.J. Srolovitz and M.G. Goldiner, *JOM* **March**, 31 (1995).
16. E. Jiran and C.V. Thompson, *J. Elec. Mat.* **19**, 1153 (1990).
17. F.A. Nichols and W.W. Mullins, *Trans. Met. Soc. AIME* **233**, 1840 (1965).
18. W. Kan and H. Wong, *J. Appl. Phys.* **97**, 043515 (2005).
19. M. McCallum, P.W. Voorhees, M.J. Miksis, S.H. Davis, and H. Wong, *J. Appl. Phys.* **79**, 7604 (1996).

IMPACT OF DEFECTS IN SILICON SUBSTRATE ON FLASH MEMORY CHARACTERISTICS

Yori Hirano[a], Ken Yamazaki[a], Fumihiko Inoue[a], Kazunori Imaoka[a], Katsuto Tanahashi[b] and Hiroshi Yamada-Kaneta[b]

[a] Spansion Inc.,
2, Takaku-Kogyodanchi, Aizuwakamatsu-shi, Fukushima 965-0060, Japan,

[b] FUJITSU LABORATORIES LIMITED,
Morinosato-Wakamiya 10-1, Atsugi 243-0187, Japan

Floating-gate type Flash memories are fabricated on Czochralski (CZ) silicon wafers. A large amount of residual misalignment in reticle shots at photolithography process step and the programming failure are observed in the low oxygen concentration wafer. The above are explained by the plastic deformation of wafer by the relaxation of internal stress during the processes.

INTRODUCTION

Flash memories are widely used for mobile phone, car navigation systems, digital still cameras and so on. Various types of Flash memories are being developed in the world. We produce two types of Flash memories; one is conventional floating gate type and the other is the NROM type[1]. The former is operated by injection of electrons through the tunnel oxide layer. The latter is a flash memory based on localized charge trapping in a dielectric layer.

Miniaturization of Flash memories proceeds by a factor of scaling to a half the die size every 3 years. One of the key technologies in the miniaturization is the photolithography. High temperature thermal processes used in Flash memory fabrication cause strain on the wafer. The pattern alignment margin in photolithography processes is significantly affected by the wafer strain. Therefore, it is important to understand how the wafer strain is produced and relaxed in the Flash memory process. In addition, sub-surface defects cause degradation in programming and/or erase characteristics of Flash memory cells. Therefore, it is also important to understand the behaviors of micro-defects in the Flash memory fabrication process.

Many types of defects are included in Czochralski (CZ) silicon wafer, such as crystal originated particle (COP)[2], oxygen precipitate (OP)[3] and dislocation[4]. These defects have serious issues for the mass production of large scale integrated devices. There have been investigations concerning the impact of defects on yield and device performance. It is reported that COPs degraded the gate oxide integrity (GOI).[5-8] OPs have many roles in device manufacturing. It is reported that the OPs suppressed the propagation of slip.[9] On the contrary; the large size OPs become the source of dislocation. Punch-out dislocations are found around the OP by transmission electron microscope observation.[3] Moreover it is well known that oxygen has the locking effect on dislocations.[4] Oxygen concentration and/or oxygen precipitates have an influence on the strength of wafer. Thus it is important to control oxygen precipitation during processes.

In this paper, we investigate the behavior of defects in the wafer during the processes of Flash memory fabrication. It is found that the residual misalignment of photolithography and electrical properties of Flash memory are closely related to the defects in silicon wafer. We have succeeded in the improvement of yield and Flash memory performance by the control of oxygen precipitation.

EXPERIMENT

The wafers used in this study are boron-doped (*p*-type), 200mm in diameter Czochralski (CZ) silicon. Oxygen concentration (Oi) is 1.1×10^{18}-1.4×10^{18} atoms/cm^3 determined by a Fourier transform infrared (FT-IR) spectrometer. The conversion factor is 4.81 (ASTM-79).

Floating-gate type Flash memory is fabricated on the above mentioned wafer. The residual misalignment data is the amount of layer-to-layer overlay mismatch between reticles at each photolithography process step

After the fabrication, Flash memory characteristics are evaluated. The program characteristic of Flash memory cells is evaluated by hot electron injection to the floating gate. Chips with programming time exceeding the spec are rejected.

Deformation of the wafer by processing is investigated by the measurement of warpage. Device patterns are stripped by dipping in 49% HF solution. Hereafter, we call this "de-processed wafer". Comparing the de-processed wafer with a bare wafer before process, we investigate whether it plastically deforms or not. Defects in wafer after processes are investigated by x-ray topography, optical microscope observation followed by preferential etching and laser scattering tomography (LST). Observations of defects by the above methods are performed in the de-processed wafer. The image of x-ray topography is observed with (220) reflection. The distribution of slip is examined by the image. Distribution of surface defects is observed by optical micrograph followed by the preferential etching in JIS-B solution of HF(49%): HNO$_3$(70%): CH$_3$COOH: H$_2$O =1: 15: 3: 3. After etching 2 μm in depth, the number of pits is counted. LST observation is performed with an infrared laser of wavelength 1.06 μm. After cleaving the sample, the incident laser is focused to a 5 μm spot on the surface. The radiation scattered by 90° is detected using an IR camera.

RESULTS AND DISCUSSION

Residual misalignment in photolithography processes by slip in wafer

Figure 1 shows the vectors of the direction and amount of residual misalignment in each photolithography process steps. Between each step the wafer is isothermally annealed for several hours above 1000°C to fabricate the devices. The step number in the figure denotes the fabrication progress. In the high ($\geqq 1.2 \times 10^{18}$ atoms/cm^3) Oi wafer, only a small amount of residual misalignment is observed throughout the fabrication process. In the low ($<1.2 \times 10^{18}$ atoms/cm^3) Oi wafer, on the other hand, the residual misalignment increases as the fabrication process proceed. A large amount of residual misalignment is observed in the wafer edge after step 4. The residual misalignment is not large in high Oi wafer. Thus, the residual misalignment depends on the Oi of wafer.

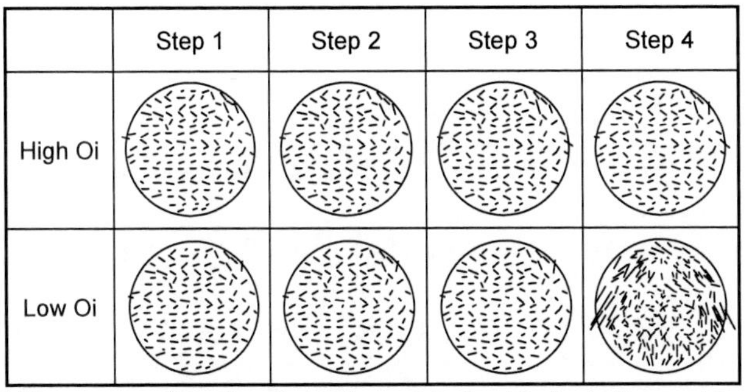

	Step 1	Step 2	Step 3	Step 4
High Oi				
Low Oi				

Figure 1 The vectors of the direction and amount of residual misalignment in a reticle shot at a photolithography process step

Generally, residual misalignment occurs due to the deformation. Therefore we investigate the deformation of wafer and defects in wafer. The warpage of wafers are measured for bare, processed and de-processed wafers. Vertical axes in Fig. 2 show the warpage of wafers. In Fig. 2, letters, a-l, are the measurement positions of warpage as shown in the schematic illustration of a wafer. Regardless of Oi in the wafer, the warpage is increased by processes, as seen by comparing the open square and open circle. This is due to the stress from the stacking thin film layers. For the high Oi wafer, the warpage of the processed wafer is recovered to that of bare wafer by de-processing. This means that the high Oi wafer elastically deformed during the process. In the low Oi wafer, on the other hand, the warpage of the de-processed wafer is not recovered to that of bare wafer. This means that the low Oi wafer plastically deformed during the processes. Figure 3 shows the images of x-ray topography for the low and high Oi wafers. As shown in Fig. 3(a), slip is not observed in the high Oi wafer. As shown in Fig. 3(b), the expansion of slip along <110> is observed from the wafer edge toward the wafer center for the low Oi wafer. Comparing between Figs. 1 and 3, it is obvious that the region of slip propagation corresponds to the area of large residual misalignment. This indicates the slip causes the residual misalignment.

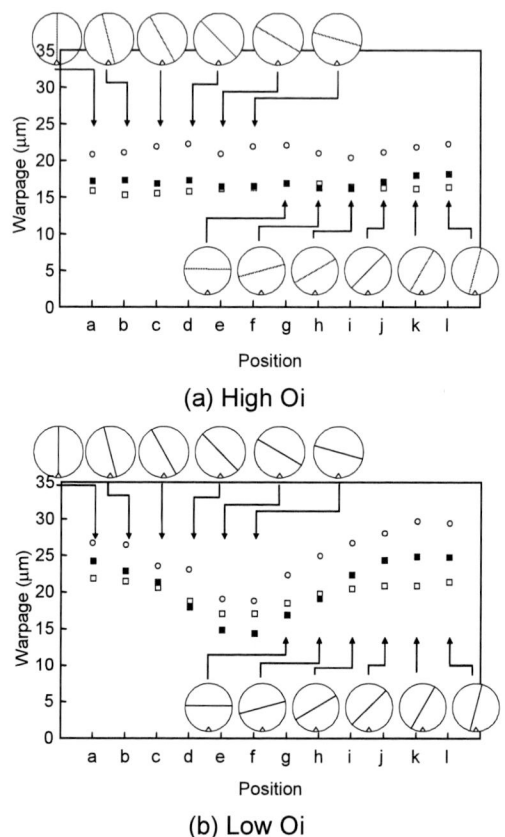

(a) High Oi

(b) Low Oi

Figure 2 Warpage of (a) high and (b) low Oi wafers. Open square, open circle and closed square denote the warpage of bare wafer, processed wafer and de-processed wafer, respectively.

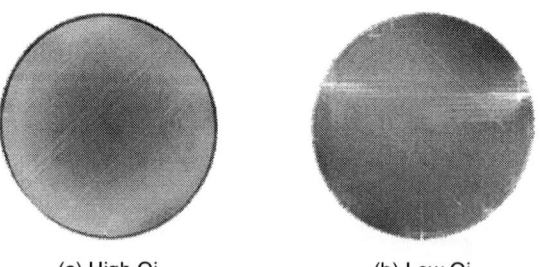

(a) High Oi   (b) Low Oi

Figure 3 Images of x-ray topography for the (a) high and (b) low Oi wafers

Programming failure by sub-surface dislocation

Programming characteristics are one of the important characteristics for Flash memory. The programming characteristics are evaluated after the process is complete. Figure 4 shows a programming failure map. The closed squares correspond to the chips with programming failures. The programming characteristic is good in the high Oi wafer. On the contrary, programming failures are observed in the center. From comparison between Figs. 1 and 4, it is found that residual misalignment and programming failure simultaneously occur in the same wafer. Moreover, it appears that the programming failure occurs in the area of small residual misalignment. In other words, the area of large residual misalignment does not show programming failure.

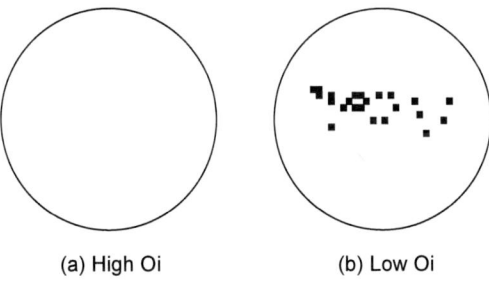

(a) High Oi   (b) Low Oi

Figure 4 Distribution of programming failure in (a) high and (b) low Oi wafer. The closed squares correspond to the chips of programming failure

The hot-electrons are injected to the floating gate through the tunnel oxide layer by applying the voltage between the floating gate and the substrate. The major reason of programming failure is the degradation of the tunnel oxide layer. Surface and/or sub-surface defects become the degradation site. There is a possibility that surface and/or sub-surface defects have a strong impact on the programming characteristic. Therefore, we observed the defects in the wafer as follows.

First, in order to observe the surface and sub-surface defects in the low Oi wafer, optical microscope observation followed by chemical etching is performed. Figure 5 shows the number of the observed pits per unit area in squared areas. Observation area of optical microscope corresponds to a quarter of the low Oi wafer as shown in schematic illustration of Fig. 5. The hatched squares are not suitable for the observation of pits due to the surface damage from another experiment. Surface pits are observed in the wafer center where the programming characteristic failed. Thus, it is found that the surface pits closely correlate to the programming failure. Next, LST observation is performed to identify the substance of pits. Figure 6 shows the LST image sub-surface region of the failure chip. The sample corresponds to the chip indicated by arrow in Fig. 5. The dark dots and the dark lines correspond to the OP and dislocation, respectively. As shown in the image, the dislocation is observed just under the surface of wafer. This indicates that the sub-surface dislocation causes the programming failure.

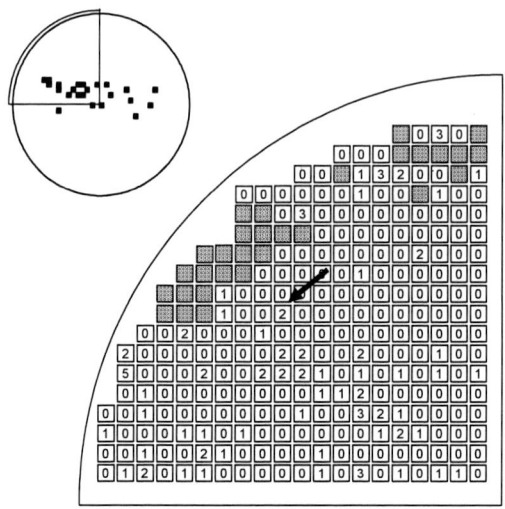

Figure 5 The number of surface pits of low Oi wafer that is observed by optical microscope after chemical etching.

surface

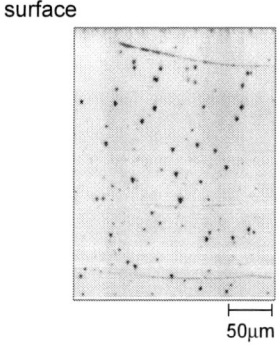

50μm

Figure 6 LST image of the sub-surface region in the low Oi wafer

Impacts of defects on Flash memory characteristic

Based on the above results, we discussed the behavior of defects during the processes. Residual misalignment and Flash memory characteristics are summarized in tables 1 and 2. Table 1 show that residual misalignment and programming failures are observed in the low Oi wafer. Distributions of residual misalignment and programming failure are contrasted. Table 2 shows that the defects are observed in low Oi wafer. The above characteristics are categorized by Oi, whether it is high or low. These clearly indicate that oxygen has strong influence on the residual misalignment and programming failure. Moreover, the origin of residual misalignment and programming failure are common.

Table 1 Residual misalignment and Flash memory characteristics in high and low Oi wafer

	Residual Misalignment	Programming characteristics
High Oi	Normal	Good
Low Oi	Large at wafer edge	Failure at wafer center

Table 2 Characteristics of defects in high and low Oi wafer

	Slip	Surface pits
High Oi	None	None
Low Oi	Large at wafer edge	Observed at wafer center

It is well known that oxygen in silicon has the locking effect for dislocations.[4] Flow stress increases with Oi of wafer. However, Oi after processing in the high Oi wafer coincides with that of low Oi wafer within 5%. Thus, there are no remarkable differences of residual Oi in the low and high Oi wafers. The locking effect of oxygen on dislocations is not the dominant factor in a residual misalignment and the programming failure.

Therefore, we must consider another effect of oxygen atoms on the strength of wafer - oxygen precipitates (OPs) formed during the processing. OPs have two roles with respect to the strength of silicon. One is the source of dislocation.[3] Flow stress decreases in the processed wafer including the excess OPs. On the other hand, OPs suppress the propagation of dislocations.[9] In this case, the flow stress of the wafer increases. The Ops are observed by LST. The vertical axis in Fig. 7 shows the density of OPs after processes as determined by the LST. As described in previous section, the residual misalignment and programming failures occurred in the low Oi wafer. Therefore, it is speculated that the density of OPs determines the residual misalignment and the programming characteristic. The slip propagation and the surface pits are observed in the low Oi wafers Figure 8 shows the LST image of the bulk in the high Oi wafer. In this image, the bowing of dislocations starting from the OP can be seen as indicated by arrow. This indicates that OPs prevent the propagation of slip. Thus, we find that the OPs have a strong effect on the suppression of slip propagation in the high Oi wafer.

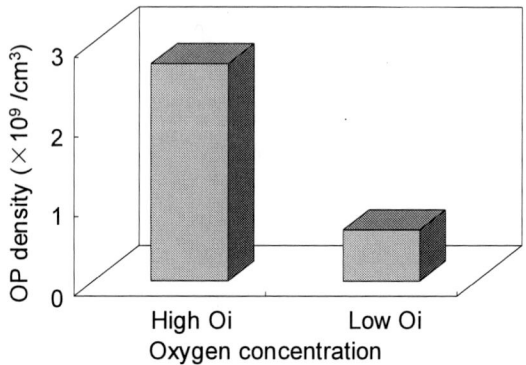

Figure 7 Density of oxygen precipitates after processing.

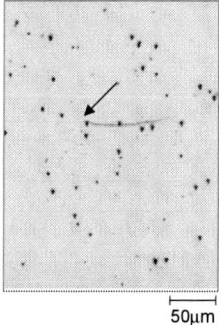

Figure 8 LST image of the bulk in the high Oi wafer

Figure 9 shows the schematic illustration of behavior of defects in the wafer during the processes of Flash memory fabrication. During the processes, internal stress occurs in the wafer. The internal stress in the wafer is relaxed by the plastic deformation. In the high Oi wafer, slip propagation is suppressed by OPs in the wafer edge. On the other hand, slip propagation is observed in the low Oi wafer as shown in Fig. 3(b). The region of slip propagation shown in Fig. 3(b) corresponds to the region of large residual misalignment shown in Fig. 1 at wafer edge in the low Oi wafer. This indicates that the slip caused by plastic deformation is the source of residual misalignment in lithography process. As a result, residual misalignment grows larger in the wafer edge where slip propagates. On the other hand, the internal stress is relaxed by the slip from the wafer front-surface in the wafer center in the low Oi wafer, which is observed as the surface pits as shown in Figs. 5 and 6. It is common that Oi is controlled for the suppression of slip propagation. In this paper, we succeed in control of slip by optimizing the density of OPs. Thus, the control of OPs is important for the yield and Flash memory characteristics.

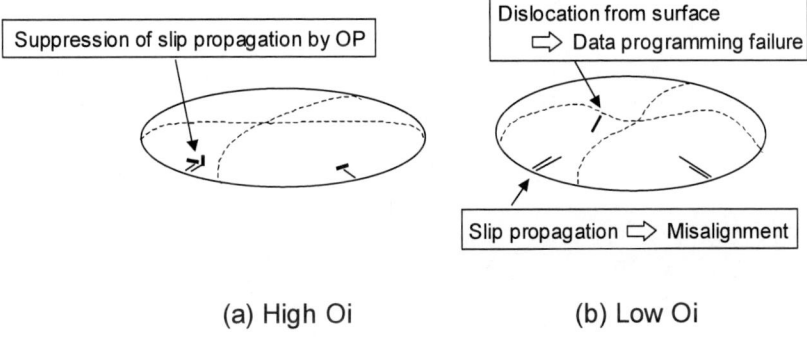

(a) High Oi   (b) Low Oi

Figure 9 Schematic illustration of behavior of defects in wafer during the processes.

SUMMARY

Flash memories are fabricated in the high and low Oi CZ silicon wafers. The residual misalignment in a reticle shot at a photolithography process step is monitored. After the fabrication of devices, the programming characteristics are evaluated. In the low Oi wafer, a large amount of residual misalignment is observed near the wafer edge. Moreover, the programming characteristic is degraded in the center of the wafer, where the residual misalignment is small. The behavior of defects in wafers by warpage, x-ray topography and LST is investigated. Residual misalignment and the programming failure are explained by the plastic deformation of silicon wafer for the relaxation of the internal stress during processes. Slip propagation can be controlled by optimizing the density of OPs.

ACKNOWLEDGMENT

The authors wish to thank Dr. Wataru Shindo, Dr. Mark Ramsbey and Daisuke Matsunaga for stimulating discussion. They are also thankful to Osamu Hideshima and Shinji Nakamura for encouragement for research.

REFERENCES

1. B. Eitan, P. Pavan, I. Bloom, E. Aloni, A. Frommer and D. Finzi, *IEEE Electron Devices Lett.*, **21**, 543 (2000).
2. J. Ryuta, E. Morita, T. Tanaka and Y. Shimanuki, *J. Appl. Phys.*, **29**, L1947 (1990).
3. H. Shimizu, *Jpn. J. Appl. Phys.*, **39**, 5727 (2000).
4. K. Sumino, H. Harada and I. Yonenaga, *Jpn. J. Appl. Phys.*, **19**, L49 (1980).
5. M. Itsumi, H. Akiya and T. Ueki, *J. Appl. Phys.*, **78**, 5984 (1995).
6. Y. Furumura, *Proc. 2nd Int. Conf. Symp. Advanced Sci. Tech. Silicon Materials*, Kona, 1996 (The Japan Society for the Promotion of Science, Tokyo, 1996) p. 418.
7. J-G. Park, H. Kirk, K-C. Cho, H-K. Lee, C-S. Lee and G. A. Rozgonyi, *Semiconductor Silicon/1994*, eds. H. R. Huff, W. Bergholz and K. Sumino (The Electrochem. Soc. Pennington, 1994) p. 370.
8. M. Miyazaki, S. Miyazaki, T. Kitamura, Y. Yanase, T. Ochiai and H. Tsuya, *Jpn. J. Appl. Phys.*, **36**, 6187 (1997).
9. K. Yasutake, M. Umeno and H. Kawabe, Appl. Phys. Lett., **37**, 789 (1980).

Low-Resistance Ti/n-Type Si(100) Contacts by Monolayer Se Passivation

J. G. Zhu, X. L. Yang, and M. Tao*

Department of Electrical Engineering, University of Texas at Arlington,
Arlington, Texas 76019, USA
* mtao@uta.edu

Low-resistance contacts fabricated by selenium passivation between Ti and n-type Si(100) substrates have been characterized by low-temperature I-V, four-point probe and circular transmission line methods. The Ti-Si contacts on Se-passivated samples demonstrate a significant reduction in Schottky barrier height over control samples in low-temperature I-V. Sheet resistance of the contacts on Se-passivated 10^{19} cm^{-3} doped n-type Si(100) substrates shows a 30% reduction as compared with control samples. Accordingly, the extracted contact resistance decreases by about one order of magnitude for samples with different Ti thicknesses and different annealing temperatures. A 125%–2900% reduction in contact resistivity is achieved by Se passivation on highly-doped n-type SOI substrates with 500 Å un-doped Si buffer layer. The reduction in contact resistance is attributed to the minimization of interface states between Ti and Si(100) surface.

INTRODUCTION

Metal-semiconductor contacts have always been an integral part of semiconductor device technology. With the continued shrinkage of feature size in Si ULSI technology, contact resistance can represent a significant fraction of the total device resistance and contributes to a loss in its performance. Fabrication of ohmic contacts by highly doped Si surface is the most widely used technique, in which the contact resistance does not strongly depend on the chosen metal. However, lower contact resistance can also be realized by lowering the Schottky barrier at the metal-Si interface. The later technique is especially useful for contacts where high doping levels are undesirable (1,2). Theoretically, the Schottky barrier height between metal and semiconductor is determined by the difference between metal work function and semiconductor electron affinity. The actual barrier height, however, is largely independent of metal work function due to Fermi level pinning by interfacial states. On the other hand, Schottky barrier height and contact resistance can be affected by both material factors (metal or silicide (3), doping level (4), and crystal structure (5)) and processing conditions (surface cleaning (6,7), oxide layer (8), annealing (9,10), and passivation (11)) because of their interface sensitivity.

Recently, Tao et al demonstrated thermally-stable potentially-negative Schottky barriers between Ti and Se-passivated n-type Si(100) surface (12,13), which should lead to low-resistance ohmic contacts to n-type Si(100) substrates. This Schottky barrier lowering effect is believed to be the result of minimized interface states when Se atoms form strong bonds with unsaturated Si atoms on the (100) surface to terminate the dangling bonds (14). The research presented in this paper is targeted to obtain direct evidence of the reduced contact resistance and contact resistivity by the Se passivation technique. Low-temperature I-V measurements reveal a significant lowering of the Schottky barrier height between Ti and Si by Se passivation. Both contact resistance and contact resistivity extracted from the four-point probe and circular transmission line methods (c-TLM) of the Ti-Si(100) contacts shows a significant reduction for the Se-passivated samples as compared to the control samples.

EXPERIMENTAL

The substrates used in the sheet resistance study were (100) oriented n-type Si wafers with resistivity of 16–24 Ω·cm P doped, 0.075–0.085 Ω·cm P doped, and <0.004 Ω·cm As doped, respectively. The substrates used for low-temperature I-V were (100) oriented n-type Si wafers with resistivity of 16–24 Ω·cm P doped. The substrates were cleaned in a solution of 2% HF in deionized water for 30 s and immediately loaded into a molecular beam epitaxy (MBE) system, which has a base pressure in the low 10^{-10} torr range. Si buffer layers of 500 Å were deposited on the substrates at 600°C after thermal desorption of H. All the samples were then annealed at 800°C for 1 h to secure a perfect crystal structure on the surface, as evidenced by sharp 2×1 reconstruction in reflection high-energy electron diffraction (RHEED). Some samples were passivated with a monolayer of Se after buffer layer growth. Other samples were used as control samples without Se passivation after buffer layer. Se passivation was accomplished in MBE by heating the Se source to 224°C while maintaining a substrate temperature of 300°C for approximately 60 s. Sharp 1×1 reconstruction was observed in RHEED after Se passivation, indicating a monolayer Se passivated Si(100) surface. After taken out of the MBE system, the wafers were cut into smaller pieces for metallization.

For low-temperature I-V measurements, Ti-Si contacts were fabricated with a lift-off process on both passivated and control samples. 500-Å Ti was deposited by e-beam evaporation and two types of contact pads were formed, 300-μm circular dots and 3430×2280-μm^2 rectangular pads. Low temperature I-V was performed on a Signatone shielded probe station with a liquid nitrogen cooled stage. For sheet resistance measurements, a layer of Ti with thickness 1000, 500, 300, and 100 Å were deposited on both passivated and control samples. These samples were then annealed in a vacuum chamber with a base pressure in the 10^{-6} torr range. The annealing was performed with 1 atm ultrahigh purity N$_2$ for 1 m at 300, 400, 500, 600, and 700°C. Finally, sheet resistance measurements were carried out using a Veeco FPP-5000 four-point probe.

The substrates used for the contact resistivity study were (100) oriented n-type SOI wafers. The Si layer was 2.5 ± 0.5 μm thick with a resistivity of 0.001–0.004 Ω·cm As doped and the buried oxide layer was 0.5 μm thick. The procedure for preparing passivated and control samples was the same as above, except that the un-doped Si buffer layer thickness was varied from 500, 50, to 10 Å. Metallization layers were 500-Å Ti for one pair of samples and 500-Å Ti at bottom plus 1500-Å Al at top for other samples. The contact resistivity was measured using the c-TLM (15). The spacing d between inner and outer electrodes was 10, 20, 30, 40, 50, and 60 μm. The current applied during measurement was 1 mA.

RESULTS AND DISCUSSION

The I-V characteristics for the Se-passivated and control samples with low 10^{15} cm^{-3} n-type doping are clearly different, as shown in Figure 1. Although both samples show linear I-V behaviors at room temperature, the control sample starts to show signs of rectification at 244 K, while the passivated sample remains ohmic till 141 K, the lowest temperature we can obtain in the probe station. This suggests that the Schottky barrier height with Se passivation is much lower than that without Se passivation. A simple-minded fitting of the I-V curves indicates a barrier height of less than ~0.1 eV for the passivated sample, depending on the area of the Schottky diode assumed. In comparison, the control sample shows a barrier height less than ~0.3 eV. The significant lowering in barrier height should lead to Ti/n-type Si(100) contacts with very low contact resistance.

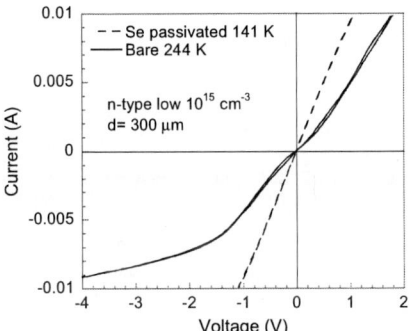

FIGURE 1. Low-temperature I-V characteristics of Ti contacts on Se-passivated and control n-type Si(100) substrates.

Figure 2 shows the comparison of sheet resistance between Se-passivated and control samples under different annealing temperatures with 100-Å Ti and moderately or low-doped n-type Si(100) substrates. The difference in sheet resistance between passivated and control is not significant. The reason is that for the moderately and low-doped substrates, the contact resistance is much smaller than the substrate series resistance, so the effect of contact resistance reduction on sheet resistance is too small to be observed. The abnormal jump in sheet resistance for the passivated sample after 500°C annealing could be caused by surface oxidation or non-uniformity in the Ti layer.

In Figure 3, the sheet resistance of Se-passivated samples is ~30% lower than that of control samples on highly-doped Si(100) substrates. It suggests that on highly-doped substrates, substrate series resistance is relatively small and contact resistance contributes a significant fraction of the overall sheet resistance. Therefore, reduction in contact resistance by Se passivation becomes evident. By assuming parallel pathways in sheet resistance measurement of the Ti-Si double-layer structure (as shown in Figure 4), contact resistance can be extracted from the measured sheet resistance with known substrate sheet resistance and Ti layer sheet resistance. Here, the substrate sheet resistance is measured from samples cut from the same wafer. The Ti layer sheet resistance is calculated using the sheet resistance data from low-doped samples for which the contact resistance is ignored due to the reason mentioned before.

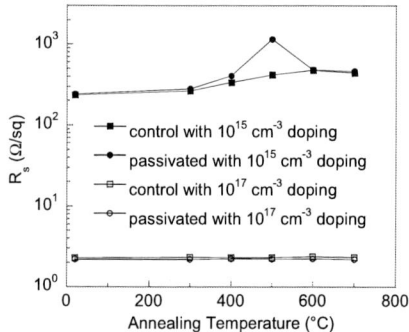

FIGURE 2. Sheet resistance as a function of annealing temperature for 100-Å Ti on Se-passivated and control n-type Si(100) samples.

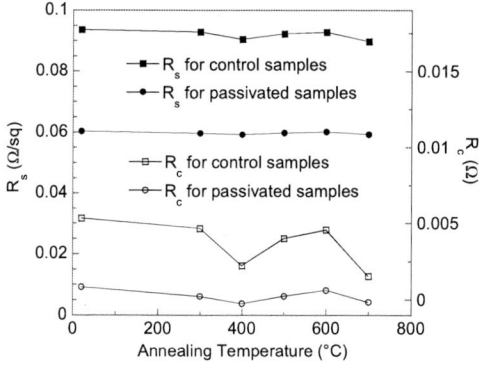

FIGURE 3. Comparison of sheet resistance and contact resistance as a function of annealing temperature for 100-Å Ti on Se-passivated and control n-type Si(100) samples with 10^{19} cm^{-3} doping.

The results in Figure 3 confirm that the contact resistance for the Se-passivated samples is always lower than that of the control samples at all annealing temperatures. Passivated samples before annealing show about one order of magnitude lower contact resistance as compared with the control samples. The variation in contact resistance for the control samples with annealing temperature is thought to be caused by silicidation. On the contrary, the contact resistance for the Se-passivated samples does not fluctuate much with annealing temperature, which suggests, and has been confirmed, that Se passivation suppresses silicidation at the metal-Si interface (16).

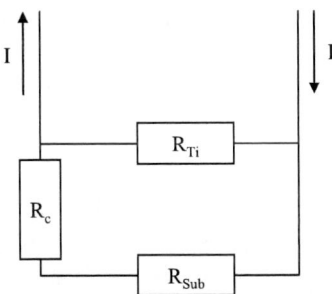

FIGURE 4. Schematic of the model for Ti-Si double-layer sheet resistance calculation.

The comparison of contact resistance as a function of substrate doping level and Ti layer thickness for Se-passivated and control samples is shown in Figure 5. On 10^{17} cm^{-3} doped n-type Si(100) substrates, contact resistances for the passivated and control samples cross each other and no consistent trend is observed. For 10^{19} cm^{-3} doped samples, contact resistance for the passivated samples is always lower then that for the control samples. These results are consistent with the sheet resistance measurements. Another interesting fact is that the contact resistance associated with thinner Ti layers is higher than that with thicker Ti layers for both passivated and control samples on the highly-doped substrates. This could be explained by interfacial reactions between the Ti layer and contaminants on the Si(100) surface since the percentage of the reacted volume for thinner Ti layers is larger.

Contact resistivity measurements for Se-passivated and control samples on SOI substrates with 500-Å Si buffer layer and 500-Å Ti layer are shown in Figure 6. The method to extract the contact resistivity from Figure 6 is well established (15)

$$R = \frac{R_s}{2\pi L}(d + 2L_T) \qquad [1]$$

and

$$\rho_c = R_s L_T^2 \qquad [2]$$

where R is the total measured resistance between the inner and outer electrodes, R_s is the substrate sheet resistance, d is the separation between the inner and outer electrodes, L is the radius of the inner electrode, L_T is the transfer length, and ρ_c is the contact resistivity, which is extracted from the experimental data using a least-square linear fitting. The passivated samples show a much lower contact resistivity than the control samples, 6.23×10^{-6} $\Omega \cdot$cm versus 1.82×10^{-4} $\Omega \cdot$cm. This is a 29 times reduction in contact resistivity between Ti and n-type Si(100) by Se passivation.

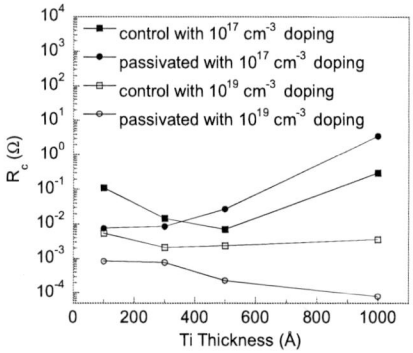

FIGURE 5. Comparison of contact resistance as a function of n-type Si(100) substrate doping level and Ti layer thickness for Se-passivated and control samples.

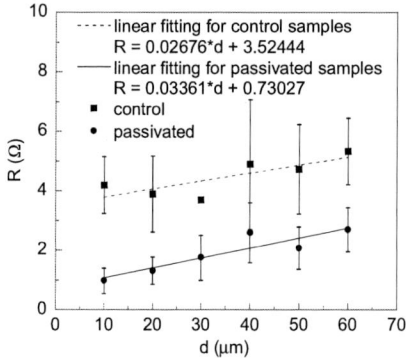

FIGURE 6. Linear fittings of the experimental data from the circular transmission line measurements of Se-passivated and control samples with 500-Å Si buffer layer and 500-Å Ti layer.

Since contact resistance depends strongly on the depletion layer thickness, an effect of the un-doped Si buffer layer thickness between Ti and the substrate is expected on contact resistivity. As shown in Figure 7, the reduction in contact resistivity is a function of the Si buffer layer thickness for both passivated and control samples with 500-Å Ti plus 1500-Å Al. When the buffer layer thickness is 500 Å, Se passivation reduces the contact resistivity by ~125% as compared to control samples in this experiment. If the buffer layer thickness is reduced to 50 and 10 Å, there is little difference between passivated and control samples. This trend could be explained by a moisture layer formed between Ti and Si when the substrates are exposed to air before Ti deposition. A transmission electron microscopy image of the interfacial layer formed between Ti and Se-passivated Si substrate is shown in Figure 8. The Se-passivated substrate was exposed to air when it was transferred from MBE after Si buffer growth and Se passivation to e-beam evaporator for Ti deposition. The sample was never annealed. The interfacial layer is ~15 Å thick after metallization, which introduces a tunneling barrier. The measured contact resistance, therefore, has two contributions, the tunneling resistance through the moisture layer, R_t, and the resistance through the Schottky barrier, R_{Sh}

$$R_c = R_{Sh} + R_t \qquad (3)$$

When the buffer layer is thick, R_{Sh} is large and the effects of Se passivation in R_{Sh} reduction are observable. When the buffer layer thickness decreases, R_{Sh} becomes smaller than R_t and the effects of Se passivation are overshadowed by R_t.

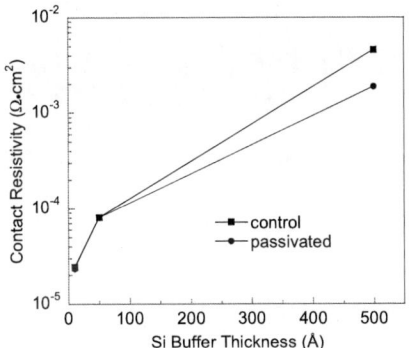

FIGURE 7. Comparison of contact resistivity as a function of Si buffer layer thickness for Se-passivated and control samples with 500-Å Ti at bottom and 1500-Å Al on top.

CONCLUSIONS

The effect of Schottky barrier lowering on contact resistance between Ti and n-type Si(100) substrates by Se passivation is studied with low-temperature I-V, four-point probe and c-LTM measurements. Low temperature I-V measurements show a significant lowering in barrier height by Se passivation as compared to the control samples. Sheet resistance of the contacts on Se-passivated 10^{19} cm^{-3} doped n-type Si(100) substrates shows a 30% reduction as compared to the control samples. Accordingly, the extracted contact resistance decreases by about one order of magnitude for samples with different Ti thicknesses and different annealing temperatures. Reduction in sheet resistance is not observed for substrates with 10^{15} cm^{-3} and 10^{17} cm^{-3} doping levels. A 125%–2900% reduction in contact resistivity is achieved by Se passivation on highly-doped n-type SOI substrates with 500-Å un-doped Si buffer layer. When the buffer layer thickness is reduced to 50 and 10 Å, the contact resistivities of the passivated and control samples reveal little difference. Possible reasons for the diminishing effects of Se passivation with lower substrate doping levels and thinner un-doped Si buffer layers are presented. The reduction in contact resistance is attributed to the minimization of interface states between Ti and Si(100) surface.

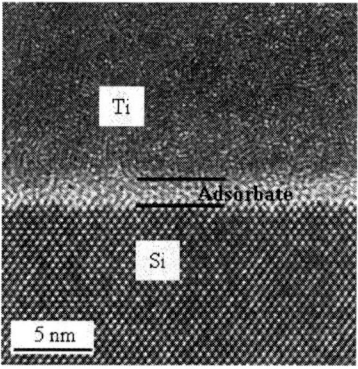

FIGURE 8. Transmission electron microscopy image of the interfacial layer between Ti and Se-passivated Si(100) substrate.

ACKNOWLDGEMENTS

This work was supported by the National Science Foundation (ECS-0322762) and the Advanced Technology Program of Texas (003656-0096-2003).

REFERENCES

1. C. Y. Ting and B. L. Crowder, *J. Electrochem. Soc.* **129**, 2590 (1982).
2. E. Sasse and U. Konig, *J. Appl. Phys.* **64**, 3748 (1988).
3. L. E. Terry and R. W. Wilson, *Proc. IEEE*, **57**, 1580 (1969).
4. C. Y. Chang, Y. K. Fang, and S. M. Sze, *Solid State Electron.* **14**, 541 (1971).
5. R. T. Tung, *Phys. Rev. Lett.* **52**, 461 (1984).
6. T. B. Hook, R. W. Mann, and E. J. Nowak, *IEEE Trans. Electron Devices*, **42**, 697 (1995).
7. T. A. Schreyer, A. J. Bariya, J. P. McVittie, and K. C. Saraswat, *J. Vac. Sci.*

Technol. A **6**, 1402 (1988).

8. S. S. Cohen, M. J. Kim, B. Gorowitz, R. Saia, and T. F. McNelly, *Appl. Phys. Lett.* **45**, 414 (1984).
9. H. R. Liauh, M. C. Chen, J. F. Chen, and L. J. Chen, *J. Appl. Phys.* **74**, 2590 (1993).
10. H. Kato and Y. Nakamura, *Thin Solid Films*, **34**, 135 (1976).
11. S. Zaima and Y. Yasuda, *Mat. Res. Soc. Symp. Proc.* **386**, 215 (1995).
12. M. Tao, D. Udeshi, N. Basit, E. Maldonado, and W. P. Kirk, *Appl. Phys. Lett.* **82**, 1559 (2003).
13. D. Udeshi, E. Maldonado, Y. Xu, M. Tao, and W. P. Kirk, *J. Appl. Phys.* **95**, 4219 (2004).
14. J. G. Zhu, M. P. Nadesalingam, A. H. Weiss, and M. Tao, *J. Appl. Phys.* **97**, 103510 (2005).
15. D. K. Schroder, *Semiconductor Material and Device Characterization*, p. 151, Wiley, New York (1998).
16. M. Tao, J. Shanmugam, M. Coviello, and W. P. Kirk, *Solid State Commun.*, **132**, 89 (2004).

410

SESSION 5

INTEGRATED METROLOGY AND
DIAGNOSTICS

Integrated Metrology and Diagnostics

Laszlo Fabry
Wacker Polysilicon, Wacker-Chemie AG
Johannes-Hess-Str 24, 84489 Burghausen, Germany

Takeo Hattori
Research Center for Silicon Nano-Science
Musashi Institute of Technology
8-15-1 Todoroki, Setagaya-ku, Tokyo 158-0082., Japan

It is beyond any doubt that Metrology & Diagnostics have played a crucial role in the rapid development of advanced semiconductor technology [1-3]. It is also beyond any doubt that it is positively necessary, but not sufficient, to establish sensitive and reliable metrology & diagnostics tools and methods in fab operations. In order to provide efficient and economic support for high volume manufacturing (HWM) lines, Metrology & Diagnostics should also be integrated in the fab operation.

In allowance for all technical aspects in efficient Metrology & Diagnostics, our session presents an overview on novel, smart, integrated and sensitive Metrology & Diagnostic methods. In cooperation with EBARA Corporation, Yamazaki et al. of Toshiba Corporation, Yokohama elaborate on the concept and discuss the details how sensitive in-line microscopic tools such as electron beam inspection, combined with projection electron microscopy (EBI-PEM), can support high volume manufacturing of post-65 nm lines. Controlling lines of post-65 nm requirements by means of conventional scanning electron microscope (EBI-SEM) proved to be insufficient due to the limited throughput of EBI-SEM. Similarly, Nutsch and Oechsner of the Fraunhofer Institute, Erlangen introduce a scenario for an integrated yield model based on defect density analyses in advanced process control (APC). Reliable defect density data must be linked to an APC network in order to support yield enhancement.

Standardized test methods are also fundamental for both efficient Metrology & Diagnostics and unambiguous technical understanding. Professor Inoue of the Osaka Prefecture University describes the achievements of Japan Electronics and Information Technology Industries Association (JEITA) Working Groups in standardization of sensitive metrology for nitrogen, carbon, bulk micro-defects and denuded zone. It is very interesting to follow up on the development of JEITA's standardization project: The process of developing standard test methods involves a deep knowledge of materials and processes beyond sensitive and reproducible instrumental analyses in order to support manufacturing lines. Validation using independent and complementary instrumental analyses is a string requirement of any standardized test. In due course, JEITA's Nitrogen Concentration Measurement Group has studied all of the available nitrogen species in CZ silicon and correlated different methods such as Fourier transform infrared spectroscopy (FTIR), secondary ion mass spectroscopy (SIMS) and charged particle activation analysis (CPAA) to provide the industry with a standardized test method for nitrogen measurement above 1×10^{14} atoms/cm³.

In cooperation between the Osaka Prefecture University and IHP/BTU Cottbus, a sensitive ultra-trace analytical method of carbon impurities has been developed. After baseline fit it has been possible to quantify carbon impurities in monocrystalline silicon down to 1×10^{14} atoms/cm³. After having correlated results of preferential etching and 90-degree laser-scattering tomography JEITA could issue a standard test specification for the quantification of bulk micro-defects (BMD) and denuded zone (DZ).

Novel engineered silicon wafers require powerful new metrology tools. Thus, T. D. Ma and a Chinese research group reports an X-ray diagnostic method for characterization of strained SiGe layers on bonded SOI. X-ray triple-axis diffractometry (TAD) and synchrotron radiation double-crystal topography (SRDT) have been applied to structural characterization of the Si capping layer and the underlying Si buffer and SOI top Si layers. High resolution reciprocal lattice mapping (HRRLM) showed that the mosaicity in SiGe layers lead to the broadening of the (004) diffraction peak. Crosshatched dislocations were spatially correlated with misfit dislocations (MD) in both SiGe and underlying Si layers.

Low frequency noise (LFN) and interface trap density measurement are important for optimizing SOI devices, as Kushner et al. of Arizona State University, Tempe present. LFN affects the quality of low-noise analog circuits such as amplifiers and the phase noise of voltage-controlled oscillators. LFN is even more important in digital circuits with reduced power supply voltage. LFN is caused by interactions of the channel carriers with oxide/semiconductor interface traps and oxide charges [4]. LFN measurements after MOSFET fabrication using Ground-Signal-Ground (GSG) on pseudo-MOSFET structures shed light on the correlation of LFN with trap densities at the buried oxide/semiconductor interface.

HfO_2/FUSI or HfO_2/poly-Si interfaces can also be tested for traps in high-k n-MOSFETs. Adding monolayers of HfO_2 increases the normalized noise in both poly-Si and FUSI devices. Differences in trap densities, as derived from the LFN spectra, can be traced back to the gate/dielectric interface. P. Srnivasan, E. Simoen, L. Pantisano, C. Claeys and D. Misra study 1/f noise in high-k n-MOSFETs in order to assess Fermi level pinning at the interface.

The session on Integrated Metrology & Diagnostics will be concluding with two presentations on general technologies such as chemimechanical polishing of copper (Cu-CMP) and organic surface contamination in cleanrooms. Professor Philipossian et al. of University of Arizona, Tucson in cooperation with researchers of IBM and Intel compared different Cu-CMP slurries such as ammonium dodecyl sulfate (ADS) and benzotriazol (BTA). The coefficient of friction (COF) is the lowest when the ADS slurry has been used. ADS, an anionic and environmentally friendly surfactant, also provides additional lubrication, thereby reducing both wear and fluctuation between the contacting surfaces.

Professor Habuka of Yokohama National University investigated real-time adsorption and desorption of water and diethyl phtalate (DEP) on silicon surfaces by means of a quartz crystal microbalance (QMC). Under steady-state conditions the adsorbed amount of water and DEP is a function of the concentration of water and DEP in the ambient nitrogen gas. However, in a non-steady state, increasing gas flow rate has enhanced both ad- and desorption rate. Thus, organic surface contamination may be enriched by increasing gas flow rate in the clean room.

References:

1. H. R. Huff, An Electronics Division Retrospective (1952-2002) and Future Opportunities in the Twenty-First Century, J. Electrochem. Soc. **149**, S35 (2002)
2. G. A. Rozgonyi, Silicon Defect Diagnostics and the "Wheel of Misfortune" **PV 77-2**, 504, The Electrochemical Society Proceedings Series, Pennington, NJ 1977
3. L. Fabry and Y. Matsushita, "Gear of Challenge" – Smart and Affordable Diagnostics, **PV 98-1**, 1459, The Electrochemical Society Proceedings Series, Pennington, NJ 1998
4. C. Claeys, A. Mercha, E. Simoen, Low Frequency Noise Assessment for Deep Submicrometer CMOS Technology Nodes, J. Electrochem. Soc. **151**, G307 (2004)

Breakthrough of in-line inspection technology in volume production for 65nm node and beyond

Yuichiro Yamazaki

TOSHIBA Corporation, Process & Manufacturing Engineering Center
8, Shisugita-cho, Isogo-ku, Yokohama, 235-8522, Japan

Optical wafer defect inspection technology for yield monitoring and systematic defect detection is widely used by the industry because the production yield of device can be directly estimated by this method. In stead of optical system, electron beam (EB) inspection system is now being introduced because the detection sensitivity of optical system has become insufficient for 65 nm node and beyond. However, low throughput of conventional EB inspection system makes it difficult to introduce the system for in-line inspection into volume production line. To overcome this problem, we have developed the prototype of EB inspection system using projection electron microscope (EBI-PEM). In this paper we present two types of applications to address the problems. In one case it is the voltage contrast (VC) defect detection technique that uses test element group (TEG) structure, while in the other case it is the capability of EBI-PEM for OPC model and reticle printability qualification in volume production line. In case of the VC inspection with TEG, EBI-PEM technique can achieve 5 times the throughput as compared to what the conventional EB inspection system can give. Furthermore, by using thin film sample to avoid the charge up effect, good sensitivity with almost 5 fold increase in throughput was obtained for OPC model and reticle printability qualification. Based on these evaluations it was confirmed that EBI-PEM is a promising candidate for in-line defect inspection system for yield monitoring and systematic defect control for post-65 nm device volume production line.

1. Introduction

With aggressive acceleration of semiconductor device shrinkage, metrology and inspection technologies are becoming increasingly more important for in-line process monitoring in volume production line (1). The throughput, accuracy, precision, and tool cost are critical for metrology and inspection. Especially, wafer defect inspection for yield monitoring and systematic defect control is an extremely important because it can directly be used to estimate production yield of device during the manufacturing processes.

Firstly, in this paper the problems of the current wafer defect inspection technologies are addressed for random defect monitoring. The current optical systems lack defect detection sensitivity required for device yield monitoring for post-65 nm node, whereas, electron beam (EB) inspection systems for in-line monitoring exhibit poor throughput to meet new challenges. Later in the paper, the requirements for monitoring of lithography induced systematic defects are also discussed.

To overcome these problems, we have proposed the electron beam inspection (EBI) technique utilizing projection microscope (PEM) called as EBI-PEM that can detect random, as well as systematic defects. Based on the POC (proof of concept) system of EBI-PEM we have developed a production worthy prototype of EBI-PEM. To verify the high throughput performance of EBI-PEM, the application of voltage contrast (VC) defect inspection with test element group (TEG) structure, and the application for the lithography induced systematic defects detection scheme are presented in this paper.

2. In-line wafer inspection technologies for production yield monitoring for post-65nm device volume production

2.1 Visible and random defect monitoring

Figure 1 shows the present in-line yield monitoring procedure using optical wafer defect inspection systems. Wafer defect inspection technologies providing the bright field and dark field optical imaging system are widely employed by semiconductor device manufacturers. The inspection, after a manufacturing process steps, is carried out to detect killer defects which typically appear in random distribution on wafer surface during the process. Using defect location data, a SEM review station is employed to classify defects under various categories. This information then provides the user with defect densities and defect classes for each process step. Based on the information on defect densities for several process steps, and by employing a statistical yield model such as "kill-ratio" model, final production yield is then estimated. However, in order to make this estimation very reliable it is necessary that sufficient defect detection sensitivity and adequate inspection productivity (or throughput) be required. In general, there always seem to be a tradeoff between detection sensitivity and throughput of a system.   Table I shows the requirements that need to be met by wafer defect inspection technology for in-line yield monitoring.

Figure 2 shows the minimum defect size that the wafer inspection systems can detect as a function of throughput of 300 mm wafer inspection.   The performances of the latest systems including optical bright field, dark field, and electron beam inspection (EBI) utilizing scanning electron microscope (EBI-SEM) are shown.   With increasing throughput, the minimum defect size increases monotonically.   Thus, as mentioned earlier, there is a trade off between the throughput and the minimum defect size that the system can detect.   When throughput is high, the minimum defect size will become larger, that would mean that the system would become less sensitive.   For in-line wafer defect inspection with minimum production loss, a throughput of more than 1WPH is required as shown in Fig. 2 by shaded region.   Until recently, the wafer inspection technologies using the optical microscope have been widely used for yield monitoring in the volume production line by detecting the random visible defect that can be detected from the top view.   With optimum investment plan based on intelligent return of investment (ROI) modeling, device makers have made the strategic decision of introducing several types of optical inspection technologies that utilize bright field and dark field inspections with visible, ultraviolet (UV) and deep ultraviolet (DUV) lights.   Furthermore, electron beam inspection system using scanning electron microscope (EBI-SEM) has also been introduced on developmental lines because of the high spatial resolution of SEM.

However, with shrinking features and their increasingly high aspect ratios, the capture rate of the visible defect has been decreasing drastically.   Even in the volume production of the present 90 nm node device, capture rate of the current optical inspection technologies is reaching only up to 30 % compared to the electric test results.   Undetectable defects of 70 % consisting of small size defects must be attributed to insufficient detection sensitivity of optical system and non-visual defect which can not be detected from the top view.   Generally, these defects are process dependent defects that are caused by limited capabilities of process tuning and small process margin.   Because the random defect has the size distribution, we can monitor the production yield correctly by detecting the defects that are typically 2 to 3 times larger than the design rule.   As the systematic defect contribution is estimated to be more significant in post-65 nm generation we have to consider the contribution of the systematic defects to the yield estimation using the results of the systematic defect detection.

2.2 Systematic defect detection for accurate in-line yield monitoring

Under this condition, device maker have to start the volume production with the systematic defects, which are ordinarily overcome during the development phase. There are two types of critical systematic defects that will be encountered in the next generation device. One is lithography process induced pattern defect and the other type relates to electric conductivity failure caused by open and short circuits which can not be detected by optical top view inspection.

Lithography process induced systematic defect has become critical because of the complex lithographic trick used in advanced design rule device. For rapid full-chip process window verification, the bright field optical inspection technology was applied to identify the random and systematic defects which occurred due to mask tolerance excursion, optical proximity correction (OPC) inaccuracies, resolution enhancement techniques (RET) design error, or unmanufacutrable layout configurations. However, the defect detection sensitivity of the optical inspection technologies for OPC model and reticle printability qualification become insufficient at smaller design rule because of the insufficient spatial resolution of optical inspection technology.

The electrical conductivity failure of interconnections and contact holes generally can not be detected by the optical view. Several applications using EB system have been proposed that include voltage contrast (VC) detection method using EBI-SEM, and absorption current measurement method using SEM. Recently, these EB applications are being introduced into the device manufacturing line as in-line monitoring system (4). However, since SEM generally consists of a single EB probing system and a single electron signal detection system, it is very difficult to improve the throughput of such a system. In fact, in SEM based inspection technology, the throughput is restricted by the number of both, EB probing system and electron signal detection system.

Table I summarizes the general requirements of wafer inspection technology for random defect detection for in-line yield monitoring, lithography induced systematic defect detection, and electric conductivity failure defect. For random and visible defect detection, critical detection sensitivity is around two times larger than the design rule, because such defect sizes predominantly influence the yield. On the other hand, in systematic defect detection, higher sensitivities are required because the systematic defect failure is not a statistical phenomena, and frequently not visible to optical techniques. Especially, lithography induced defect is very critical because even small pattern defect can be fatal. We have to monitor the pattern failure of less than half the design rule size within a whole chip. In case of an electric conductivity failure, defect detected is generally non-visual. EB inspection and metrology system should be necessary for both, the development phase and volume production phase.

3. Development of EBI-PEM prototype for high throughput in-line EB inspection

EB inspection system has to be introduced into in-line inspection system in conjunction with optical inspection technology in order to increase the capture rate of the systematic defects including small size visible defects and non-visual defects. However, ROI of the in-line inspection system including present EBI-SEM can not meet the requirements for post-65 nm device production line because the throughput of EBI-SEM is insufficient for in-line use. It is well known, that the throughput of EBI-SEM is dominated by image blur due to the space charge effect. Hence, by increasing the electron beam current to achieve the higher throughput, the image blur comes into play and the inspection sensitivity degrades.

To overcome the throughput problem with the EB inspection systems, we have developed high throughput EB inspection system based on projection electron microscope (EBI-PEM) (5, 6, 7, 8). In comparison with EBI-SEM, the EBI-PEM has both, faster mapping rates due to parallel detection and higher resolution because of direct imaging produced by projection optics. This imaging concept is almost similar to the concept of a bright field optical inspection system. Firstly, we developed the proof of concept (POC) system (5, 6). The basic concepts of optical system and the optical performances for practical semiconductor devices had been proved using the POC system. On the basis of the results of the POC system, EBI-PEM production prototype was developed and is now being improved to catch up with the requirement for post-65nm generation (7, 8). The performances of the system will be described in future reports.

The system configuration of the EBI-PEM prototype is schematically illustrated in Figure 3. The system consists of an electron optical system providing illumination optics and projection optics, a 300 mm wafer stage, and 32 pipeline image-processing engines with defect detection algorithm. A rectangular electron beam generated by the illumination optics illuminates the wafer on the stage. Secondary electron (SE) image emitted from the wafer surface is projected on to an electron image sensor. The wafer stage moves continuously along the short axis direction of the illumination beam (X direction) for scanning motion, and steps in the long axis direction (Y direction) for stopping motion. The electron image is acquired by a time-delay and integration (TDI) sensor that is synchronously linked with the stage motion in X. The optical image on the fluorescence screen is transferred by the optical lens system to the TDI sensor. Image signals detected by the TDI sensor with 32 taps are transferred to the image processing engines and handled for defect detection in parallel. Details of the electron optics are shown in Fig.4. The illumination optics produces the rectangular illumination beam using an electron gun and an electrostatic quarupole triplet. The illumination beam is bent by a beam separator and is normally irradiated on the wafer, which is biased to negative potential. Due to the acceleration filed between the wafer and the projection optics, the secondary electron is accelerated and transferred to the projection optics. The SE image is magnified and forced on to the electron image sensor by the projection optics.

4. Verification of EBI-PEM for in-line systematic defect monitoring

Performances of EBI-PEM for two types of systematic defect inspections were verified. One is a VC defect detection scheme using test element group (TEG) structure, the other is a capability of EBI-PEM for OPC model and reticle printability qualification.

4.1 Voltage contrast defect detection using TEG structure pattern

To improve the EBI-SEM throughput for VC defect detection, the inspection using TEG was proposed and being gradually introduced into production line monitoring. As shown in Fig.5 (a), the conventional TEG structure for open and short failure detection is comb-pattern with floating electrodes and grounded electrodes. The floating electrodes at a common potential and the grounded electrodes are provided alternately. By scanning the electron beam at the edge of the pattern, open and short defects can be detected with reasonable throughput by detecting the abnormality of VC periodicity with the image comparison algorithm. Also, to detect the electric conductivity open failure of contact hole, the TEG structure provid the two layered island Cu electrodes in which the upper electrodes are connected to the lower electrodes through contact holes as shown in Fig.5 (b). Electrically isolated lines are produced by connecting in a direction through contact holes, where one end of each of the line electrodes is grounded. By scanning the electron beam at the edge of the TEG pattern, image contrast of the electrode changes drastically at the line where there is an open contact hole, and thus the failure line electrode can be focused. By scanning on the detected electrode using EBI or review SEM, the open contact hole location can be identified, and the failure analysis at the location can be made with reasonable throughput.

In this application, TEG area is very important function.   The entire inspection area of TEG is determined by the defect density at the field, which is modified with the product generation of development phase, ramp up phase, and volume production phase.   On the other hand, the size of the unit patterns which construct the TEG is dominated by the target throughput of the EB inspection and defect capturing performance.   To improve the capturing accuracy, it is better to reduce the size of the unit pattern because the signal to noise (S/N) ratio of EB induced VC abnormality and defect capturing rate in defect review using review SEM increase with reducing the unit size.   Therefore, the unit pattern size is becoming less than 10 μ m square with design rule shrinkage.   When the unit pattern size becomes small, high throughput inspection system should be necessary to meet the suitable inspection time with the suitable whole TEG area. By adopting EBI-PEM in this TEG application, higher throughput can be obtained.

Figure 6 shows the detection number of real open and short failures produced in the 100 μ m design rule TEG under several CMP process conditions.   There is a discrepancy in the defect number between the two systems.   Because the image generating mechanism of EBI-PEM is different from that of EBI-SEM, detection number can not relate to each other.   However, by changing the process condition, the defect number of EBI-PEM with 540MPPS data rate is proportional to the data of conventional EBI-SEM with 100MPPS, where its correlation coefficient R^2 is 0.99.   EBI-PEC can achieve the good correlation with conventional EBI-SEM with more than 5 times the throughput.

4.2 Defection of lithography induced defect

It is well known that in the next generation of semiconductor manufacturing, lithography induced defects will become more critical.   The most serious issue is going to be pattern defect due to the proximity effect.   Hot spots, which are pattern defects cased by the optical proximity effect, have to be eliminated during the developmental phase using several procedures.   One is the inspection based procedure in which the hot spots are detected by the wafer inspection followed by correcting the OPC data.   The other is simulation based procedure, in which the hot spots are estimated by the lithography simulation.   In both case, high sensitive pattern defect inspections are compared with the ideal pattern such as the design date of mask pattern.   Recently, EB inspection system providing die to data base algorithm have been proposed for the hot spot detection in which the SEM image compared with the design date of mask pattern to detect the pattern fidelity like an EBI-SEM.   For OPC verification in the development line, we can accept the performance of the throughput for EBI-SEM at the least.

However, in the next generation semiconductor volume production line, OPC management on volume production line will become very important because the complicated OPC features.   In that case we have to monitor the OPC qualification by inspecting the one-die area on the wafer with a reasonable throughput.   To achieve the target performances for hot spots managements as shown in Table I , we have to improve both, the inspection sensitivity and the throughput.

In case of EB inspection, sample charging is quite serious for stable inspection.   To reduce the influence on the EBI-PEM inspection due to the surface charging, we proposed an imaging condition using reflection electrons, because the refection electron with higher emission energy is robust for the surrounding inhomogeneous surface potential due to the charging (8).   However, the reflected electron imaging exhibits low signal to noise ratio because of the low emission yield of the reflection electron as compared with the secondary electron imaging.

We developed the new method to inspect after the transferring a pattern on SiO_2 thin film grown on Si wafer.   By optimizing the thickness of SiO_2 film and electron beam condition of EBI-PEM, we could minimize the influence of charge up, and obtain a high contrast EBI-PEM image.

Figure 7(a) shows the relative image shift of the SiO_2 thin film line and space pattern as a function of the SiO_2 film thickness at kinetic energy of incident electron E_L of 3keV and 1keV.   The relative image shift means the ratio of relative line-pattern-width at pattern edge area with respect to the dense pattern area and becomes 100% when there no distortion at the edge area.   This relative pattern shift is caused by image distortion due to the surface charging.   In the pattern edge area, the charging causes the inhomogeneous surface potential and induces the image distortion.   Then, the charging contribution can be monitored qualitatively by this relative pattern shift.   As shown in Fig. 7(a), the relative pattern shift at landing energy of 3 keV is grater than the shift at 1.0 keV, and with increasing the film thickness, the pattern shift enhances.   The wafer surface charging is strongly dependent on the penetration of the incident electrons, which is proportional to the incident electron energy.   Therefore, the image distortion due to the surface charging decreases with increasing the E_L.

Also, to obtain the higher quality image of the SiO_2 thin film pattern, we have to maximize the signal intensity with respect to the film thickness and the landing energy as shown in Fig. 7(b).   The signal intensity of the SiO_2 pattern at E_L =1.0keV is more than three times greater than the intensity at 3.0keV, and takes its maximum at SiO_2 film thickness of 10nm.   Because the low landing energy electrons distribute within the film and do not penetrate to the Si substrate, the secondary electron emission yield at E_L=1.0keV is enhanced in comparison with that at 3keV. Furthermore, the surface charging and the SE emission yield penetration of the incident electron, which are dependence on the penetration of the incident electron, induce the modulation of the signal intensity. With increasing the film thickness, the surface charging is enhanced, and the SE emission yield of SiO_2 surface is degraded. These two contributions cause the maximum relative intensity at SiO_2 film thickness of 10 nm as shown in Fig. 7(b). Based on these results, the optimum condition was obtained; E_L of 1.0keV and SiO_2 film thickness of 10nm.

Under the optimized condition, we evaluated the capture rate of programmed defect wafer in which 56 programmed defects per one die were created as shown in Fig.8. The size of programmed defect was measured on SEM image. There are two types of defect size definitions. Open and short circuits defect is mostly affected by the size of defect in the plane in the pattern. Line and space thinning is mostly affected by the size of defect vertical to the pattern. Figure 8 shows the capture rate of 5 dies with total defect of 280 as function of the defect size. Almost 100% capture of the short and open defects was achieved. On the other hand, the averaged capture rate of the line and space thinning defects decreases monotonically with reducing the defect size. The defect acquisition rates of 60 nm and 35 nm are 80% and 50%, respectively with data rate of 40 MPPS. Inspection throughput of this data rate was almost same as that of the conventional EBI-SEM system.

5. Conclusions

In semiconductor device volume production line, production yield of the device is quite important for precise yield estimation during the intermediate processes. Until recently, we have been applying the conjunction tool strategy using dark field (DF), bright field (BF), and electron beam (EB) inspection technologies. In the post-65nm device volume production line, we will be facing the limitation in the inspection sensitivity and increasing the cost of in-line inspection tool. To overcome these problems, we have developed the high throughput EB inspection system EBI-PEM.

To verify the performances of EBI-PEM for in-line inspection in next generation volume production, two types of applications are presented. One is a VC defect detection using TEG structure, and the other is a capability of EBI-PEM for OPC model and reticle printability qualification in volume production line. In case of the VC inspection with TEG, inspection at data rate of 500MPPS or more was achieved with the almost same detection sensitivity as with conventional EBI-SEM system with 100MPPS. Then, EBI-PEM can achieve 5 times the throughput as compared to EBI-SEM for VC defect inspection. Furthermore, by using thin film sample to avoid the charge up effect, it can be achieved that the defect capture rates of 60 nm and 35 nm are 80% and 50%, respectively with data rate of 40 MPPS. The inspection throughput of this data rate was almost same as that of the conventional EBI-SEM system. Based on these evaluations, we can confirm that EBI-PEM is a promising candidate for in-line defect inspection system for post-65 nm device volume production yield monitoring and systematic defect control.

Acknowledgments

The author would like to acknowledge Mr. M. Shiozaki and Dr. N. Hayasaka for helpful discussions and kind support to this work. Also, the author acknowledges Mr. I. Nagahama, Mr. A. Onishi, Dr. T. Kaga, and Mr. N.Noji for valuable technical support to this work.

References

1. ITRS Road Map 2005
2. W. D. Meisburger et *J. V.Sci. Technol.* **B9**, 3010(1991).
3. M. Brodsky, S. Halle, V. Jophlin-Gut, L. Liebmann, D. Samuels, G. Crispo, K. Nafisi, V. Ramani, and I. Peterson, *Proc.SPIE, vol.5756*, 51(2005).
4. H. Hayamashi, Y. Yamazaki
5. T. Kitamura, K.Kubota, T.Hasebe, F.Sakai, S. Nakazawa, N.Vohra, M.Yamamoto, and M.Inoue, *Proc.SPIE, vol.5756*, 73(2005).
6. M. Miyoshi, Y. Yamazaki, T. Nagai, I. Nagahama, and K. Okumura, *J. Vac. Sci. Technol.* **B17**, 2799(1999).
7. M. Miyoshi , Y. Yamazaki, I. Nagahama, A. Onishi, and K. Okumura, *J. Vac. Sci. Technol.* **B19**, 2852(2001).
8. Y. Yamazaki, I. Nagahama, and A. Onishi, *Proc. SPIE, vol.5041*, 212(2003).
9. Y. Yamazaki, N. Noji, M. Miyoshi, and K. Okumura, *Nikkei Microdevices*, Apr, 95(2003).
10. T. Satake N. Noji, T.Murakami, M.Tsujimura, I. Nagahama, A. Onishi, and Y. Yamazaki, *Proc. SPIE, vol.5375*, 1125(2004).
11. I. Nagahama, A. Onishi, Y.Yamazaki, T.Satake, and N.Noji, *Proc. SPIE, vol.5375*, 921(2004).

12. T. Satake N. Noji, T.Murakami, M.Tsujimura, I. Nagahama, Y. Yamazaki, and A. Onishi, *Proc. SPIE, vol.5752*,1219(2005).
13. A. Onishi, I. Nagahama, and Y.Yamazaki, *Proc. SPIE*, (2006).

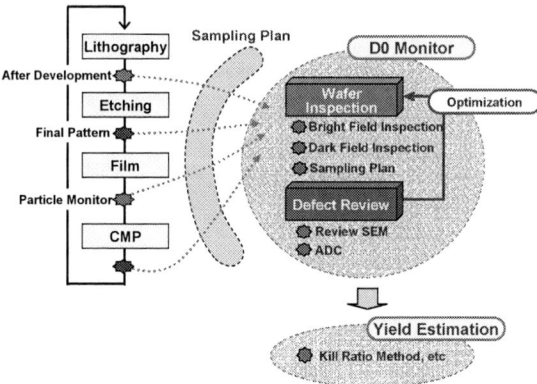

Figure 1 In-line yield estimation and yield control procedure using bright field and dark field optical inspection technologies.   Based on the defect density obtained by the wafer defect inspection,

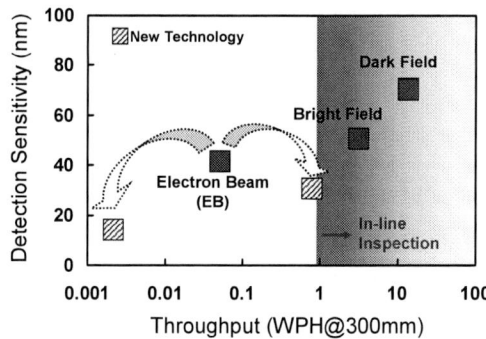

Figure 2 Performance chart of the in-line inspection technologies. Dark field and bright field optical inspection technology are widely used for in-line yield monitoring.

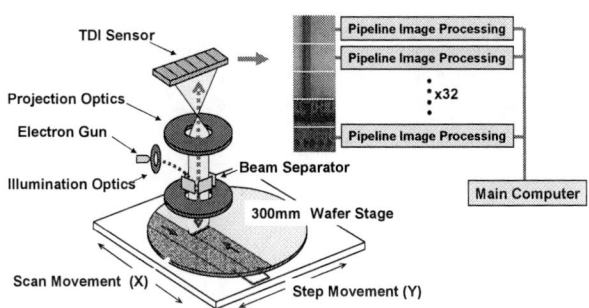

Figure 3 System configuration of the prototype of EBI-PEM which consists of 300mm wafer stage, 32 pipe-line image processing units

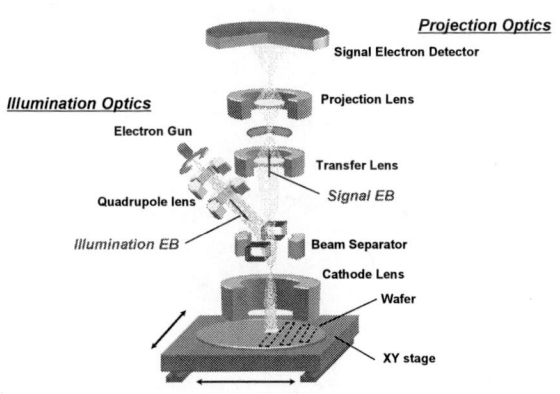

Figure 4 Optics configuration of the prototype EBI-PEM. System consists of illumination optics, bema separator, illumination optics and electron beam image sensor.

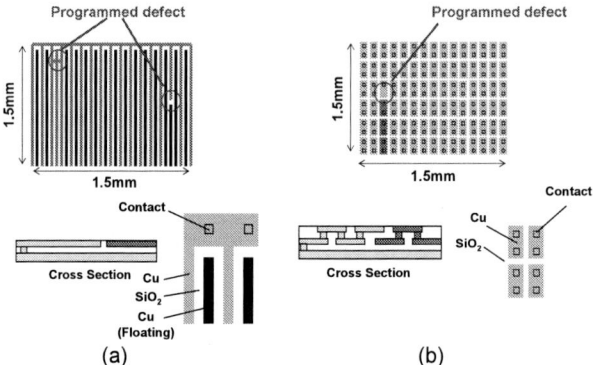

Figure 5 TEG structures for the voltage contrast (VC) defect detection. (a) TEG structure for open and short defect detection. Floating electrode and grounded electrode are provided periodically. By scanning the electron beam at the edge of the pattern, open and short defect can be detected using voltage contrast abnormally. (b) TEG structure for contact open defect detection.

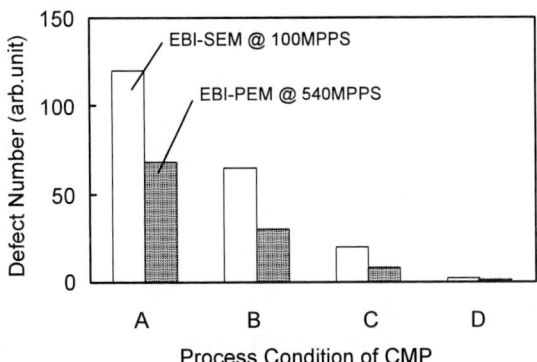

Figure 6 Inspection performances of VC defect detection with the TEG for 65 node logic device. Kinetic energy of incident electron, data rate, pixel size, and current density are 0.6keV, 540MPPS, 50nm, and 15mA/cm^2, respectively. At the several CMP process conditions, the defect number detected by EBI-PEM is proportional to the number of the EBI-SEM. Good correlation of R^2=0.99 was achieved with more than 5 times high throughput compared with the EBI-SEM.

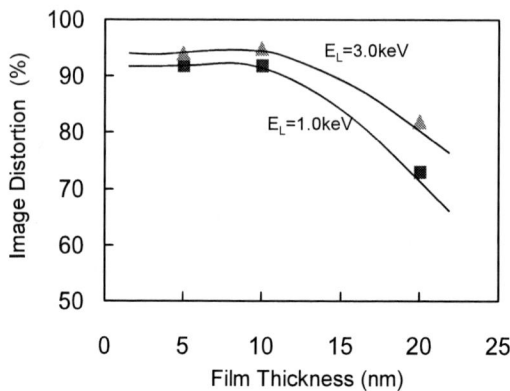

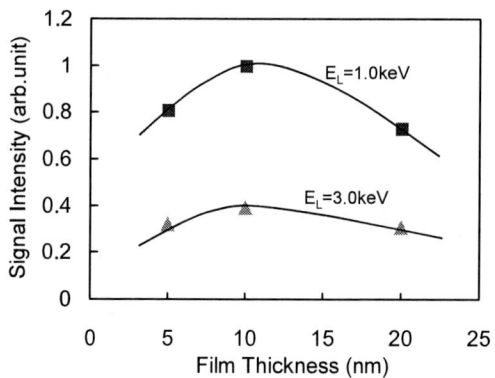

Figure 7(a) Image distortion of the TEG pattern with respect to the SiO_2 film thickness. (b) Signal intensity of the TEG with respect to SiO_2 film thickness.

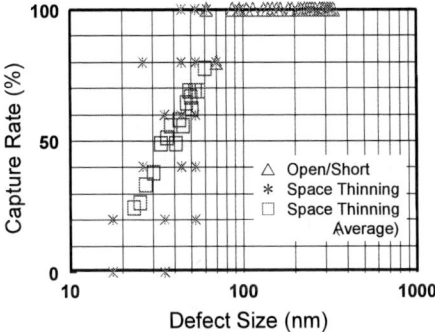

Figure 8 Inspection performances of EBI-PEM for programmed defects on SiO$_2$ thin film pattern at the beam incident energy of 1kV, date rate of 40MPPS.

TABLE I. Requirements for in-line inspection for yield monitoring and systematic defect control. .

	In-line yield monitoring	Lithography induced defects control	Electric conductivity defects control
Kind of defect	Particle induced defect (random)	Pattern defect (systematic)	Electrical failure (systematic)
Min. Sensitivity	More than DR	1/2 DR	Open & short failure
Throughput	1WPH		1WPH
Use Case	Volume production	Development Volume production	Development Volume production
Inspection Area	Full wafer	Several dies	Full wafer

TABLE II. Specifications of EBI-PEM prototype

	Performance
Minimum Sensitivity (nm)	30 nm
Date Rate	600 MPPS
TDI size	2048 x 512 pixel
TDI tap	32 taps
Throughput (WPH) at 300mm wafer	1WPH@100nm pixel size
Inspection Algorithm	Die to Die, Cell to Cell, Die to Any Die

ECS Transactions, 2 (2) 433-452 (2006)
10.1149/1.2195679, copyright The Electrochemical Society

Scenario for a Yield Model Based on Reliable Defect Density Data and Linked to Advanced Process Control

A. Nutsch, R. Oechsner

Fraunhofer Institute Integrated Systems and Device Technology, Schottkystrasse 10, 91058 Erlangen, Germany

Corresponding author contact: andreas.nutsch@iisb.fraunhofer.de, +49 9131 761 115

Currently, yield enhancement and APC (advanced process control) systems operate independently within the manufacturing process. Yield enhancement has several requirements such as the root cause analysis of yield loss, the reliability of defect density data, and the demand for a link between yield models and APC. For root cause studies chemical analysis uses different preparation techniques to describe the nature and location of contamination or defect densities, e.g. copper bulk contamination or wafer edge effects. The reliability of defect density data and the matching of defect inspection tools improve by the reduction of noise. In combination with a process model this is the prerequisite to set up a random defect-based yield scenario which describes single processes. The combination of this scenario with methods for data acquisition and tracking from APC enhances the prediction of yield.

I. Introduction

The continuous reduction of critical dimensions for micro and nano electronics is driving the technology development of yield-relevant contamination and process control, as well as the defect inspection for ULSI (ultra large scale integrated) devices of the latest technology generations e.g. 65 nm or beyond. Leading-edge manufacturing uses APC (advanced process control) environments for reduction of process variations and improvement of yield. Currently, yield enhancement and APC systems operate independently within the manufacturing process. Yield enhancement is based on defect density engineering whereas APC requires control algorithms for processes. Therefore, contamination control, defect inspection, and process models are the base for high-yielding manufacturing. Yield enhancement for semiconductor manufacturing comprises tools for the root cause analysis of yield loss and for improvement of the reliability of defect density data. Potential obstacles are: a) the impact of the edge and backside of the wafers on processes; b) the noise in view of detection of defects, and c) the challenge to link yield modeling and APC. A scenario based on yield models and reliable defect density data combined with methods for data acquisition and tracking from APC environment enhances the prediction of yield.

Yield loss due to random mechanisms can be caused by particulate contamination, process cross contamination due to wafers, process variations, or faults caused by the processes and materials such as e.g. pinholes, short cuts, and open circuits. Models associate this defect density and with the corresponding yield loss. Therefore, methodologies are required to study the sources and impact of defect densities e.g. wafer edge and backside contamination and defect densities, bulk contamination, and the change of wafer shape

433

through polishing processes. Models describing such impacts of processes on the wafer are the base for a reliable combination with APC.

Yield management systems in combination with APC require consistent and comparable defect density data. Therefore, the reliability and the matching of defect inspection tools of one vendor or different vendors are fundamental. The transfer of defect densities of single tools to an in-line tool yield in combination with APC enhances yield prediction for single wafers during the manufacturing process.

This work summarizes and proposes new methodologies for contamination studies which are able to distinguish between bulk and back side contamination. Root cause analysis like, for example, the impact of wafer back side and the impact of edges on processes is necessary for both, yield enhancement and for APC. Furthermore, APC requires modeling of single processes for appropriate control and adjustment of the parameters. The reliability of defect densities data was studied in detail. Beyond this, the results were used to develop a methodology to match defect inspection tools. Reliable defect density is required for input to a random yield model. This yield model describes components of single process yield and can be combined with APC applications. The single process yield is the base to calculate and predict the overall line yield.

II. Impact of wafer back side and edge on processes

Yield enhancement requires a reliable root cause analysis which is capable to determine the nature and location of contamination. In-line defect density control and process control use test wafers and product wafers. These wafers as well as silicon wafer starting materials require excellent topography of the surface and lowest possible defect densities in terms of contamination, flatness and surface roughness. This continuous quality improvement of the wafer surface is demanded by the smaller and smaller feature sizes of the technology generations below 65 nm. The number of polishing processes increases with each technology generation. Defect and contamination reduction as well as flatness improvement are essential, especially at the wafer edge.

Wafers run through a sequence of processes e.g. deposition, lithography, etching, deposition, planarization, etc. during the manufacturing of ULSI devices. Since batch processing is widely used and wafers pass through handlers and storages, cross contamination due to surface edge and back side is crucial. The wafer edge, bevel and the wafer backside are often unaccounted during standard inspection and analysis. Routinely, defect detection or analytical methodology controls the front side of a wafer. All other areas of the wafer are disregarded due to standard procedures until yield loss or contamination issues occur.

For example particle type contamination at the back side might reduce the yield as shown in the literature (1). Additionally, particulate contamination from polishing processes and flakes from the wafer edge have a high potential to cause yield loss during ULSI manufacturing. Another source for contamination is the bulk of the wafer. For example, copper once completely cleaned from the wafer surface might have diffused at raised temperatures or even at room temperature from the bulk to the wafer surface. This causes precipitates resulting in surface defect densities. This contamination also increases the risk of cross contamination.

A methodology to study nature and location of contamination as well as a suitable detection methodology for copper bulk contamination is presented in the following sections. The importance of the wafer edge for processes was emphasized by adding the specification for inspection tools in the recent edition of the ITRS 2005 (2).

CMP (chemical mechanical planarization/polishing) is considered to be crucial during the manufacturing of integrated circuits both in terms of the defect density and concerning the topography on the surfaces. Polishing processes are very important for the wafer geometry at the wafer edge. A correlation between process, consumables and wafer shape is shown. Such correlations are important in view of APC. They are the base for the success of a model-based APC application.

Determination of nature and location of contamination

For yield loss root cause analysis, it is mandatory to identify the nature as well as the location of contamination. Chemical trace analysis results in information on the nature of contamination. For wafer preparation different tools e.g. VPD (vapor phase decomposition), PEM (pack extraction method), and etching are available (32-35). The methods listed above study different areas of the wafer, e.g. the surface, the whole wafer or thin surface layers, respectively. Additionally, thin layers can be dissolved by usage of suitable chemicals to obtain depth information. Therefore, it is possible to use these chemical analysis steps to study different areas of the wafer. If the analysis steps are applied in the right orders it is possible to distinguish between e.g. front side surface contamination and edge and back side contamination.

An example to study the nature and location of contamination for bare silicon wafers is given below by application of three different methods for contamination analysis. VPD and PEM methods extract impurities in the following ways:

1) On surfaces of the silicon wafers: The wafers are put into diluted HCl (mixture HCl : $H_2O = 1:100$) within a sealed polypropylene (PP)-bag (HCl-PEM).
2) In the native oxide of the silicon wafers: The wafers are put into HF vapor dissolving the native oxide. Contaminants on the surface are collected by using H_2O droplets (VPD).
3) In the bulk the silicon wafer: The wafers are etched with a mixture of HNO_3 : HF : $H_2O = 1:1:100$ in a sealed PP-bag (HF/HNO_3-PEM).

The obtained liquids are analyzed by AAS (atomic absorption spectroscopy) or ICP-MS (inductively coupled plasma mass spectroscopy). Additionally TOF (time of flight) and TXRF (total reflection x-ray fluorescence) can be used. These methods give a depth profile of impurities. The TOF SIMS method is surface-sensitive, analyzing only the first monolayers of the sample surface. The TXRF determines surface trace contamination, enabling the measurement on all kind of surfaces e.g. polished, etched, ground or lapped. Measurement of thin layers such as oxide or alumina is possible, too. However this method loses sensitivity with increasing surface roughness (3).

The TXRF uses an X-ray beam with an incident angle smaller than the critical angle of the silicon surface which excites the contamination within the surface to fluorescence. Increasing the incident angle of the beam above the critical angle resulted in an increased penetration depth of the X-rays. The measured peaks of the elements as a function of the incident angle enable very sensitive depth profiling of sub surface contamination or shallow implanted layers (4, 5). An example for the study of back side contamination using the above mentioned methodology is given in (6).

Detection of copper bulk contamination

Copper bulk contamination causes, with a time delay, surface defect densities and degrades electrical properties of the silicon, e.g. lifetime and mobility of minority carriers. Copper contamination is observed for example after the storage of silicon wafers. After a longer storage period, the defect densities or increased haze values might be detected. Surface defect densities or haze are related to copper precipitates on the silicon wafer surface (28, 29). These storage phenomena can be explained as copper diffuses at room temperature within the silicon bulk. Cleaning changes the boundary conditions of the diffusion processes. Therefore, this diffusion is enhanced due to high differences in concentration of clean surface and of the bulk. The clean surface acts as an infinite sink for copper.

The methodology discussed in the previous section is not reliable for detection of copper bulk contamination. For detection of crucial copper bulk contamination out diffusion supported by temperature can be used. The cleaning changes the boundary conditions of the diffusion processes. As a result high differences in concentration of clean surface and of the bulk are achieved. Therefore, this enhances the out-diffusion due to temperature. The clean surface acts as an infinite sink for copper.

The following steps are used to perform the copper out-diffusion:
a) Firstly, the initial surface contamination of silicon wafers regarding copper is determined. This step cleans also the front side of the wafer.
b) Secondly, the silicon wafer is exposed to temperature.
c) Thirdly, the copper contamination is determined again on the wafer front side.

Step b) is crucial for the reliability of this methodology. The temperature and the exposure time used for this diffusion process have to be determined. Therefore, this study needs a model. The model assumes that the copper concentration is constant within the bulk of the wafer. The concentration is dependant on the process time t, the wafer thickness W and the temperature dependence T of the diffusion coefficient D. Using the assumption that the Cu concentration is lowered by 4.65×10^{-5}, the following simplified expression

$$\frac{D(T) \cdot t}{4 \cdot W^2} = 1$$ [1]

can be used for calculation of the required time and temperature. For example, wafers with a diameter of 300 mm have a thickness of 775 μm. D is given by the relation

$$D(T) = 3 \cdot 10^{-4} \cdot e^{\frac{-0.18 eV}{kT}} \left[\frac{cm^2}{s} \right]$$ [2]

This model does not consider the dependence of copper diffusion on dopants. As the solubility at temperatures between 200 and 400 °C is negligible copper diffused to the surface is not expected to diffuse back into the bulk. The required time at a specific temperature can be determined from figure 1. At 300 °C, e.g., the process time should be approx. 45 min.

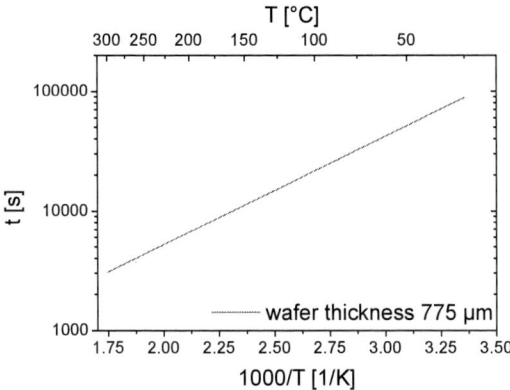

Figure 1: Logarithmic plot of time versus reciprocal temperature for out-diffusion of copper. The assumptions for this plot are to lower the bulk copper concentration by 4.65×10^{-5} in the center of the 775 μm thick 300 mm wafer.

The reliability of the presented methodology was proven experimentally with silicon wafers. These silicon wafers had a diameter of 300 mm and were boron-doped with an epitaxial p layer of approx. 2 μm thickness. The wafers had an initial copper contamination below detection limit of 0.01×10^{10} atoms/cm². After temperature exposure of 45 minutes at 300 °C the copper surface contamination was 0.3×10^{10} atom/cm². For stored wafers with an initial copper contamination of 0.3×10^{10} atoms/cm², the post temperature treatment copper values were found to be 200×10^{10} atoms/cm². The results show impressively that the methodology can be used to study copper contamination of the bulk of silicon wafers. For example abrasive polishing processes for silicon wafers are identified as possible sources of copper contamination due to the ionic transfer of Cu into the bulk silicon. Subsurface damage and temperature assist the transfer process and the diffusion process.

Inspection and characterization of the wafer edge

The methodologies presented above allow studying the nature and location of contamination mainly on bare silicon wafers. Within the semiconductor manufacturing process, the defect inspection is usually used for yield and process control. The continuous control guarantees a good interaction and compatibility of the processes. The inspection of bevel and wafer edge on multilayer product wafer becomes a big challenge as more and more defect/process problems have their origin in those areas of the wafer. Defect inspection concepts or technologies are under development or have to be realized within the next years. Currently, it is a key challenge to find a method for the root cause inspection of wafer edge and backside (30). Important criteria beside coverage of all areas, sensitivity and speed are ADC (automated defect classification and characterization) and optical review capability on the tool as well as a standard result file allowing SEM (secondary electron microscope) review.

Impact of polishing processes on the wafer edge

Beside the defect densities, there is the wafer geometry and shape which impacts subsequent processes. Especially, polishing processes modify the properties of wafer geometry at the edge. A very sensitive monitor for the study of polishing processes and their impact on the surface and wafer edge is the bare silicon wafer. Therefore, silicon wafers from the manufacturing process were studied. During these processes, either polishing on both sides or on one side is applied. The double-side polishing process is a batch process for 15 silicon wafers with a diameter of 300 mm at once. The subsequent single polishing process is for single wafers. This process has a lower silicon removal compared to the double side polishing process. The purpose of these two to three polishing process steps is to reduce subsurface damage. Additionally, the polishing improves the flatness due to planarization and the roughness by two to three orders of magnitude (9). During manufacturing of integrated circuits the single-side polishing is used basically for planarization.

The above-mentioned silicon polishing processes were studied with highly sensitive surface inspection and height profiling. For determination of haze maps dark field narrow detectors of a defect inspection tool were used. The haze maps of bare silicon wafers after a double-side abrasive polishing process and an intermediate abrasive polishing process were studied. It was observed that the haze has been usually higher in the wafer center and lower at the wafer edge (see figure 2). At the same time the shape of the wafer changed. This was indicated by the total thickness variation of the wafer. High precision measurement of geometry at the edge of the wafer showing an increase of site flatness substantiated these results. Between radiuses of 110 mm to 150 mm the wafer thickness decreases by several 100 nm.

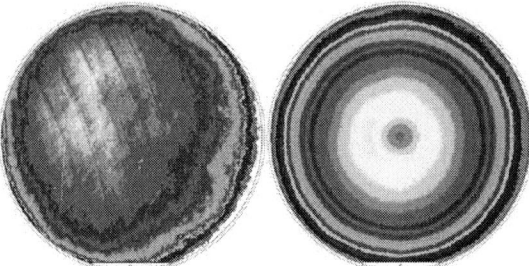

Figure 2:   left side: The haze map of a double side polished wafer is shown. The haze range is from 0.6 at the wafer edge to 1.1 ppm in the center of the wafer.
right side: The haze map of a single side polished silicon wafer is shown. The haze range is from 2.5 ppm at the edge to 5 ppm in the center of the wafer.

The data presented above were explained by studying the slurry, the corresponding removal rate and emerging haze in detail. The processes used colloidal silica slurries which were pH stabilized in the alkaline range (10). The stabilization avoids dehydration of the slurry particle surface and therefore prevented a subsequent agglomeration. The size of the SiO$_2$ particles was in the range of 15 nm to 55 nm. When silicon is polished the colloidal silica particle rolls onto the surface and gets larger and more aqueous as the outer shell is taking up oxide and water (11). The subsequent oxidation of the surface is due to the interaction with atmospheric oxygen.

The removal rate can be described in the following way. During the polishing with colloidal silica the silicon surface is treated in two ways. One mechanism on the surface is the oxidation through the environmental oxygen and the subsequent removal of the oxide with the slurry particles. This mechanism can be described using a modified Langmuir Hinshelwood model (12, 13). The other mechanism is the etching of the silicon surface due the alkaline stabilization of the slurry. Therefore the removal rate has two components, a chemical portion and a mechanical portion. The chemical portion of the removal rate is mainly influenced by pH value and temperature, whereas the mechanical portion of the removal rate is influenced by the polishing pressure and efficiency of the slurry particles. Increasing the temperature enhances the chemical processes within the polishing process. The etching increases the removal rate and shortens process time. Increasing the pressure on the other hand enhances the mechanical component of the polishing process. Therefore increased pressure and/or temperature increase the removal rate. Commonly, the removal rate non-uniformity is caused by pressure differences between center and edge of the wafer.

The haze is correlated to the removal rate of the slurry in combination with the pad in a way that lower removal rates result in lowest haze of the surface. This was experimentally proven by testing of different final polishing slurries.

In the above described observation during the polishing processes the haze in the center was higher than at the edge. This indicated that the chemical etching process was enhanced in the wafer center due to higher temperature. The raised temperature is due to the frictional forces between pad and wafer. This result would suggest that the wafer is getting thinner in the center of the wafer but the experimental observation showed the opposite. Assuming that the slurry particles get bigger, softer and contain more water at the surface, the removal rate is expected to decrease with the time of a slurry particle remaining on the wafer surface. This would explain the observed phenomena of higher haze in the center of the wafer and higher removal rate at the edge of the wafer. At the edge of the wafers, the slurry particles get in contact with a high mechanical component capable of removing silicon dioxide very efficiently. The loss in the efficiency of slurry particles is reduced when the slurry is supplied through the polishing pad as it is done in the double-side polishing process. Therefore, the differences of the haze non-uniformity between double-side polishing and single-side polishing can be explained. Additionally, the chemical component of the alkaline slurry causes etching of the wafer surface in the wafer center enhanced by the higher temperature in the center of the wafer. The etching non-isotropy of the stabilization causes the surface roughness.

III. Defect Detection

In addition to the root cause analysis, yield enhancement requires the reliable detection of defect densities on wafer surfaces. Furthermore, the consistence of defect density data within semiconductor manufacturing sites and therefore, the matching of defect inspection systems are mandatory. The impact of noise on the performance and the matching of systems for inspection of non-patterned wafers are studied in detail. An advanced methodology for the study of defect inspection at 50 nm defect size for leading-edge technologies is described. The methodology was approved through experimental results achieved on defect inspection systems designed for bare silicon wafers. This methodology is applicable for non-patterned and patterned defect inspection.

SSIS (scanning surface inspection systems) are capable to detect light scatter events on the wafer surface. These light scatter events are classified as different defect types by the system software. This chapter will focus on two types of defects that cause surface defect densities on non-patterned silicon wafers. These two types are particles and COP (crystal originated point defects). Particles are solids and agglomerations of substances such as silicon compounds, organics, or metal compounds lying on the wafer surface. Additionally, the formation of metal precipitates on the wafer surface might occur. In contrast, COPs are small holes in the wafer surface (14). The origin of the COPs is the silicon crystal (15, 16). The separation of COPs and particles requires the detection of characteristic light scatter distributions. COPs are usually detected by the light scatter intensity measured with a high angle position of the detector. The light scatter intensity at oblique angles is much weaker (17). Therefore, defect inspection systems are equipped with multiple detector design. The reliable classification of COPs and particles is mandatory for the characterization of polishing and cleaning processes. The etching of the wafer surface, either during polishing or cleaning, dishes these COPs. Very small COPs are important in view of highly sensitive defect inspection and reliable process control.

<u>Study of defect density data</u>

The defect inspection on non-patterned wafers was studied in detail by tracking single defects through a sequence of scans. For this purpose, an algorithm for the comparison of defect maps was designed. The defect maps are obtained from a sequence of 50 or 100 measurements. This is to obtain statistical significant data. During the comparison of defect maps through the sequence of scans single wafers are tracked regarding size and coordinates. Figure 3 shows the scheme for the algorithm.

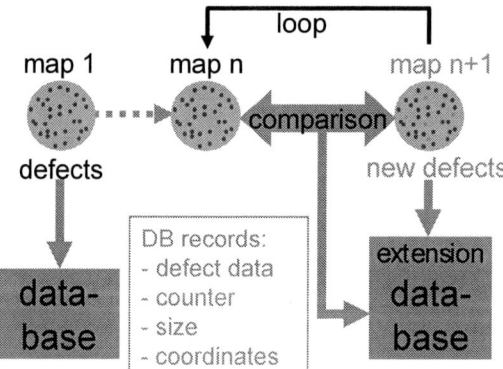

Figure 3: Scheme of an algorithm for evaluation of multiple scans of a defect inspection system on a silicon wafer.

The algorithm used a defined search radius (e.g., 100 or 300 μm) for the recovery of defects on the follow-up maps. Size and coordinates of each defect were recorded in a database. This database was extended by new defects subsequently. As a result, the data base enables the calculation of capture rates, sizing accuracy, and coordinate accuracy.

Experiments were performed using the SSIS for non-patterned wafers with a diameter of 300 mm from different vendors. A laser beam scans the surface of a spinning wafer. The wafer surface defect densities are measured in terms of the intensity of scattered light detected by photomultipliers. Tool A and C had multiple detectors and tool B had two detectors. Tool A and B, installed at different facilities, were compared for matching. These systems had comparable thresholds of 50 nm defect size. Tool C had a threshold of 40 nm defect size and was used for principle studies of defect capture.

For the tools A, B, and C, this so-called 'super map' showed a standard deviation for sizing of detected defects better than 2% and for coordinates better than 30 μm. The capture rate for single defects was shown to be as high as 100%. The capture rate decreased at the sensitivity limit of the system. Decreasing the sensitivity limit shifted the cut-off of the capture rate to smaller sizes and vice versa (see figure 4).

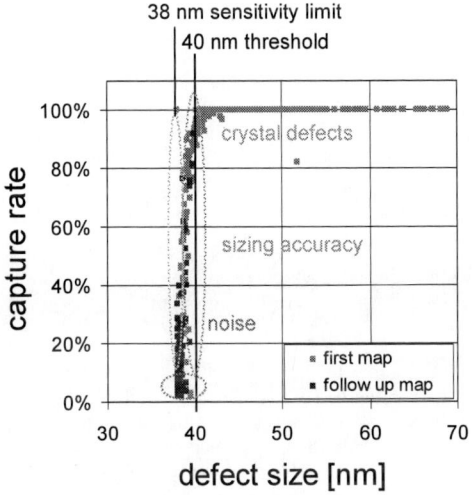

Figure 4: Capture rate versus defect size for multiple defect inspection of a bare silicon wafer with a diameter of 300 mm. The tool detected 238 defects larger than 40 nm, and the sensitivity limit was 38 nm.

Additionally, the capture rate for COP was studied. The capture rate for COPs above the sensitivity limit was 100%. The capture rate for COP at the sensitivity limit decreased more significantly than for particles. This was observed for both tools, A and B. The standard deviation for sizing of detected COPs was found to be 3 %.

The significant decrease of the capture rate at the sensitivity limit for defect detection is due to: a) sizing accuracy, b) noise due to the detection units, c) micro surface roughness, and d) crystal defects. The impact of sizing accuracy is observed through the cut-off of the capture rate at the sensitivity limit. Noise is indicated by capture rates below 10%. In principle, it is impossible to differentiate between noise and micro roughness of the wafer surface. COPs are physically very small but classified as bigger particles with a low capture rate. The obviously stronger decrease in the capture rate at the sensitivity limit was explained by less sensitive detection of light scatters perpendicular to the wafer surface, e.g. due to higher noise (18). The lower sizing accuracy also reduced the capture rate for crystal defects. In summary, tool A had a higher sensitivity to COPs compared to tool B.

Matching of defect inspection tools using defect maps:

Reliable tool matching requires several prerequisites e.g. identical system specifications such as: a) precise size calibration, b) equal adjustment of sensitivity limits and thresholds and c) high capture rates for defects.

The previously described algorithm enables the calculation of reliable defect density data. This data enables a map-based matching for defect inspection tools. This map-based matching shows the following advantages: Firstly, noise data included in the measurement is suppressed using the above described 'super maps'. Secondly, the use of coordinates enabled matching of different classified defects, e.g., COPs, and particles and also differences in sizing information.

To generate reliable maps, the above described algorithm applied to multiple scans on a wafer sample was used. This algorithm evaluates the coordinate and sizing information of defect maps regarding the reproducibility of the scans. This information was used to calculate a 'super map' used for the matching of defect inspection tools. During matching, the coordinates and the sizing information of defect maps were compared. A coefficient for matching of defect inspection systems can be calculated from the comparison of defect maps using equation 1.

$$P = \frac{N_{matched}}{(N_{Matched} + \Delta N_{map1} + \Delta N_{map2})} \qquad [3]$$

In this equation, P is the matching coefficient, $N_{matched}$ the number of defects detected on both tools, N_{map1} the number of defects detected only by tool 1, N_{map2} the number of defects detected only by tool 2. Tool-to-tool matching was applied by using the following methodology. The evaluation algorithm, described above, applied to the data of e.g. 50 scans on the same wafer, generates a 'super map' with high data reliability. The 'super maps' generated with each tool from the same samples were compared.

For the experiments the wafer samples were measured at tool B. The wafer samples were transferred to tool A in another facility and measured again. Contamination due to transport was negligible. The experiments were performed with wafers of different qualities e.g. 'high COP' and 'low COP' silicon wafers. The map to map repeatability of the defect inspecion for tool A was determined as 93% and significantly higher compared to tool B showing a repeatability of 82%. The match between tool A and B was found to be 84% (see figure 5). Tool B showed an increased noise at the sensitivity limit, identified as false defects. They were effectively eliminated by the described methodology according to which defects with a capture rate below 10% were classified as noise. Tool A showed a higher sensitivity for detection of COPs than tool B (see figure 6). Concerning the comparison of both tools, all light scatters of tool B matched those detected by tool A. Even particles classified by tool B and not identified by tool A matched as soon as the COP data was displayed.

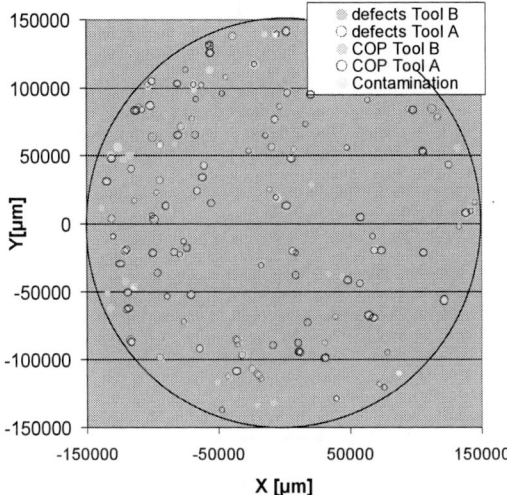

Figure 5: Overlay of defect maps at 60 nm of two different defect inspection tools measuring the same wafer. The diameters of the circles represent the defect size. The filled and open circles are the defects detected by tool B and tool A, respectively. The match between the tools was found to be 84 %.

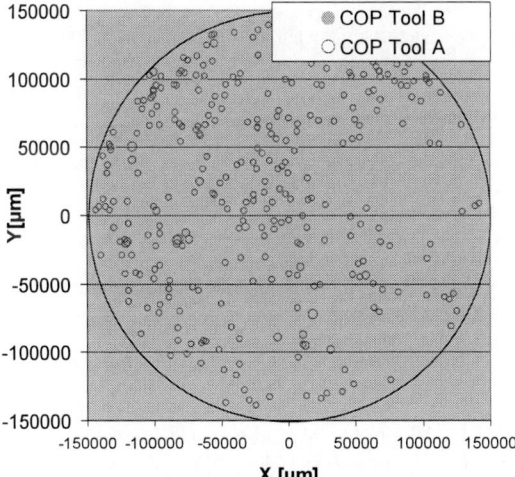

Figure 6: Overlay of defect maps for COP from defect inspection tool A and B. The overlay shows that tool A displays more defect data than tool B. This is explained by the higher sensitivity for COPs of tool A.

The separation of particles and COPs is crucial for the matching and requires the detection of characteristic light scatter distributions. COPs are usually detected by the light scatter intensity measured with a high angle position of the detector. The light scatter intensity at oblique angles is much weaker. It has been proven that tool A with multiple detectors is more effective for defect inspection than tool B with only two detectors. Additionally, tool A is more sensitive and therefore more reliable for COP detection.

Binomial model for defect detection

Regarding defect inspection expenses it is useful to identify a statistical model for description of the defect inspection. This reduces the number of measurements for example in the case that the number of defects could be used instead of repeating the experiment for calculation of standard deviations. A binomial distribution described by the following formulas

$$B(n,x) = \binom{n}{x} \cdot p^x \cdot (1-p)^{n-x} \qquad [4]$$

$$p = \frac{1}{2}\left(1 \pm \sqrt{1 - 4 \cdot \frac{\sigma^2}{\mu}}\right) \qquad [5]$$

$$\sigma^2 = n \cdot p \cdot (1-p) \qquad [6]$$

$$\mu = n \cdot p \qquad [7]$$

could be used to describe the defect inspection. In these equations, σ is the variance, μ the expected number of defects, n the number of defects, x the expected number of detected defects and p the capture rate. For scanning the wafer surface, the scan path starts in the center of the wafer and spirals out. Along the scan path, the probability of detecting defects is usually very high. Repeating scans on one wafer was expected to show a binomial frequency plot versus defect counts. This was experimentally proven by several tools. Equation 2 was used for evaluation of results. Tool A showed capture rates of 92 % for defects at 50 nm. Tool B showed capture rates of 83 % for defects at 50 nm. Tool C had the best sensitivity showing capture rates of 92 % for defects at 40 nm. The result for tool C is shown in figure 7. These experiments demonstrated the validity of the binomial model for defect detection.

The binomial distribution was found to describe the defect inspection. As shown previously the repeatability is strongly influenced by the sizing accuracy and noise signals. They are always present at the detection threshold and they do not vanish at adjusted lower sensitivities. The probabilities for defect counts due to noise and sizing accuracy are described with Gaussian distributions. Therefore, it is obvious that the results of a defect inspection system have to be corrected for missed defect counts due to sizing accuracy or additional counts due to noise. The frequency plot of the defect counts is a useful and fast tool for characterization and monitoring of defect inspection systems without using the methodology of defect tracing. Particle counts above the maximum number of particles are a very sensitive indicator for noise when observed in the binomial frequency plot versus defect counts. This is due to the Gaussian distribution of noise caused by the detection units or by surface micro roughness. Additionally, the sizing accuracy at the detection threshold causes differences in the defect counts for multiple scans, resulting in a lower capture rate when calculated only with the binomial distribution.

The binomial model described above was proposed as a SEMI standard (31) for particle per wafer pass (PWP) testing. The model enables the calculation of the real number of particles and the variance from only one single scan. This simplifies evaluation and comparison of data as results are statistically proved. Furthermore, the methodology enables cost-efficient defect inspection for semiconductor manufacturing as the number of repeated scans can be reduced.

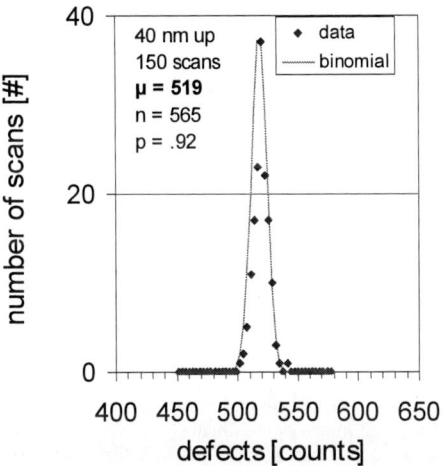

Figure 7: Frequency plot of number of scans with tool C displaying a number of defects. The data is in good agreement with a binomial distribution.

IV. Yield Models and Process Control

Models for yield improvement and cost optimization of semiconductor manufacturing processes have been developed and discussed over the last 40 years (19). This part of the paper intends to show a scenario for a single process yield model. The model requires reliable data and information from root cause analysis to predict the yield for single wafers. The wafer sort yield of a semiconductor manufacturing site is calculated from two components the random yield and, the systematic yield. The focus here will be on random yield.

Yield loss due to random mechanisms can be ascribed for example to particulate contamination or process induced defects. Yield models correlate these defect densities with random yield. Particulate contamination can be expressed in so-called defect budgets of semiconductor manufacturing equipment. The determination of particulate defect budgets can be easily done by determination of PWP (particle per wafer pass) values for single equipment tools. But it is difficult to scale the defect budget information of single processes to a common defect size as every defect type scales according to specific size distributions. Furthermore, the transfer of this information between different manufacturing sites or technology generations is difficult, too. In order to obtain the yield impact of the defect budgets, it is necessary to determine the according kill ratios.

Semiconductor manufacturing equipment causes process induced defect densities. Therefore, process models enable the adjustment of parameters according to measured process variations or defect densities. Based on the availability of process induced defect densities the calculation of yield prediction from yield models is possible from current and historic defect density data. Furthermore, this information in combination with APC enhances the prediction and the improvement of yield as APC allows forward adjustment of process parameters.

Yield models

The development and application of yield models is described within the overview of Cunningham (20). The usage of the yield models depends strongly on the manufacturing process. The Poisson model is applicable for smaller chips and defects with random spatial distribution at low propabilities. The negative binomial model is valid for large area integrated circuits and the occurrence of spatially clustered defects. The negative binomial yield correlation was calculated by compounding the Poisson probability distribution function with a gamma distribution for spatial clustering of defects (21). The negative binomial yield is described by the following equation:

$$Y(D) = \left(1 + \frac{A \cdot D}{\alpha}\right)^{-\alpha}$$

[8]

In this expression α is a clustering parameter, A the chip area and D the defect density. The negative binomial model approximates to the other commonly used yield models by adjustment of the cluster parameter (22). Therefore, the usage of this model has the best potential to obtain a holistic yield model. To obtain, for example, the Seed's yield expression, the clustering paramenter is set to one (23). A cluster parameter of one describes a high degree of defect clustering due to exponential spatial defect distribution. Negligible clustering of defects would be expressed by a large clustering parameter resulting in an approximation to Poisson's spatial defect distribution. Therefore, the distributions are adjusted according to the requirements of the desired application. A yield model requires different input data to enable the calculation of the yield. These data are the critical chip area and the killer defect density which has to be correlated to the chip area, to the defect density and yield loss. The knowledge of this data enables e.g. the determination of the cluster parameter, which depends on the application in terms of the product or technology. On the other hand, the data can be used to calculate and predict yield when the cluster parameter was determined from production data and the defect density is measured in-line. To obtain highly integrated circuits a sequence of processes is necessary. Each of these processes has a yield portion. Calculating the product sum of this yield per process results in the line yield:

$$Y_{line} = \prod_{i=1}^{k} \left(1 + \frac{A \cdot D_i}{\alpha_i}\right)^{-\alpha_i}$$

[9]

In this expression, A is the critical chip size, D_i the defect density and α_i the cluster parameter for each process level i. k is the total number of process modules. If required each process level consider a specific critical chip size by using instead of A the process level specific area A_i. This expression can be further adjusted o real conditions by introducing different yield impacting defect densities for each processes step. Therefore, the following expression is obtained to describe the yield of a manufacturing line:

$$Y_{line} = \prod_{i=1}^{k} \prod_{j=1}^{n} \left(1 + \frac{A \cdot D_{ij}}{\alpha_{ij}}\right)^{-\alpha_{ij}}$$

[10]

In this expression j describes the impact of different random yield impacts e.g. contamination, particulate contamination or random defects through process variations for each process level. If required each process level consider a specific critical chip size by using instead of A the defect type level specific area A_j. The obtained formula is able to describe the yield at a process level.

Defect budgets of equipment

The defect density of particulate contamination as well as the according cluster factor is determined by checking the PWP of the semiconductor manufacturing equipment with either bare wafers or patterned wafers. The methodologies described above, guarantee a cost-efficient and reliable determination of defect densities. There are two requirements for the defect budget data.

Firstly, the data requires transferability to new technology generations. For scaling the defect density data to the next technology generations, it is suggested to use half of the defect densities of the previous technology generation. Simultaneously, the defect control size is adjusted to the new technology generation.

Secondly, the data requires comparability for tools of a manufacturing site or even of different manufacturing sites. The comparability of the data is guaranteed when it is scaled to a common defect size even if the data have been determined with different methodologies or at different defect sizes. This requires the knowledge of the function which relates defect size and defect density. The defect density as a function of the defect size depends on the process. Therefore, a general expression is used to describe the defect density D as a function of the defect size x for specific process levels

$$D(x) = \sum_{i=1}^{m} \frac{C_i}{x^{n_i}}$$

[11]

where C_i is a coefficient, and n_j is a power of the defect size (24). M is the number of different components required to describe the defect density as a function of the defect size. A more complicated defect distribution using a discrete exponential distribution is given in (25, 26). Assuming that each defect type has a specific scaling function this size distribution would enable to determine the defect mechanisms from defect density data. Additionally, each defect type scales with a specific power n_i the defect density. For example within (27) or (24) it was found that n equals relatively often three. For only one defect type these assumptions would result in the following simplified formula

$$D(x) = \frac{C}{x^3}.$$

[12]

Formula [12] is useful to scale defect density data e.g. to common defect sizes or for comparison of equipment. For example the scaling of defect density data enable the prediction of yield for different technologies from a baseline process.

For example, the calculation of defect density data from defect budgets surveys for the ITRS (International Technology Roadmap for Semiconductors) requires the usage of common defect sizes as well as the transfer to other technology generations. An example for scaling of defect density data is shown in figure 8. The defect density data lines are plotted as a function of the defect size. The defect budgets of equipment are scaled by moving on one of these lines to the according common defect size. The common defect size has to be defined. If it is necessary to transfer data between technology generations, the following procedure is suggested: The defect density is assumed to be half of the previous technology generation. This assumption is useful, for example, for predicting tolerable defect densities for the ITRS technology generations or initial yield models of new technology generations until new defect density data is available. Therefore, this defect density size model can be used to enable the comparison of defect density data within a defect budget survey at different semiconductor manufacturing sites and concerning different technologies.

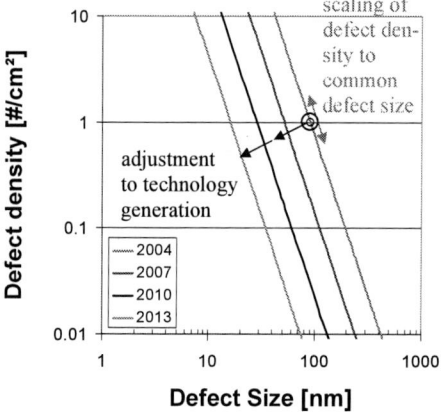

Figure 8: Visualization of the scaling of defect budgets to common defect sizes at for semiconductor manufacturing equipment. The figure visualizes the transfer of defect budgets to future technology generations according to the ITRS.

Example for APC: polishing processes

APC is required to achieve fast yield learning cycles and to reduce costs during semiconductor manufacturing. Process readjustment techniques use e.g. run-to-run, feedforward and feedback control mechanisms (36). Run-to-run control mechanisms determine the process parameters that are dependent on the wafer history and the process model. The control mechanism defines, for example, the process parameters through preceding measurements, e.g. for layer thickness, surface flatness and macroscopic defects for a polishing process. The process equipment, for example, allows modifying the process parameters according to the incoming inspection and a process model or optimizing the process results and reducing the process variability. An example for a process model and related determination of defect density data for polishing of wafers was given above. Thus the process time, pressure uniformity and therefore, the topography and the flatness of the wafer is optimized during the process. On the other hand, control mechanisms adjust the process parameters for the wafers that are being processed. These are, for example the control of the polishing pad thickness or any kind of end-point detection terminating the process. The polishing process is followed by an end-inspection of the wafers. The data of the subsequent inspection could be used for adjustment of post-treatment of the wafers, resulting in a reduction of defect density. If necessary, a post-treatment of the corresponding wafer is initiated automatically by the APC system. By input check and final check, the APC expert system learns the process behavior. Using this information, the system adjusts the process parameters for the following wafers. An APC system can be trained and therefore learn to handle wafers according to individual wafer status information. By using sensor technology and control algorithms, the semiconductor manufacturing equipment becomes more tolerant regarding input quality of wafers and according processes.

Scenario for combination of yield model and APC system

The yield model described above allows calculating the yield data during the process sequence as soon as the defect density data is available and linked to the wafer in the process. Therefore, the current yield status is available for each wafer for the different process steps. This is shown schematically in figure 9. Historic defect density data or defect budgets allow interpolating the expected yield for a wafer according to formula 9. An yield management system provides this data to the wafers. Furthermore, the APC system in combination with a yield and process model uses advanced control mechanisms to improve the yield of a wafer.

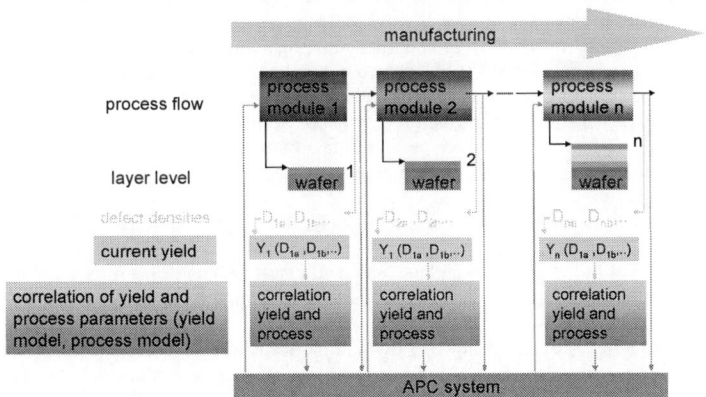

Figure 9: The scheme shows the link between a defect density control and the APC System using according yield models and process models.

The presented scenario requires as a prerequisite single wafer tracking. This enables calculation of the currently expected yield of the wafer. Historical data provided by the control system allows interpolating the expected yield of the wafer. Therefore, it is possible to link a specific yield to each wafer passing through process equipment. This enables monitoring of the single process yield improving simultaneously yield learning cycles. Furthermore, yield impact caused by different defect mechanisms could be separated by the analysis of in-line yield data and by study of the defect density and defect size correlation of each process module. The yield correlation of defect densities caused by specific processes enables the adjustment of APC control mechanisms. It is possible to e.g. adjust process parameters in order to reduce defect densities for the following wafers, enable corrective process adjustment for the subsequent process module, or introduce additional process modules as cleaning. Thus, the yield is improved and yield learning cycles are shortened. Especially, APC has the capability to enhance the flexibility during manufacturing.

V. Conclusion

A scenario to achieve a holistic yield model linked with methods for data acquisition and APC is based on three components: a) On a yield cause root analysis based on chemical analysis methodologies and appropriate process models, b) The generation of reliable defect data through the reduction of noise in the defect inspection, c) A yield model which takes into account the differing impacts of single processes and the therefore allows the improvement of wafer yield during processing.

The combination of sophisticated chemical analysis methodologies such as VPD (vapor phased decomposition) and PEM (pack extraction method) with corresponding analysis allows determining the nature and location of contamination. Especially, the study of the wafer edge and back side is enabled by this methodology. For the study of bulk contamination, the copper diffusion was analyzed. The description of copper diffusion by a model allows detection of crucial copper contamination in the bulk of silicon wafers. This bulk contamination causes, with a time delay, surface defect densities and impacts electrical properties of the silicon, e.g. lifetime and mobility of minority carriers.

The polishing process employing colloidal silica for semiconductor surfaces and materials is widely used. Edge effects like haze and non-uniform removal rate during chemical mechanical polishing are explained by interaction of the used colloidal slurry with the surface. A model for the polishing process using colloidal silica slurry was determined. The presented model is capable to perform a qualitative description of surface roughness and wafer shape. The spatial removal rate on the wafer surface during the polishing process depends on the local temperature and pressure. Additionally, the chemical component and the efficiency of the slurry are important. The slurry particle surface loses efficiency during the contact to the wafer surface. Therefore, at the center of the wafer the slurry efficiency was found to be low and the temperature to be higher. This resulted in a reduction of the removal rate which explains edge effects during polishing of wafers with colloidal silica slurries.

The methodologies of defect tracking by multiple scans allow system evaluation, qualification, and tool-matching using bare silicon wafers in contrast to state-of-the-art methods using wafers contaminated with PSL (polystyrene latex spheres). A reliable matching methodology for tools at differing sites and from different vendors was established. Yield management systems require defect maps displaying the defects, avoiding background noise from detection units or substrates. Noise caused by detection units and wafer surfaces will be a key challenge for detection of defects with a size below 50 nm. Also, noise is essential for tool-to-tool matching either within a semiconductor manufacturing site or between different sites. The studies showed that the tool-matching by comparing defect maps is very reliable.

Models for yield improvement and cost optimization of semiconductor manufacturing processes based on defect densities have been developed for 40 years. The negative binomial yield model was identified as the most common applicable yield model from the various available yield models. The yield model was transferred to a single process level, which allows in combination with methods for data acquisition, single wafer tracking and APC the prediction of yield. Additionally, APC allows feedback from the yield model results to the process chain. The system has the capability to link specific yield estimation to single wafers initiate corrective actions within this scenario.

VI. Acknowledgments

The authors thank L. Pfitzner and P. Pichler from Fraunhofer IISB, I. Thurner from Infineon Technologies AG; M. Ikeno, H. Otha, and F. Supplieth from Hitachi High Technologies; L. Fabry from Wacker - Chemie AG; H. Shimizu from Nihon University for very fruitful discussions and the great support. Results of this paper were partly obtained during joint research projects with Hitachi High Technologies. This work was partly funded by the European Commission within the project SEANET No. 0279280 (Framework Program 6).

References

1. V. Parks and F. Koschinsky, ISSM 2002, Tokyo, Japan, Conf. Proc. p. 475 (2002)
2. ITRS 2005 Edition, to be printed at the beginning of 2006
3. A. Nutsch, V. Erdmann, G. Zielonka, L. Pfitzner, H. Ryssel, Spectrocimica Acta Part B, **56**, pp. 2301-2306, (2001)
4. N. Streckfuss, Erlanger Berichte Mikroelektronik, 3/95, Verlag Shaker (1995)
5. C. Steen, A. Nutsch, P. Pichler, H. Ryssel, Preprint submitted to Elsevier Science 26 October 2005, (2005)
6. A. Nutsch, H. Shimizu, A. Englmüller, L. Fabry, Conf. Proc. of the 2003 IEEE Int. Symp. on Semiconductor Manufacturing ISSM, pp. 229, (2003)
7. H. Bracht, Mat. Sci. in Semiconductor Processing **7**, 113-124 (2004)
8. A. A. Istratov, E. R. Weber, Physics of Copper in Silicon, J. Electrochem. Soc., **149** (1), pp. G21-G30 (2002)
9. L. Pfitzner, E. Bär, J. Frickinger, J. Nguyen, A. Nutsch: Bd. **327**, pp. 136 - 152, ISBN 3-86012- 218-5, (2004)
10. S. H. Li, B. Tredinnick, M. Hoffmann, Chemical Mechanical Polishing in Silicon Processing, Semiconductor and Semimetals Vol. 63, 139, Academic Press (2000)
11. T. Woignier, J Phalippou, J. of Non-Cryst. Solids, 93, pp. 17 (1987)
12. Z. Li, L. Borucki, I. Koshiyama, A. Philipossian, J. Electrochem. Soc., **151**(7), pp. G482-G487 (2004)
13. C.L. Burst, D. G. Thakurta, W. N. Gill, R. J. Gutmann, J. Electrochem. Soc., **149**(2), pp. G 118-G127 (2002)
14. J. Ryuta, E. Morita, T. Tanaka, and Y. Shimanuki, Jpn. J. Appl. Phys., **29**, pp. L1947–L1949, (1990)
15. H. Nishikawa, T. Tanaka, Y. Yanase, M. Hourai, M. Sano, and H. Tsuya, Jpn. J. Appl. Phys., **36**, pp. 6595–6600, (1997)

16. J. Ryuta, E. Morita, T. Tanaka, and Y. Shimanuki, Jpn. J. Appl. Phys., **31**, pp. L293–L295, (1992)
17. W.-P. Lee, H.-K. Yow, T.-Y. Tou, 'IEEE Trans. on Semiconduct. Manufact., **17** (3), pp. 422-431, (2004)
18. H. Altendorfer, G. Kren, C.T. Larson, S.E. Stokowski, Solid State Technol., pp. 93-97, August (1996)
19. B.T. Murphy, Proc. IEEE, **52**, pp. 1537-1545 (1964)
20. J. A. Cunningham, IEEE Trans. on Semiconduct. Manufact., **3** (2), pp. 60-71, May (1990)
21. B. Meister, IBM J. Res. Develop., **27** (6), pp. 545-548, (1983)
22. C.H. Stapper, IEEE Trans. on Semiconduct. Manufact., **4** (4), pp. 294-297, May (1991)
23. R.B. Seeds, IEEE Int. Conv. Rec., **6**, pp. 61-66, Apr. (1967)
24. H.G. Parks, E.A. Burke, IEEE Proc. of Int. Nat'l Manufact. Sci. Symp., pp. 131 – 135, (1989)
25. H. Sato, M. Ikota, A. Sugimoto, H. Masuda, IEEE Trans. on Semiconduct. Manufact., **12** (4), pp. 409 - 417, November (1999)
26. A.V. Ferris-Prahu, IEEE Transactions o Electron. Dev., **ED-32** (9), pp. 1727-1985, Sep 1985,
27. R. Glang, Proc. IEEE 1990 Int. Conf. on Microelec. Test Structure, **3**, pp. 57-60, March (1990)
28. M. Hourai, K. Murakami, T. Shigematsu, N. Fujino, T Shiraiwa, Japan. J. Appl. Phys. **28** (12), pp. 2413-2420, (1989)
29. B. O. Kolbesen, H. Cerva, Phys. Stat. Sol. (B) **222**, 303, (2000)
30. J. D. Morillo, T. Houghton, J. M. Bauer, R. Smith, R. Shay, 2005 IEEE/SEMI Advanced Semiconductor Manufacturing Conference, (2005)
31. Semi Standard SEMI E146-0306, 2006 edition of SEMI® Standards
32. A. Shimazaki, H. Hiratsuka, Y. Matsushita and S. Yoshii: Extended Abstract 16th Conf. on Solid State Devices and Materials (Kobe,1984) (Tokyo: Business Center for Academic Societies, Japan) pp.281-284.
33. L. Fabry, S. Pahlke, L. Kotz, E. Schemmel and W. Berneike, "Crystalline defects", ed. by B. O. Kolbesen, P. Stallhofer, C. Claeys and F. Tardif, PV 93-15, Electrochem. Soc. Inc., Pennington, NJ, p. 232 (1993).
34. S. Kiyota and S. Ishiwari, Clean Technology, 4, p. 35 (1994).
35. H. Shimizu and S. Ishiwari, Semicond. Sci. Technol., 15, pp. 776-781, (2000)
36. L. Pfitzner, Proc. of 209th meeting of the Electrochem. Soc. in Denver May 7-12, 2006, '10th Symposium of Silicon Materials Science and Technology', (2006)

ECS Transactions, 2 (2) 453-460 (2006)
10.1149/1.2195680, copyright The Electrochemical Society

Standardization of Measurement of Nitrogen Concentration in CZ Silicon Crystals

N. Inoue[1], K. Masumoto[2], M. Shinomiya[3], K. Kashima[4], K. Eifuku[5], M. Koizumi[6], T. Takahashi[7] T. Takenawa[8], A. Karen[9], K. Shingu[10], H. Yagi[11] (*)

Nitrogen Concentration Measurement Group,
Japan Electronics & Information Technology Association (JEITA)
11, Kanda-Surugadai 3-chome, Chiyoda-ku, Tokyo 101-0062, Japan

ABSTRACT

Standards for the measurement procedures of nitrogen concentration in CZ silicon crystals were established for infrared absorption spectroscopy (IR), secondary ion mass spectroscopy (SIMS) and charged particle activation analysis (CPAA) for concentration above 1×10^{14} atoms/cm^3. Good agreement was obtained between the laboratory within the same method and between the three methods. Conversion coefficient from (weighted sum of) absorption coefficients to nitrogen concentration agreed well with the previous one.

INTRODUCTION

Nitrogen doping is widely used in growing CZ silicon crystals for advanced devices, especially for annealed wafers. It is important to control and characterize the nitrogen concentration. Concentration of light elements in silicon crystals such as oxygen and carbon is usually determined by infrared (IR) absorption by local vibration modes and the measurement procedures are established in the ASTM standards. In case of nitrogen, however, there is no standard for IR and SIMS is the only one whose measurement procedure is established as a standard test method [1]. We have conducted a project including three measurement methods, IR, SIMS and CPAA, and reported the preliminary result 4 years ago [2]. In this paper we summarize our recent results.

SAMPLES

Samples from several vendors were used for the round-robin measurement. CZ and FZ, including non-doped crystals were prepared. Estimated nitrogen concentrations were between 5×10^{13} and 4×10^{15} atoms/cm^3. Samples for IR, SIMS and CPAA were prepared side by side. For IR measurement, they were polished on both sides with final thickness 10 and 2 mm. CZ samples were annealed at 650°C after measurement was done on as-grown specimen. For SIMS measurement, some samples were cut from mirror polished wafers, others were mainly 2 mm thick. For CPAA measurement, 2 mm thick samples were also used, as the acceleration voltage was increased to increase signals.

INFRARED ABSORPTION

453

In the IR measurement, wavenumber resolution was 2 cm^{-1}. The calibration method of apparatus was established to improve the detection limit by absorbance units of 10^{-4}. This is to measure the same sample repeatedly and confirm that the variation in the differential absorbance spectrum between the repeated measurements is sufficiently below the peak height of the nitrogen absorption line. There are 7 absorption lines observed in CZ silicon [3] and their origins were confirmed as NN, at 963 and 776 cm^{-1}, and NNOi at 801, 996 and 1027 cm^{-1}, and NNOiOi at 810 and 1018 cm^{-1} [4, 5], respectively. Therefore it is necessary to measure at least three peaks to determine the total nitrogen concentration. Measurement accuracy was evaluated and improved by comparing the absorbance of the different peaks from the same nitrogen complexes: The ratio between the absorbance of the peaks from NN, at 963 and 776 cm^{-1}, and that between those from NNOi, at 801, 996 and 1027 cm^{-1}, and that between those from NNOiOi, at 810 and 1018 cm^{-1}, respectively, should be constant over all the samples. It was evaluated for the results from all participating organizations. It was almost constant over all samples in the case of the absorption by NN, as an example shown in Fig. 1(a). In case of NNOi and NNOiOi, the ratio was not ideally constant, as shown also. It was due mainly to the overlapping absorption by oxygen dimer at 1013cm^{-1} [6] which can not be removed by the donor killer annealing for the case of the nitrogen absorption lines especially that at 1018 cm^{-1}. Therefore the line at 810 cm^{-1} was recommended instead of the 1018 cm^{-1} line which has been used previously. As already pointed out, there is a steep valley at 799 cm^{-1} between phonon absorption peaks which is almost overlapping the absorption line at 801 cm^{-1}. Therefore the absorbance at 801 cm^{-1} line generally includes some error. The results of some organization were nearly constant, but those from others were less constant.

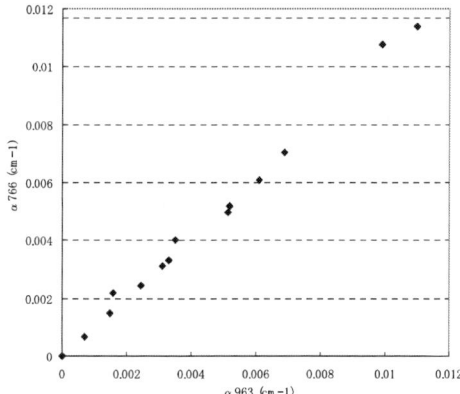

Figure 1 An example of relation among the absorbance of different line from the same configuration, NN, 963 and 766 cm^{-1}.

To obtain the total nitrogen concentration, weighted sum of absorbance from three main structures must be used [2]. The weights were determined by measuring the complimentary change among the absorptions by the heat treatment at various temperatures, tentatively as $\alpha_{sum} = \alpha_{963} + 1.5 \times \alpha_{996} + 0.5 \times \alpha_{810}$. The conversion coefficient of the weighted sum of the absorption coefficients to the nitrogen concentration determined by SIMS was nearly equal to the conventional one for FZ Si [7], as will be discussed later. There was good agreement between some organizations and acceptable for others.

A nearly equal result was obtained by the measurement on 2 mm thick samples. In case of as-grown samples absorption by the thermal donors and oxygen dimmers interferes the measurement. Therefore donor-killer annealing [8] is recommended.

It is difficult to measure the absorbance of weak absorption lines, for example, at 810 and 1018 cm^{-1}. And 996 and 1018 cm^{-1} lines suffer from overlapping by oxygen dimer absorption even the donor-killer annealing has been performed. Tanahashi et al. proposed to measure absorbance of 963 line after annealing the samples at 600 and 750 °C, respectively and calculate [N] = $\{\alpha_{963}(750) - \alpha_{963}(600)\}$ x $4.07x10^{17}$ [9]. The preliminary evaluation of this method has also been performed.

One difficulty with the IR nitrogen measurement is that the background differential absorbance is wavenumber dependent. Thus, the conventional baseline procedure is not always applicable. It is likely that because nitrogen absorption is so much smaller than the phonon absorption that the differential phonon absorbance spectrum is left. A procedure of spectral analysis was established to determine the background absorbance at the nitrogen peak accurately: One is to delete the thermal donor absorption in case of as-grown samples and the oxygen dimer absorption by using the standard spectrum created by nitrogen-undoped sample and nitrogen & oxygen undoped reference. The other is to fit a reduced phonon spectrum to the background spectrum rather than to draw a straight baseline. Also, it was found that a low-pass filter reduced the background differential absorbance and gave the better result. These two methods are discussed in detail in a separate paper [10].

SIMS

SIMS measurement was done by two organizations according to the SEMI standard [1] in which the raster change method [11] had been adopted as based on the JEITA proposal. Interlaboratory agreement between the two organizations was examined as shown in Fig. 2. Linear relationship was obtained. There was a slight difference between the results from the two organizations. It was confirmed that there was no difference in the relative sensitivity factor [11] between the two organizations.

Nitrogen is precipitated in as-grown wafers in case of high concentration samples as well as in annealed wafers. Precipitated nitrogen is observed as spikes in a depth profile and bright spots in an image as an example shown later. It is difficult to determine the total nitrogen concentration when the nitrogen has been precipitated. A simple way to include precipitated nitrogen concentration is to add sum of height of spikes to average concentration. The result is compared as an example shown in Fig. 3. In the high concentration regime, [N] with spike is slightly higher than [N] without spike. The difference is distinct for $[N]>1x10^{15}/cm^3$.

It is useful when the precipitated nitrogen concentration can be determined by SIMS. The preliminary study was performed as shown below. Fig. 4(a) shows an example of depth profile taken by Cameca IMS-4F using a primary ion of Cs at 14.5 KV, 110 nA. From the integrated intensity over the scanned area of 30μϕ and depth of 1.2 – 1.3 μm, average [N] was estimated to be $2x10^{11}$ atoms/cm². As a result, number of nitrogen atoms in the area (= in a precipitate in case of one precipitate per image) was estimated to be $2x10^{11}$xπx$(15x10^{-4})^2$=$1.4x10^6$. Fig. 4(b) shows an example of image using 42Si-N-ion. The average precipitate density was estimated to be $1x10^9/cm^3$. As a result, precipitated nitrogen concentration was estimated to be about $(1.4x10^6 x 1x10^9/cm^3=)$ $1x10^{15}$ atoms/cm³. This proved to be a reasonable value according to an estimation by IR.

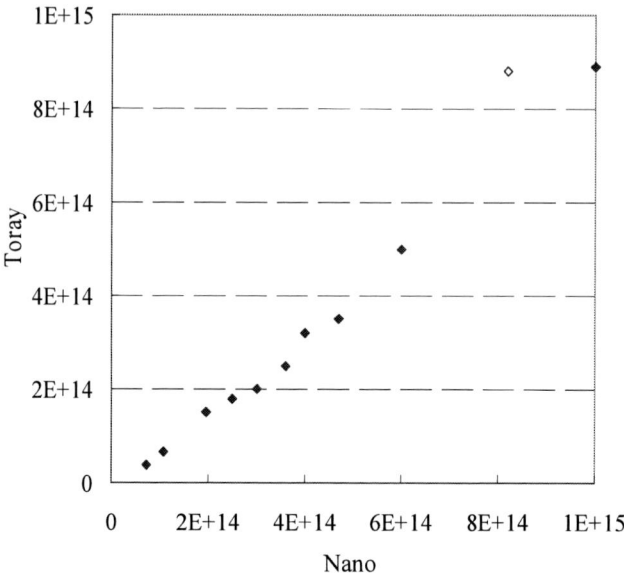

Figure 2 Comparison between the SIMS results from two organizations.

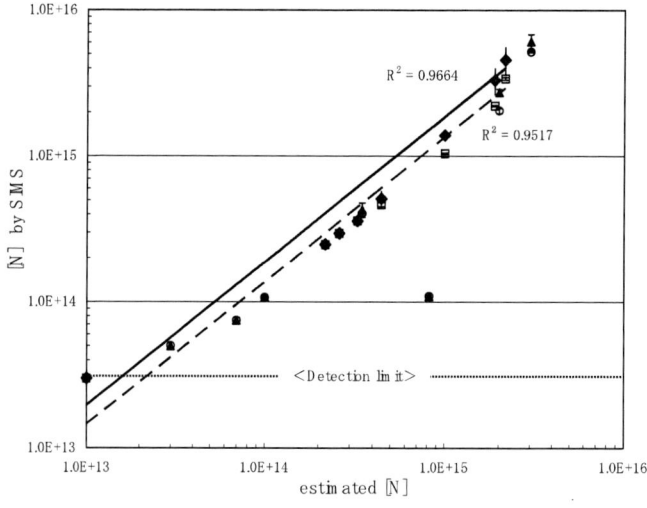

Figure 3 Comparison of results with spikes (solid line) and without spikes (broken line).

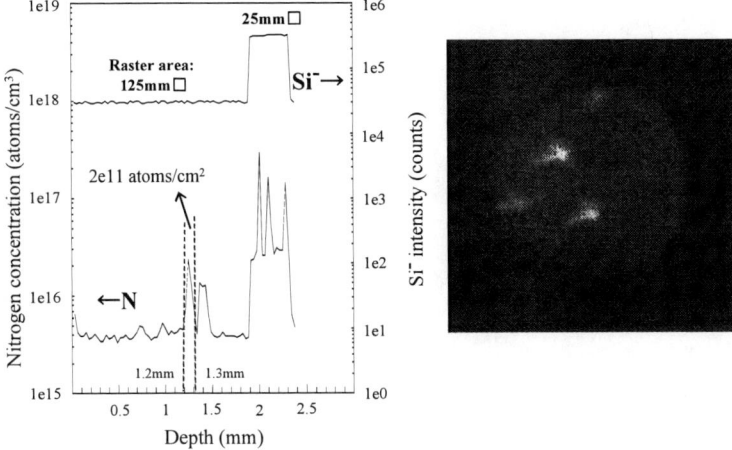

Figure 4 SIMS measurement of precipitated nitrogen (a) depth profile, (b) N-Si secondary ion image.

CPAA

Charged particle activation analysis (CPAA) utilizes the radioactive reaction $^{14}N(p, \alpha)^{11}C$. The basic procedure and preliminary result has been reported already [12]. Two organizations participated the measurement using the following two fixed-energy cyclotrons giving about 20 MeV proton beams, respectively, (1) CYPRIS 370V cyclotron of Sumitomo Heavy Industry Examination and Inspection (SHIEI), and (2) the Shimadzu MCY-1750 cyclotron of Nishina Memorial Cyclotron Center (NMCC), Japan Radio Isotope Association (JRIA). The incident proton energy was reduced to 15 MeV by aluminum absorber foils. The beam current and bombardment time were set 5μA and 20 min. The nitrogen concentration in silicon is so low that the radioactivity of ^{11}C needs to be separated with other interfering radioactive species before the measurement. It must be done quickly because the half life of of ^{11}C is 20.4 min half-life. We intended to use two methods of chemical separation, dry fusion and wet chemistry, in order for the cross-check of their reliability. Detailed procedure of chemical separation after irradiation was established for the wet process this time. For ganma ray measuremet at positron emission, two types of detectors were used, NaI (Tl) scintillation detectors (5 in. ϕ) at SHIEI and BGO scintillation detectors (3 in. ϕ) at NMCC. The $^{11}B(p, n)^{11}C$ reaction makes a serious interference in the CPAA of nitrogen. Since boron content of silicon can be known from electric conductivity fairly accurately, we can correct for the interference of boron by the use of the ratio of ^{11}C activity formed from elemental boron to that formed from AlN under equal bombardment conditions. Fig. 5 shows the comparison between the results from two organizations. Good interlaboratory agreement has been obtained.

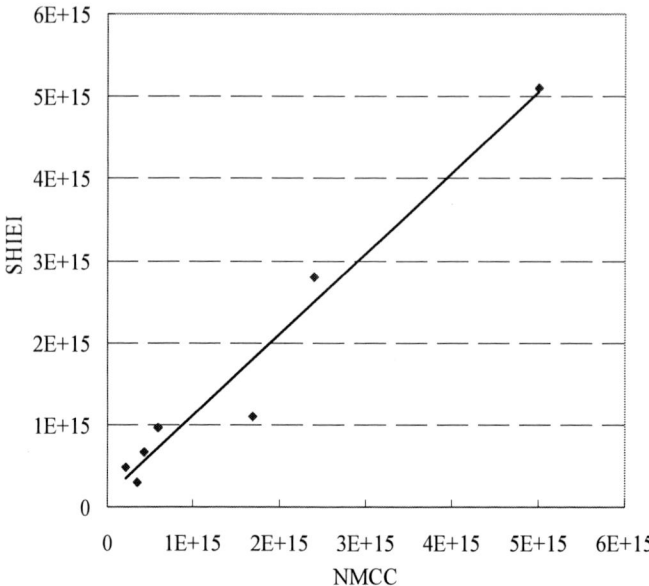

Figure 5. Relation between the CPAA results from two organizations.

COMPARISON AMONG THE THREE METHODS

Results of IR, SIMS and CPAA were compared in detail, and it was found that the agreement was good for most data. Fig. 6 shows the relation between IR and SIMS, IR and CPAA and SIMS and CPAA, respectively.

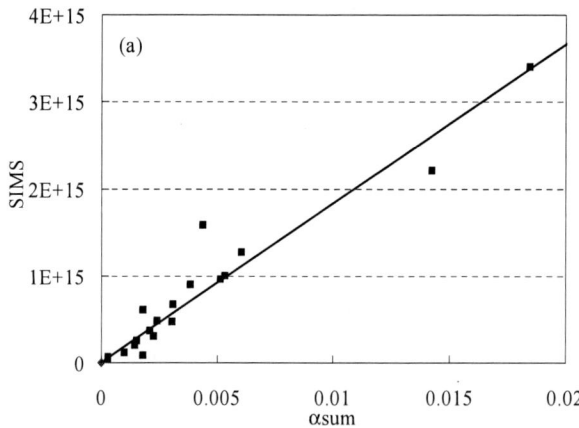

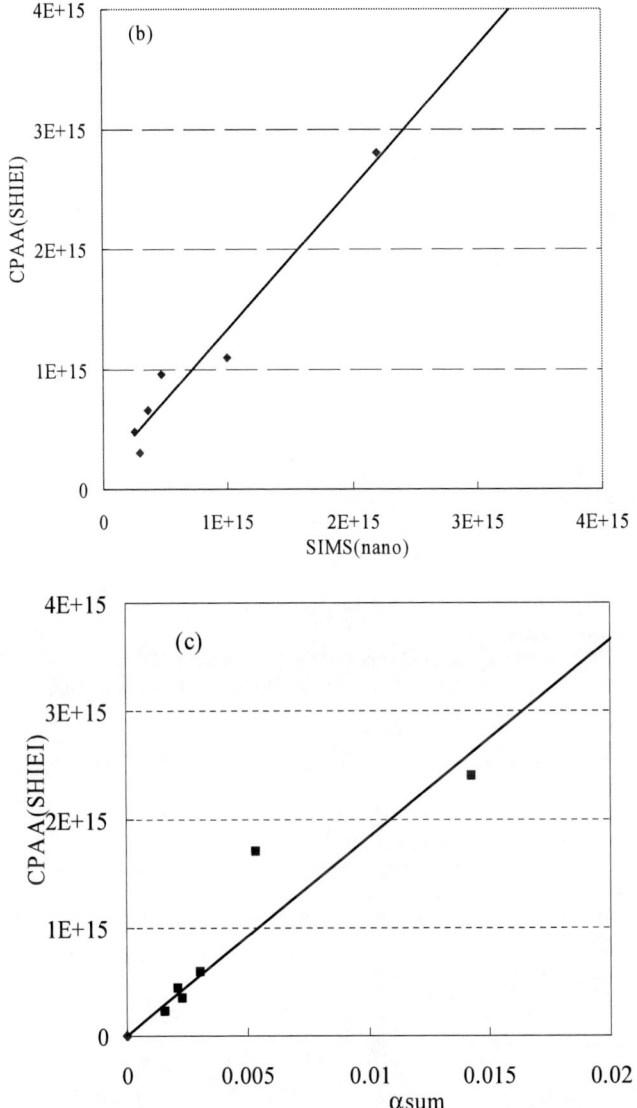

Figure 6. Example of relation between the three methods, (a) IR and SIMS, the line shows the previous calibration, (b) SIMS and CPAA, (c) IR and CPAA.

Below 1×10^{14} atoms/cm^3, sensitivity (signal to noise ratio) of the three methods must be improved. Nitrogen does not form NN pair but mainly forms NO pair accompanied by oxygen. For IR measurement, new peaks were found [13], candidates for the absorption related to shallow thermal donors which were dominant in low concentration regime. Sample dependence and annealing behavior were examined in detail.

SUMMARY

In summary, nitrogen concentration measurement was done by IR, SIMS and CPAA using the established procedure by participating organizations. Good agreement was obtained among the organizations and methods. Conversion coefficient from weighted sum of IR absorption coefficient of $\alpha 963+1.5 \times \alpha 996+0.5 \times \alpha 810$ to nitrogen concentration was nearly equal to the previous one of 1.83×10^{17} atoms/cm^2. As a result, the standard measurement procedures were established for the three methods for nitrogen concentration above 1×10^{14} atoms/cm^3 and issued from JEITA.

Acknowledgments

The authors are grateful to JEITA working group members and people preparing or measuring samples. This work is partially supported by the JSPS grant No. 0016560014.

References

[*] Following organizations prepared the samples or performed the measurement (alphabetical order). Sample suppliers are marked with*. Accent Optical Tech[8]., Komatsu Electronic Metals*[5], NanoScience (Charles Evans)[10], MEMC*[7], Osaka Prefecture Univ.[1], Rad. Sci. Center KEK[2], SHI Exam. & Inspect.[11], Shin-Etsu Handotai*[3], SUMCO*[6], Toray Research Center[9], Toshiba Ceramics*[4].

1. Test Method for Measuring Nitrogen Concentration in Silicon Substrates by Secondary Ion Mass Spectrometry, SEMI MF2139-1103.
2. N. Inoue, K. Shingu and K. Masumoto, Semiconductor Silicon 2002, p. 875.
3. P. Wagner, R. Oeder and W. Zulehner, Appl. Phys. A , 46, 73 (1988).
4. T. Matsumoto, Y. Yamanaka, H. Harada and N. Inoue, Materials Science and Engineering, B91-92, 144 (2002).
5. A. Karoui, F. S. Karoui, G. A. Rozgonyi, M. Hourai and K. Sueoka, Semiconductor Silicon 2002, p. 670.
6. T. Hallberg et al., Mater. Sci. Forum, 258-263(1997)361.
7. Y. Itoh, T. Nozaki, T. Masui and T. Abe: Appl. Phys. Lett., 47(1985)488.
8. H. Shirai and K. Kashima, Materials Science and Engineering, B95(2002)33.
9. K. Tanahashi and H. Yamada-Kaneta, Japan. J. Appl. Phys., 42(2003)L223.
10. M. Nakatsu, N. Inoue and V. D. Akhmetov, Semiconductor Silicon 2006.
11. N. Fujiyama, A. Karen, D.B. Sams, R.S. Hockett, K. Shingu and N. Inoue, Appl. Surf. Sci., 203-204(2003)457.
12. K. Masumoto, T. Nozaki, H. Yagi, Y. Minai, Y. Saito, S. Futatsugawa and N. Inoue, Proc. Diagnostics on Semiconductors, (Electrochem. Soc. Proc. Vol. 2004-05) 69.
13. N. Inoue, M. Nakatsu and H. Ono, to be published in Physica B (2006).

Infrared absorption measurement of carbon concentration in silicon crystals

[a]N. Inoue, [a]M. Nakatsu and [b]V. D. Akhmetov

[a]RIAST, Osaka Prefecture University, 1-2, Gakuencho, Sakai, Osaka, 599-8570, Japan
[b]IHP/BTU Joint Lab, BTU Cottbus, Konrad-Wahsmann-Alle 1, 03046 Cottbus, Germany

Abstract

Sensitivity and accuracy of carbon concentration measurement by infrared (IR) absorption spectroscopy are improved. Unnecessary high energy light input was cut by a low pass filter. All the measurement conditions for samples and references were kept as reproducible as possible by using the sample changer and measure them alternately for many times. By keeping the temperature both of the sample and of the reference as similar as possible the accuracy was improved. In the analytical procedure, we use a reduced phonon spectrum fitting instead of straight baseline. Standard carbon spectrum fitting to a small carbon peak make it possible to determine the carbon concentration accurately. As a result, we can measure differential carbon concentration down to about 1×10^{14} atoms/cm^3.

Introduction

The detection limit of carbon concentration by infrared absorption spectroscopy (IR) has been thought to be 2×10^{15} atoms/cm^3 in the SEMI standard [1] and most modern crystals are considered to contain carbon below the detection limit. We have applied the technique developed for the nitrogen concentration measurement [2] and made it possible to detect carbon down to about 1×10^{14} atoms/cm^3. To improve the signal-to-noise ratio, low-pass filter was used. A sample changer was adopted to eliminate the differential background absorption. Effect of sample temperature on the accuracy was examined. New arithmetic procedure has been developed to eliminate the background absorption.

Experimental and arithmetic procedure

Samples were prepared from various Czochralski-grown silicon crystals. In addition, axial distribution of carbon was also examined on one crystal. The samples were 2 mm thick double side mirror polished. IR measurement was done using Fourier transform IR (FTIR) apparatus at room temperature with a wavenumber resolution of 2 cm^{-1}. In the impurity concentration measurement by IR absorption, differential absorption between samples and a reference without impurities is usually applied. We used samples and a reference that had similar characteristics, for example, sample and reference were provided from the same wafer vendor or from the same crystal. Effect of low-pass filter on the accuracy and sensitivity was examined. Arithmetic procedure on the obtained spectrum was developed as will be discussed below. The effect of sample temperature on the measurement error was also examined.

Results and discussion

Effect of low-pass filter

Carbon atoms occupy substitutional positions in monocrystalline Si and the carbon IR absorption peak is located at 605cm⁻¹. Figure 1 shows an example of transmitted intensity spectrum around the carbon absorption peak. One of the problems with the carbon concentration measurement is that the carbon absorption peak is located on a shoulder of a big double-phonon absorption peak at 610 cm⁻¹. The absorbance of the latter is about 0.7. In the case of $[C] = 1 \times 10^{15}/cm^3$, the absorbance is 0.0011 (equivalent to the phonon absorption of 2μm thickness difference). It is to be noted that we have to draw a baseline over absorbance range of 0.7.

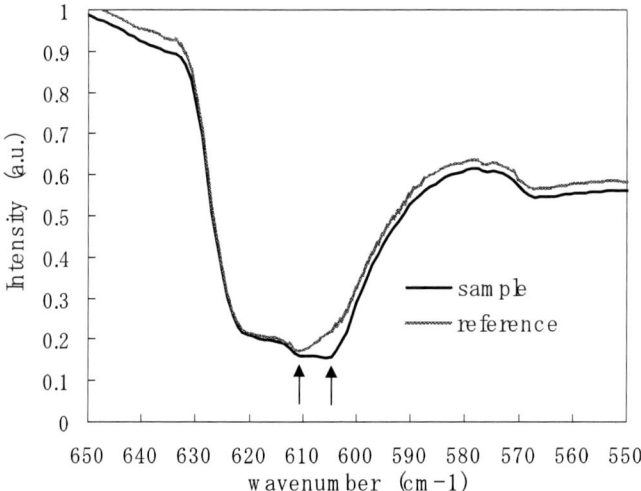

Figure 1. Transmitted intensity spectrum from sample and reference. The carbon absorption peak is located at 605cm⁻¹ and the phonon absorption peak is at 610cm⁻¹.

Moreover, it is easily expected that the background absorption difference between the sample and reference easily exceeds the absorbance by nitrogen. This makes drawing a baseline very difficult in case of low carbon levels

To overcome these problems, improvement of signal to noise ratio is helpful. One method for this is to use a low-pass filter. Figure 2 shows an example of transmission intensity spectrum from a commercial FTIR. Wavenumber range is 4000-400cm⁻¹, and carbon absorption peak is located at 605cm⁻¹. Most of the energy input to a detector is not necessary to measure the carbon peak but increase the noise. By using a low-pass filter to cut unnecessary high energy, most of the superfluous lights was cut off as an example shown in figure 2 (the thick line).

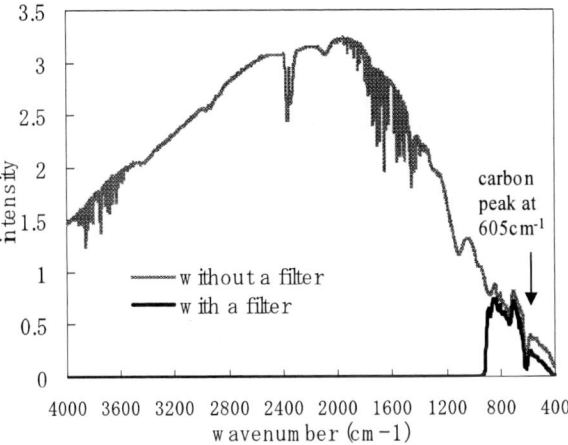

Figure 2. Transmitted light intensities without a filter (gray and thin) and with a filter (black and thick).

Figure 3 shows an example of differential absorption spectrum with or without a filter, respectively. carbon concentration was about $1.4E+15/cm^3$. In the case of without using a low-pass filter, the apparent difference of phonon absorption remained and formed irregularity near 625 cm^{-1} as indicated by an arrow. In the case of using a filter, the background differential absorption got more even and a straight baseline could be drawn.

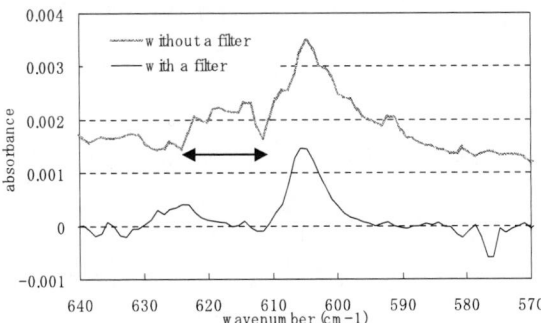

Figure 3. Improvement of background spectrum by the use of filter. Up: a differential absorption spectrum without a filter. Down: a differential absorption spectrum with a filter.

Figure 4 depicts the differential absorption spectra between sample and reference with and, resp., without a low-pass filter. The carbon concentration was even lower than in the sample shown in figure 3. Without a filter, the background spectrum was straight within only a narrow wavenumber range as shown in figure 4, upper spectrum. This was due to the apparent background spectrum difference between the sample and reference. Therefore, we could draw only a short baseline. This resulted in rather poor accuracy. When we used a filter as shown in figure 4, down spectrum, on the other hand, the background absorption was straight over a wider range and a longer and reliable baseline could be drawn.

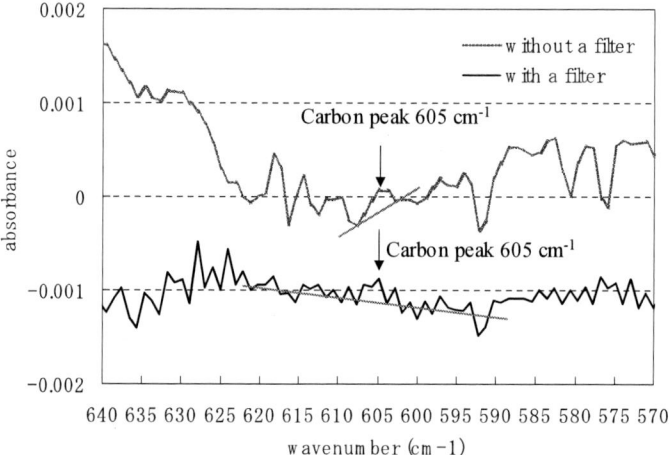

Figure 4. The differential absorption spectra between sample and reference without low-pass filter (upper spectrum) and with a low-pass filter (lower spectrum).

Effect of measuring condition

Next, we examined the conditions of the measurement. In the differential absorption measurement between sample and reference, we can not cancel the background absorption if the sample and the reference are measured under different conditions. Therefore, conditions may not vary between sample and reference measurements. A suitable solution had been developed by one of the coauthors [3]. An important feature is to use a sample changer and to measure the sample and reference alternately for many times. It was confirmed that by using this method, the background differential absorption was much reduced than in the case of without using the sample changer. As a result, it became easier to draw a baseline for low [C] samples.

Usually, we have to calculate the results from data taken on different days. Thus, we examined the influence of variations of data on different days. So we examined how the data obtained on the different day reduced the accuracy. We measured the sample and reference on the same day and different day and compared. Figure 5 shows an example, in which the same sample was used as the sample and the reference (no differential carbon concentration). The smooth background was obtained in a measurement that was carried out on the same day. The accuracy was reduced by carrying out measurements on reference and sample on different days. However, we can accurately measure carbon on the same day.

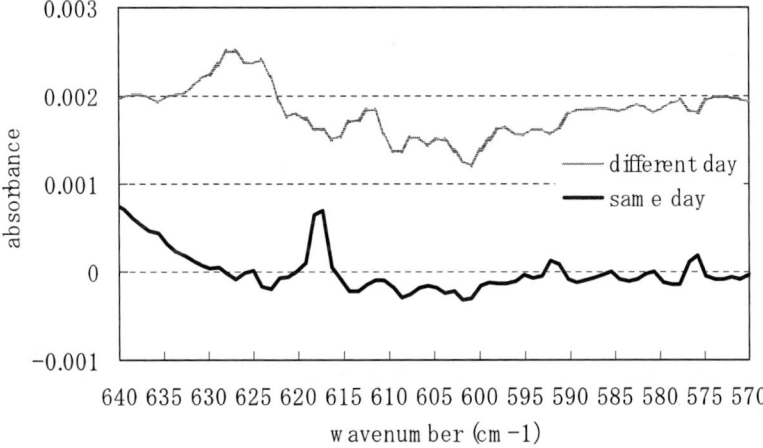

Figure 5. the differential absorption spectrums by the measurements on the different day (up) and on the same day (down). Vertical axis is shifted for clarity

It the FTIR measurement, it is recommended to keep the temperature around the instrument constant, for example, at 22°C. The reason is that the FTIR is a delicate optical instrument. It is to be noted that the phonon absorption has temperature dependence. So if the sample temperature is different from the reference temperature, the background differential absorption should rise. When the impurity concentration is high, the influence of temperature is small. But in the case of low impurity concentrations the influence of temperature on the measurement may be large. One possible cause of reduced accuracy by the measurement on the different day may be the temperature difference between sample and reference measurement conditions. Thus, we studied how temperature variations influence the background. The sample temperature and room temperature were measured during the IR measurement. Figure 6 shows an example of differential absorption spectrum in the case of temperature difference of 3.8 °C (in this case, the same specimen was used for clarity). The background is irregular especially at the carbon peak and 3 points marked in the figure. Figure 7 shows the dependence of difference absorbance at there on the temperature. The background gets more irregular, as the temperature difference gets larger. As a result, it was difficult to fit and draw a straight baseline.

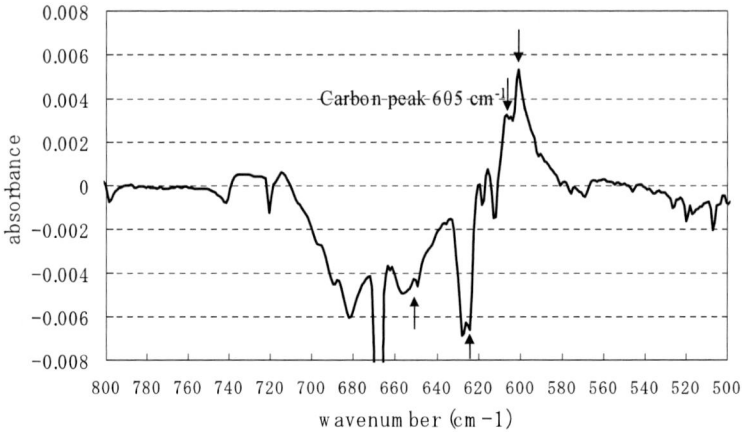

Figure 6. The differential absorption spectrum in the case of different temperature between sample and reference.

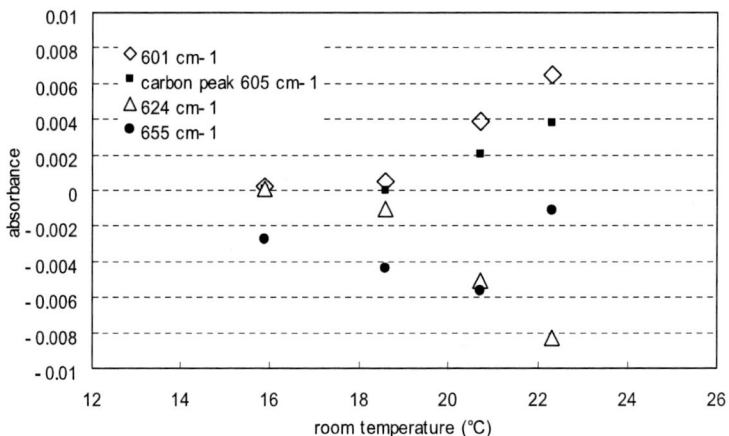

Figure 7. The dependence of differential absorbance at the carbon peak and 3 points indicated in Fig. 6 on the temperature. (the temperature of reference is 15.9 °C)

To examine if the main origin of reduced accuracy in the measurement on the different day, we measured on the different day while we changed the sample temperature. (in this case, the same specimen was used also) Figure 8 shows the temperature dependence of the differential absorbance at carbon peak and 3 typical points. Even the measurement was done on the different day, the difference was small in case that the temperature difference was small. As a result, the straight baseline could be drawn over these 4 points and we could accurately measure by keeping the temperature of sample and reference nearly equal.

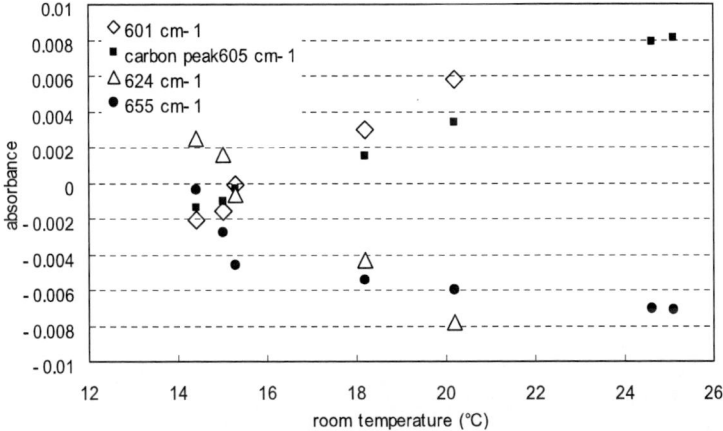

Figure 8. The temperature dependence of the differential absorption at carbon peak and 3 typical points in case of measurements on a different day of sample and reference measurement. (The temperature of reference was 15.9 °C)

Cancellation of background absorbance by arithmetic procedure

Phonon spectrum fitting to the background absorption. In case of carbon concentration approaching to 1×10^{14} atoms/cm^3, the peak height is nearly equal to the residual background difference. It is difficult to draw a straight baseline even when using a low-pass filter. The origin of the remained background differential absorption is the phonon absorption. It is therefore possible to use the phonon absorption to fit the remained background absorption instead of drawing a straight baseline. Figure 9 shows an example of differential absorbance spectrum. There is a plateau on the higher wavenumber side of the carbon absorption. The thin line shows the reduced reference absorbance spectrum (essentially composed of phonon absorption) fitted to the differential spectrum. It was obvious that the reduced phonon absorption spectrum fits better to the differential spectrum than a straight baseline. In this case, applying a straight baseline would underreport the background differential absorption.

ECS Transactions, 2 (2) 461-470 (2006)

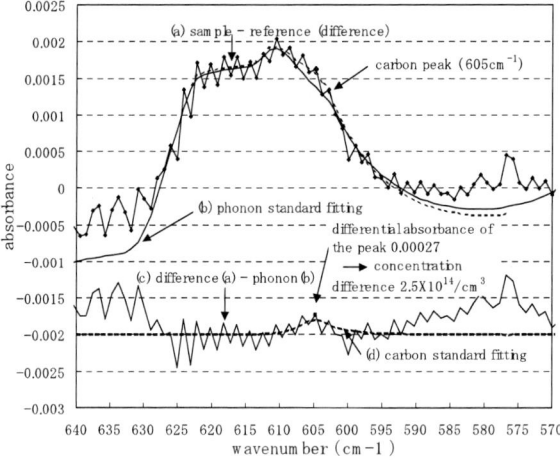

Figure 9. An example of differential absorbance spectrum with a straight baseline and a phonon absorption spectrum fitted to the background ([C] = 1.4×10^{15} atoms/cm^3).

Standard carbon absorption spectrum fitting In case of very low [C], the wavenumber range of the carbon peak is narrow. As a result, we may use very small number of data points which include relatively big error. Especially it includes much error when we use only one peak data. In case of using an interference filter, the fringe modifies the peak height also. To overcome this problem, we fit a carbon standard spectrum to the differential spectrum, instead of using a peak data.

Figure 10. An example of differential absorbance spectrum and arithmetic procedure. (a) sample – reference (differential absorbance), (b) reduced phonon absorbance fitted to the difference, (c) difference (a) – phonon (b), (d) reduced carbon standard fitting.

468

The carbon standard spectrum was prepared from high concentration sample. In the high concentration sample, we can draw a long straight baseline easily. Figure 10 shows an example of simultaneous fitting both of the phonon absorption spectrum and the carbon standard spectrum to a differential absorption spectrum in case of a low concentration sample. It is obvious that we can measure the weak carbon peak more accurately by using this arithmetic procedure.

As a result, in the case of 2mm thick sample it is possible to measure differential concentration of about $1\times10^{14}/cm^3$. By the same procedures, detection limit and accuracy of measurement of wafers were also improved. The result was compared with that from the 2 mm thick samples from the same origin in figure 11. It is possible to measure down to $5\times10^{14}/cm^3$.

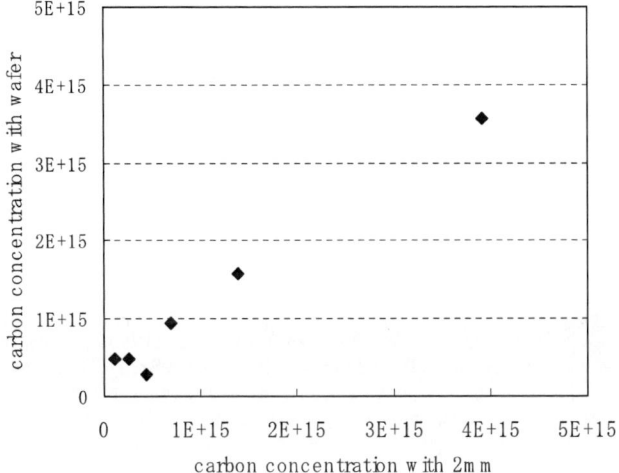

Figure 11. Comparison the carbon concentration with 2mm thick samples to wafers on the measurement.

Figure 12 demonstrates the accuracy of our method by showing the axial distribution of carbon concentration in lowest carbon concentration crystal. This compares the results by the conventional procedure and by the filter and fitting procedure. In the latter case, smoother axial distribution was obtained down to below $10^{14}/cm^3$ of difference to the concentration near the top of the crystal (used as a reference). Curved line shows the estimated concentration distribution using the normal freezing equation corrected with contribution from atmosphere. The result suggested that the incorporation of carbon was dominated by segregation. However, a contribution of atmospheric carbon via inert gas / handling has also been observed.

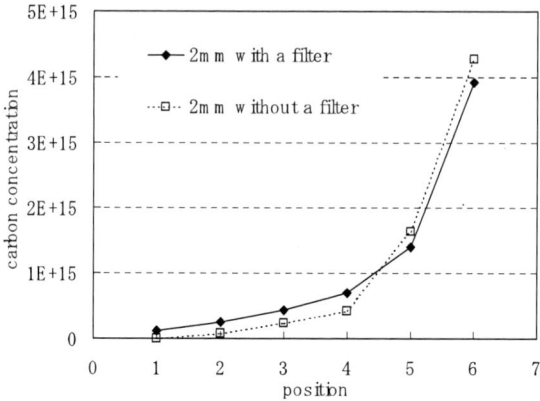

Figure 12. Axial distribution The horizontal axis shows the approximate position along the crystal.

Summary

In summary, high sensitivity and high accuracy infrared absorption measurement of carbon concentration in silicon is established. The advantage of using a low-pass filter, sample changer and keeping the temperature of sample and reference is demonstrated. By fitting reduced phonon spectrum and carbon standard absorption spectrum to differential absorption spectrum, differential carbon concentration can be determined accurately down to 10^{14} atoms/cm^3.

References

1. The test methods for substitutional atomic carbon content of silicon by infrared absorption, SEMI MF1391-0704.
2. N. Inoue, K. Masumoto and K. Shingu, Semiconductor Silicon 2002, p. 875. N. Inoue, A. Karen and H. Yagi, Semiconductor Silicon 2006.
3. V. D. Akhmetov, O. Lysytskiy and H. Richter, High Purity Silicon VIII (Electrochem. Soc. Proc. Vol. 2004-05) p. 109

ECS Transactions, 2 (2) 471-484 (2006)
10.1149/1.2195682, copyright The Electrochemical Society

Standardization of test methods for bulk micro-defects and denuded zones in annealed CZ Silicon

Ryuji Takade[1)2)3)] , Naohisa Inoue[2)4)], ,Kazuo Moriya[1)2)5)] ,Kazuhiko Kashima[1)2)6)] , Kenji Nakashima[2)7)], Masahiro Kato[1)8)], Satoru Kitagawa[1)9)] , Toshiaki Ono[1)10)], Hiroki Urushido[1)11)] ,Nonuhito Nango[2)12)] ,Vladimir Akhmetov[13)]

1)JEITA Defect Characterization WG, 3-11, Kanda-surugadai, Chiyoda-ku, Tokyo 102-8471, Japan

2)Defect Character. Subc., JSPS 145 Comittee, 1-6, Ichibancho, Chiyoda-ku, Tokyo 101-0062, Japan

3)Toshiba Ceramics CO., Ltd., Higashi-kou, Seiro-machi, kitakannbara-gun, Niigata, 957-0197, Japan

4) RIAST, Osaka Perfecture University, 1-2, Gakuen-cho, Sakai-Shi, Osaka 599-8570, Japan

5)Mitsui Kinzoku, 1333-2, Haraichi, Ageo-shi, Saitama 362-0021, Japan

6)Toshiba Ceramics Co., Ltd. 1-6-3, Osaki, Shinagawa-ku, Tokyo, 141-0032, Japan

7)TOYOTA CENTRAL R&D LABS., Inc. 41-1 Yokomichi, Nagakute, Nagakutecho, Aichi, 480-1192, Japan

8)Shin-Etsu Handotai Co., Ltd. 13-1, Isobe 2-chome, Annaka-shi, Gunma, 379-0196, Japan

9)Komatsu Electronic Metals Co.,Ltd. 3-25-1, Shinomiya, Hiratsuka, Kanagawa, 254-0014, Japan

10)Sumitomo Mitsubishi Silicon Corporation, 2201, Oaza-Kamioda, Kouhoku-cho, Kishima-gun, Saga 849-0597 Japan

11)MEMC Japan Ltd. 11-2, Kiyohara Industrial Park, Utsunomiya, Tochigi, 321-3296, Japan

12)RATOC SYSTEM ENGINEERING Co., Ltd. Kirin 1st Bldg. 540 Tsurumaki-cho, Waseda, Shinjuku, Tokyo, 162-0041, Japan

13) IHP/BTU Joint Lab, Konrad-Wachsmann-Allee 1, 03046 Cottbus,

The requirement to standardize measurement methods for BMD (**B**ulk **M**icro **D**efect) density and DZ (**D**enuded **Z**one) CZ (**Cz**ochralski) silicon has lead to the establishment of a SEMI standard for annealed CZ silicon wafers. Therefore, it was decided that we should aim at standardizing the preferential-etching and 90-degrees laser-scattering tomography techniques as a collaborative work between JEITA (**J**apan **E**lectronics and **I**nformation **T**echnology Industries **A**ssociation) and JSPS (**J**apan **S**ociety of **P**romotion of **S**cience) 145th Committee. In this work a set of "round robin" tests were curried out. Regarding standardization, both the preferential etching and the 90-degrees laser-scattering methods were examined. This resulted in a standardized measurement protocol for BMD density and DZ width, which has become known as the JEITA standard "**EM 3508**"[1].

Introduction

The single crystal silicon usually employed for semiconductor device substrates includes a small quantity of grown-in defects that are formed during the crystal growth process. These defects can affect the device performance over long-term thermal cycles or improve the wafer performance by

471

their ability "gettering" impurities [2]. The knowledge of these defects is essential for the user. Historically, many different etching solutions have been used in preferential-etching methods in order to reveal defects [3-9]. Due to historical references, both engineers and researchers apply "personalized" BMD test methods. The most popular etching solution has been the so-called "Wright solution", which contains Cr^{6+} ions [5]. In Japan, e.g., an etchant using Cr^{6+} had been also defined as the standard etchant until the current etchant, without containing Cr, was specified in 1996. Since that time, a system without Cr, standardized as **JIS-H0609** [10], has been applied. Thermal-process-induced oxygen precipitates in a silicon single crystal can be revealed as etch-pits by using this etching solution. Oxygen precipitates increase during the long thermal processes which are generally required in device fabrication, and are derived from oxygen atoms that are trapped within the crystal during CZ crystal-pulling. For about 15 years, a laser-scattering technique has been used for the measurement of oxygen precipitates. This method, known as infrared (IR) tomography, applies a laser at an IR wavelength, and is now widely used in silicon wafer fabrication [11]. Other methods employ surface-scattering and light transmission. However, there is no existing standard test document for these methods. Over this period of time, there has been a need for the standardization of annealed wafers to a SEMI standard, together with a need to establish detailed standard test methods for BMD and DZ width. To this end, we initiated a working group in December 2003. During developing a standard test method, we studied a preferential-etching technique and a laser-scattering process as examples of two typical measurement methods, and we described a set of instructions for each method. In fact, the preferential-etching technique already had a standard document, known as JIS-H0609. However this has been tested but found to be inadequate for DZ measurements. A better choice was thought to involve a laser-scattering method combined with preferential-etching. In this report, we summarize the activity of the JEITA and the JSPS 145th Committee for Standardization.

Comparison of the preferential-etching and laser-scattering methods.

In this study, we report two of the methods available for detecting micro defects in silicon. As described below, we tried to determine what we believed to be the benefits of each method. Examples of preferential-etching techniques for silicon single crystal evaluation are shown in table 1.

Table 1. The preferential-etching solution for silicon single crystals

Solution	Composition	Etching rate(μm/min)	Detected Defects
Sirtl[3]	HF(49%):CrO$_3$(5M)=1:1	~1	Oxidation induced Stacking Fault (OSF)・Dislocation
Secco[4]	HF(49%):K$_2$Cr$_2$O$_7$(0.15M)=2:1	~1.5	OSF
Wright[5]	HF(49%):HNO$_3$(70%):CrO$_3$(5M):Cu(NO$_3$)$_2$:H$_2$O:CH$_3$COOH=60ml:30ml:30ml:2g:60ml:60ml	~1	OSF・BMD・Dislocation
Dash[6]	HF(49%):HNO$_3$(70%):CH$_3$COOH(99.9%)=1:3:12	~0.03	OSF
Schimmel[7]	HF(49%):CrO$_3$(1M)=2:1	~1.75	Dislocation
Yang[8]	HF(49%):CrO$_3$(1.5M)=1:1 ⇒ Y3 solution	~1.5	BMD
Sato[9-10]	HF(49%):HNO$_3$(70%):CH$_3$COOH:H$_2$O=1:12.7:3:3.7 ⇒ JIS"A"solution	~1.8	OSF・BMD
Modified-Dash[10]	HF(49%):HNO$_3$:H$_2$O:AgNO$_3$=1:2.5:10.5:0.005mol/l ⇒ JIS"E"solution	1.4~3.3	OSF・BMD

The specific etching solutions are selected depending on the type of crystal defects to be tested. The appropriate choice to reveal the large dislocations created during crystal puling would be e.g. the Sirtl etch. Small defects on the surface of a silicon wafer are revealed by using the Dash etching solution, which has a low etch-rate. The Wright etch, which contains Cr^{6+}, is used for the etching of silicon in about 60% of cases. Thus, it is very important to understand different types of defect that are present in the silicon in order to select the appropriate etching solution. Methods that utilize the laser-scattering technique are shown in Table 2. The laser-scattering techniques include the 90 degrees scattering method, the Brewster-angle incidence method, particle counting, etc.

Table 2. Techniques for laser-scattering from defects in / on a silicon wafer

Method	Model and Function			Country
	Manufacture	Model	Functions	
Laser Scattering Methods	Mitsui Kinzoku	MO-411	Obtains fault-scattering image of bulk defects by finely focusing a laser on a small region of the crystal	Japan
	Raton System Engineering	MILA-EXP	and scattering with the laser. Detectable densities: 10^4 to 10^{10}/cm^3. Density measurements and size evaluations are made through image processing.	Japan
Total Reflection Methods	Raton System Engineering	MILA	By focusing the laser at a slant to the cleavage plane, total reflection is obtained at the wafer surface and defects immediately below the surface are detected. These methods can observe defects just below the surface at detection sensitivities equivalent to laser-scattering methods.	Japan
Grazing Incident Methods (Brewster's angle)	Mitsui Kinzoku	For bulk defects. MO-511	Obtains a set of bulk longitudinal defect image slices in a non-distractive manner to get bulk defect distributions. Detectable densities: 10^4 to 10^{10}/cm^3.	Japan
		For near surface defect. MO-521	Gives the placement of defects near surface the crystal surface from the measurement-position dependency in an non-distractive manner. Detectable densities: 20 to 10^8 /cm^3. Totally automated non-destructive measurement of defects near the crystal surface.	
		LSTD Scanner MO-6		
	Hitachi High Technologies	OSDA	Inspection device for surface-layer defect.	Japan
	Ratoc System Engineering	MILSA	Non-destructive inspections for near surface defect.	Japan
Double Dark-Field Method	Hitachi DECO	LS6600	Inspection device for surface faults	Japan
	Topcon	WM-5000	Inspection device for surface faults	Japan
	KLA-Tencor	SP1-TB1	Inspection device for surface faults using grazing-incident method.	USA
		SP1	Inspection device for surface faults using a perpendicular laser.	
Perpendicular Examination methods	Semilab	SIRM300	Inspection device for bulk defects with a confocal, reflective differential-interference system together with back scattering.	Hungary
	ACCENT	OPP300	Defect measurement using transmission differential-interference measurements.	USA
	Laser Tech	MAGICS	Measures surface-layer defects, faults, and surface-shape transformations with a reflective, confocal microscope.	Japan

We performed a survey of these, as follows. A questionnaire was sent to 22 member organizations of the JEITA Silicon Technical Committee, which consists of cross-industrial participants, and a Silicon Measurement Committee. The questionnaire was designed to elucidate the following information: Which of the organizations used a preferential-etching method, and which ones a laser-scattering method. When preferential-etching was being used, we asked who used a non–Cr etching solution to JIS-H0609, and who used Wright etch or M-Dash (cf. comments to table 1). These were further subdivided according to the etch-depth that was used, which could be between 2 and 5μm. Regarding the processes used for counting BMD, all of the organizations that were questioned utilized a microscope, while a few organizations also utilized automated counting using an optimized microscope and a CCD system. Regarding the laser-scattering method, we found that all of the organizations utilized the 90 degrees laser-scattering method [11-12]. At that method, we decided to ignore standardization of "transmission type" systems, because this approach now appeared to be phasing out of commercial use. The questionnaire allowed us to formulate a set of measurement conditions and to carry out a "round robin" measurement exercise.

Round robin test conditions
The tests were intended to uncover gaps in the test methods and to correlate the results from each organization in order to achieve standardization. An additional aim was to compare commonly applied preferential etching with 90 degrees laser-scattering tomography. Six wafer vendors, two equipment manufacturers and one overseas research organization group participated in the group experiment. However, only the silicon vendors participated in the round of etching trials. 200mm diameter p-type CZ silicon sample wafers were prepared. The resistivity of all of the samples that were tested was approximately 1 Ωcm. The BMD density levels were 1×10^8, 5×10^8, and $9 \times 10^9 / cm^3$, and the DZ widths were targeted at 7, 10, 30 and 200μm. All of the samples were annealed at 780°C for 3 hours +1000°C for 16 hours in dry O_2 to facilitate the oxygen precipitation, after which all of the samples were separated into halves by cleaving. One half was assigned to the scattering method, while the other half underwent preferential-etching by etching off 5 μm of material with modified Dash solution, following the procedure in the JIS H609 Standard Document. The cleaved surfaces were cleaned with clean room wipers containing an organic solvent such as the ethanol to avoid any measurement error. Regarding the measurement locations, we identified three positions at 99mm, 100mm and 101mm from the edge of the wafers. BMD density measurements were taken at a depth of 200μm from the wafer surface, where an exposed surface was observed. For the DZ width measurements, we performed observations on the wafer surface and then recorded first, second and third defects at depths that were noted as DZ1, DZ2 and DZ3, respectively. For the 90 degrees scattering method, we decided to study the same positions that were examined for the etching method, only using the other halves of the wafers.

Results and discussion
BMD density results, comparing the etching and 90 degrees laser-scattering methods
Fig.1 indicates the results from each laboratory in this test. Seven samples were divided into two groups comprising three levels of samples for BMD density use and four levels of samples for DZ

width use. The Y-axis indicates the average of three values for the wafers from each laboratory, and the X-axis indicates the grand average for each of the samples from all laboratories. The dotted line in Fig.1 shows a 1:1 relationship. The actual BMD levels were between 1×10^7 and 1×10^{10} /cm^3, as expected. From this result, we concluded that there was no difference between laboratories in terms of BMD counting and observation using optical microscopy with an etching technique. Thus, it was understood that there should be no concerns about the standardization of BMD measurements when carrying out preferential-etching. In this experiment, laboratory D' had made measurements using an automated system, combined with an optical microscope and a CCD camera. This system was in good agreement with normal manual inspection and management under favorable conditions. The results of the 90 degrees laser-scattering method experiments are shown in Fig 2. The Y- and X-axes are the same as those used in Fig.1, including the average of three points and the grand averages. In this case, laboratory *I'* had performed measurements using the back scattering method (SIRM method)[13]. As in the preferential-etching results, the results showed good agreement between the different laboratories, and therefore we need not be concerned about the standardization of BMD density using the 90 degrees laser-scattering method. Fig.3 compares the results of the etching and laser-scattering methods, including the grand average values from all laboratories.

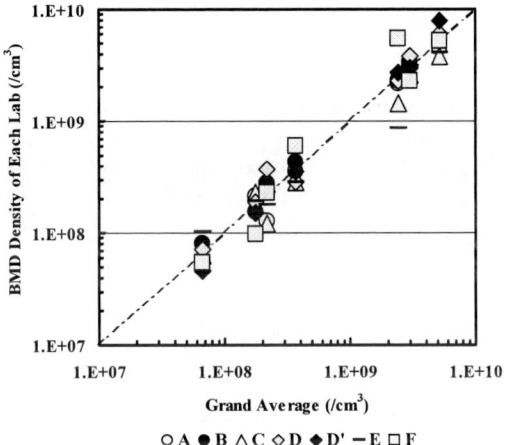

O A ● B △ C ◇ D ◆ D' — E □ F

Fig 1 Comparison of results from each measurement laboratory using the preferential-etching method.

In this case, the Y-axis indicates the BMD density values obtained from the 90 degrees laser-scattering method, whilst the X-axis indicates the BMD density from the preferential-etching method. The results demonstrate good correlation between the two methods, and therefore it was believed to be possible to create a joint standard for the measurement of BMD.

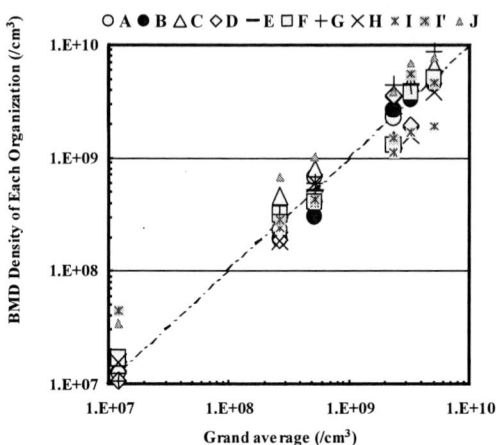

Fig 2 Comparison of results from each measurement laboratory using the 90 degrees laser-scattering method.

However, we found that the measurement results were slightly different at extremely low BMD densities below 10^8/cm^3.. It seems that large statistical errors were caused by narrow fields of observation. When using the preferential-etching technique with manual inspection, the operator automatically changed the magnification of the observation microscope to view a wider field area, whereas the 90 degrees laser-scattering method was performed at a fixed magnification and field of view.

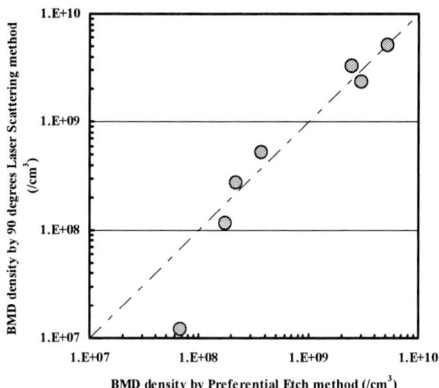

Fig 3 Comparison of the BMD density measurement by the 90 degrees laser-scattering and the preferential-etching methods.

These differences disappear at large BMD densities, above $10^{10}/cm^3$. Thus, we recommend that the standard method should also adopt changing the magnification and the field of view like during manual measurement with preferential etching.

Results of DZ width measurements

The samples for DZ width measurements included four different levels of DZ width but DZ widths on 200μm wafers were excluded from the analysis because the measurements were not sufficiently reproducible at such low BMD densities. Fig.4 shows DZ3 values (representing the third defect measurement from the ML surface) using the preferential-etching method. The Y-axis indicates the DZ width from each laboratory, and the X-axis represents the grand average from all laboratories. The standard deviation of the results from each laboratory was 6.1 μm. The reason for the large standard deviation may stem from the variation in DZ1 values, representing the first defect position from the ML surface.

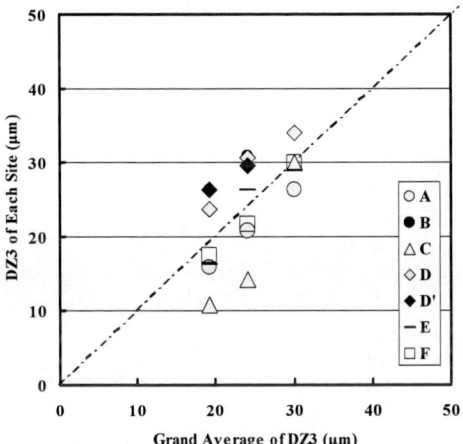

Fig 4 Comparison of DZ3 values from each laboratory using the preferential-etching method

Additionally, dirty wafer surfaces, which were a result of handling during shipments in the "round robin" test, also, contributed to the errors. Furthermore, the variations occurred to a different extent for each laboratory. Nonetheless, it was still possible to classify each one of levels of samples. In a similar fashion, we show results from the 90 degrees laser-scattering experiments in Fig.5. This result indicates the present normal equipment status in each laboratory. In this figure, only two sample levels are shown, because the width on the third level was greater than 50μm. The standard deviation of these results was 4.9 μm. One conclusion of these experiments was that we were now in a position to classify each level of sample.

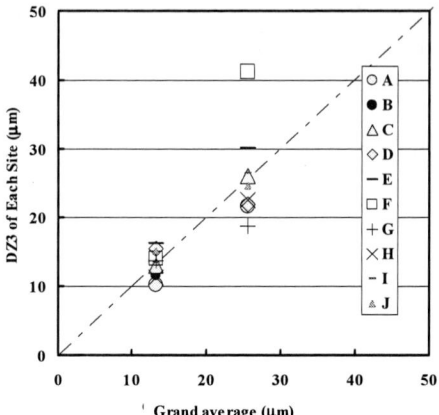

Fig 5 Comparison of DZ3 values from each laboratory
using the 90-degree laser-scattering

A comparison of the two measurement methods is shown in Fig.6. This figure shows the average of DZ1, DZ2 and DZ3 values obtained at each laboratory. The data were in good agreement over the range of 10μm to 25μm. However, there was an increasing gap between the etching and laser-scattering methods at larger DZ widths. It was shown that the two methods were sufficiently correlating. However, we have recommended that the relationship between the preferential-etching and 90 degrees laser-scattering methods needs to be confirmed.

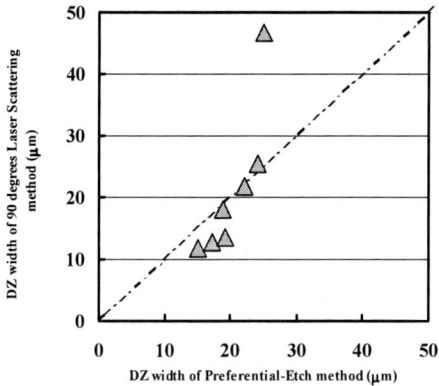

Fig 6 Comparison of DZ width measurements using the 90 degrees
laser-scattering method and the preferential-etching method.

Influence of the field-of-view at low BMD densities or wide DZ widths

The origin of the difference between the two methods in Fig.6 is believed to stem from the large deviation in the location of the measured positions with decreasing BMD counts in a particular field-of-view. Therefore, we used the 90 degrees laser-scattering method to examine the extent of changes in the DZ width with increasing field-of-view. The result, shown in Fig.7, indicates good agreement between the two measurement techniques in a "normal" DZ width-range of 25μm. However, deviations occurred when the DZ width was around 40μm. In contrast, when the field-of-view was enlarged in order to observe an increasing number of defects such as the defects in the tested circle shape, the measured values approached those obtained when using preferential etching. The dimensions of the field-of-view were originally 200μm x 166μm. This was then increased by a factor of between 3- and 8-times. Thus, in the case of very low levels of BMD density, good agreement between both methods was obtained by enlarging the field-of-view. With this modification, the measurements of BMD densities were in good agreement between the participating laboratories having returned results that were strongly correlating with both the etching and laser-scattering methods. Thus, we decided that it was possible to standardize both of the correlating methods. Regarding the measurement of DZ width, we found that each level of DZ width could be classified. However, it cannot be said that there was absolute quantitative agreement. Nevertheless, we judged that standardization would still be viable.

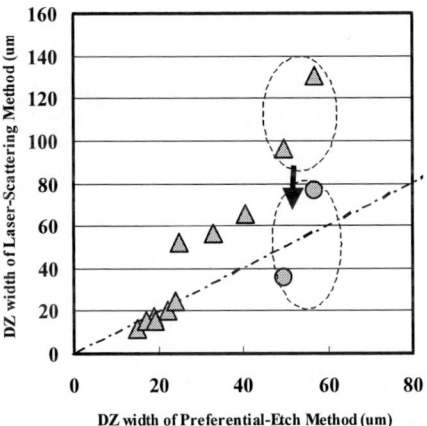

Fig 7 DZ width after increasing the measurement area using
the 90 degrees laser-scattering method

A future problem

The definition and identification of an appropriate method for determining DZ width

The definition of DZ width and the identification of a measurement method have been noted as problems for the future. Essentially, the mean of the DZ measurements did not relate to crystalline defects in the near-surface region. However, DZ in this standard had been described as the DZ mean of thermally-induced oxygen precipitates (BMD). In general, the width of the DZ would be required to be several μm on an IC device. However, the measured DZ widths were very diverse in this study. From this standpoint, the requirement for a short DZ width was low in comparison with the BMD density required to effect a gettering capability for heavy metal impurities. The probability of void defects (so called COP: Crystal Originated Particle) occurring was very high in comparison with BMD in the near-surface region of a silicon wafer, down to several μm in depth, whereas the COP density was very low, at about below 10^6 /cm^3. In this study, we separated the DZ (detected BMD-free zone) and COP free zones. In this case, the actual DZ width was determined from the measurement of the first defect as an etch-pit or a scattering point from the polished surface. Decisions regarding the DZ width were mainly taken from DZ1 to DZ3, which had been performed using the old method by most wafer venders. Other techniques used the average of DZ1-DZ3. However, the first defect measurement was a basic one that could be used for any DZ width decision. The DZ width could be in error if the first defect position was misjudged. For example, we can evaluate accurately in those cases where the DZ width is associated with a high BMD density. However, in cases with a low BMD density corresponding to a large DZ width, there was a significant difference between the preferential-etching and 90 degrees scattering methods. In the first trial of this large-scale round robin test for defect measurements, we ascertained the status of each site. This permitted an understanding of the problems that are encountered when measuring DZ width.

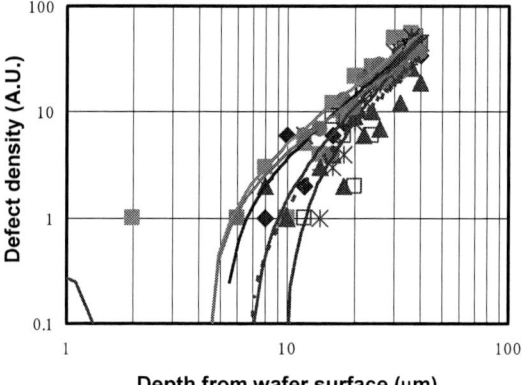

Fig 8 Approximation by survey and the secondary function of the depth profile for defect density. By increasing the measurement area, the values approach those obtained from the etching results.

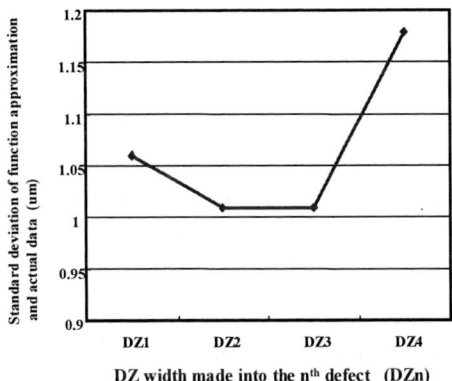

Fig 9 Standard deviation of the defect position approximating the BMD density profile with the secondary function, and the actual values of DZ 1 to 4

For example, it is necessary to scrutinize the definition of DZ width. On the other hand, we had examples of good repeatability of results for the determination of DZ width using the DZ3 position. Here, Fig. 8 shows the BMD density depth profiles of 6 pieces of DZ silicon wafers. We made a curve-fitting approximation using a quadratic function for the BMD depth profiles of these wafers. We then simulated their DZ widths by using a curve-fitting formula. Assuming that the DZ width was associated with the first, second, third and fourth defects from the polished surface, we then show a comparison between the measured values and assumed values, with standard deviations, in Fig.9. We found that the standard deviations were lowest in the DZ2 and DZ3 cases. This being the case, we thought that a good way to specify DZ width would be to apply DZ3 data without a distribution function for the defect density. In other words, the first-counted defect results in a large variation in the expected position as a result of statistical defect-distributions in a silicon wafer, because the DZ1 defect was not a BMD induced by thermally-induced oxygen precipitation, but was more likely to be a void-defect with a low density, below 10^6 /cm^3. Again, it can be assumed that a large variation is associated with a narrow field-of-view. In addition, Inoue et al. [14] have reported that the 90 degrees laser-scattering method can be used to measure BMD depth-profiles automatically, without operator error. Furthermore, they reported different detection limits for the preferential-etching and 90 degrees laser-scattering methods. Regarding DZ width, they recommended an analysis technique for DZ width in which the first step was to evaluate the BMD depth profile in the silicon bulk, and, then to curve-fit the defect results at different depths into the polished surface, defining the DZ width as the position where the defect density was 1/e or 1/10. We judged that such a depth value should be defined as a 'defect growth layer' width property rather than a DZ width.

Influence of detection limits

In this analysis, we had assumed large differences in the measurement data between the preferential-etching and 90 degrees laser-scattering methods, but these differences were actually slight. Because of this slight discrepancy between two the methods, we believe that, due to small errors in counting, the annealed wafers for the 'round robin' exercise possessed larger-sized defects than the as-grown wafers. We needed to know the real target BMD size and the detection limit of each method and to then optimize the detection limit when it was found to be necessary. From the 90 degrees scattering results, the average scattering intensities of the BMD measurements in these tests were between 800,000 and 5,000,000 counts. However, the detection limits of the laser-scattering method were assumed to be several tens of counts. Therefore, the 90 degrees scattering method should be able to detect defects of 1/10 the size in the test samples, where we have assumed the size of the BMD to be proportional to the 6th power of the BMD diameter. Although the size was considered to be about 10 nm at present, direct observation by TEM (Transmission Electron Microscope) is fraught with difficulties. On the other hand, it has been reported that direct observation of the BMD can be carried out by using RIE (Reactive Ion Etching) [15-16]. Here, we studied information about BMD size for a test sample by using RIE and TEM observations. In details, the measurement of BMD size from the surface to a depth of 20μm was carried out using the RIE method, and further size information was obtained at depths of over 100μm by direct TEM observation. The BMD sizes were defined as the average of the values between "a" and "b", as shown in Fig.10. Fig.10 shows four typical structures of oxygen precipitates within a silicon wafer; octahedral, transformation of void defect 1 or 2, and platelet, respectively [17].

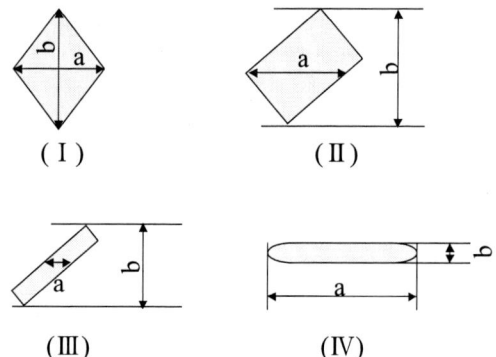

Fig.10. Definition of BMD size from TEM result

(I) Octahedral, (II) Void transformation defect 1, (III) Void transformation defect 2, (IV) Platelet

So far, we have mostly observed defects of types (I) or (II) during TEM observations. Fig.11 shows the depth profiles of BMD size obtained by the RIE and TEM methods. In this figure, the circle symbols show the data obtained from RIE, and the triangle symbols show the data obtained from the

TEM method. From this result, we estimated the detection limit for BMD size that can be derived from DZ width information; this turned out to be approximately 60nm by preferential-etching and 41nm by the 90 degrees laser-scattering methods.

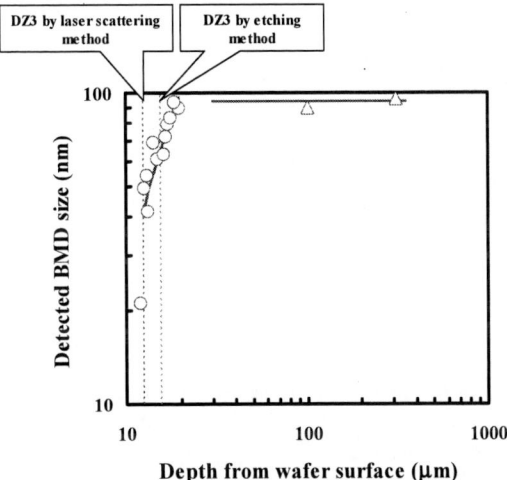

Fig.11 Depth profile of BMD size using one test sample by RIE and TEM observations.

The sample was determined to have a DZ width of 17.5 μm by the preferential-etching method and of 13.4μm by the 90 degrees laser-scattering method, respectively. Therefore, the BMD value for the laser-scattering method of 41nm is very reasonable; it was previously reported as being 30-40nm by Moriya et al. [11]. We believe that the same detectable levels of BMD size should be achieved between the preferential-etching and laser-scattering methods for a variety of silicon crystals. It should be noted that, the smaller a device structure is, the more it is affected by crystallographic defects. Recognizing this connection, BMD and DZ could be easily discriminated by the conventional preferential-etching method. However, exact measurement is difficult; it is very time-consuming, but information regarding crystal defects in the device layers near the wafer surface has recently become important. In order to acquire information about crystal defects close to the detection limit using the laser-scattering method, it is believed that examination using methods involving direct observation, such as TEM and RIE, are required for comparison.

Conclusion

A large-scale experiment to evaluate defect measurements has been carried out in a 'round robin' exercise as a first trial, and has evaluated the conditions presently employed at each of the sites involved in the trial. As a result of the experiments, several problems became clear. For example, it is particularly necessary to consider the definition of DZ width. There is also difficulty in establishing a definitive standard, since the detection limit depends on the depth from the surface. Recently, Inoue et al. have developed test methods using a simple laser-scattering-method, and noted that they were not operator dependent, although there were differences in the detection limits. Regarding the determination of DZ width, their method required some knowledge of the BMD density distribution in the bulk. Their method allows the calculation of DZ width as a function of the 1/e value. This can be obtained from the defect curve, which decreases toward the polished wafer surface. Accurate and meaningful results will depend on further advances in measurement technology in the future.

Acknowledgement

The authors would like to thank all participating laboratories involved in the inter-laboratory studies for all measurements and for constructive discussions. In particular, we would like to recognize Dr. M. Miyazaki of SUMCO, Mr. K. Takanashi, K. Sato, M. Nakase, Dr. T. Takahashi and K.Nakai of JEITA for their support. Also, the authors thank Miss N. Ohmori of Toshiba Ceramics for technical support with TEM observations.

Reference

[1]   http://www.jeita.or.jp/english/standard

[2]   T.Y.Tan, E.E.Gardner and W.K.Tice: Appl.Phys.Lett. **28**, 564 (1976)

[3]   E.Sirtl and A.Adler: Zeischrift for Metalkunde., **52**, 529 (1961)

[4]   F.Secco, d'Aragona: J.Electrochem.Soc., **119**, 948 (1972)

[5]   M.Wright: J.Electrochem.Soc., **124**, 757 (1977)

[6]   W.C.Dash: J.Appl. Phys., **27**, 1193 (1956)

[7]   D.G.Schimmel: J.Electrochem.Soc., **126**, 480 (1979)

[8]   H.Yang: J.Electrochem.Soc., **131**, 1140 (1984)

[9]   S.Nomura,R.Takeda and S.Takasu: Electrochem.Soc.,1990 Spring meeting in Montreal, Extended Abstract p430.(1990)

[10]   JIS H0609-1999: "Test Methods of crystalline defects in silicon by preferential etch techniques", Japanese Standards Association, 1st English Edition (2004)

[11]   K.Moriya,K.Hirai, K.Kashima and S.Takasu, J.Appl.Phys., **66**. 5267 (1989)

[12]   K.Moriya and T.Ogawa,: J.J.Appl.Phys **22**, L207 (1983)

[13]   G.R.Booker, Z.Laczik and P.Kidd, Semicond. Sci. Technol., **7**, A110 (1992)

[14]   N.Inoue, K.Moriya, K.Kashima, R.Takeda, V.Akhmetov, O.Lysisky and K.Nakashima: in proc. of "The 4th International Symposium on Advanced Science and Technology of Silicon Matrials" p123 (2004)

[15]   K.Nakashima, Y.Watanabe, T.Yoshida and Y.Mitsushima: J.Electrochem.Soc., **147**, 4294 (2000)

[16]   K.Nakashima, T.Yoshida and Y.Mitsushima: J.Electrochem.Soc. **152,** G339, (2005)

[17]   H.Fujimori, H.Matsushita, I.Oose and T.Okabe: J.Electrochem. Soc., **147**, 3508 (2000)

ECS Transactions, 2 (2) 485-490 (2006)
10.1149/1.2195683, copyright The Electrochemical Society

X-ray Reciprocal Space Mapping and Synchrotron Radiation Topography of Strained Si/Si$_{1-x}$Ge$_x$ on Bonded SOI

T. D. Ma [a, b], H. L. Tu [a], G. Y. Hu [b], B. L. Shao [b], A. S. Liu [b]

[a] National Engineering Research Center for Semiconductor Materials, General Research Institute for Nonferrous Metals, No. 2 Xinjiekouwai Street, Beijing 100088, P. R. China
[b] National Center of Analysis and Testing for Nonferrous Metals and Electronic Materials, General Research Institute for Nonferrous Metals, No. 2 Xinjiekouwai Street, Beijing 100088, P. R. China

> X-ray triple-axis diffractometry (TAD) and synchrotron radiation double-crystal topography (SRDT) have been utilized to investigate strained Si/Si$_{1-x}$Ge$_x$ on bonded SOI. The structural characteristics of the Si capping layer and the underlying Si layers (the Si buffer and SOI top Si layer) have been described using high resolution reciprocal lattice mapping (HRRLM). The mosaicity in the SiGe layer leads to (004) diffraction peak broadening asymmetrically. The crosshatched contrast spatially correlated with the misfit dislocations (MDs) has been observed both in the SiGe layer and the underlying Si layers. The granular pattern due to the mosaic blocks is also shown in the topograph of the SiGe layer. It is argued that the compressive strain in the SiGe layer has been primarily released at the interface between the SiGe layer and the underlying Si layers.

Introduction

Strained Si on relaxed SiGe layers has been extensively studied in recent years for its promising applications in microelectronic devices. (1) The enhanced carrier mobilities in strained Si channel layer allow the development of high performance metal-oxide-semiconductor field-effect transistors (MOSFETs). (2, 3) The combination of strained Si with SiGe on SOI offers a new opportunity to develop MOSFETs with higher electron and hole mobilities. (4) Understanding of the relaxation mechanism in strained Si/Si$_{1-x}$Ge$_x$ on SOI is extremely important, because substantial relaxation usually degrades the electrical and optical performance of the devices (5). The use of TAD has been introduced to provide an essentially delta-shaped reciprocal space probe (6, 7), which allows the nondestructive and precise determination of the strain in the active layers and the composition of the virtual substrates through HRRLM (8-11). Double-crystal reflection topography has matured as a highly deformation sensitive technique for strained epitaxial layers. (12) It has the advantage that the substrate and epilayer can be imaged independently and defects in each layer can be identified. (13) The early application of SRDT was in the study of MDs in GaAlAs epitaxial layers on GaAs (14) and SRDT dislocation images were compared with theoretical simulations (15). The combination of different x-ray topography techniques and reciprocal space mapping has been used to monitor the early stages of relaxation in silicon p-on-p+ epitaxial structures. (16) It is somewhat interesting that relatively little attention has been paid to combining these two techniques to investigate the strain relaxation and the spatial distribution of defects in strained Si/Si$_{1-x}$Ge$_x$ on bonded SOI.

Expeimental

A specimen was grown in the UHVCVD system. The starting bonded SOI substrate was chemically cleaned for 15 minutes in a boiled solution of H_2SO_4: H_2O_2= 4:1, then rinsed in de-ionized water for at least 10 minutes. Native oxide on the wafer surface was etched in 10 % HF solution for 30 seconds and then ferried into the loading chamber. When the vacuum reached 10^{-5} Pa, the wafer was put into the growth chamber. The growth chamber was pumped with a 1000 l/s turbo-molecular pump and a base pressure of 5×10^{-7} Pa was obtained. The chamber pressure was maintained below 0.13 Pa during the epitaxial process. As soon as the base vacuum was approached, the temperature would be raised to 750-800 °C at a high ramp-up rate. After the wafer was heated in the vacuum for 5 minutes, the temperature would descend to the growth temperature, for Si, 600 °C and for SiGe, 550 °C. Pure SiH_4 and 15 % GeH_4 diluted in H_2 were used for SiGe epitaxial growth. The flow rates of SiH_4 and GeH_4 were 10 sccm and 2 sccm respectively during the growth. The final structure from top to bottom was composed of a 20 nm-thick Si capping layer, a 35 nm-thick $Si_{0.81}Ge_{0.19}$ layer, a 15 nm-thick Si buffer layer, and a bonded SOI wafer with a 85 nm-thick top Si (001) layer. The specimen was in-situ thermally treated at 750 °C for 30 minutes to achieve partially relaxed SiGe layer. TAD measurements were performed using a Philips X'Pert diffraction system equipped with a beam monochromator consisting of a four-bounce Ge (220) crystal that gave an incident Cu $K_{\alpha1}$ beam with a divergence of 0.002 ° and wavelength dispersion of 2.3×10^{-5}. (17) HRRLM was obtained by performing $\omega/2\theta$ scans (rotation of the sample and detector; ω: the angle of incidence, 2θ: the diffraction angle) for different offsets given by ω scans (rotation of only the sample) with a three-bounce Ge (220) analyzer crystal between the specimen and the detector. The (004) and (113) diffractions were measured in the triple axis mode. SRDT was performed at the topography station at the 4W1A beam line of Beijing Synchrotron Radiation Laboratory (BSRL). Two parallel Si (111) crystals were used as the monochromator in double-crystal diffraction optics and (004) diffractions were examined. The storage ring was running at 2.2 GeV with beam current varying between 50 and 100 mA. The topographs were recorded on Fuji films and developed on one side of the exposed plate.

Results and Discussion

The crystallographic misalignment within strained $Si/Si_{1-x}Ge_x$ on bonded SOI allows the direct measurements of the normal and in-plane lattice parameters of the Si layers over the buried oxide and the SiGe layer using TAD. (18) A set of $\omega/2\theta$ scans (radial from (000)) at different ω offsets (transverse in reciprocal space) have been measured and plotted as isointensity lines around (004) and (113) reciprocal lattice points of the Si layers over the buried oxide and the SiGe layer. The calculation results with respect to the TAD measurements are listed in Table 1.

Two diffraction peaks are found in (004) and (113) HRRLM of the Si layers over the buried oxide (see Figure 1a and 1b), which have been identified from the Si capping layer and the underlying Si layers (18). On the basis of the comprehensive analyses, it has been deduced that the tensile strain in the Si capping layer is 0.04 % and both the tensile strain and Ge diffusion from the SiGe layer lead to 2θ shift of the underlying Si layers in (004) HRRLM.

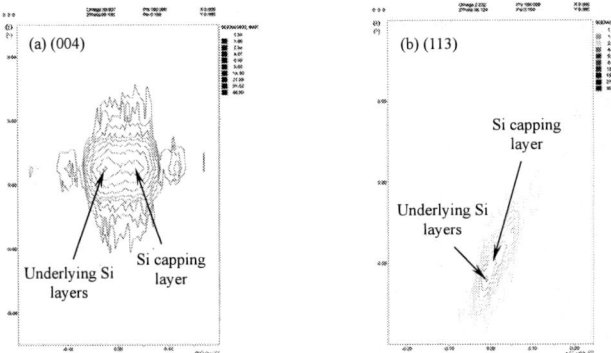

Figure 1. (004) and (113) HRRLM show the two diffraction peaks from the Si capping layer and the underlying Si layers.

(004) and (113) HRRLM (see Figure 2) of the SiGe layer indicate that the relaxation percentage of the SiGe layer is 4.7 % and the lattice constant of the unstrained SiGe layer corresponds to Ge mole fraction 18.7%. The diffracted intensity in the tail of the SiGe layer is not symmetric with intensity broadening toward the positive ω angle (see Figure 2a). This is assumed due to the formation of MDs at the lower interface of the SiGe layer, which induces the increase in the maximum angular tilt between the mosaic blocks during relaxation process (19). The elongation of the SiGe (004) peak along the 2θ/ω scan direction (see Figure 2a) suggests that the strain variation is a dominant effect over the mosaicity.

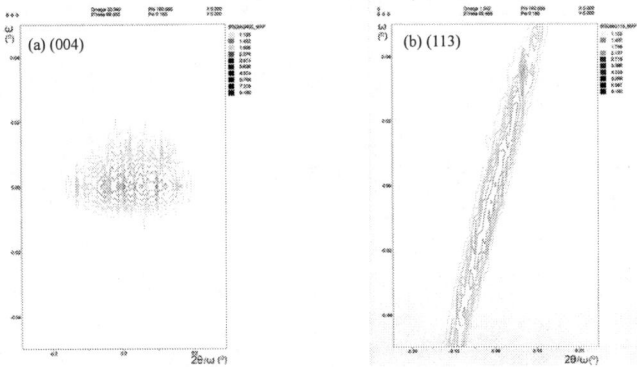

Figure 2. (004) and (113) reciprocal space maps of the SiGe layer: (004) HRRLM indicates the strain variation in growth direction is dominant effect over mosaicisity.

Figure 3a and 3b are assigned to the Si capping layer and the underlying Si layers respectively. Further details about the identification process can been found in Ref. 18. Very faint crosshatched contrast can be observed in Figure 3a and the crosshatched contrast is clear and parallel to the <011> directions in the Figure 3b. The crosshatched contrast has been proved to correlate spatially with the presence of MDs. (20) Since larger lattice distortion can lead to sharper crosshatched contrast, more MDs exist in the underlying Si layers than in the Si capping layer. Not only crosshatched contrast but also granular pattern can be observed in the SRDT topograph of the SiGe layer (see Figure 3c). The granular pattern is attributed to the mosaic blocks that lead to intensity broadening asymmetrically as shown in Figure 2a. The different contrasts in the SiGe layer and the underlying Si layers have been due to the individual dislocation distribution, and the strain relaxation in the SiGe layer does not predominantly depend on the compliant substrate effect (21). The MDs distribution both in the SiGe layer and the underlying Si layers demonstrates the partial relaxation of the compressive strain in the SiGe layer occurs at its lower interface.

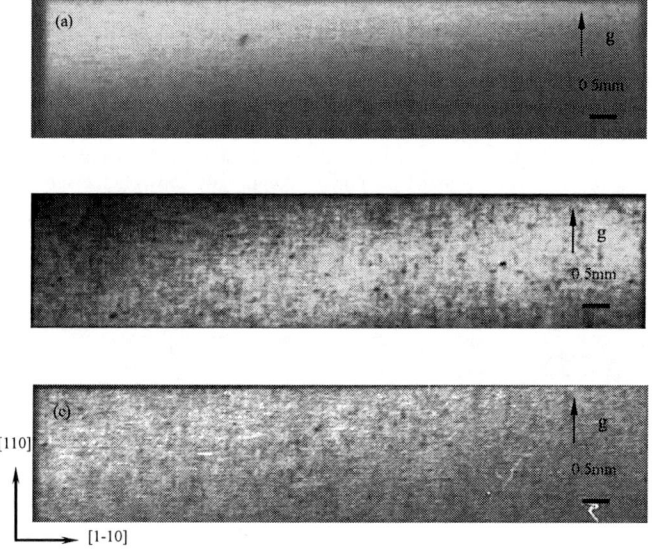

Figure 3. (004) SRDT topographs of strained Si/SiGe on bonded SOI: (a) the topograh of the Si capping layer showing faint crosshatched contrast; (b) the topograph of the underlying Si layers showing sharp crosshatched contrast; (c) the topograh of the SiGe layer showing the crosshatched contrast and granular pattern. The arrow with "g" indicates the projection of 004 scattering vector.

Table I. Out-of-plane and in-plane lattice constants, unstrained lattice constants, relaxation percentage and tensile strain measured by TAD.

Measured structures	a_v(nm)	a_p(nm)	a_0(nm)	ε_p(%)	R(%)
Si capping layer	0.54286	0.54328	0.54304	0.04	-
SiGe layer	0.55041	0.54330	0.54733	-	4.7
Underlying Si layers	0.54324	0.54336	0.54329	0.013	-
Bulk Si substrate	0.54309	0.54309	0.54309	-	-

Conclusions

Strained $Si/Si_{1-x}Ge_x$ on bonded SOI has been characterized using TAD and SRDT. The strain status and diffusion within the Si layers over the buried oxide have been depicted. The mosaic blocks are due to the strain relaxation at the interface between the SiGe layer and the underlying Si layer, which lead to the diffracted intensity in the tail of the SiGe layer broadening asymmetrically. The crosshatched contrasts have been observed both in the underlying Si layers and in the SiGe layer by SRDT. The granular pattern in the topograph of the SiGe is assumed due to the mosaic blocks. All experimental results demonstrate that the compressive strain in the SiGe layer has primarily been released at the interface between the SiGe layer and the underlying Si layers.

Acknowledgments

This work is supported by the contract No. 50502008 from National Natural Science Foundation of China.

References

1. J. Welser, J. L. Hoyt, and J. F. Gibbons, IEDM Tech. Dig., p. 1000, 1992.
2. T. Mizuno, S. Takagi, N. Sugiyama, H. Satake, A. Kurobe, and A. Toriumi, IEEE Electron Device Lett. 21, 230 (2000).
3. K. Rim, J. L. Hoyt, and J. F. Gibbons, IEEE Trans. Electron Devices 47, 1406 (2000).
4. J. Welser, J. L. Hoyt, and J. F. Gibbons, IEEE Electron Devices Lett. 15, 100 (1994).
5. R. Hull, J. C. Bean, C. Buescher, J. Appl. Phys. 66 (1989) 5837.
6. P. F. Fewster, J. Appl. Cryst. 22, 64 (1989).
7. P. F. Fewster, J. Appl. Cryst. 24, 178 (1991).
8. J. M. Li, H. L. Tu, G. Y. Hu, C. Q. Wang, A. S. Zheng and J. Y. Qian, Mater. Resear. Bull., 38 (2003) 675.
9. M. O. Tanner, M. A. Chu, K. L. Wang, M. Meshkinpour, M. S. Goorsky, J. Cryst. Growth, 157 (1995) 121.
10. G. M. Cohen, P. M. Mooney, E. C. Jones, K. K. Chan, P. M. Solomon, and H.-S. P. Wong, Appl. Phys. Lett., 75 (1999) 787.
11. E. M. Rehder, C. K. Inoki, T. S. Kuan and T. F. Kuech, J. Appl. Phys., 94 (2003) 7892.
12. S. J. Barnett, (and 16 others) J. Phys. D 28, A17-22 (1995).
13. Z. H. Mai, C. Y. Wang, L. S. Wu, C. Jiang and J. M. Zhou, Science China (Series A), 37 (1994) 1125.
14. J. F. Petroff, M. Sauvage, P. Riglet and H. Hashizume, Phil. Mag. A, 42 (1980) 319.

15. P. Riglet, M. Sauvage, J. F. Petroff and Y. Epelboin, Phil. Mag. A, 42 (1980) 339.
16. M. S. Goorsky, P. Feichtinger, H. Fukuto, and G. U'Ren, Phil. Trans. R. Lond. A 357, 2777 (1999).
17. W. J. Bartels, J. Vac. Sci. Technol. B1, 338, 338 (1983).
18. Tongda Ma, Hailing Tu, Guangyong Hu, Beiling Shao, Ansheng Liu, J. Cryst. Growth (to be published)
19. K. Durose, A. Trunbull, and P. Brown, Mater. Sci. Eng. B 16, 96 (1993).
20. K. Sawano, S. Koh, Y. Shiraki, N. Usami and K. Nakagawa, Appl. Phys. Lett., 83 (2003) 4339.
21. T. D. Ma, H. L. Tu, B. L. Shao, A. S. Liu, and G. Y. Hu, Materials Science and Engineering B 124-125, 148 (2005)

SOI Low Frequency Noise and Interface Trap Density Measurements
With the Pseudo MOSFET

V.A. Kushner, J. Yang, J.Y. Choi, T.J. Thornton, and D.K. Schroder

Department of Electrical Engineering and Center for Solid State Electronics Research,
Arizona State University, Tempe, AZ 85287-5706, USA

Low frequency noise (LFN) is important in analog and digital
circuits. In analog circuits it affects the performance of low-noise
amplifiers and the phase noise (1) of voltage-controlled oscillators
(2). In digital circuits it becomes more important as the supply
voltage is reduced and it degrades substrate noise coupling. Low-
frequency noise is due to interactions of the channel carriers with
oxide/semiconductor interface traps and oxide charges and is very
dependent on the quality of the oxide/semiconductor interface and
noise measurements can give important information about such
interfaces and defects (3). Silicon-on-insulator devices have two
oxide/semiconductor interfaces and the bottom interface is
generally worse than the top interface. Most LFN measurements
are made after MOSFET fabrication, but it is desirable to
characterize such materials without fabricating devices. In this
paper we discuss silicon-on-insulator (SOI) low-frequency noise
and interface trap density measurements using a Ground-Signal-
Ground (GSG) pseudo-MOSFET structure with minimum
fabrication.

Advantages and Fabrication of GSG Pseudo MOSFET

The pseudo MOSFET (Ψ-MOSFET) is a simple, yet powerful, device to characterize
various aspects of SOI wafers and is routinely used for incoming wafer inspection to
determine material parameters (3-7). It comes in the *point contact* and the *mercury probe*
(HgFET) configurations. The point-contact Ψ-MOSFET simply requires two probes on
an SOI wafer. However, the contact geometry is poorly defined leading to questions in
the interpretation of the I_d-V_g data. The HgFET has the advantage of well-defined
source/drain contacts, but it has an Hg/Si interface and all the vagaries that accompany
metal/Si contacts, where barrier heights change with time due to surface state changes (5),
(7). The Ψ-MOSFET is routinely used for current-voltage, mobility, threshold voltage,
etc. measurements. However, to our knowledge is has not been used for noise
measurements. Yet, it is a convenient test structure for such measurements.

Motivation for Ground-Signal-Ground Ψ-MOSFET

Attempts to measure LFN of *point contact* Ψ-MOSFETs have not been successful
because of repeatability problems. It has not been possible to place the point probes
consistently on the same spots and apply the same pressure (6). Thus the results have had
significant variance from one measurement to the next one with the same operating point.
Mercury probes also have not been very suitable for LFN measurements due to their
dependence on the sample surface conditions (7).

For reproducible measurements we use deposited metal electrodes on the SOI wafer forming Schottky barrier source/drain contacts, and using the substrate as the gate. This allows us to use GSG probes for noise and frequency response measurements. Titanium was chosen for the contacts on p-type SOI wafers, because it has almost equal electron and hole barrier heights. Thus the titanium GSG Ψ-MOSFET allows us to analyze all regions from accumulation to deep inversion, which is not possible with conventional MOSFETs.

Fabrication of GSG Ψ-MOSFETs

The GSG Ψ-MOSFET is fabricated on an SOI substrate. GSG pads provide a stable silicon contact and good shielding from environment noise. The silicon island structure has a well defined device region compared to the traditional point probe pseudo MOSFET.

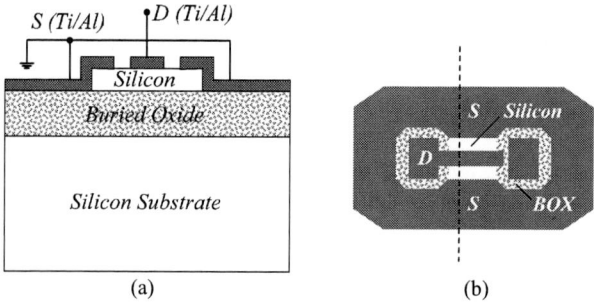

(a)   (b)

Figure 1. (a) Cross section through the dashed line on (b); (b) top view of the GSG Ψ-MOSFET. The unconnected island on the right is a gate contact artifact of the GSG FET mask.

The cross section of device structure is shown in Fig. 1(a). The fabrication is very straightforward. A p-type boron-doped bonded wafer, doped to 5×10^{14} cm^{-3}, is used. The silicon film and buried oxide thicknesses are 191 nm and 392 nm. To define the silicon island, the wafer is patterned with photoresist and dry etched with reactive ion etching (RIE). The silicon island has a dimension of 400 µm by 110 µm. After forming the silicon island, the device is coated by photoresist then a GSG pattern, shown in Fig. 1(b), is transferred to the device by lithography. The device is cleaned in 5% buffered oxide etch (BOE) for 1 minute to remove the native oxide and to terminate the contact region silicon surface with hydrogen bonds. Immediately after the HF rinse, 150 nm Ti and 150 nm Al films are deposited in sequence with electron beam evaporation. Titanium is selected as the contact metal due to its work-function near the mid-gap of silicon. Biasing the substrate with different polarity voltages, allows both electron and hole characteristics to be extracted. The Al layer provides better step coverage and acts as a protection layer for the Ti film as well. The GSG pad is then formed by the liftoff process after metal deposition. As the last step, the device was sintered for 30 seconds to get a good Ti/Si contact. To avoid forming titanium silicide, the sintering temperature was set to 450°C.

A micrograph of the finished device is shown in Fig. 2. High precision optical measurements on pseudo n-MOSFETs yield a gate length of 25 µm and a gate width of 760 µm (two 380 µm fingers).

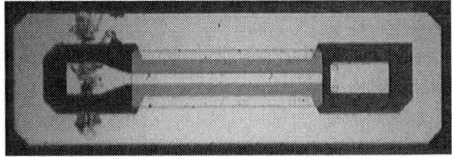

Figure 2. Micrograph of the pseudo n-MOSFET with 25 µm gate length and 760 µm gate width. The device structure is a GSG "two finger gate" pseudo n-MOSFET.

DC Measurements

Measurement Setup and Results

Measurements are made with the HP-4156B semiconductor parameter analyzer utilizing one GSG probe with three electrodes (Fig. 3). The two outer electrodes are shorted and grounded, connecting to the source of the Ψ-MOSFET. The center electrode carries the signal and connects to the drain. The gate voltage V_g is applied to the substrate through the probe station chuck. All voltages are referred to the grounded GSG Ψ-MOSFET source.

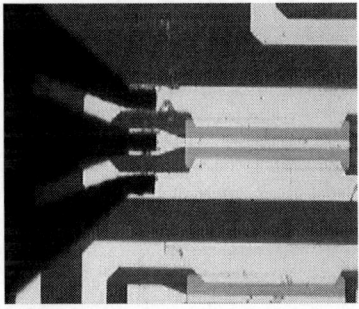

Figure 3. Ground signal ground (GSG) probe on the source (two outside contacts) and the drain (the contact in the middle).

The device has good control in both accumulation (Fig. 4(a)) and inversion (Fig. 4(b)). The significant range of currents in both regions is possible through the Ti drain and source contacts with the Schottky barrier height (0.56 to 0.6 eV). However, it leads to some complications. The OFF current changes significantly with an increase of the absolute drain voltage in the range of 1 to 5 V: from 1.2 nA up to 15 nA for negative drain bias (Fig. 5(a)) and from 1 nA up to 120 nA for positive drain bias (Fig. 5(b)). The maximum ON current reaches a few milliamperes in the accumulation regime (Fig. 5(a)) and a few hundred microamperes in the inversion regime (Fig. 5(b)), but is very different for positive and negative drain voltages (Figs. 6 and 7). These DC parameters make the GSG Ψ-MOSFET a suitable device for noise spectroscopy. However, caution needs to be exercised to account for the influence of the Schottky junctions, because such junctions introduce additional noise.

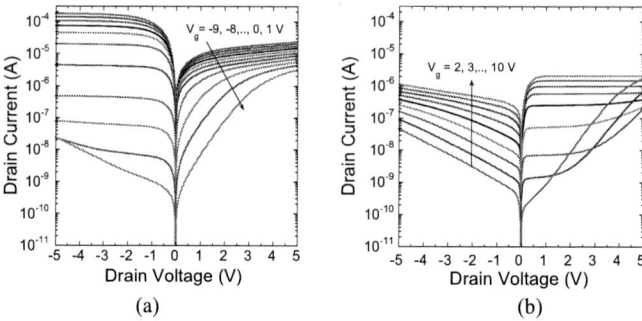

(a)   (b)

Figure 4. GSG n-MOSFET drain current versus drain voltage in (a) accumulation, (b) depletion and inversion regions.

(a)   (b)

Figure 5. GSG n-MOSFET drain current versus gate voltage for (a) negative, (b) positive drain voltages.

Basic GSG Ψ-MOSFET Model

Traditional point or mercury Ψ-MOSFETs often demonstrate behavior characteristic of long-channel MOSFETs because of almost ohmic drain and source contacts, achieved by point contact damage in the point-contact (4) or by low barrier height for either holes or electrons in the HgFET (7). Contrary to that, the GSG Ψ-MOSFET DC characteristics are more complex (Figs. 4-7), because for any bias conditions the drain current has to flow through one forward biased Schottky diode, the silicon film, controlled by the gate (substrate voltage in our case), and through the reverse-biased Schottky diode. Also the source contact area is twice as large as the drain contact area, making the GSG Ψ-MOSFET highly asymmetrical. Edge effects may cause some other secondary effects in spite of the large contact areas.

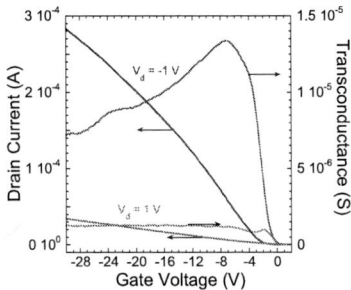

Figure 6. The reverse-biased Schottky junction at $V_d = 1V$ causes a significant decrease in the drain current and transconductance in the accumulation region.

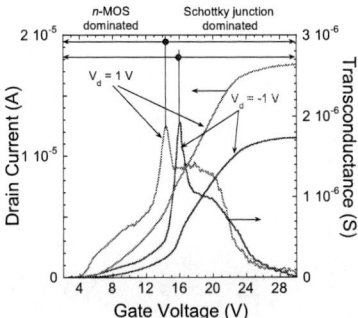

Figure 7. The asymmetrical device structure and the Schottky junctions cause a threshold voltage shift and a drain current reduction for $V_d = -1V$ versus $V_d = 1V$ in the depletion and inversion regions.

In order to simplify the device analysis we can assume that the device behaves as a long channel MOSFET in series with a reverse-biased Schottky diode. This means that the drain current is dominated by the lowest of the MOSFET or the Schottky contacts currents for any given bias conditions. Hence, we approximate the drain current of the GSG Ψ-MOSFET as

$$I_d = \frac{I_{MOS} I_{Schottky}}{I_{MOS} + I_{Schottky}}$$ [1]

where the MOSFET current $I_{MOS}(V_g, V_d)$ in accumulation or strong inversion follows the simple MOSFET equation (10)

$$I_{MOS} = \mu_{eff} C_{BOX} \frac{W}{L} \left(V_g - V_{th,FB} - \frac{V_d}{2} \right) V_d$$ [2]

and the Schottky diode current $I_{Schottky}$ is given by (11)

$$I_{Schottky} = AA^*T^2 \exp\left(-\frac{q\phi_b}{kT}\right)\left[\exp\left(\frac{qV_d}{nkT}\right)-1\right] \qquad [3]$$

where μ_{eff} is the effective mobility, C_{BOX} the buried oxide capacitance/unit area, W the gate width, L the gate length, V_d the drain voltage, V_g the gate voltage, V_{th} the threshold voltage for the inversion region, V_{FB} the flatband voltage for the accumulation region, A the area of the Schottky contact, A^* Richardson's constant, T the temperature, ϕ_B the effective Schottky barrier height, and n the ideality factor.

GSG Ψ-MOSFET Noise Spectroscopy

<u>Low Frequency Noise Measurements</u>

The probe station for the D.C. measurements was also configured for low frequency noise measurements. Of particular interest is the low frequency $1/f$ noise behavior which is associated with traps at the Si/BOX interface. The measurement configuration is shown schematically in Fig. 8. It can be used for the direct measurement of the drain current noise spectral density, S_{Id}. However, by applying a small signal voltage to the gate of the device the overall transfer function of the device plus current amplifier can be measured. From the known response of the current amplifier the transfer function of the MOSFET can then be deduced and used to determine the gate-referred noise voltage spectral density S_{Vg}.

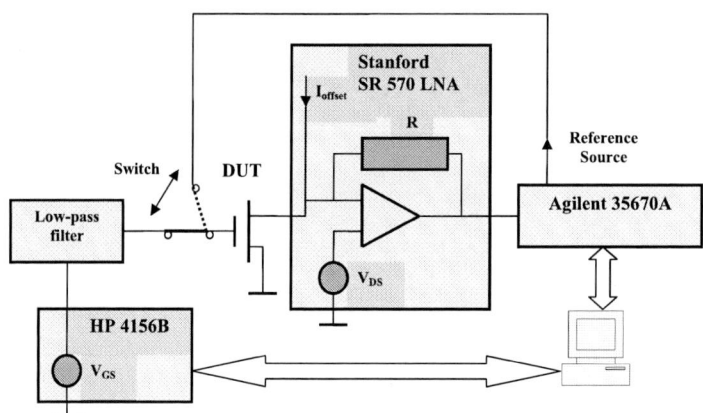

Figure 8. Schematic illustration of the setup used to measure the LFN spectrum (adapted from (12)).

The low-frequency drain current noise spectral density (S_{Id}) follows the $1/f$ relation in the accumulation as well as in the inversion regions of the bottom Si/BOX interface (Figs. 9 and 10). S_{Id} is proportional to I_d^2 in inversion (Fig. 11), which is consistent with n-MOSFETs (8). However S_{Id} is proportional to $I_d^{1.75}$ in the accumulation region (Fig. 11), that can be explained by volume conduction in weak accumulation, with interface traps having less influence on the active carriers.

The gate-referred noise voltage power spectrum S_{Vg} is determined from S_{Id} and the GSG Ψ-MOSFET transconductance g_m (8) (Figs. 12 and 13). S_{Vg} is almost independent of the drain current in the accumulation region and shows a weak dependence in strong accumulation (Fig. 12). The same is found for the depletion and weak inversion regions. However in strong inversion the device transconductance decreases drastically (Fig. 7) due to the influence of the reversed biased drain Schottky junction. As a result we see a significant increase in the drain referred noise spectrum (Fig. 13).

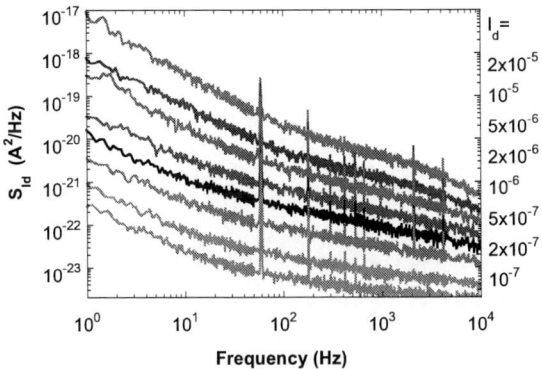

Figure 9. S_{Id} vs. frequency for Ψ-n-MOSFET in accumulation at $V_d = -1$ V.

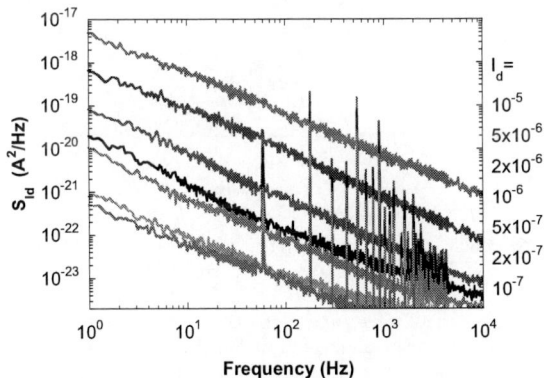

Figure 10. S_{Id} vs. frequency for Ψ-n-MOSFET in inversion at $V_d = 1$ V.

Careful examination of the plots in Figs. 9 and 10 show that in the accumulation region (Fig. 9) the slopes of curves between 10 and 100Hz change from 1/f to less steep slopes. Also the plots for the inversion region keep 1/f slope throughout the entire range of frequencies. Such phenomena may be caused by the generation-recombination (G-R) noise component due to defects in the depletion region of the gate Schottky contact or because of interface traps (13, p. 212 and 203).

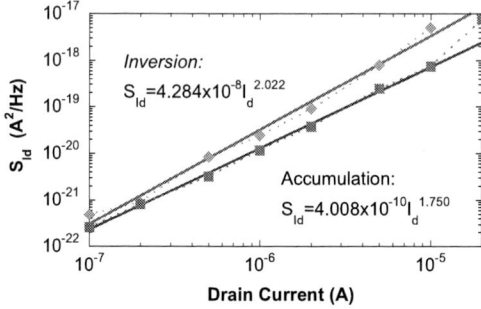

Figure 11. S_{Id} vs. I_d at $f = 1$Hz in accumulation and inversion of the Si/BOX interface.

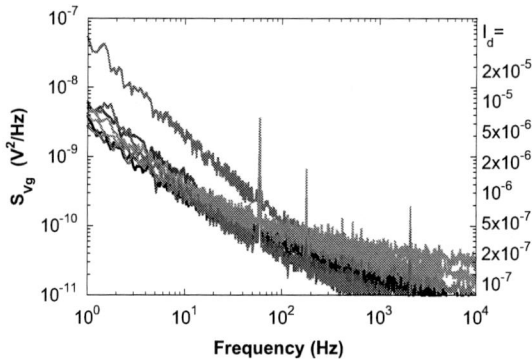

Figure 12. S_{Vg} vs. frequency for Ψ-n-MOSFET in accumulation at $V_d = -1$ V. The increase of the gate-referred noise spectral density occurs in strong accumulation.

In order to extract the defect parameters, the LFN measurements are done for different drain voltages in the accumulation region (Figs. 14, 15). The drain current is held constant at 1μA by adjusting the gate voltage. The plot on Fig. 14 confirms our hypothesis about the generation-recombination process in the depleted region of the drain Schottky junction: the G-R Lorentzian noise peaks (13, p. 201) dominate for reverse-biased (positive) drain voltages and not for negative V_d. Nevertheless, we cannot rule out the possibility of interface trap participation. The noise spectrum curves do not change with the increase of the drain voltage from 1 to 5 V (Figs. 14, 15). This is because the silicon film is fully depleted under the gate, the drain current is kept constant and the Ψ-MOSFET channel is uniform in the source to drain direction.

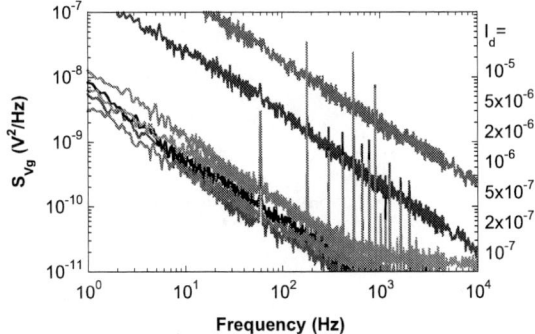

Figure 13. S_{Vg} vs. frequency for Ψ-n-MOSFET in inversion at $V_d = 1$ V. The increase of the gate-referred noise spectral density occurs in the strong inversion.

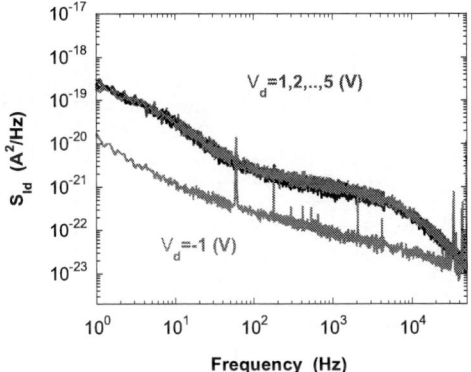

Figure 14. S_{Id} vs. frequency for Ψ-n-MOSFET in accumulation at constant $I_d = 1$ μA and $V_d = -1, 1, 2..., 5$ V.

The $f*S_{Vg}$ versus frequency plot on Fig. 15 helps to extract the zero-frequency plateau values of the Lorentzians (13, p. 206), since it is easier to find maxima of the curves on Fig. 15 than the corner frequencies on Fig. 14.

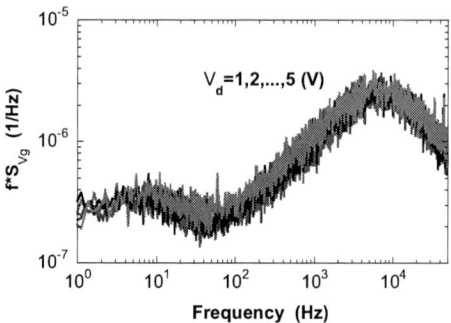

Figure 15. f^*S_{Vg} vs. frequency for Ψ-n-MOSFET in accumulation at constant I_d=1 μA and V_d=1, 2..., 5 V.

Analysis of LFN Data

In order to derive meaningful parameters from the measured data we need to determine the GSG Ψ-MOSFET LFN sources and the types of noise. The current noise power spectral density $S_{Id}(f)$ can be written as the sum of flicker noise, generation-recombination noise, shot noise and thermal noise (8)

$$S_{Id}(f) = \frac{C_1}{f} + \frac{C_2/f_0}{1+(f/f_0)^2} + C_3 + C_4 \qquad [4]$$

where C_1 is the amplitude of the flicker noise, C_2/f_0 the zero-frequency plateau value, $C_3=2qI_{Schottky}$ the Schottky contact shot noise spectrum and $C_4=8kTg_m/3$ denotes the MOSFET thermal noise. The depletion and inversion plots in Fig. 10 show only a $1/f$ noise component. The generation-recombination component is present on the accumulation plot in Fig. 9, but the Lorentzian plateaus are obscured by the $1/f$ component, making it impossible to extract C_2/f_0 (13). The G-R defects parameters can be extracted from plots in Fig. 14 and Fig. 15 where the G-R component of LFN dominates.

The $1/f$ LFN behavior in the linear part of the inversion region can be explained by the unified correlated model (9) where the carrier number and mobility fluctuations are taken into account in a correlated fashion

$$S_{Vg}(f) = \frac{kTq^2}{\gamma f WL C_{BOX}^2}(1 + \alpha\mu_{eff}N)^2 N_T(E_F) \qquad [5]$$

where γ is the attenuation coefficient, μ_{eff} the effective carrier mobility, α the Coulombic scattering parameter, N the number of channel carriers/unit area, C_{BOX} the buried oxide capacitance per unit area, WL the gate area and $N_T(E_F)$ the oxide-trap density (cm^{-3}eV^{-1}) (9).

The input voltage referred noise spectral density S_{Vg} is calculated from the frequency response data. The interface and border trap density at the quasi-Fermi level $N_T(E_{Fn})$ is extracted as $N_T(E_{Fn}) = 9.4\text{x}10^{18}$ cm^{-3}eV^{-1} for weak inversion and $4.1\text{x}10^{18}$ cm^{-3}eV^{-1} for accumulation, which corresponds to $1.9\text{x}10^{12}$ cm^{-2}eV^{-1} and $8.2\text{x}10^{11}$ cm^{-2}eV^{-1} for the 2 nm Si/BOX interface. The $N_T(E_{Fn})$ values are higher than for conventional SOI MOSFETs, which are typically 1×10^{17} cm^{-3}eV^{-1} (14), because the SOI Si film/buried oxide interface is worse than the MOSFET gate oxide/Si interfaces, leading to a higher trap density.

Interface trap densities can also be extracted from the G-R noise component of LFN (Fig. 15), if we assume that the Si/BOX interface traps cause the Lorentzians (11, p. 204). The density of the noisy interface traps can be approximated at the gate bias Fermi level by

$$N_{it}(E_{Fn}) = \frac{S_{Vg}(0) f_0 C_{BOX}^2 WL}{2\pi q^2} \quad [6]$$

where $S_{Vg}f_0$ are the maxima of the curves on Fig. 15. This gives $N_{it}(E_{Fn})=3.2 \times 10^{10}$ cm^{-2}eV^{-1} at $f_0=6.625$ Hz from the first maximum and $N_{it}(E_{Fn})=2.2 \times 10^{11}$cm^{-2}eV^{-1} at $f_0=7074.5$ Hz from the second maximum. These numbers are four times lower than the ones from the 1/f noise component, indicating that the G-R noise is most likely caused by defects in the depletion region of the Schottky drain junction. However, there may be an alternative explanation. LFN behavior, similar to Fig. 14, is observed in reversed biased Schottky diodes (15). In order to determine the source of G-R noise components the LFN measurements need to be done on Ψ-MOSFET with a partially depleted channel under the drain Schottky contact. This also gives the vertical distribution of defect densities in the device channel. Such measurement involves low drain and high gate voltage biases. This has not been done due to equipment limitations. Another approach involves the use of the probe station with a temperature controlled chamber (13, p. 214).

Conclusions

A new approach for SOI wafer characterization has been proposed allowing noise measurements with the pseudo MOSFET structure. This approach allows SOI wafer noise measurements before conventional device fabrication. A new ground-signal-ground (GSG) Ψ-MOSFET has been developed and characterized. Consistent and repeatable low-frequency noise measurements have been made on the Ψ-MOSFET for the first time. DC and noise measurement data have been explained and the Si/SiO$_2$ interface trap densities at the pseudo-Fermi levels have been extracted. The GSG Ψ-MOSFET noise spectroscopy can be further expanded through controlled temperature measurements, providing an opportunity for the trap density/energy level spectroscopy, similar to the DLTS measurements (11).

References

1. B. Razavi, *IEEE Journal of Solid-State Circuits*, **31**, 331-343 (1996).
2. M. Borremans, B. de Muer, M. Steyaert, *Electronics Letters*, **36**, 857-858 (2000).
3. H. Wong, *Microelectron. Reliab.* **43**, 585-599, (2003); C. Claeys, A. Mercha, and E. Simoen, *J. Electrochem. Soc.* **151**, G307-G318 (2004).
4. S. Cristoloveanu, D. Munteanu, and M. S. T. Liu, *IEEE Trans. Electron Dev.* **47**, 1018-1027 (2000).
5. H.J. Hovel, *Solid-State Electron.* **47**, 1311-1333 (2003).
6. S. G. Kang, D. K. Schroder, *IEEE Trans. Electron Dev.* **49**, 1742-1747 (2002).

7. J. Y. Choi, S. Ahmed, T. Dimitrova, J. T. C. Chen, and D. K. Schroder, *IEEE Trans. Electron Dev.* **51**, 1164-1168 (2004).
8. A. van der Ziel, Noise in Solid State Devices and Circuits, Wiley, NY (1986).
9. K. K. Hung, P. K. Ko, C. Hu, and Y. C. Cheng, *IEEE Trans. Electron Dev.* **37**, 654-665 (1990).
10. S. Cristoloveanu and S. S. Li, *Electrical Characterization of SOI Materials and Devices.* Boston, MA: Kluwer (1995).
11. D. K. Schroder, *Semiconductor Material and Device Characterization,* 2nd ed., p. 169, Wiley, New York (1998).
12. A. Blaum et al., *A New Robust On-Wafer 1/f Noise Measurement and Characterization System, Proc. IEEE Int. Conf. on Microelectronic Test Structures,* **14**, p. 125-130 (2001).
13. N. B. Lukyanchikova in *Noise and Fluctuations Control in Electronic Devices,* 201-203, American Scientific Publishers (2002).
14. S. Haendler, J. Jomaah, G. Ghibaudo, F. Balestra, *Microelectronics Reliability,* **41**, 855-860 (2001).
15. H. Ouacha, M. Mamor, M. Willander, A. Ouacha, F. D. Auret, *J. Appl. Phys.,* Vol. **87**, 3858-3863 (2000).

ECS Transactions, 2 (2) 503-513 (2006)
10.1149/1.2195685, copyright The Electrochemical Society

1/f noise as a tool to assess Fermi Level pinning (E_F) at the HfO$_2$/ poly-Si and FUSI interface in high-κ n-MOSFETs

P. Srinivasan[1,2*], E. Simoen[1], L. Pantisano[1], C. Claeys[1,3] and D. Misra[2]

[1] IMEC, Kapeldreef 75, B-3001 Leuven, Belgium
[2] Department of Electrical Engineering, NJIT, Newark, NJ, USA
[3] Electrical Engineering Department, KU Leuven, Belgium

Evidence is provided that 1/f noise may be useful in the analysis of the traps responsible for Fermi level pinning at the HfO$_2$/poly-Si or HfO$_2$/FUSI interface in high-κ n-MOSFETs. As reference devices, transistors with 1.5 nm SiON gate dielectric have been used. It is shown that adding a few (5, 10, 20) monolayers of HfO$_2$ enhances markedly the normalized noise magnitude in both poly-Si and FUSI devices. The 1/f noise characteristic behaves according to the number fluctuations theory and the results are interpreted in terms of trapping and de-trapping of channel carriers by defects in the gate dielectric layer. Differences in trap densities, derived from the low-frequency noise spectra are noticed at the gate/dielectric interface, which can explain the Fermi-level pinning in these devices. Additionally, it is shown that the correlated mobility fluctuations derived from the 1/f noise at larger gate voltage overdrives correlate well with the low-field mobility of the n-MOSFETs, demonstrating that the same traps in the gate dielectric are also partly responsible for the mobility degradation.

Introduction

The scaling requirements have brought the gate oxide thickness (t_{ox}) at its useful limits imposed by direct tunneling, in the range 1.2-1.5 nm. This bears serious consequences for the off-state current and necessitates the introduction of so-called high-κ dielectrics for low off-state power applications in the near future. Another parameter that is expected to undergo fundamental changes is the low-frequency noise, usually dominated by 1/f noise for sufficiently large-area transistors. It is believed that the 1/f noise will reduce according to a power law t_{ox}^n, with n somewhere between 1 and 2, as has been experimentally verified [1-2]. It is generally accepted that the dominant 1/f noise mechanism is related to trapping and de-trapping: in the frame of the number fluctuations model, based on the original McWhorter theory [3], the current fluctuations arise from carrier tunneling events to and from traps in the gate oxide. For the state-of-the-art deep submicron transistors, with a physical oxide thickness in the range 1.2-1.5 nm, the tunneling distance becomes of the same order or even larger than t_{ox}. In that case the devices are probed very near to the gate-dielectric/gate-electrode interface.

---

* Contact Author: e-mail.purushothaman@ieee.org

Two types of gate electrodes – poly-Si and FUlly Silicided (FUSI) gates with SiON as the gate dielectric, have been studied. The latter have recently attracted considerable interest as candidate for mid-gap metal gate [4-8]. Among the different silicides, NiSi offers some clear advantages due to its low sheet resistance, process temperature and silicon consumption in the source/drain regions. The impact of the gate electrode-SiON interface on 1/f noise is studied using a thin layer of HfO_2 between the poly-Si gate and the SiON layer. In a similar study Hobbs *et al.* [9] concluded that higher V_T shifts originate at the HfO_2/poly-Si interface, which are mainly due to Fermi-level pinning below the poly-Si conduction band. It has been shown that not only the static parameters, i.e. threshold voltage and transconductance, but also the 1/f noise is affected by the used gate electrode material [10]. Here the origin of these observations is further elaborated. Due to Fermi-level pinning at the HfO_2/poly-Si interface, charges present in the layer are observed as a variation in trap density at the depth corresponding to this interface. A similar observation with reduced effect is seen for a HfO_2 /FUSI interface.

Experimental

The 1/f noise performance was investigated in n-MOSFET devices fabricated using a conventional CMOS process flow. The starting gate dielectric is a 1.5 nm thick SiON layer and ALCVD was used to deposit 5, 10, and 20 cycles of HfO_2 on the top of the SiON as shown in Figure 1. More information on the volume of deposited Hf can be found from the Rutherford Backscattering Studies (RBS) from [11]. The equivalent Oxide Thickness (EOT) was 1.5 nm +/- 0.2 nm for all studied devices. It is to be noted that 10 cycles corresponds to an oxide thickness of ~ 1 nm. The samples were subjected to a Forming Gas Anneal (FGA) at 520°C for 20 min. To understand and verify the Fermi-pinning effect on 1/f noise at the gate-SiON interface, two types of gate material are being used: n-type polysilicon (poly Si) or FUlly nickel SIlicided (FUSI) (undoped).

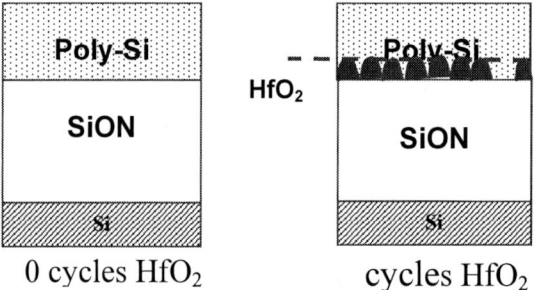

0 cycles HfO_2 cycles HfO_2

Figure 1. Schematic of gate stack with and without the growth cycles of HfO_2.

The noise has been measured on 10 µmx1 µm and 10 µmx0.25 µm poly-Si devices, while only 10 µmx0.25 µm n-MOSFETs were used for FUSI gates. On-wafer noise measurements were performed in linear operation at a constant drain voltage $|V_{DS}|$=0.05 V for gate voltages $|V_{GS}|$ between 0.3 – 1.2 V, in steps of 50mV using BTA9812 hardware and NoisePro software from Cadence.

Results

Figure 2a and 2b shows the transconductance (g_M) characteristics of, poly and FUSI gates with and without HfO_2 cycles respectively. A 10 ~ 15% reduction in mobility is observed when a few cycles of HfO_2 are present at the gate-SiON interface in these devices. The ~0.6 V increase in the threshold voltage for SiON in Fig. 2b compared to Fig. 2a is related to the 0.6 eV lower work function of the FUSI gate [6]. For the FUSI gates, a significant increase of the inversion capacitance density (C_{eff}) and reduced poly-depletion effects were observed.

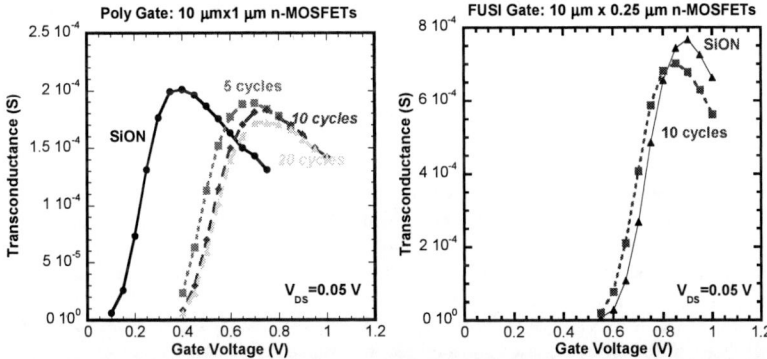

Figure 2. Input characteristics in linear operation (V_{DS}=0.05 V) for (a) 10 μmx1.0 μm poly-silicon gate and 10 μmx0.25 μm FUSI gate (b) n-MOSFETs with and without a few cycles of HfO_2 on top of a 1.5 nm SiON gate dielectric.

Figure 3 shows the drain current spectral density (S_I) [12] of pure SiON and devices with 5, 10 and 20 cycles of HfO_2 for $|V_{DS}| = 0.05$ V and $|V_{GS} - V_T| = 0.10$ V for poly-Si gates. The drain current spectral density S_I obtained on SiON devices with few cycles of HfO_2 showed a similar $1/f^\gamma$ behavior across all the frequencies with a frequency exponent $\gamma \sim 1$. A mixed behavior was noticed for 1.5 nm EOT SiON devices whereby the device exhibits two different slopes – a higher value of γ ($\gamma \sim 1.2$) for the low frequency part f < 100 Hz, and a lower value ($\gamma \leq 1$) at higher frequencies. A similar nature was also observed in FUSI gate devices with SiON and 10 cycles of HfO_2. Assuming that the current fluctuations are due to trapping, this indicates a higher trap density close to both interfaces.

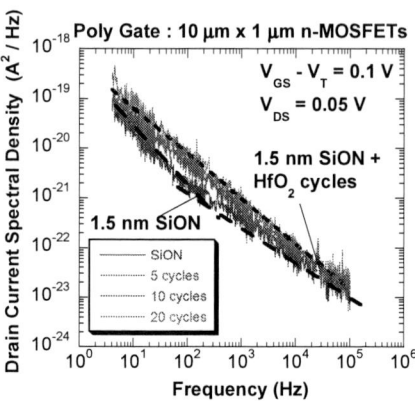

Figure 3. Drain current spectral density S_I Vs frequency f of devices with SiON and 5, 10 and 20 cycles of HfO$_2$. Two different slopes are observed for pure SiON devices

The values of the frequency exponent γ were also plotted for various gate voltage overdrives as shown in Figure 4. In the case of SiON with few cycles, γ is found to be constant ($\gamma \sim 1$) over the measured gate voltage overdrives, while for SiON, γ varies significantly from 0.9 to as high as 1.8. Excess noise peaks are observed as indicated in Figure 4 by numbers. These peaks have a frequency exponent $\gamma > 1$, which would mean that a greater number of low-frequency traps are away from the substrate-dielectric interface. In the case of a few cycle devices, the frequency exponent γ is less than 1 for $V_{GS} > V_T$ indicating that the trap distribution is increasing towards the substrate-dielectric interface.

As the normalized spectral density (S_I/I_D^2) and the $(g_M/I_D)^2$ ratio are found to be parallel to each other, the McWhorter theory [3] related to carrier trapping can be used to analyze the noise origin. Though this emphasizes that carrier number fluctuations due to tunneling to and from the traps are the cause for the observed 1/f noise, it is probable that scattering-related events could also contribute to the source of 1/f noise. Both approaches are further used to get a better insight into the underlying fundamental differences in these devices.

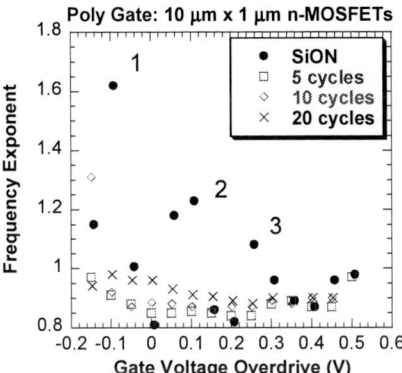

Figure 4. Frequency exponent γ Vs gate voltage overdrive for poly-Si devices with pure SiON and with a few cycles of HfO$_2$ on top of SiON. Numbers indicate excess noise peaks in SiON.

Discussion

The above differences observed are explained using two possible fluctuation mechanisms related to trapping and scattering – number fluctuation theory and correlated mobility fluctuation theory. The first theory assumes that channel carriers are the origin of 1/f noise while the second theory takes into account the scattering related events also.

(i) Number fluctuation theory based approach

The low-frequency contribution generated by deeper lying traps can be studied by plotting the trap density values with depth as shown in Fig. 5a and 5b, based on the simplest approximation, where the tunneling distance z is related to the noise frequency f by:

$$\frac{1}{2\pi f} = \tau_0 \exp(\alpha_t z) \tag{1}$$

with τ_0 the time constant at the Si/SiO$_2$ interface ($\sim 10^{-10}$ s) and α_t the electron wave function attenuation parameter, usually taken 10^8 cm^{-1}.

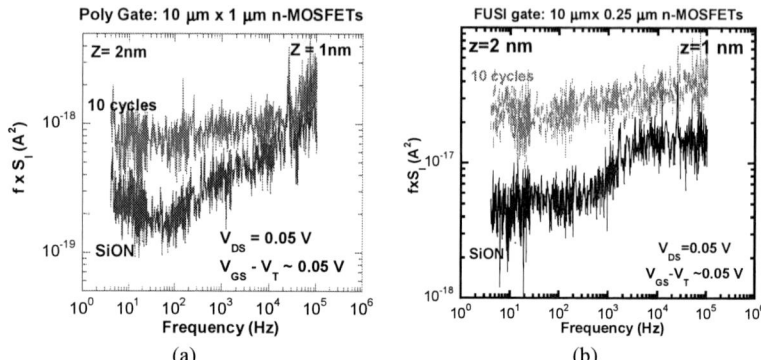

(a)   (b)

Figure. 5 Normalized low-frequency noise spectrum f x S_I versus frequency f in linear operation (V_{DS}=0.05 V) and at V_{GS} -V_T ~ 0.05 V, for (a) 10 µmx1 µm poly-Si n-MOSFETs and (b) 10 µmx0.25 µm n-MOSFETs, with SiON and SiON plus 10 cycles HfO$_2$, respectively

In Fig. 5a and 5b, the current noise spectral density S_I multiplied by the frequency f is represented versus f for poly and FUSI gate devices with and without 10 cycles of HfO$_2$ at V_{GS} -V_T ~ 0.05 V. By representing the spectra in this way, a few features become more obvious. First, it is clear that the highest noise is found in case of 10 cycles for both FUSI and poly gates. Next, it is also obvious that the frequency exponent (γ) changes with f; in the case that γ=1, one would expect a horizontal curve, which confirms the results obtained in Fig. 4. In the context of the McWhorter model, one can say that the variation of γ is due to a non-uniform density of oxide traps. As shown in Fig. 6a, the trap density reduces towards the gate-dielectric interface, compared with the Si/SiON interface. Putting this in perspective, the tunneling depth at the lowest (~4 Hz) and highest (100 kHz) frequency, calculated from Eq. (1) is indicated in Fig. 5a and 5b. It is clear that the low-frequency part of the spectra corresponds to a depth that is larger than the physical oxide thickness, but in the same range of the electrical or effective thickness.

The trap density calculations are based on the relationship with S_{VG} as

$$f. \ S_{VG}= q^2kTN_t \ / \ WLC_{EOT}^2\alpha_t \tag{2}$$

where N_t is the volume trap density (1/ cm^{3}eV), α_t is the the electron wave function attenuation parameter (1/cm), C_{EOT} is the gate oxide capacitance per unit area associated with EOT, i.e., equal to ε_{ox} /EOT, with ε_{ox} the permittivity of SiO$_2$.

A decaying profile is noticed near the interfacial layer as shown in Figure 6a, which is in agreement with the well-established fact that there exists a highly defective transition layer close to the Si/SiO$_2$ interface. The spectra of Fig. 5a suggest an increased trap density in the SiON layer when approaching the silicon interface. There are two possible reasons for that: it is known that the presence of nitrogen introduces additional noise centers [12-14], so that Fig. 6a would indicate an increasing nitrogen profile towards the silicon interface. Alternatively, it is known that the transition layer of 0.6 nm between the silicon substrate and the bulk oxide is highly defective and consists of suboxides [15,16]. This again could be the origin of a higher 1/f noise at higher frequencies in the case of the poly or FUSI gate transistors [17].

Around 1.7 nm, the volume trap density of the SiON device increases to as high as 10^{17} cm^3. This is close to the gate-SiON interface and suggests that there is an increasing trap density probed in the transition layer of the poly-Si/SiON interface [18]. Unlike the SiON case, the trap values are higher and are found to be constant throughout the oxide depth, in devices with few cycles HfO_2 as shown in Fig. 6a.

Due to Fermi-level pinning at the HfO_2/poly-Si interface, oxide charges present at this interface could translate into a more or less constant trap density in the oxide as seen in Fig. 6a. These additional charges are not present in the pure SiON layer and therefore one sees lower trap density values for pure SiON device. Further confirmation can be seen from the static DC characteristics where a shift in threshold voltage $V_T \sim 0.2$ V is observed in Fig. 6b.

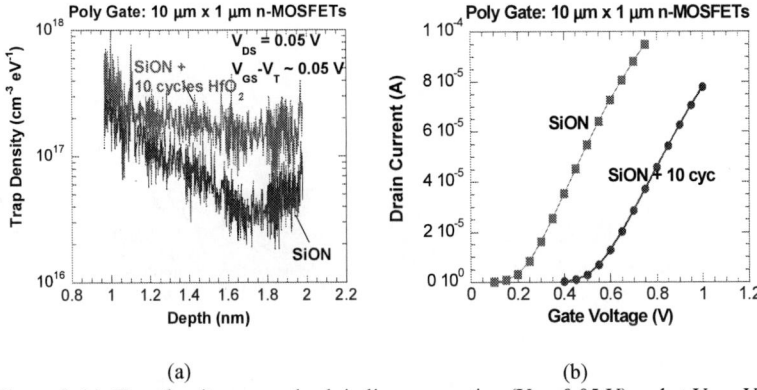

(a)   (b)

Figure 6. (a) Trap density versus depth in linear operation (V_{DS}=0.05 V) and at V_{GS} - V_T ~ 0.05 V, for a 10 µmx1 µm poly-Si n-MOSFET with and without 10 cycles HfO_2 on top of a 1.5 nm SiON gate dielectric. (b) Threshold Voltage V_T shifts observed between SiON and SiON with 10 cycles of HfO_2.

When regarding the low-frequency part, corresponding with the layer near/at the gate-oxide interface, it is clear that the density of fluctuation sources is a strong function of the gate material used, with the lowest value for the FUSI pure SiON devices. Translated to a density of traps, it means that there exist about three times more traps close to the gate-oxide interface for the FUSI + 10 cycles HfO_2 transistor, compared with its FUSI pure SiON counterpart. Moreover, a Fermi-level pinning effect has, for example, also been reported for the PtSi/HfO_2 gate stack [7], whereby it is shown that the presence of Si atoms at the interface causes the pinning effect, i.e., by creating a high density of interface traps. From Fig. 7, it is seen that the trap density increases based on number of cycles deposited, confirming higher Fermi-level pinning effect at the interface. A similar trap density increase at f=10 Hz, i.e., close to the gate-dielectric interface was obtained for a FUSI gate with 10 cycles of HfO_2.

In the literature, different types of traps have been proposed to explain the Fermi-level pinning [9]. They all rely on an oxygen deficit, leading either to the formation of Si-Hf bonds at the interface or to V-O centers. Although 1/f noise cannot identify the defects sites responsible for the increased trap density, the fact that there appears to be a continuous increase of N_{ot} into the bulk of the SiON layer supports the second hypothesis. Alternatively, one could suppose that the traps at the gate-oxide interface correspond to Si-Hf bonds while deeper in the material additional V-O centers are being created during the HfO$_2$ deposition, possibly by an out-diffusion of oxygen to the surface.

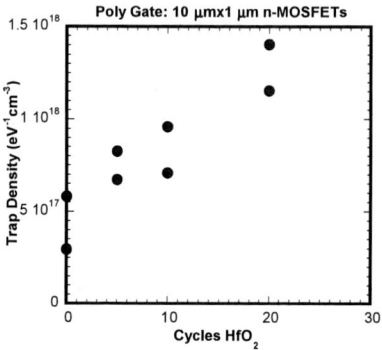

Figure. 7 Trap Density Vs cycles HfO$_2$ for poly-gate 10 μm x 1 μm n-MOSFETs.

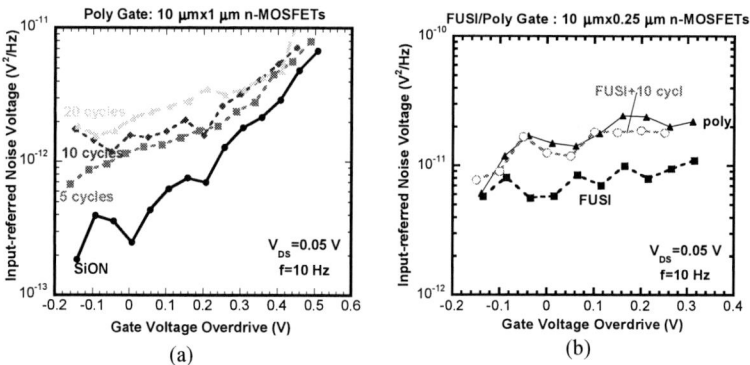

Figure 8. Input-referred noise spectral density S_{VG} at 10 Hz versus gate voltage overdrive $V_{GS} - V_T$ in linear operation (V_{DS}=0.05 V) for n-MOSFETs corresponding with (a) a poly silicon gate and SiON or SiON plus a few cycles HfO$_2$, and (b) a poly-Si gate with SiON and a poly-Si or FUSI gate with SiON plus a few cycles of HfO$_2$.

Comparable trap densities were found for FUSI and poly-Si gate devices, close to the gate and near the silicon interface. This is due to similar values of input-referred noise S_{VG} observed for FUSI and poly-Si gate in Fig. 8b, when account is made of the polysilicon depletion effect on the effective capacitance.

(ii) Correlated mobility fluctuation theory based approach

For larger gate voltage overdrives, the input-referred noise spectral density of the studied devices can be described by the correlated-mobility fluctuations model [19], described by:

$$S_{VG} = S_{VFB} \left[\ 1 \pm \alpha_{sc}\mu_0 \ (V_{GS} - V_T) \ \right]^2 \tag{3}$$

S_{VFB} is the voltage spectral density at flat-band condition (weak inversion), responsible for the current fluctuations through trapping and release of channel carriers and μ_0 the low-field mobility. The effective mobility was calculated from the channel transconductance values measured during noise and is plotted along the electric field and the injected charge Q_{inj} on the primary and secondary X-axes respectively as shown in Fig. 9a. It is seen that the peak mobility μ_{peak} for devices with few cycles reduced to ~ 80% of the pure SiON value. In that case, the noise increase for a few cycles has to be related with the mobility degradation, which was evaluated by estimating $q\alpha_{sc}N_{ot}$ as a figure of merit, where α_{sc} is the scattering coefficient and N_{ot} the oxide trap density.

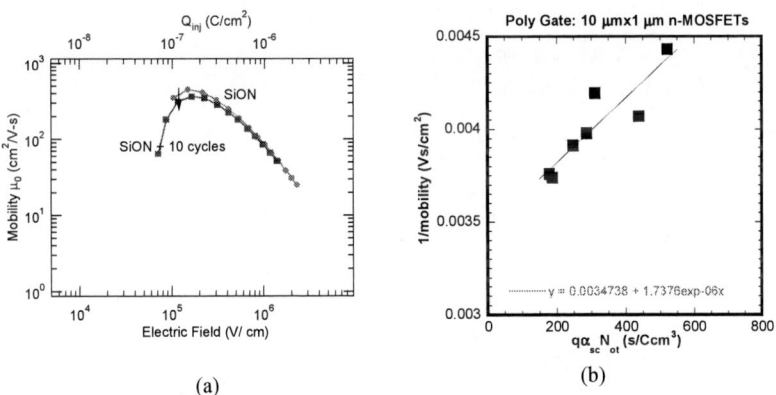

(a)

(b)

Figure 9. (a) Mobility Vs Electric Field (1 X-axis), Injected Charge (2 X-axis) for poly-Si devices with pure SiON and SiON with 10 cycles of HfO$_2$ (b) 1/mobility versus $q\alpha_{sc}N_{ot}$ – A figure of merit parameter.

Figure 9b shows the correlation between 1/mobility and the figure of parameter - $q\alpha_{sc}N_{ot}$ for poly-Si gate devices. For n-MOSFET, it is seen that $q\alpha_{sc}\mu_0$ reduces from its original value corresponding with pure SiON due to the addition of few cycles of HfO$_2$. Kuga *et al.* [20,21] assumed that the mobility is limited by the coulomb scattering of channel carriers by the trapped charge at the interface and in the dielectric (μ_{cit}) with μ_{cit} = $1/\alpha_{sc}N_{IT} = \mu_{c0} \sqrt{N}/ N_{IT}$, where N_{IT} is the number of occupied traps per unit area and μ_{c0} is a fitting parameter.

Conclusion

The effect of the gate-dielectric interface on 1/f noise using few cycles of HfO_2 has been studied. The presence of a small amount of HfO_2 at the top interface, thought responsible for the Fermi-level pinning, gives rise to a strong increase in the 1/f noise. Based on the experimental behavior reported here, the most straightforward explanation is that the existence of a high density of gate interface traps is causing a higher 1/f noise.

Acknowledgments

The authors would like to thank A. Mercha and V. Subramanian for stimulating discussions. P. Srinivasan wishes to acknowledge NSF (Award # ECS-0140584) for his financial support.

References

1. M.J. Knitel, P.H. Woerlee, A.J. Scholten and A.T.A. Zegers-Van Duijnhoven, in *IEDM Tech. Dig.*, p. 463 The IEEE (New York), (2000).
2. R. Kolarova, T. Skotnicki and J.A. Chroboczek, *Microelectron. Reliab.*, **41**, 579 (2001).
3. A.L. McWhorter, in *Semiconductor Surface Physics*, University of Philadelphia Press, Philadelphia, p. 207 (1957).
4. J. Kedzierski, D. Boyd, C. Cabral, Jr., P. Ronsheim, S. Zafar, P.M. Kozlowski, J.A. Ott and M. Ieong, *IEEE Trans. Electron Devices*, **52**, 39 (2005).
5. J. Yuan and J.C.S. Woo, *IEEE Electron Device Lett.*, **26**, 87 (2005).
6. W.P. Maszara, *J. Electrochem. Soc.* **152**, G550 (2005).
7. M. Kadoshima, K. Akiyama, N. Mise, S. Migita, K. Iwamoto, N. Yasuda, K. Tominaga, M. Ikeda, H. Satake, T. Nabatame and A. Toriumi, *Jpn. J. Appl. Phys.* **44**, 2267 (2005).
8. K. Sano, M. Hino, N. Onishi and K. Shibahara, *Jpn.. J. Appl. Phys.*, **44**, 3774 (2005).
9. C.C. Hobbs, L.R.C. Fonseca, A. Knizhnik, V. Dhandapani, S.B. Samavedam, W.J. Taylor, J.M. Grant, L.R.G. Dip, D.H. Triyoso, R.I. Hegde, D.C. Gilmer, R. Garcia, D. Roan, M.L. Lovejoy, R.S. Rai, E.A. Hebert, H.-H. Tseng, S.G.H. Anderson, B.E. White and P.J. Tobin, *IEEE Trans. Electron Devices*, **51**, 978 (2004).
10. P. Srinivasan, E. Simoen, R. Singanamalla, Y. Hong Yu, C. Claeys and D. Misra, *Solid-State Electronics*, submitted.
11. L.-A. Ragnarsson, L. Pantisano, V. Kaushik, S.I. Saito, Y. Shimamoto, S. De Gendt and M. Heyns, *IEDM Tech. Dig.*, p. 87-89 (2003).
12. R. Jayaraman and C.G. Sodini, *IEEE Trans. Electron Devices,* **36**, 1773 (1989).
13. P. Morfouli, G. Ghibaudo, T. Ouisse, E. Vogel, W. Hill, V. Misra, P. McLarty and J.J. Wortman, *IEEE Electron Device Lett.*, **17**, 395 (1996).
14. C. Claeys, E. Simoen and A. Mercha, *J. Electrochem. Soc.*, **151**, G306 (2004).

15. H. Ono, T. Ikarashi, K. Ando and T. Kitano, *J. Appl. Phys.*, **84**, 6064 (1998).
16. H. Yamada, *J. Appl. Phys.*, **91**, 1108 (2002).
17. A. Szewczyk, T. Ernst, C. Leroux, G. Ghibaudo and J.A. Chroboczek, in *Proc ESSDERC 2001*, Eds H. Ryssel, G. Wachutka and H. Grünbacher, Paris: Frontier Publishing, p. 455 (2001).
18. E. Simoen, A. Mercha, L. Pantisano, C. Claeys and E. Young, *Solid State Electronics*, **49,** 702 (2005).
19. K.K. Hung, P.K. Ko, C. Hu and Y.C. Cheng, *IEEE Trans. Electron Devices,* **37**, 654 (1990).
20. J. Koga, S. Takagi and A. Toriumi, *Proc. Int. conf. Solid State Devices and Materials,* p. 895 (1994).
21. J. Koga, S. Takagi and A. Toriumi, *IEDM Tech. Dig.*, p. 475 (1994).

514

Study of Inhibition Characteristics of Slurry Additives in Copper CMP Using Force Spectroscopy

A. Philipossian[a,b], H. Lee[a], S. V. Babu[c], U. Patri[c], Y. Hong[c], L. Economikos[d], M. Goldstein[e], Y. Zhuang[a,b] and L. Borucki[b]

[a] Department of Chemical & Environmental Engineering, University of Arizona, Tucson, AZ 85721, USA
[b] Araca Incorporated, Tucson, AZ 85750, USA
[c] Center for Advanced Materials Processing, Clarkson University, Potsdam, NY 13699, USA
[d] IBM Corporation, Hopewell Junction, NY 12533, USA
[e] Intel Corporation, Santa Clara, CA 95052, USA

Using a reference slurry, ammonium dodecyl sulfate (ADS), an anionic and environmentally friendly surfactant, was investigated as an alternative to BTA for its inhibition and lubrication characteristics. Results demonstrated that the inhibition efficiency of ADS was superior to that of BTA. Coefficient of friction (COF) was the lowest when the slurry contained ADS. This suggested that adsorbed ADS on the surface provided lubricating action thereby reducing the wear between the contacting surfaces. Temperature results were consistent with the COF and removal rate data. ADS showed the lowest temperature rise again confirming the softening effect of the adsorbed surfactant layer and less energy dissipation due to friction. Spectral analysis of shear force showed that increasing the pad-wafer sliding velocity at constant wafer pressure shifted the high frequency spectral peaks to lower frequencies while increasing the variance of the frictional force. Addition of ADS reduced the fluctuating component of the shear force and the extent of the pre-existing stick-slip phenomena caused by the kinematics of the process and collision event between pad asperities with the wafer. By contrast, in the case of BTA, there were no such observed benefits but instead undesirable effects were seen at some polishing conditions. This work underscored the importance of real-time force spectroscopy in elucidating the adsorption, lubrication and inhibition of additives in slurries in CMP.

1. Introduction

Chemical Mechanical Planarization (CMP) of copper has emerged as a critical component for the fabrication of integrated circuits.[1] Several studies have been carried out to study the effect of inhibitors on copper CMP.[1-5] Deshpande *et al* examined the effect of benzotriazole (BTA) as inhibiting agent in the presence of hydrogen peroxide and glycine at low pH.[1] Attempts have been made to elucidate the mechanism of copper removal in the presence of oxidizer, complexing agent and inhibitor.[4] Recent work [2] has shown that the inhibition characteristic of copper dissolution by sodium dodecyl sulfate (SDS) was better than that of BTA. It has been demonstrated that ammonium dodecyl sulfate, an environmentally friendly surfactant, can be used as an efficient corrosion inhibitor for Electrochemical Mechanical Planarization of copper in solutions containing hydrogen peroxide and glycine solutions.[3] Surfactants have also been investigated for their effect on dishing and polishing speed.[5]

In this study, using a reference slurry, ammonium dodecyl sulfate (ADS), an anionic and environmentally friendly surfactant, was investigated as an alternative to BTA for its inhibition and lubrication characteristics. Measured parameters included real-time mean and fluctuating component of frictional force (both in terms of average shear force and variance) at the pad wafer interface, pad surface temperature and average copper removal rate. Spectral analysis (in frequency domain) of the shear force data was employed to determine the inhibition and lubrication of the additives in the slurry

2. Experimental Procedure

Three different types of slurries were tested, which were based on a reference slurry to which 3.0 mM concentration of ADS or BTA was added. The reference slurry was comprised of 1 percent by weight glycine, 5 percent by weight H_2O_2 and 3 percent by weight fumed silica at a solution pH of 4.0. The pH of the slurry was adjusted by the addition of dilute perchloric acid ($HClO_4$). Slurries, prepared with UPW and reagent grade chemicals, were tested for copper removal rate, thermal and frictional characteristics. In all cases, Rohm and Haas IC1000 K-groove pad was used to polish 200 mm electroplated blanket copper wafers. Copper removal rate was calculated by measuring wafer weight before and after polishing using a microbalance. Pad surface temperature was recorded by an infrared camera (Agema 550 infrared thermal imaging camera) which took three thermal images every second. Pad temperatures at five locations outside the leading edge of the wafer carrier were averaged throughout the polishing process to report the mean pad leading temperatures. The polisher and associated accessories are described elsewhere.[6] Shear force was measured with a precision of $\pm$ 10 percent using load cells which were attached to strain gauge amplifiers that would send voltage to a data acquisition board. Experimental procedure consisted of 30-minute initial diamond conditioning (100-grit TBW diamond disc) with UPW at a pressure of 0.4 PSI, rotational velocity of 30 RPM and oscillation frequency of 20 per minute. Conditioning was followed by a 5-min pad break-in with the reference slurry. The same rotational velocity and oscillation frequency were used for in-situ pad conditioning. Two sliding velocities, 0.31 and 0.94 m/s, were used in the study. Both wafer and pad rotated in the counter-clockwise direction during polishing. Polishing pressure was kept constant at 2 PSI. Slurry was injected at the center of the pad and the flow rate was maintained constant at 200 cc/min. Two wafers were polished at each polishing condition and average values are presented everywhere.

3. Results and Discussion

3.1. Removal Rate Analysis

The effectiveness of ADS in suppressing copper removal by mechanical polishing is compared with that of BTA as a function of polishing power. Fig. 1a illustrates the mean removal rate of copper for reference slurry with and without BTA or ADS at additive concentration of 3.0 mM. This representation is followed in all the figures. The reference slurry exhibits the highest removal rate as expected. Addition of ADS or BTA reduces the removal rate thereby confirming the inhibition characteristics of the additives. The removal rate with ADS containing slurry is lower than that of BTA for both concentrations and polishing conditions establishing ADS to be a superior inhibitor than BTA. It has been reported that adsorption of ADS, which exhibits adsorption properties on metals very similar to those of SDS, is expected to follow the simple Langmuir or Frumkin type behavior on copper [7] and an electrostatic physisorption mechanism is likely to dominate over the chemisorption process.[8] The ionic head group of ADS ensures strong interaction with the copper and the hydrophobic tail acts as an effective barrier to diffusion of oxidants to the reacting surface. BTA, being a non-linear molecule may not form a complete closed packed adsorbed layer as ADS and hence provides deeper interaction zone. Based on removal rate results it appears that at 3.0 mM concentration of ADS the surface coverage is complete and there is sufficient availability of species for adsorption on the metal/metal oxide surface which results in significant suppression of removal of copper from the surface.

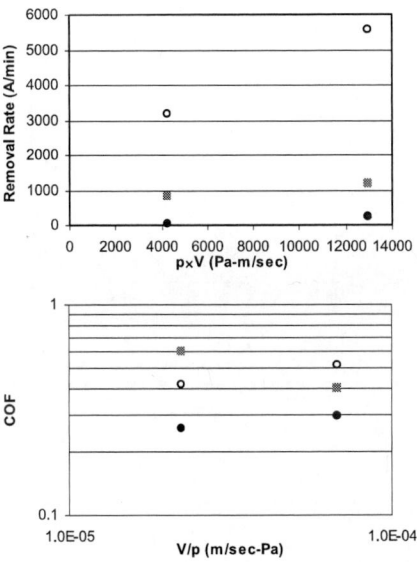

Fig. 1. Removal rate (top, a) and COF (bottom, b). Open circles represent reference slurry, filled circles represent reference slurry with 3.0 mM ADS concentration and squares denote reference slurry with 3.0 mM BTA

3.2. Frictional and Thermal Analysis

Fig. 1b shows the mean coefficient of friction (COF) vs. V/p, a representative of the Sommerfeld number [9], where V is the sliding velocity and p is the polishing pressure. Results indicate that COF remains relatively constant for reference and ADS containing slurry, indicating boundary lubrication, and decreases for the slurry containing BTA where a transition from boundary lubrication to partial lubrication may be taking place.[9] It is possible that at lower velocity where boundary lubrication exists and pad-wafer interface is devoid of BTA containing liquid layer, the adsorbate films cannot equilibrate under sliding conditions. Therefore the COF determined is not the true COF of adsorbed BTA. At some higher transition velocity, the fluid film thickness increases which ensures the availability of BTA which replenishes a friction-damaged adsorbate film under the conditions of repetitive sliding contact, as observed at 3.0 mM BTA concentration, causing the COF to drop. The key point to note is the reduction in COF with the addition of 3.0 mM concentration of ADS. This suggests that adsorbed ADS provides lubricating action at the pad wafer interface by reducing the shear force and protecting the wafer and pad from chemical and mechanical wear. This lubricating effect is probably caused by the formation of low shear strength interface between the opposing surfaces.[10] On the other hand, at lower velocity addition of BTA to the reference slurry increases the shear force between the pad and the wafer surface and therefore results in higher energy dissipation due to friction. At higher velocity some benefits are observed with BTA but not as significant as with ADS. It is obvious that molecular structure or shape of the adsorbate has a strong influence on the effectiveness of lubrication. Presence of polar group, as in ADS, ensures strong adhesion to the metallic copper surface and long linear hydrophobic chain provides closed packed monolayers.

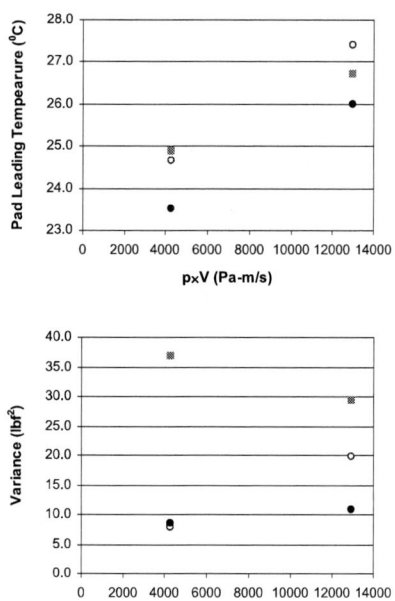

Fig. 2. Pad leading temperature (top, a) and Variance of shear force (bottom, b). Open circles represent reference slurry, filled circles represent reference slurry with 3.0 mM ADS concentration and squares denote reference slurry with 3.0 mM BTA

Fig. 2a shows the measured mean pad leading temperature for slurries at different polishing conditions. As one can notice, the mean pad leading temperature increases with the polishing power as expected. It is interesting to note that ADS containing slurry shows the lowest mean pad leading temperature. Therefore addition of ADS lowers the friction force at the pad wafer surface due to its lubricating effect and consequently offers relatively lower temperature rise during polishing. This confirms the consistency in removal rate, COF and temperature results for ADS which exhibits its effect by formation of adsorbed barrier layer supplemented by reduction in both mechanical and chemical (by lowering the temperature) contributors to the removal rate of copper. In the case of BTA at lower velocity, it appears that the inhibition characteristics of BTA is predominantly by preventing the diffusion of the chemicals and oxidizers through formation of a barrier layer at the reacting surface and not by reduction in mechanical or thermal effects. Borucki et al have discussed the effect of viscous shear forces to the total shear force in the nanolubrication layer[11] which may exists for the ADS containing slurry. It was hypothesized that thermal softening of the asperity tip can increase the viscous element of the COF by decreasing pad modulus. The viscous contribution to the COF may therefore not only be partially contributory in creating a temperature increase but will also be affected by it. This effect may, to some extent, be observed in our case of reference and ADS containing slurry where we observe a slight increase in COF with increasing velocity.

3.3. Shear Force Variance and Spectra

In addition to removal rate, COF and temperature data, the variance of the shear force as well as its spectral characteristics may be used to further explore the adsorption, lubrication and inhibition of additives during CMP. The unidirectional shear force can be separated into two components, a mean component and a fluctuating component. Fig. 2b summarizes the mean variance corresponding to all process conditions based on measured values of shear force (a total of 50,000 measured values per run). Results indicate that at higher sliding velocity, addition of BTA increases the variance of the shear force. On the other hand, addition of ADS reduces the extent of irregularity in the shear force. This signifies the dynamic nature of adsorption during the polishing process. The rate limiting step in the formation of an adsorbate film under sliding conditions is believed to be re-adsorption and typical uniqueness of the additive is required for this process to occur within the time available between successive sliding contacts.[10] ADS reduces this unsteady condition and hence lowers the variance of shear force in the polishing process. However, in the case of BTA it appears that it is the intrinsic characteristic of the molecule itself that creates the instability in the process. Hence BTA containing slurry exhibits large oscillations in the friction force demonstrating higher variability and unevenness of the system. Similar trends are noticed for BTA at lower polishing condition. At this condition, the variance of shear force for reference slurry is sufficiently low and therefore addition of ADS does not produce any further benefit. Variance has also been demonstrated to be a useful parameter in determining the lubrication regime as suggested from consistency in trends in COF and variance results.[12]

The Fast Fourier Transform function is then utilized to convert measured shear force from time domain to frequency domain.[13] Normalized spectral amplitudes plots for different slurries at lower polishing power are shown in Fig. 3. The slurries at lower

ECS Transactions, 2 (2) 515-522 (2006)

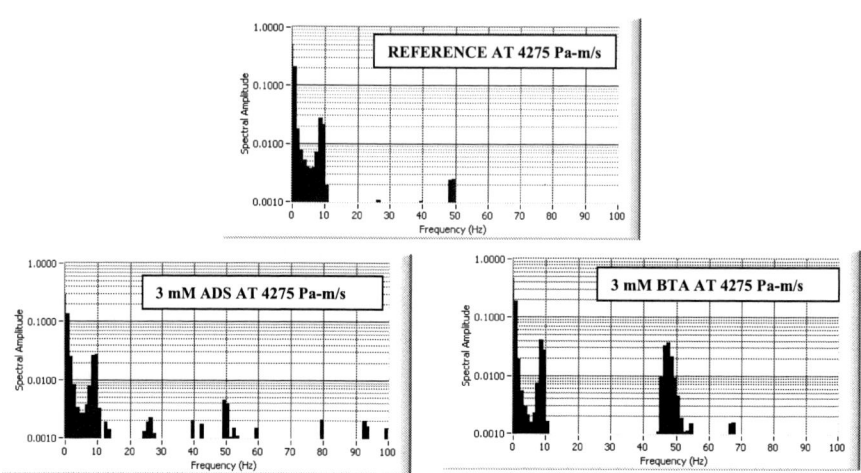

Fig. 3. Stick-Slip phenomena distribution in frequency domain for investigated slurries at relative pad wafer sliding velocity of 0.31 m/s

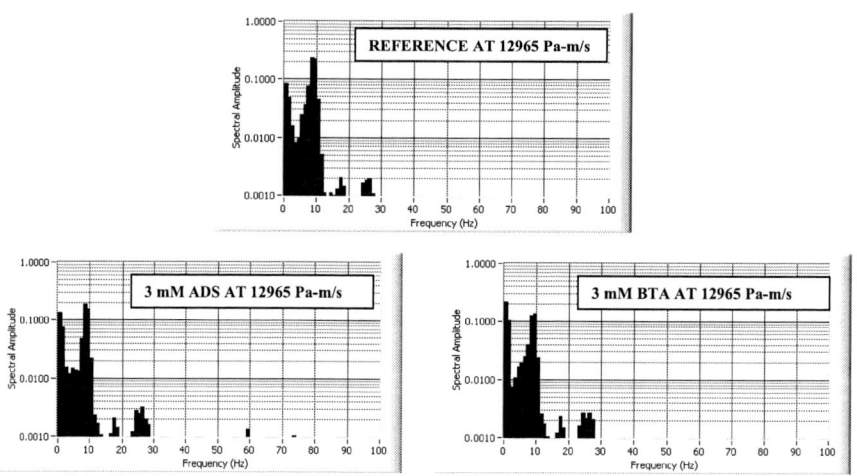

Fig. 4. Stick-Slip phenomena distribution in frequency domain for investigated slurries at relative pad wafer sliding velocity of 0.94 m/s

polishing power exhibit dominant peaks between 0-12 Hz and 40-60 Hz. Some of these peaks are assumed to be associated with stick-slip phenomena occurring due to fundamental processes at the microscopic (such as adsorption or desorption of surface active agents at an interface) and macroscopic (such as collision of wafer's advancing edge with neighboring grooves) level. Though these peaks normally signify elementary occurrences in the process, growth of these peaks can stipulate enhanced vibrations in the polishing process causing increased friction, wear and instability. On the other hand, suppression of these peaks suggests subdued vibrations thus resulting in smoother polishing. The emergence of peaks at lower frequencies (less than 12 Hz) is caused by the kinematics of the process including the rotational velocity of the platen, wafer and conditioner as well as by the collision event of groove and pad asperities with polished wafer. The other frequencies with lower spectral amplitude may be considered as noise of the polish process that has no significant implication.

The high frictional characteristic of the BTA is evident from Fig. 3 where addition of BTA transforms high frequency spectral peaks. In the presence of BTA these spectral peaks (40-60 Hz) grow and become outsized indicating high variance, shear force and unsteadiness in the process. Conversely, addition of ADS does not significantly alter the spectra. However, one can notice development of low amplitude peaks at different frequencies which may be reflective of different mechanistic microscopic behavior exhibited by ADS, due to its inherent surface active attributes. Currently there is not enough information to predict these phenomena and further investigation will be required to explore the details. Normalized spectral amplitudes plots at higher polishing power are shown in Fig. 4.Increasing polishing power shifts the spectral peaks from higher frequency to lower frequency regime. Therefore, on comparing spectral peaks at the two different polishing powers, one can observe an increase in population of peaks at frequency lesser than 12 Hz at higher polishing power. As mentioned earlier, these peaks below 12 Hz, are caused by the kinematics of the process and therefore it is typical to detect an increase in peak masses with an increase in the polishing power or relative pad wafer sliding velocity. However, unfortunately these lower frequency high amplitude peaks restrain the appearance of other peaks at higher frequency. Therefore the spectral peaks which were present at higher frequency in lower velocity spectra disappeared in higher velocity spectra.

4. Conclusions

Results demonstrate that the inhibition efficiency of ADS is superior to that of BTA. COF obtained from frictional force measurement is the lowest when the slurry contains ADS. This suggests that adsorbed ADS on the surface provides lubricating action thereby reducing the wear between the contacting surfaces. ADS containing slurry shows the lowest temperature rise again confirming the softening effect of the adsorbed surfactant layer, thereby resulting in less energy dissipation due to friction. Spectral analysis (in frequency domain) of the shear force data is employed to determine the inhibition of the additives in the slurry. It is noticed that increasing the pad-wafer sliding velocity at constant wafer pressure shifts the high frequency spectral peaks to lower frequencies and also increases the variance of the frictional force at the pad-wafer interface. Addition of ADS reduces the fluctuating component of the shear force and the extent of the pre-existing stick-slip phenomena caused by the kinematics of the process and collision event between pad asperities with the wafer. By contrast, in the case of BTA, there are no such observed benefits but instead undesirable effects are seen in some polishing conditions.

This work underscores the importance of real-time force spectroscopy in elucidating the adsorption, lubrication and inhibition of additives in slurries in CMP.

Acknowledgments

The authors wish to express their gratitude to the NSF/SRC Engineering Research Center for Environmentally Benign Semiconductor Manufacturing for financial support.

References

1. S. Deshpande, S. Kuiry, M. Klimov, Y. Obeng, and S. Seal, *J. Electrochemical Soc.*, **151**, G788 (2004).
2. Y. Hong, U. Patri, S. Ramakrishnan, and S. Babu, *Mater. Res. Soc. Symp. Proc.*, **867**, W1.10 (2005).
3. Y. Hong, D. Roy, and S. Babu, *Electrochem. Solid-State Lett.*, **11**, G297 (2005).
4. T. Du, Y. Luo, and V. Desai, *Microelectronic Engineering*, **71**, 90 (2004).
5. P. Bernard, P. Kapsa, T. Coud, and J. Abry, *Wear*, **259**, 1367 (2005).
6. D. Rosales-Yeomans, L. Borucki, T. Doi, L. Lujan and A. Philipossian, *22nd VMIC conference* (2005).
7. V. Kolev, K. Danov, P. Kralchevsky, G. Broze, and A. Mehreteab, *Langmuir*, **18**, 9106 (2002).
8. Q. Luo, D. Campbell, and S. Babu, *Langmuir*, **12**, 3563 (1996).
9. A. Philipossian and S. Olsen, *Jpn. J. Appl. Phys.*, **42**, 6371 (2003).
10. G. Stachowiak and A. Batchelor, *Engineering Tribology*, Butterworth-Heinemann, Inc., Woburn (2001).
11. L. Borucki, A. Philipossian and Y. Zhuang, *22nd VMIC conference* (2005).
12. A. Philipossian, D. Rosales-Yeomans and T. Doi, *Submitted to Jpn. J. Appl. Phys.* (2005).
13. E. Brigham and H. Oren, *The Fast Fourier Transform and Its Applications*, Prentice-Hall, Inc., Inglewood Cliffs (1988).

ECS Transactions, 2 (2) 523-536 (2006)
10.1149/1.2195687, copyright The Electrochemical Society

INFLUENCE OF GAS VELOCITY AND HUMIDITY ON DIETHYL
PHTHALATE ADSORPTION AND DESORPTION ON SILICON SURFACE

Hitoshi Habuka, Masaki Tawada, Kisho Suzuki, Takashi Takeuchi and Masahiko Aihara

Department of Chemical Engineering Science, Yokohama National University,
Yokohama 240-8501, Japan

The real-time adsorption and desorption of water and diethyl
phthalate on silicon surface were studied using a quartz crystal
microbalance in ambient nitrogen, having various values of
humidity, gas velocity and diethyl phthalate concentration. In a
steady state, the adsorbed amount of water and diethyl phthalate is
considered to depend only on each concentration in the ambient
nitrogen. However, in a non-steady state, the gas velocity at the
quartz crystal microbalance enhances both the adsorption rate and
the desorption rate. Therefore, the gas velocity at the measurement
point is concluded to be one of the key parameters that influence
the organic contamination measurement.

INTRODUCTION

In order to produce highly reliable gate oxide films on a silicon wafer surface for
microelectronics manufacturing, the advanced technology to build and maintain clean
environment in a clean room is necessary. However, airborne molecular contamination in
a clean room is widely known to cause many problems, such as degradation of the oxide
film quality, an error in the gate oxide film thickness measurement and an increase in the
leakage current due to silicon carbide formation during annealing (1-16).

In order to study the complicated time-dependent behavior of the organic
contamination, a real-time measurement of the airborne molecular contamination in a
clean room is necessary and expected. Therefore, the measurement using a quartz crystal
microbalance (QCM) has been evaluated (17-23). Although this method can measure
only the total weight of the contaminants and is theoretically unable to perform the
species identification, the information under the controlled environment is expected to
clarify the airborne molecular contamination behavior on a silicon surface in a steady
state and in a non-steady state. Therefore, Okamura *et al.* showed the influences of
inorganic compounds and the behavior of the slower desorption compounds of dibutyl
phthalate (DBP) (19-21).

Subjects to be studied further are the influences of humidity, gas velocity and
contaminant concentration in the gas phase. Particularly, the influence of gas velocity
should be clarified, because not only the QCM sensor, but also the wafers are usually
exposed to air at the position having various air velocities depending on the arrangement
of the wafers and the equipment (15, 16, 25, 26).

Therefore, the adsorption and desorption of water and diethyl phthalate (DEP), widely known contaminant (3, 5) on silicon surface, are studied in detail using the quartz crystal microbalance method in ambient nitrogen having wide ranges of humidity, average gas velocity and DEP concentration, in addition to our previous studies (22, 23). The behavior of the adsorbed water and DEP amount is evaluated using the model of multicomponent organic species adsorption-induced contamination (MOSAIC), particularly focusing on the influence of the gas velocity (5, 6).

EXPERIMENTAL

Figure 1 shows the quartz crystal microbalance system used to measure the adsorption and desorption of water and DEP on a silicon surface. The QCM sensor (Halloran Electronics Co., Ltd., Tokyo) is made of a circular-shaped quartz plate, AT-cut, with a diameter of 8 mm. Its front and back sides have attached circular silicon electrodes, formed by sputtering (19), with the diameter and the thickness of 4 mm and 50 nm, respectively. Because the silicon electrode surface was oxidized during the storage in air (room temperature, humidity of 40 % in a clean box) for more than one month after its preparation, this sensor is expected to have the same function of silicon wafer surface having native oxide film. The amounts of the adsorbed and the desorbed chemical compounds on the silicon surface were evaluated using the decrease and the increase in the frequency of the QCM sensor, respectively. The frequency of the QCM sensor was adjusted to 25 MHz. The decrease in the frequency of 1 Hz shows an increase in the surface concentration of 0.5 ng/cm^2 in this study.

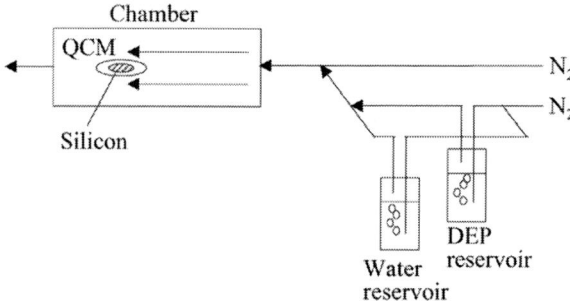

FIGURE 1. The quartz crystal microbalance system used to measure the adsorption and desorption of water and DEP on the silicon surface.

This system used nitrogen as the ambient gas. A part of the nitrogen gas was introduced into two liquid reservoirs, one of which had deionized water and the other contains DEP liquid. Water and DEP were vaporized by bubbling nitrogen gas through them. The vaporized amounts of water and DEP were obtained by measuring their weight decrease after the measurement. The relative humidity was obtained by dividing the gas phase concentration of water by its saturated value at the temperature of the QCM system. The concentrations of water and DEP in the gas phase were adjusted by adding the nitrogen gas.

The gas mixture was introduced into the chamber. The QCM sensor was positioned at the center position of the downstream edge of the chamber. The gas mixture was lead through a small stainless tube and uniformly spread at the entrance of the chamber using a shower-head type gas distributor. Thus, the gas mixture was flowing in a direction parallel to the surface of the QCM sensor in the chamber, as shown in figure 1. The gas supply tube and the chamber were made of stainless steel. The two liquid reservoirs were made of a fluorocarbon resin.

The relative humidity was adjusted to 0-40 %. The average gas velocity in the chamber was adjusted from 0.1 to 0.4 m/s. The DEP concentration was 0.01-0.04 g/m^3. Although the very high gas phase DEP concentration is used to perform the measurement, its surface concentration, comparable to the actual organic contamination amount, was assumed to show the behavior quite similar to an actual airborne molecular contamination due to phthalate esters. The temperature and the pressure of the measurement system shown in figure 1 were maintained at 298 K and atmospheric pressure, respectively.

RESULTS AND DISCUSSION

Entire behavior of water and DEP

First in this section, the adsorption and desorption of water and DEP, were measured and described using the QCM method. Figure 2 shows the typical change in the surface concentration of water and DEP on the silicon surface of the QCM sensor.

In figure 2, by adding the water vapor to the nitrogen atmosphere, the surface concentration immediately increased to 110 ng/cm^2 and reached a steady state. After, the further addition of DEP vapor, the surface concentration immediately increased to near 150 ng/cm^2; then it reached a steady state. By means of terminating the supply of the DEP vapor, the surface concentration decreases. After reaching a steady state, the surface concentration is found to recover to the same value as that before supplying DEP vapor. Finally, by terminating the supply of water vapor, the surface concentration became the initial value of this measurement.

Therefore, in figure 2, the increase and decrease in the surface concentration were considered to correspond to the start and termination of the supply of water vapor and DEP vapor, respectively.

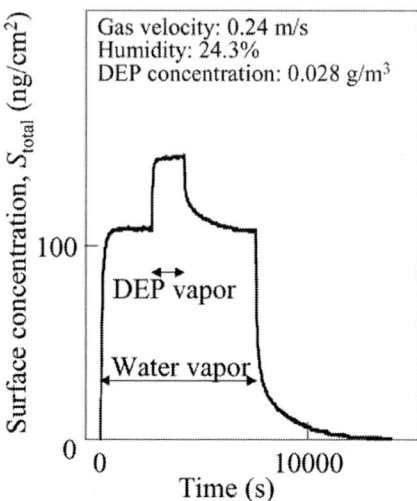

FIGURE 2. Change in total amount of water and DEP by adsorption and desorption on the silicon surface of QCM at 298 K. Arrows show the period of the water vapor supply and DEP supply.

Water adsorption

The amount of water adsorbed on a silicon surface was studied by means of exposing the QCM sensor to ambient nitrogen having various values of humidity and gas velocity.

Figure 3 shows the surface concentration of water on the silicon surface at 298 K. The amount of water is considered to have the largest value of the compounds adsorbed on silicon surface, because it exceeds 100 ng/cm² at humidity of 40%, As shown in figure 3, the surface concentration of water adsorbed on the silicon surface of the QCM sensor at various gas velocities is considered to follow a linear relationship indicated by the solid line and did not show a relationship with the gas velocity. Therefore, the surface concentration of the adsorbed water on the silicon surface in a steady state was considered to depend only on the humidity, with a negligible dependence on the gas velocity around the QCM sensor. The increasing and decreasing rate of the water amount was also evaluated. However, its dependence on the gas velocity was not clear. Therefore, the gas velocity is concluded to only negligibly affect the adsorption and desorption rate of water.

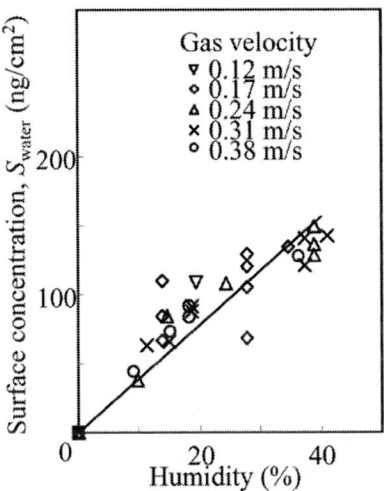

FIGURE 3. Surface concentration of water on the silicon surface of QCM at 298 K at various gas velocities and at the relative humidity of less than 40 %.

<u>DEP in a steady state</u>

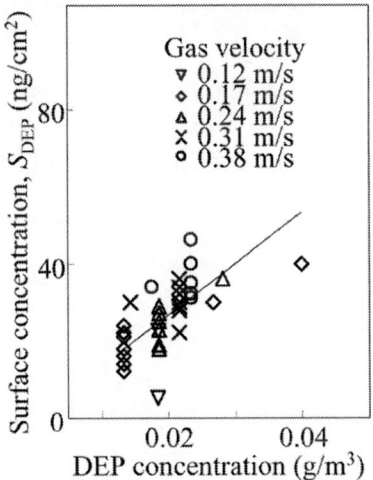

FIGURE 4. Surface concentration of DEP on the silicon surface of QCM at 298 K, and at the DEP gas concentration of less than 0.04 g/m³, at various gas velocities.

Figure 4 shows the surface concentration of DEP on the silicon surface at the temperature of 298 K, the DEP gas concentration of less than 0.04 g/m^3 and the gas velocities of 0.12 - 0.38 m/s. The solid line in figure 4 shows that the surface concentration of DEP is proportional to the DEP concentration in the gas phase. A clear dependence of the adsorbed DEP amount on the gas velocity does not appear in figure 4. Therefore, the surface concentration of DEP in a steady state is considered to be independent of the gas velocity, similar to that of water.

Water and DEP on Silicon Surface

Based on the information of the adsorbed amount of water and DEP, the condition of water molecule adsorbed on silicon surface is discussed, taking the Takahagi et al.'s notable study (24) into account. They reported that there were three layers of water on silicon wafer surface, classified from the viewpoint of chemical condition of water molecule.

(i) The first layer is Si-OH layer formed on the surface of silicon oxide film.
(ii) The second layer is that consisting of water molecule having hydrogen-bond with the first layer.
(iii) The water layer is formed on the second layer, by physisorption as a top layer.

Although an amount of the first and the second layers does not depend on humidity, the amount of the third water layer is proportional to humidity (24). Because the amount of water adsorbed on the QCM sensor as obtained in this study is proportional to humidity like the top layer found by Takahagi et al., the water adsorption measured by the QCM method could be considered to correspond to the top layer formation.

The silicon surface condition is further described using figure 5 which shows a schematic of the silicon surface condition induced by supplying water vapor and DEP vapor.

First, by supplying water vapor to the silicon surface of the QCM sensor in figure 5(a). Water is considered to form a top layer with its molecules, as shown in figure 5(b); its amount reached a steady state due to the adsorption and desorption. Both adsorption and desorption are simultaneously occurring on the native oxide surface. The top layer is considered to be quickly produced in a clean room, because silicon wafers immediately after cleaning are usually exposed to air having a humidity of 40-50 %. The top water layer has an interaction with the organic molecules causing their adsorption, in other words, the contamination.

Next, DEP vapor comes after the water layer formation and is considered to adsorb on the water layer, as shown in figure 5(c). It then desorbs from there in figure 5(d). If a very strong interaction, such as mixing, between the DEP molecule and water molecule exists, a complicated trend in the adsorption or desorption behavior may appear, particularly, for a very high gas phase concentration of DEP as in this study. However, only a simple adsorption and desorption behavior is observed in figure 2. Therefore, very simple adsorption on DEP is considered to occur.

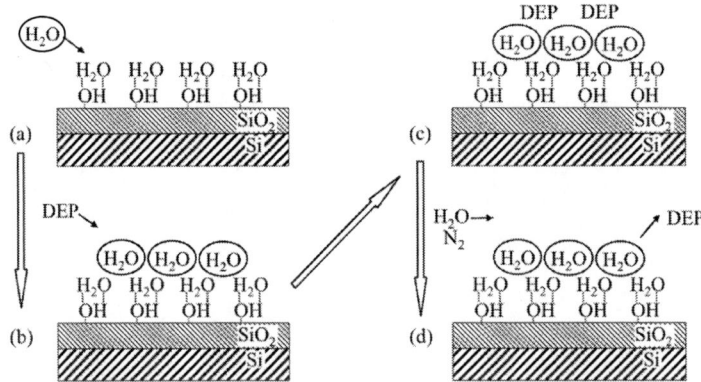

FIGURE 5. Silicon surface condition for adsorption and desorption of water vapor and DEP vapor. (Solid line: covalent bond, dotted line: hydrogen bond and circled H$_2$O: physisorbed water.)

DEP in a non-steady state

Because the airborne molecular contamination has been proved to be a time-dependent phenomenon (2, 3), the influence of the gas velocity on the increasing and decreasing rates of the DEP amount is described in this section.

Figure 6 shows the surface concentration ratio of DEP, $S_{DEP}/S_{steady\ state}$, increasing with time after the initiation of the DEP vapor supply. The DEP surface concentration is normalized using that in the steady state.

Immediately after initiating the DEP vapor supply, the surface concentration of DEP increases. It gradually reaches a steady state corresponding to the DEP concentration in the gas phase. In figure 6, the increase in DEP at the gas velocity of 0.38 m/s is quicker than the others at the slower gas velocities. Overall, the increasing rate seemed to become greater with the increase in the gas velocity.

Figure 7 shows the decrease in the surface concentration ratio of DEP on the silicon surface at 298 K after the termination of the DEP vapor supply. The surface concentration ratio of DEP is obtained by dividing the surface concentration of DEP, S_{DEP}, by that in the steady state, $S_{steady\ state}$, before terminating the DEP vapor supply.

Immediately after terminating the DEP vapor supply, the surface concentration of DEP decreases as shown in figure 7. Within 40-60 seconds, the half amount of adsorbed DEP is desorbed. In this figure, the desorption rate at the high gas velocity becomes greater than that at the low gas velocity. At 80 seconds, the surface concentration ratio of DEP at the gas velocity of 0.38 m/s is nearly 65 % of that at 0.12 m/s.

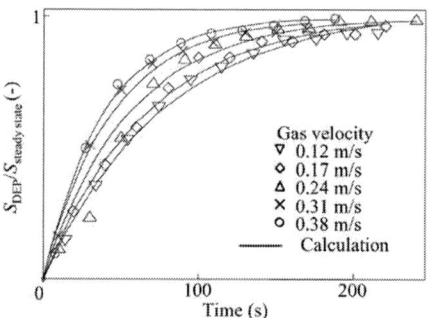

FIGURE 6. Surface concentration ratio of DEP, $S_{DEP}/S_{steady\ state}$, increasing with time after the initiation of the DEP vapor supply, at various gas velocities. Solid lines are the calculations.

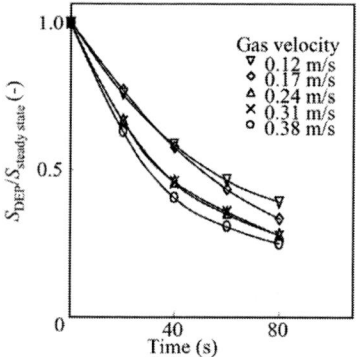

FIGURE 7. Decrease in the surface concentration ratio of DEP on the silicon surface at 298 K, $S_{DEP}/S_{steady\ state}$, after the termination of the DEP vapor supply, at various gas velocities.

In order to clearly evaluate the influence of the gas velocity, the rate constants of adsorption and desorption were obtained.

The adsorption and desorption of DEP molecules have been assumed to show a single-component behavior on the water layer (22). From the MOSAIC model (5), the single-component adsorption and desorption can be described in equation [1].

$$\frac{dS_{DEP}}{dt} = (S_e - S_{DEP})k_{ad,DEP}C_{DEP,s} - k_{de,DEP}S_{DEP} \qquad [1]$$

Here, t is time (s), $k_{de, DEP}$ is the desorption rate constant of DEP (1/s), and S_{DEP} is the surface concentration of DEP (ng/cm^2). $C_{DEP, s}$ is the gas concentration of DEP at the silicon surface (ng/cm^3), $k_{ad, DEP}$ is the adsorption rate constant of DEP on the silicon surface (cm^3/(ng s)), and S_e (ng/cm^2) is the effective higher limit of the concentration of the organic compounds adsorbed on the silicon surface.

The DEP desorption rate is described in equation [2].

$$\frac{dS_{DEP}}{dt} = -k_{de,DEP} S_{DEP}$$ [2]

Using equation [2] and figure 7, the desorption rate constant of DEP, $k_{de, DEP}$, is depicted by triangles in figure 8. This figure shows that $k_{de, DEP}$ increases with the gas velocity of the ambient nitrogen.

Next, S_e is evaluated prior to obtaining $k_{ad, DEP}$. Since the adsorbed amount of both DEP and water can be assumed to depend on the net surface area of the silicon, S_e is simply estimated following the assumption that the net surface is proportional to the adsorbed water amount on the QCM sensor surface. In our previous studies (22, 23), the relationship between the adsorbed amount of water and humidity is as follows:

Adsorbed water amount = 1.5 x Humidity (%) (22). [3]

Adsorbed water amount = 3.9 x Humidity (%) (23) [4]

S_e for the QCM sensor used in this study was estimated to be 110 ng/cm^2 and 280 ng/cm^2.

Using the values of S_e and $k_{ad, DEP}$ obtained in this study, the product of $k_{ad, DEP} C_{DEP}$ is obtained for each of the measurements in figure 6 so that the calculation agrees with the values of $S_{DEP}/S_{steady\ state}$, described by equation [5] derived from equation [1].

$$\frac{S_{DEP}}{S_{steady\ state}} = 1 - \exp\{-(k_{ad,DEP} C_{DEP,s} + k_{de,DEP} S_{DEP})\}$$ [5]

The solid lines in figure 6 are the calculations which agree with the measurement. Additionally, the change in the DEP's increasing rate with the gas velocity can be reproduced.

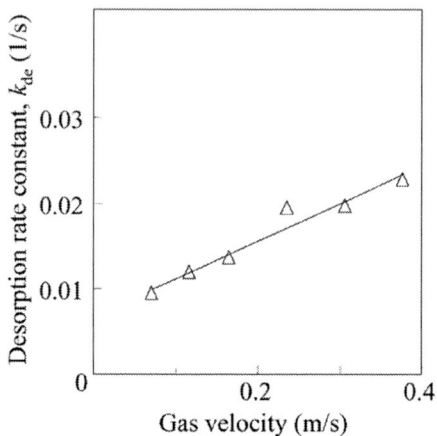

FIGURE 8. Change in the desorption rate constant of DEP with the gas velocity.

The value of $k_{ad, DEP}$ was obtained as shown in figure 9. This figure shows the increase in $k_{ad, DEP}C_{DEP, s}/C_{DEP, g}$ at 298 K with the gas velocity. The gas concentration (g/m^3) of DEP at the inlet of the chamber, $C_{DEP, g}$, is used for estimating $k_{ad, DEP}$. Because an actual value of $C_{DEP, s}$ is usually very difficult to be measured, and because the DEP consumption rate is very smaller than its feed rate, $C_{DEP, s}$ can be assumed to be comparable to $C_{DEP, g}$.

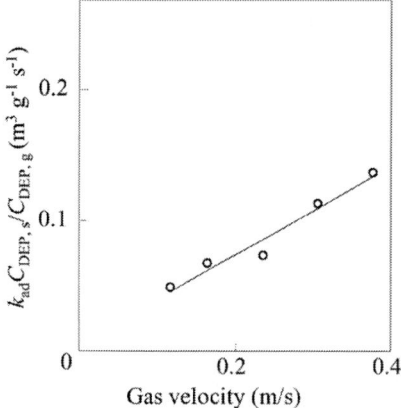

FIGURE 9. Change in $k_{ad} C_{DEP, s}/C_{DEP, g}$ at 298 K with the gas velocity. $C_{DEP, s}$ and $C_{DEP, g}$ are the gas concentration (g /m³) of DEP at the silicon surface and at the inlet of the chamber, respectively.

In multicomponent system

The influence of the increase and decrease in the adsorption and desorption rate on the surface concentrations of the organic compounds in a multicomponent system was evaluated.

Here, two organic compounds, C1 and C2, are assumed. C1 has a hundred times higher $k_{ad}C$ and k_{de} values than those of C2. By maintaining this relationship, when a set of $k_{ad}C$ and k_{de} values is 200% (Double) and 50% (Half) of the standard (Std.), the surface concentrations and the abundance of C1 and C2 are calculated using the MOSAIC model (5, 6). Figure 10 (a) shows the changes in the surface concentrations of C1 and C2. S_e is 280 ng/cm^2. $k_{ad}C$ and k_{de} used in this calculation are listed in Table 1. The surface concentrations of C1 for the Double, Std. and Half appear to overlap each other. They immediately increase after the initiation of the exposure (0 s). After reaching their peaks, their concentrations gradually decrease. Simultaneously, the surface concentration of C2 gradually increases and approaches a steady state. For C2, Double becomes a steady state earlier than Std. and Half. Therefore, figure 10 (a) concludes that the higher $k_{ad}C$ and k_{de} values can give an earlier approach to the steady state.

Table 1 Rate parameters used in figure 9

Compound	Line	C1		C2	
		$k_{ad}C$ (s^{-1})	k_{de} (s^{-1})	$k_{ad}C$ (s^{-1})	k_{de} (s^{-1})
Double	--------	0.004	0.04	0.00004	0.0004
Standard (Std.)	————	0.002	0.02	0.00002	0.0002
Half	-··-··-	0.001	0.01	0.00001	0.0001

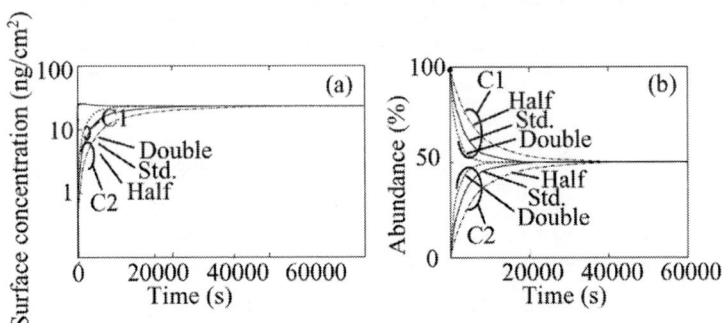

FIGURE 9. Change in (a) the surface concentrations, and (b) the abundance of the organic compounds having a fast rate (C1) and slow rate (C2) of adsorption and desorption. Double is the behavior for 200% of the medium rates (Std.); Half is that for 50% of the Std.

In order to study further the influence of the increase and decrease in the $k_{ad}C$ and k_{de}, the abundances of C1 and C2 are obtained, as shown in figure 10 (b), which shows that both C1 and C2 reach a steady state earlier when the $k_{ad}C$ and k_{de} values are higher. Additionally, the abundance of C2 having a small k_{de} rapidly increases when $k_{ad}C$ and k_{de} become high, corresponding to increasing gas velocity above and around the silicon surface. Therefore, the values of the surface concentration and the abundance are influenced by the $k_{ad}C$ and k_{de} values, even when they proportionally increase or decrease following the change in the gas velocity.

This result predicts that the typical organic contaminants, such as dibutyl phthalate and bis(2-ethylhexyl)phthalate, having very small $k_{ad}C$ and k_{de} values (6), can easily increase under the higher air velocity.

CONCLUSIONS

The real-time adsorption and desorption of water and DEP on silicon surfaces were studied using a quartz crystal microbalance in ambient nitrogen. The humidity, average gas velocity and DEP concentration were 0-40 %, 0.1-0.4 m/s and 0.01-0.04 g/m^3, respectively.

(1) In a steady state, the surface concentration of water and DEP, S_{water} and S_{DEP}, respectively, are proportional to each gas phase concentration. S_{water} at fixed humidity reaches the same value under various gas velocities; S_{DEP} becomes steady state at the value corresponding to the DEP gas concentration, with negligible relationship with the gas velocity. Therefore, the influence of the gas velocity on S_{water} and S_{DEP} in a steady state is considered to be negligible.

(2) The increase and decrease of the surface concentration of DEP become faster with increasing gas velocity. Then, the rate parameters of the adsorption and desorption of DEP are obtained using the single-component equation derived from the multi-component organic species adsorption-induced contamination (MOSAIC) model. The rate parameters of the adsorption and desorption are shown to increase with the gas velocity.

(3) The time-dependent organic contamination behavior of two virtual organic compounds is predicted by the calculation using MOSAIC model. One of the organic compounds had the large adsorption and desorption rate parameters, and the other had the small adsorption and desorption rate parameters. The surface concentration of the organic compound having the larger adsorption and desorption rate parameters increases earlier than that having the smaller parameters. Because the rate parameters include the influence of the gas velocity, according to our present study, this result predicts that organic contamination on a silicon wafer under fluent air flow increases faster than that under stagnant air flow.

ACKNOWLEDGMENTS

The authors would like to thank Dr. S. Okamura of National Institute for Materials Science, Profs. K. Okuyama and M. Shimada of Hiroshima University, for their discussion. They also thank Mr. N. Enomoto of Halloran Electronics Co., Ltd. for his technical suggestion.

REFERENCES

1. Y. Shiramizu and H. Kitajima, *Extended Abstracts of the 1995 Int. Conf. Solid State Devices and Materials*, p. 273 (Osaka, Japan, 1995).
2. T. Wakabayashi, M. Saito, T. Hayashi, Y. Wakayama and S. Kobayashi, *Extended Abstracts of IEE Japan*, p. 55 (Sendai, Japan, Nov., 1996).
3. Y. Sakamoto, K. Takeda, T. Nonaka, T. Taira, T. Fujimoto, N. Suwa and K. Otsuka, *Proc. 18th Annual Tech. Meeting Air Cleaning and Contamination Control*, p. 15 (Tokyo, Apr., 2000).
4. T. Takahagi, S. Shingubara, H. Sakaue, K. Hoshino and H. Yashima, *Jpn. J. Appl. Phys.*, **35** ,L818 (1996).
5. H. Habuka, M. Shimada and K. Okuyama, *J. Electrochem. Soc.*, **147**, 2319 (2000).
6. H. Habuka, M. Shimada and K. Okuyama, *J. Electrochem. Soc.*, **148**, G365 (2001).
7. C. L. Tsai, P. Roman, C. T. Tu, C. Pantano, J. Berry, E. Kamieniecki and J. Ruzyllo, *J. Electrochem. Soc.*, **150** (2003) G39.
8. N. Muenter, B. O. Kolbesen, W. Storm and T. Muller, *J. Electrochem. Soc.*, **150**, G192 (2003).
9. K. Saga and T. Hattori, *in Analytical and Diagnostic Techniques for Semiconductor Materials, DEvices, and Processes and ALTECH 2003*, B. O. Kolbesen, C. L. Cleays, P. Stallhofer, F. Tardif, D. K. Schroder, T. J. Shaffner M. Tajima and P. Rai-Choudhury Editors, PV **2003-03**, P. 136, The Electrochemical Society Proceedings Series, Pennington, NJ (2003).
10. S. D. Gendt, D. M. Knotter, K. Kenis, M. Depas, M. Meuris, P. W. Mertens and M. M. Heyns, *Jpn. J. Appl. Phys*, **37**, 4649 (1998).
11. F. Sugimoto, S. Okamura, T. Inokuma, Y. Kurata and S. Hasegawa, *Jpn. J. Appl. Phys.*, **39**, 2497 (2000).
12. J. J. Guan, G. W. Gale, J. Bennett, *Jpn. J. Appl. Phys.*, **39**, 3947 (2000).
13. K. Kawase, J. Tanimura, H. Kurokawa, K. Wakao, M. Inoue, H. Umeda and A Teramoto, *J Electrochem. Soc.*, **152**, G163 (2005).
14. A. E. Braun, *Semicond. Int. Jap. Ed,* **2**, 17 (Feb., 2005).
15. M. Tamaoki, K. Nishiki, A. Shimazaki, Y. Sasaki and S. Yanagi, *Proc. IEEE/SEMI Advanced Semicond. Manuf. Conf.*, p. 322 (Nov., 1995).
16. T. Fujimoto, *Proc. SEMICON Europe 2000* (Organic Contamination Workshop), p. 55, (Munich,2000)
17. M. Shimada, K. Okuyama, S. Honda and H. Habuka, *Kuukiseijo*, **40**, 24 (2002).
18. S. Okamura and R. Takasu, *Proc. 21th Annual Tech. Meeting on Air Cleaning and Contamination Control*, p. 37 (Tokyo, Apr., 2003).
19. S. Okamura, M. Shimada and K. Okuyama, *Jpn. J. Appl. Phys.*, **43**, 2661 (2004).
20. S. Okamura, M. Shimada and K. Okuyama, *Jpn. J. Appl. Phys.*, **43**, 4135 (2004).

21. S. Okamura, M. Shimada and K. Okuyama, *Jpn. J. Appl. Phys.*, **43**, 5496 (2004).
22. H. Habuka, K. Suzuki, S. Okamura, M. Shimada and K. Okuyama, *J. Electrochem. Soc.*, **152**, G241 (2005).
23. H. Habuka, M. Tawada, T. Takeuchi and M. Aihara, *J. Electrochem. Soc.* **152**, G862 (2005).
24. T. Takahagi, H. Sakaue and S. Shingubara, Jpn. J. Appl. Phys., **40**, 6198 (2001),
25. S. Ishiwari, H. Kato and H. Habuka, *J. Electrochem. Soc.*, **148**, G644 (2001).
26. H. Habuka, S. Ishiwari, H. Kato, M. Shimada and K. Okuyama, *J. Electrochem. Soc.*, **150**, G148 (2003).

SESSION 6
STRAINED SILICON

538

Strained Silicon

Eicke Weber
University of California, Berkeley
Berkeley, CA

Rick Wise
Texas Instruments, Inc.
Dallas, Texas 75243

Since the 90nm node, straining the silicon lattice to enhance hole and electron mobility in the channel has been used in CMOS technology. The resulting improvement in device performance has been so significant that the industry has quickly and pervasively implemented material, process, and integration schemes to introduce strain. However, despite the proven performance enhancement, physical measurement of the strain with resolution relevant to the transistor, dimensions where the strain techniques are incorporated, has remained challenging. Although other material parameters which impact the level of strain imparted to the channel (e.g., film thickness, film composition) may be measured and combined with numerical simulations to predict strain profiles in devices, experimental verification of the strain is desired from both the development and production control standpoint. X-ray diffraction and micro-Raman spectroscopy have been applied to this challenge but have limitations in terms of spatial resolution. Three research groups in the Silicon Wafer Engineering and Defect Sciences (SiWEDS) consortium have used the convergent beam electron diffraction (CBED) technique of transmission electron microscopy (TEM) in attempts to directly measure strain in the channel region of devices fabricated from the same characterized wafer. Another focus of the SiWEDS consortium relevant to the semiconductor industry's adoption of strained silicon is the study of dislocations (threading and misfit) generated in heterostructures.

Surface relaxation of the strained layer during preparation of the thin foils required for TEM was an issued encountered by the three research groups. In the first paper, Huang and Kim, University of Texas at Dallas, found that by adding a thin oxide layer on the top surface of the wafer prior to sample polishing, the relaxation could be minimized and the sample thickness and uniformity better controlled. The sample thickness strongly influences the intensity of the High Order Laue Zone (HOLZ) lines formed by electron diffractions from lattice planes in the upper Laue zones. Since the CBED technique of strain measurement is based upon matching experimental and simulated HOLZ patterns, the accuracy is largely determined by the clarity of the HOLZ lines. Energy filtering was used to minimize broadening of the HOLZ lines caused by inelastic electron scattering. Using CBED, they were able to measure strain in the channel region beneath the gate of a CMOS transistor and show the gradient of decreasing strain with increasing distance into the substrate (away from the gate). Using <230> and <340> zone axes to obtain the CBED patterns, they report a strain measurement in the 10^{-3} range with an error as low as ~10%.

In their CBED work, Zhao et al., North Carolina State University, used Hough transformations were used to detect HOLZ lines. The HOLZ line positions were simulated using kinematic theory and lattice parameters refined by minimizing the fitting function, χ^2. For thin samples (~ 300 nm), they observed that strain relaxation occurs by deforming the TEM foil. Areas of high strain gradient (beneath the gate) and non-uniform strain distribution (deformed areas of the foil) cause significant HOLZ line splitting/ blurring and make direct measurement of strain with CBED very difficult. They discuss the use of finite element modeling combined with

HOLZ line splitting information as a possible way to overcome these limitations and determine the initial strain state of the structure.

Zhang et al., University of California, Berkeley, show that for a structure of blanket strained silicon on relaxed silicon germanium, lattice constant measurements by CBED were in good quantitative agreement with those obtained by Raman spectroscopy. As with other papers in this session, they also encountered the limitations of CBED for measurement of locally strained regions caused by relaxation of the TEM foil and abrupt strain gradients. The new LIBRA[TM] TEM in the National Center for Electron Microscopy (NCEM) used for this research offers several features of value for sub-100nm strain detection, such as a monochromator and in-column corrected Omega filter. Strain down to 2×10^{-4} can be detected in this way. The authors were able to determine stress levels in the channel direction of a 35 nm physical gate length PMOS device at depths of 15 nm and 25 nm beneath the gate. They propose amorphizing the surfaces of the TEM sample as a way to shield the scattering of electrons from the partially relaxed crystalline columns (thus reducing HOLZ line blurring).

In the last paper of the session, Lu et al., North Carolina State University, have studied threading dislocations in strained silicon / silicon germanium heterostructures and the electrical activity of these defects using minority-carrier transient spectroscopy (MCTS). Following the method of earlier researchers, they illustrate the two-step defect etch technique to delineate threading dislocations from misfit dislocations which lie in the plane of the strained Si / SiGe interface. A short duration etch (diluted Secco or Schimmel solution) is used to open "pipes" in the threading dislocations down to the underlying SiGe. An etch solution (containing HF, H_2O_2, HAc) selective to SiGe is then used to undercut the relaxed SiGe beneath the strained Si film making the threading dislocation defects easily resolved by an optical microscope. They also showed that a one-step defect etch technique could be effectively used if the SECCO etchant was chilled to 2 °C. The MCTS studies revealed a dominant deep level. The authors studied the dependence of the signal height on the filling pulse duration and found it to be logarithmic. Such a behavior is typical for trap filling at extended defects such as dislocations, allowing the authors to conclude that this observed deep level is indeed caused by the dislocations present in the structure.

ECS Transactions, 2 (2) 541-547 (2006)
10.1149/1.2195689, copyright The Electrochemical Society

**Probing Nanoscale Local Lattice Strains in Advanced Si CMOS Devices by CBED:
A Tutorial with Recent Results**

J. Huang[a], P. R. Chidambaram[b], R. B. Irwin[b], P. J. Jones[b], J. W. Weijtmans[b],
E. M. Koontz[b], Y. G. Wang[b], S. Tang[b], R. Wise[b], and M. J. Kim[a]

[a] Department of Electrical Engineering, The University of Texas at Dallas, Richardson,
Texas, 75083
[b] Texas Instruments, MS 3739, 13560 N. Central Expressway, Dallas, Texas, 75243

The experimental methodology to characterize the nanoscale local
lattice strain in advanced Si CMOS devices by using Focused Ion
Beam (FIB) system and Convergent Beam Electron Diffraction
(CBED) is discussed. Through both high spatial resolution of
Transmission Electron Microscopy (TEM) and high strain
sensitivity of the CBED technique, compressive lattice strains in
the order of 10^{-3} from the nanoscale Si PMOS channel region are
detected. The one-dimensional quantitative strain-mapping is
performed by obtaining and simulating high quality CBED patterns
with different zone axes such as <230> and <340>.

Introduction

As the continuous scaling-down of Si-based MOSFETs is challenged to meet the
trend of the International Technology Roadmap for Semiconductors (ITRS), strained Si
technology is of great interest for providing the electron or hole mobility enhancement
(1-3). In order to better understand and engineer the strain incorporated in the nanoscale
region, local strain characterization with high resolution and sensitivity is essential.
However, conventional lattice strain measurements using micro-Raman spectroscopy (4)
or x-ray diffraction (5) can not be used due to their lack of spatial resolution needed for
the characterization of nanoscale devices. An advanced transmission electron microscopy
(TEM) technique, Convergent Beam Electron Diffraction (CBED), is a powerful method
for measuring local changes in lattice parameter due to strain (6,7). The CBED technique
provides strain-sensitive three-dimensional structural information with a spatial
resolution of approximately 1 nm with a sensitivity on the order of 10^{-4}, based on the
strain induced shift of High Order Laue Zone (HOLZ) lines (8). This technique allows
the strain components to be determined at each point of the sample from the analysis of
the corresponding CBED pattern. As a result, it is now possible to map out process-
induced and/or new device structure-induced strains in the active region of a local
isolated device with nanometer spatial resolution. However, few reports have been found
to use this technique on nanoscale strained Si CMOS devices (9,10). Most studies have
been focused on measuring the strain in Shallow Trench Isolation (STI) structures which
have relatively larger dimensions (11-13). In this paper, the local strain measurements on
advanced PMOS devices with sub-100 nm gate length are reported.

Experimental

541

The TEM sample is prepared by first mechanically cutting and polishing a Si (100) device wafer to be about 30 μm thick. A dual-beam Focused Ion Beam (FIB) system is then used to ion-mill the sample until the area of interest is about 300 nm thick. The thickness and milling process is monitored by in-situ Scanning Electron Microscopy (SEM). Pt is deposited prior to the ion milling to protect the device from the ion beam damage. A final cleaning step is applied by using low ion beam energy to reduce the damage and contamination on the sample. The surface relaxation induced during the sample preparation can be minimized by depositing a thin oxide layer on the surface of the wafer prior to the sample polishing. This method allows not only to do site-specific analysis for nanoscale device samples, but also to precisely control the sample thickness and uniformity, which are essential to obtain high quality CBED patterns. If the sample is too thin, the intensity of the HOLZ lines will be reduced due to the lack of the electron diffractions. But if the sample is too thick, the HOLZ lines will become broad. Both of the cases will degrade the accuracy of the strain measurement. Low power plasma cleaning is applied to remove the surface contamination after the sample preparation.

In this experiment, a 200 kV field emission gun (FEG) TEM is used. The probe size can be as small as approximately 1 nm, which can be utilized to measure point-to-point strain gradients in the device. The CBED patterns are obtained by focusing the electron beam on the area of interest. The diffraction pattern will appear as an array of disks. In the central disk, the HOLZ lines are formed by electron diffractions from lattice planes in the upper Laue zones, so that the HOLZ line positions are very sensitive to the small change of the lattice constant. As the specimen is tilted to a certain zone axis to acquire the CBED pattern, the unwanted specimen-tilt projection effect is induced. This effect becomes larger as the tilt angle and specimen thickness increase, which reduces the analyzable specimen area. As shown in Figure 1, the projection effects and CBED patterns from three different zone axes such as <230>, <340> and <560> are compared, where the specimen is tilted at 11.3°, 8.1° and 5.2° respectively off the <110> axis orientation. It is obvious that the projection effect is minimum at <560> zone axis (Z.A.), however, it is found that there are too much dynamical effects in the <560> CBED pattern so that the matching process becomes rather computing intensive. As illustrated in Figure 2, the Kinematically simulated pattern does not match with the experimental one. Since both <230> and <340> zone axes have been proven to be free of dynamical effects and the Kinematical approximation would provide the simulation with good accuracy, these two zone axes are chosen in this experiment (14). In order to improve the accuracy of the measurements, the clarity of the HOLZ line needs to be improved. When the electron beam goes through the sample, especially the sample with a few hundred nanometers thickness, the inelastic scattering effect will result in the HOLZ line broadening. A Gatan Imaging Filter (GIF) can be used to reduce the inelastic scattering. Figure 3 shows room temperature energy-filtered CBED pattern, comparing to the unfilted pattern. The improvements due to energy-filtering are clearly observed.

The strain tensor is determined by finding the best fit between the experimental pattern and the simulated one. A JEMS simulation software package is used to generate the HOLZ lines by either the Kinematical or Dynamical approach (15). Since the strain tensor has six components, it is impractical to simulate a CBED pattern with all six variables at the same time. Depending on the crystal orientation of the specimen, it is found that the independent strain tensor components can be reduced from six to three (7,16). Here it is assumed that only ε_x, ε_z, and ε_{xy} are allowed to change while ε_y, ε_{xz}, and

ε_{YZ} are constant. As the electron acceleration voltage also has an effect on the position shift of the HOLZ lines, the effective voltage is determined by retrieving CBED pattern from the unstrained Si and matching it with the simulated pattern that only changes with the voltage parameter. Then this effective voltage is used for the strain measurement.

Results and Discussion

Figure 4(a) is a cross-sectional TEM image of a Si PMOS transistor with a 37 nm gate length and the SiGe layer incorporated in the recessed Si at the drain extension location (17). A compressive strain gradient that decays from the gate-channel interface toward Si substrate and a tensile strain underneath the drain are observed, which is expected as the SiGe has a larger lattice constant than Si and is of finite depth. The CBED patterns taken from locations marked by circles in Figure 4(a) are shown in Figure 4(b), (c) and (d). The Kinematical simulation is first used in pattern matching to obtain the strain values. Then the full dynamical CBED pattern is generated to confirm the results. As a result, a compressive strain gradient in ε_x from about -2.8×10^{-3} to -1.3×10^{-3} and a tensile strain of approximately $+3.4 \times 10^{-3}$ are determined, which is consistent with the ANSYS based stress simulations. The negative and positive signs refer to the compressive and tensile strain respectively. A corresponding 35% drive current improvement is gained in this device and the detailed electrical measurement results are described elsewhere (3).

Using the nanometer size electron probe in TEM, a one-dimensional strain map in the active device region can be constructed by scanning along a line normal to the gate-channel interface to obtain CBED patterns from each individual point. In a Si PMOS device with a patterned gate length of 65nm, an epitaxial SiGe layer at the source/drain is chosen to compressively strain the channel in order to increase the hole mobility. Both <230> and <340> zone axes are used to obtain the CBED pattern. Figure 5(b) demonstrates this strain mapping, while Figure 5(a) shows a <340> Z.A. pattern taken from a single spot that is about 40 nm away from the gate-channel interface and superimposed by the matched simulation. The strain tensor components extracted from this CBED pattern are determined to be $\varepsilon_x = -6.8 \times 10^{-3}$, $\varepsilon_z = -2.8 \times 10^{-3}$ and $\varepsilon_{XY} = +0.9 \times 10^{-3}$. A compressive strain gradient in ε_x that decays from the gate-channel interface toward Si substrate is observed, which is also consistent with the fact that the SiGe has a larger lattice constant than Si. The measurement error due to the pattern matching can be as small as 3×10^{-4} for both <230> and <340> zone axes, which is acceptable for measuring the strain in the range of 10^{-3}.

Conclusions

TEM/CBED technique has been proven to be a powerful tool for high-resolution local strain measurement in the nanoscale Si CMOS devices. The process in preparing the site-specific and thickness controlled TEM sample by FIB has been proven to be effective. <230> and <340> Z.A are chosen to obtain the CBED patterns. As a result, the compressive strain gradients in the order of 10^{-3} are observed from sub-100 nm Si PMOS devices. With the help of nanometer electron probe used in FEG TEM, one-dimensional strain-mapping in the nanoscale active device region is possible.

Acknowledgments

This project is supported by Semiconductor Research Cooperation.

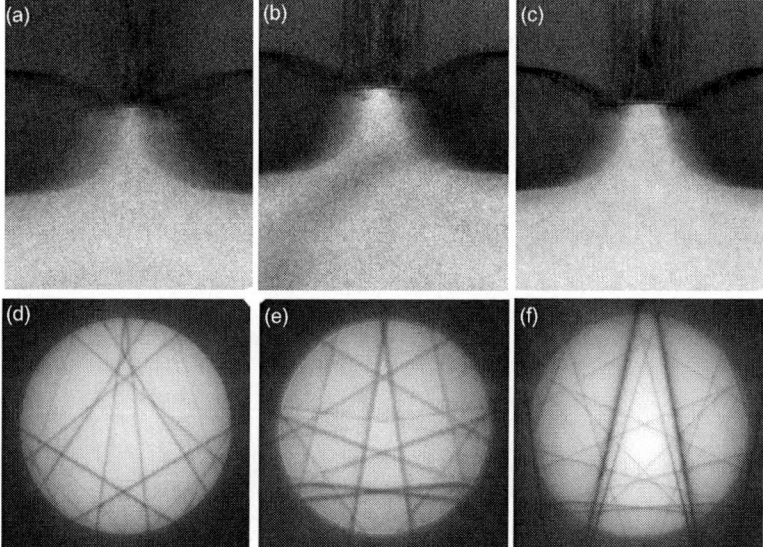

Figure 1. (a), (b) and (c) show the specimen-tilt projection effect from <230>, <340> and <560> Z.A., respectively. (d), (e) and (f) are the corresponding CBED patterns at these three zone axes.

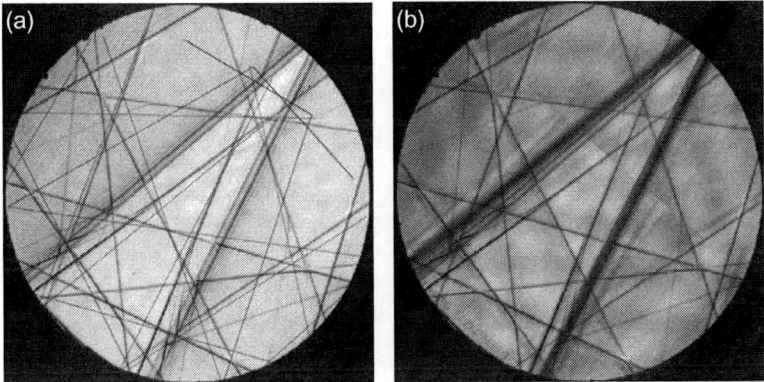

Figure 2. Experimental CBED patterns taken at <560> Z.A. are superimposed by Kinematical (a) and Dynamical (b) simulations. The simulation performed by Kinematical approximation does not fit with the experimental pattern.

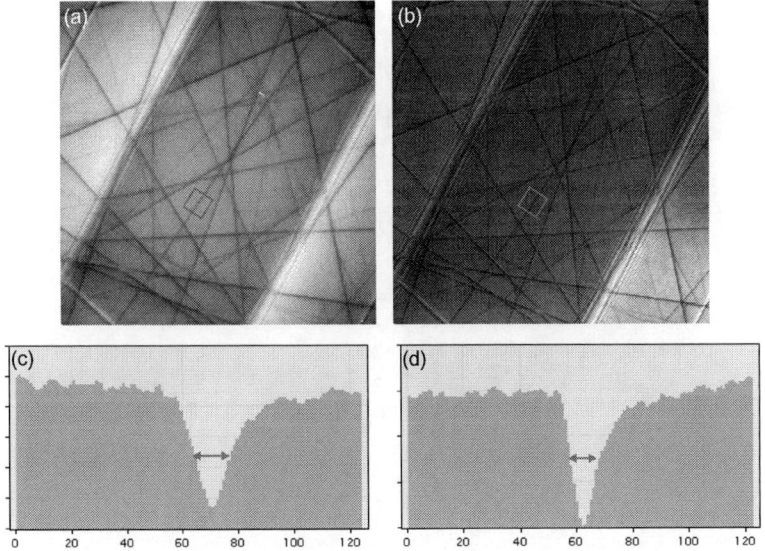

Figure 3. Experimental <230> CBED patterns: (a) room temperature unfiltered and (b) energy-filtered. Intensity line scans across the marked areas in (a) and (b) are shown in (c) and (d), respectively. Note the narrow line profile in the filtered one shown in (d).

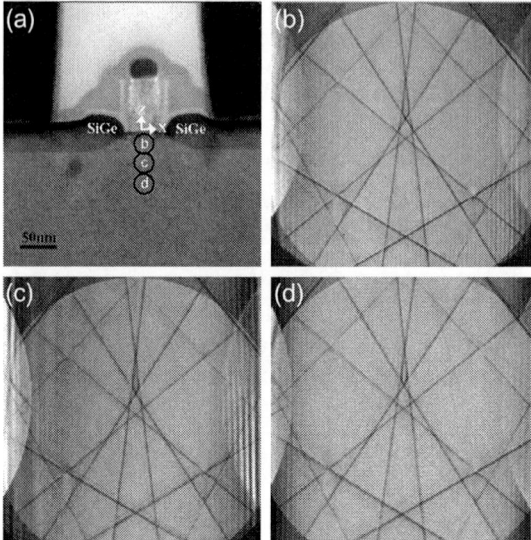

Figure 4. (a) TEM image of a Si PMOS transistor with a 37 nm gate length and the SiGe layer in the extended drain location. (b), (c) and (d) are experimental <230> Z.A. CBED patterns taken from the regions shown in (a), superimposed by dynamically simulated ones. The strain analysis indicates a compressive strain gradient in ε_x that decays from the center channel region: (b) -2.8×10^{-3}, (c) -2.0×10^{-3}, and (d) -1.3×10^{-3}.

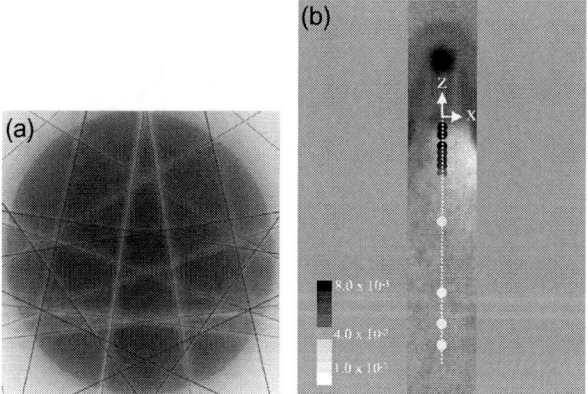

Figure 5. (a) <340> Z.A. CBED pattern is taken from a spot that is about 40 nm away from the gate-channel interface, superimposed by the kinematically simulated one. (b) Cross-sectional TEM image of a Si PMOS transistor with one dimensional strain map obtained by acquiring CBED patterns from the spots marked in the image, where the compressive strain gradients in ε_x are detected.

References

1. K. Rim, S. Koester, M. Hargrove, J. Chu, P.M. Mooney, J. Ott, T. Kanarsky, P. Ronsheim, M. Ieong, A. Grill and H. Wong, *Symp. VLSI Tech. Dig.,* p.59-60 (2001).

2. T. Ghani, M. Armstrong, C. Auth, M. Bost, P. Charvat, G. Glass, T. Hoffmann, K. Johnson, C. Kenyon, J. Klaus, B. McIntyre, K. Mistry, A. Murthy, J. Sandford, M. Silberstein, S. Sivakumar, P. Smith, K. Zawadzki, S. Thompson and M. Bohr, *Tech. Dig.- Int. Electron Devices Meet.,* p.978-980 (2003).

3. P.R. Chidambaram, B.A. Smith, L.H. Hall, H. Bu, S. Chakravarthi, Y. Kim, A.V. Samoilov, A.T. Kim, P.J. Jones, R.B. Irwin, M.J. Kim, A.L.P. Rotondaro, C.F. Machala and D.T. Grider, *Symp. VLSI Tech. Dig.,* p.48-49 (2004).

4. I. De Wolf, H. Norstrom and H.E. Maes, *J. Appl. Phys.,* **74**, 4490 (1993).

5. Y. Ando, J.R. Patel and N. Kato, *J. Appl. Phys.,* **44**, 4405 (1973).

6. J.M. Zuo, *Ultramicroscopy,* **41**, 211 (1992).

7. R. Balboni, S. Frabboni and A. Armigliato, *Philos. Mag.,* A **77**, 67 (1998).

8. P.M. Jones, G.M. Rackham and J.W. Steeds, *Proc. R. Soc. London, Ser.,* A **354**, 197 (1977).

9. A. Toda, N. Ikarashi, H. Ono, S. Ito, T. Toda and K. Imai, *Appl. Phys. Lett.,* **79**, 4243 (2001).

10. S.L. Toh, K.P. Loh, C.B. Boothroyd, K. Li, C.H. Ang and L. Chan, *J. Vac. Sci. Technol.,* B **23(3)**, 940 (2005).

11. M.Y. Kim, J.M. Zuo and G.S. Park, *Appl. Phys. Lett.,* **84**, 2181 (2004).

12. A. Armigliato, R. Balboni, S. Frabboni, A. Benedetti, A.G. Cullis, G.P. Carnevale, P. Colpani and G. Pavia, *Mater. Sci. Semicond. Process.,* **4**, 97 (2001).

13. C. Stuer, J. Van Landuyt, H. Bender, I. De Wolf, R. Rooyackers and G. Badenes, *J. Electrochem. Soc.,* **148**, G597 (2001).

14. A. Armigliato, R. Balboni and S. Frabboni, *Appl. Phys. Lett.,* **86**, 063508 (2005).

15. P. Stadelmann, *Microsc. Microanal.,* **9**, 60 (2003).

16. Deliverable D2, *STREAM Project,* p. 25 (2000), http://stream.bo.cnr.it

17. P.R. Chidambaram, B.A. Smith, L.H. Hall, H. Bu, S. Chakravarthi, Y. Kim, A.V. Samoilov, A.T. Kim, P.J. Jones, R.B. Irwin, M.J. Kim, C.F. Machala and D.T. Grider, in *SiGe: Materials, Processing, and Devices/*2004, D. Harame, J. Boquet, J. Cressler, D. Houghton, H. Iwai, T.-J. King, G. Masini, J. Murota, K. Rim and B. Tillack, Editors, PV 2004-07, p.123, The Electrochemical Society Proceedings Series, Pennington, NJ (2004).

ECS Transactions, 2 (2) 549-558 (2006)
10.1149/1.2195690, copyright The Electrochemical Society

Local Strain Measurement on Strained-Si/SiGe Heterostructures Using Convergent Beam Electron Diffraction (CBED) Analysis

W. J. Zhao[a], G. Duscher[a,b], and G. Rozgonyi[a]

[a] Department of Materials Science and Engineering, North Carolina State University, Raleigh, North Carolina 27695, USA
[b] Solid State Division, Oak Ridge National Lab, Oak Ridge, Tennessee 37831, USA

> Local strain variation in microelectronic device structures has been investigated with convergent beam electron diffraction (CBED) The analysis has been improved with a series of tasks: high order Laue zone (HOLZ) line detection by the Hough transform, HOLZ line position simulation with the kinematic theory, lattice parameter refinement by examination least square fitting, and theoretical strain state calculation with finite element (FE) modeling. The goal of this paper is to demonstrate the procedure of this strain measurement analysis. The actual strain variation in a TEM foil of a CMOS device structure was shown, followed by the FE in a strain Si/SiGe heterostructure.

Introduction

With the complexity of semiconductor devices increasing and geometric sizes decreasing, the local strain and the related stress, for example, for sSi/SiGe enhanced mobility devices, becomes one of the major issues in semiconductor technologies. Understanding and controlling strain is important for exploring strain generation mechanism, and selecting optimal processing and device operating conditions. However quantification of local strain variation with a nano-meter or sub nano-meter spatial resolution is still a challenge. With a potential strain sensitivity of 2E-4 and a nanometer spatial resolution, convergent beam electron diffraction (CBED) is believed to be the method of choice for investigations of strains at the smallest scale for future microelectronic devices, being far superior to X-ray diffraction or micro-Raman spectroscopy (1-5).

The CBED technique is based on analyzing the shift in higher order Laue zone (HOLZ) lines in a CBED pattern because the positions of HOLZ lines are very sensitive to the lattice parameters. Ideally, unknown lattice parameters (a, b, c, α, β, γ) can be determined simultaneously from a single CBED pattern. This involves careful selection of the "most sensitive" elements in an experimental pattern and then refining the fit with the simulated pattern. Different goodness-of-fit functions have been defined based on the selected elements, either distances between points or the areas enclosed by the points (6-9). This procedure can be complicated especially in cases where several lattice parameters are unknown. A selection of a certain number of intersection points or areas of HOLZ lines is inevitably lack of accuracy because not all of the HOLZ lines are included. When dealing with strain mapping in devices, a large number of CBED patterns have to be processed. Gathering and analyzing these CBED patterns is very time consuming and, therefore, a high level of computer assisted automation is required.

549

One of the big problems in strain measurements with CBED is sample deformation during TEM sample preparation. This is caused by the relaxation of the initial strain resided in the bulk, especially near the interfaces, when the sample is getting thinned. This deformation both complicates the original strain state characterization in bulk and degrades the quality of a CBED pattern and, therefore, makes it hard to detect HOLZ lines (10-12).

In this paper we developed a program with MATLAB and ANSYS for local strain CBED measurements. The HOLZ lines in the experimental patterns can be detected by the Hough transformation. The HOLZ line positions at different zone axis are determined with kinematic simulations. The effective voltage, the effective camera length and lattice parameters are optimized by examining the goodness-of-fit function. Lastly, we examine strain variation as a function of depth along a strained Si channel in a SiGe COMS device structure by comparing the splitting of HOLZ lines. A finite element simulation has started to determine the initial strain state in the bulk structure. The strain distribution in a strained Si/SiGe heterostructure is shown.

Line Detection with the Hough Transform

The Hough transform is widely used to detect straight lines, circles or any parameterized curve (13). Its main advantages are that it is relatively unaffected by noise or missing portions of the boundaries of objects of interest. The line detection program in this paper is based on the work by R. C Gonzalez et. al.(14). The underlying principle of the Hough transform is explained briefly in the following.

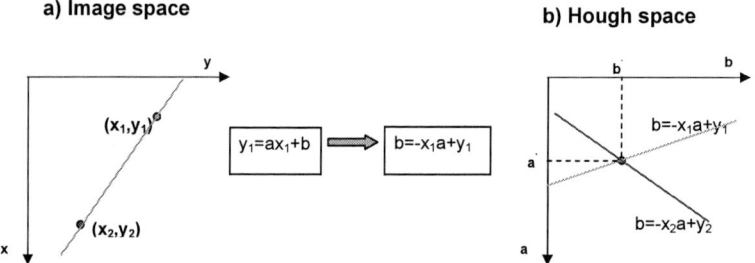

Figure 1. Schematic example of the Hough transform of a straight line in image space into a point in Hough space. Any point, for example (x_1,y_1), along a straight line in image space a), can be simply represented by a equation: $y_1=a*x_1+b$; By the equation transform, the point (x_1,y_1) can be represented by a straight line with a slope of $-x_1$ and an intercept of y_1 in Hough space b).

Figure 1 shows a schematic example of the Hough transform of a straight line in image space into a point in Hough space. Assume there is a straight line in image space to be detected (Fig. 1 a)). Any point, for example (x_1,y_1), along this line corresponds to a straight line in Hough space based on the following linear equation transform:

$$y_1 = ax_1 + b \Rightarrow b = -x_1 a + y_1 \qquad [1]$$

also shown in Figure 1 b). Transforming all of the points along this straight line results in an accumulation of counts (intensity) at a specific point (a`, b`) in Hough space.

The limitation with this parameterization in Cartesian coordinate is the divergence of the slope for vertical lines (15,16). To avoid this divergence problem, a parameterization (θ,ρ) in Polar coordinate is used. A line therefore can be represented by:

$$x\cos\theta + y\sin\theta = \rho \qquad [2]$$

where θ is the angle between the line normal to the x axis and ρ is the distance from the origin (center of the image in a CBED pattern) to the straight line, as shown in Figure 2 a). A point along a straight line will be Hough transformed into a sinusoidal curve, as shown in Fig. 2 b), instead of a straight line.

a) Image space

b) Hough space

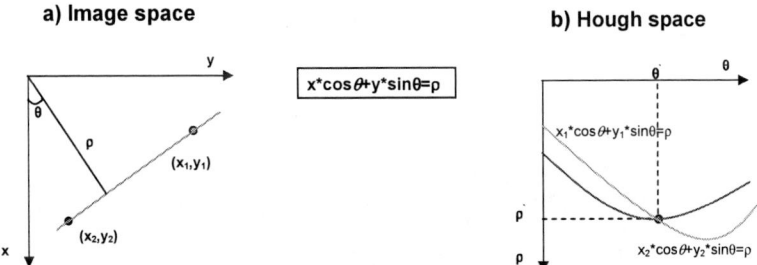

Figure 2. Schematic representation of Hough transform in Polar coordinate. Any point, for example (x_1,y_1), along a straight line in image space a), can be simply represented by a equation: $x_1*\cos\theta+y*\sin\theta=\rho$; By the equation transform, the point (x_1,y_1) can be represented by a sinusoidal curve, instead of a straight line, and the straight line in image space can be transformed in a specific point $(\theta`,\rho`)$ in Hough space b).

In the line detection program, each pixel in the image is transformed into Hough space and contributes a count to each element (θ,ρ) along the sinusoidal curve. Points along a straight line, $x_i*\cos\theta_i+y_i*\sin\theta_i=\rho_i$, give more counts to the element (θ_i,ρ_i) than any other elements since the sinusoidal curves all pass through the element (θ_i,ρ_i). Therefore, the peaks can be found by voting with respect to counts. Elements of (θ,ρ) with counts above a threshold value are treated to correspond to straight lines in image space.

Figure 3 show the HOLZ line detection with the Hough transform programmed on MATLAB: a) is a typical CBED pattern on [023] zone axis; the red lines superimposed on the CBED pattern in b) indicate the line positions measured with the Hough transform; and c) is the corresponding Hough transform of a) in (θ,ρ) parameterization. Each bright point marked with a number is treated to correspond to a HOLZ line in b). The precision of θ is set to be 1 degree and that of ρ is 1 pixel.

In practice, the accuracy of a line position determination depends on the line width and the signal-to-noise ratio. These can be improved by cooling down the sample with liquid nitrogen, energy filtering out the inelastic scattering electrons, and masked the background and the intersections of HOLZ lines in a CBED pattern. However, nonuniformity strain distribution within the electron interaction volume can broaden the

HOLZ lines. This is hard to avoid for a TEM foil, as will be discussed more in the experimental part of this paper. This broadening is a major issue for CBED strain measurement.

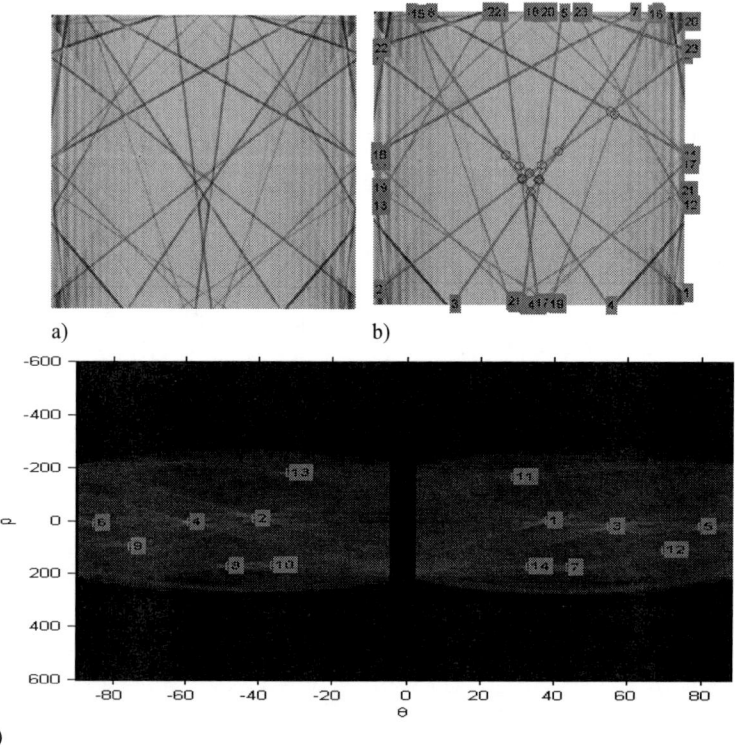

c)

Figure 3. Illustration of HOLZ line detection with the Hough transform programmed on MATLAB: a) is a typical CBED pattern on [023] zone axis; b) shows the detected HOLZ lines (in red) superimposed on the CBED pattern; c) is the corresponding Hough transform of the image in a) in (θ,ρ) parameterization. Each marked spot corresponds to a HOLZ line in the CBED pattern in b).

Kinematic Simulation of HOLZ Lines

Theoretical HOLZ lines positions for a selected zone axis need to be simulated and will be used to compare with the experimental ones for lattice parameter determination. The most precise simulation for HOLZ line positions is based on dynamical theory (17). However the dynamical simulation is very time-consuming, therefore not suitable in automation.

The kinematic approximation means that the HOLZ line positions are determined by Bragg law only. This theory can be found in most of the textbooks about transmission electron microscope. It's been found that in many situations the HOLZ line positions predicted by kinematic theory are acceptable. The effect of the dynamic shifts can also be reduced by applying an effective high voltage first proposed by Lin at el. (18), using an effective distance between the dynamic and kinematical dispersion surface suggested by Zuo (19) or applying an effective dynamic shit for each individual HOLZ line used by Kramer & Mayer (**Error! Bookmark not defined.**). Since the kinematic simulation is very fast it is suitable for automatic strain analysis. In this paper, we use MATLAB (20) to reprogram the codes from J. C. H. Spence and J. M. Zuo (21).

Goodness-of-Fit of Effective Voltage and Strain

In the literature, a number of "most sensitive" elements are carefully chosen and then either the distances between points or the areas of the triangles formed by the points are calculated for the goodness-of fit functions. In this paper, we use a different method. In this method we use all of the detected HOLZ lines to calculate $\chi 2$, based on the following equation:

$$\chi^2 = \sum_n (e_i^{Theory} - c \times e_i^{Exp})^2 \qquad [3]$$

where e is the chosen element, n is the number of chosen elements, and c is a variable magnification factor for fine-tuning the camera length L. This method will be shown to be more sensitive because more HOLZ lines are involved in minimizing $\chi 2$.

Experimental

The sample investigated in this paper is a MOSFET structure with SiGe source and drain regions. The aim of our work is to map the local strain along and across the Si channel. The TEM sample was prepared to be 300nm with focused ion beam (FIB) technique. The advantage of this technique is that the sample thickness can be well controlled resulting in a uniform thickness over a large area. This makes it easier to obtain good quality CBED patterns.

The CBED images were taken on a JEOL 2010F scanning transmission electron microscope equipped with a Gatan GIF 200 energy filter, scanning unit (STEM mode) and a slow scan rate 1024×1024 CCD camera. The microscope was operated on 200keV. An energy-selecting slit with a width of 10eV was inserted and centered around the zero-loss peak. The microscope was set up in scanning TEM mode where a fine probe of a few Angstroms was obtained. CBED images were taken in two different modes: spot mode and scanning mode. In spot mode, the electron probe was held stationary at a point and CBED patterns were acquired like in conventional TEM mode. In scanning mode, a series of CBED images were consecutively captured while the probe was scanning across the sample, simultaneously a Z-contrast image was acquired. This method allows for a high confidence in strain mapping because we know exactly where the CBED patterns were taken.

First the thickness of the TEM sample was determined in (004) two beam condition based on the following equation (22):

$$(\frac{S_i}{n_i})^2 + (\frac{1}{n_i})^2 (\frac{1}{\xi_g})^2 = \frac{1}{t^2} \qquad [3]$$

with S_i the excitation error at the ith minimum, t the effective specimen thickness along the beam direction, ξ_g the extinction distance, and n_i the nth intensity fringe. Figure 4 a) shows the (004) two beam condition and b) shows the contrast profile used for intensity fringe determination. The thickness of the TEM sample was found to be around 319.4nm from the plot in Fig. 4 c), within the thickness range for good quality CBED HOLZ line images.

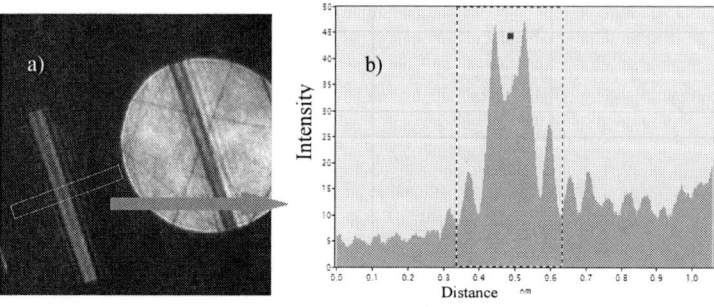

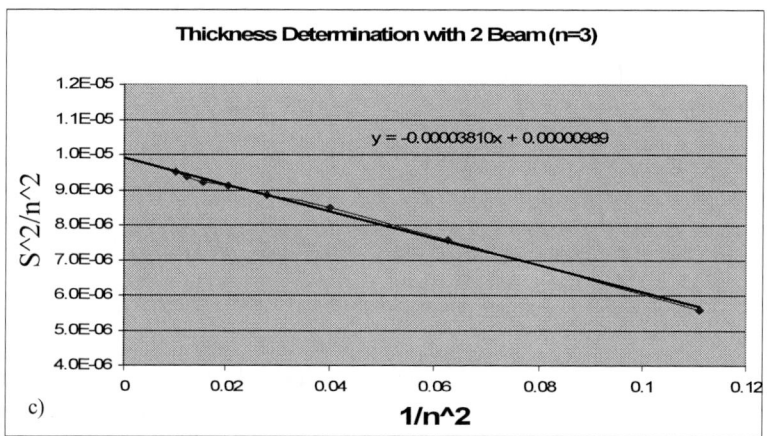

Figure 4. Thickness determination with two beam theory: a) is (004) two beam condition; b) is contrast scan along (004) beam rocking curve; c) is the graph of $(S_i/n_i)^2$ against $(1/n_i)^2$ for thickness determination.

The actual or effective accelerating voltage was calibrated before the strain quantification. The CBED pattern was taken from the Si substrate, far away from the strained Si region. The HOLZ lines were detected by Hough transform followed by minimizing the χ^2 value using with the voltage the only fitting parameter. The effective accelerating voltage was determined to be 199.5keV for the usual settings in the here used TEM.

Figure 5 a) is a Z-contrast image of the sample. Brighter areas are SiGe source and drain and the dark area is Si. The strain variation measurement was done in the scanning TEM mode by continuously scanning the electron beam along the strained Si channel region, marked as a rectangular box in Fig. 5 a). Figure 5 b) is a blowup of the rectangular box from Fig. 5 a). As a result totally 17 CBED patterns and Z-contrast images were acquired simultaneously, with each CBED pattern taken from a 8nm×40nm area, as shown in Fig. 5 b) and c). Figure 5 c) is a typical CBED pattern on [230] show the splitting HOLZ lines due to sample deformation during TEM sample preparation, as explained by L. Clement et. al. (**Error! Bookmark not defined.**). The splitting variation long the vertical distance to the channel was plotted on Fig. 5 d). The splitting of HOLZ line (-1,1,11) for each spectrum was converted to a value in degree based on the convergence angle, as plotted in Fig. 5 d) against vertical distance.

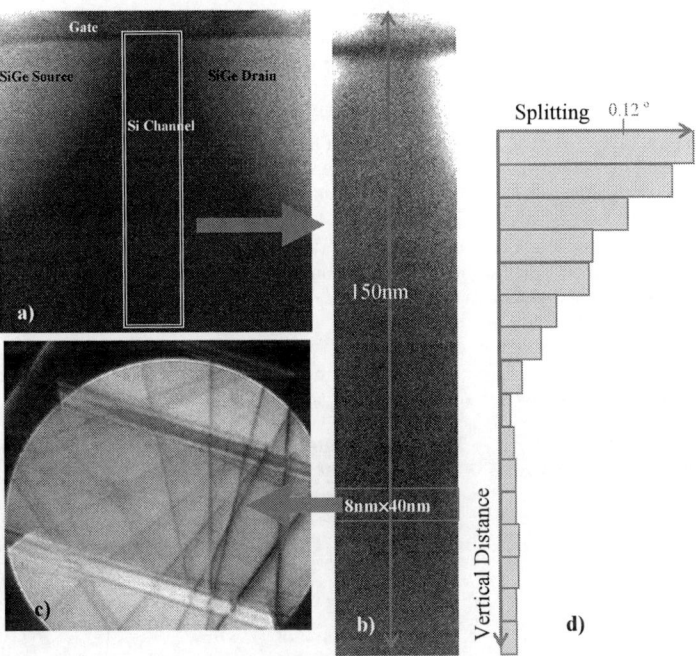

Figure 5. Strain variation along vertical distance (150nm) to the Si channel: a) is a Z-contrast image showing gate, SiGe drain, source and Si channel. The rectangular box shows the line scanning region with 150nm long; b) is the blowup of the rectangular box in a); c) is a [230] CBED pattern taken a 8nm×40nm area show in b); and d) shows the splitting in degree of (-1,1,11) HOLZ line with the vertical distance.

The splitting is so strong in the area under the gate that it's hard to measure. Therefore, the first few spectra were not included in the line-splitting profile in Fig 5. d). We can clearly see, that the splitting decreased with distance away from the gate. As we know, if the strain is uniform within the electron interaction volume, sharp HOLZ lines should be obtained. Due to TEM sample preparation, the sample is so thin that the initial lattice mismatch strain cannot be maintained any more. This causes strain relaxation by deforming the TEM foil. Therefore, the strain distribution is not uniform within the deformed areas, such as film surface and interfaces, where we are interested in knowing the strain state. Areas with greater strain gradient give more severe splitting, such as the area right under the gate, as shown in Fig.5 b) and d). This makes it impossible to measure the strain directly with CBED. However Clement et al. have shown that with the help of finite element modeling, the initial stain state can be calculated with the information of HOLZ line splitting (**Error! Bookmark not defined.**).

Finite Element Simulation

As stated earlier, finite element modeling can be used to help understand the strain state. We first started with a simple structure: strained Si on Ge heterostructure. The FE simulation was performed in ANSYS (23). The geometry of the sample and the frame of references are shown in Fig. 7 a). The thickness of Ge layer was chosen to be 100nm and of Si layer to be 5nm. This thickness ratio would be sufficiently representative of the substrate behavior as an *infinitely thick* layer (24). The strain due to lattice mismatch between Si and Ge was simulated by applying a uniform temperature change to the structure (**Error! Bookmark not defined.**,25,26):

$$\varepsilon = \alpha \times \Delta T \qquad [4]$$

where α is the effective coefficient of thermal expansion. Note that ΔT is introduced here just for generation of lattice mismatch strain in the heterostructure. There is no boundary condition applied to the sample.

Our preliminary FE simulation shows that the normal stress in the Si layer distributes uniformly near the center of the sample while it gets less and less uniform towards the edges. The normal stress accumulates with the highest value in the center, while it gets smaller away from the center.

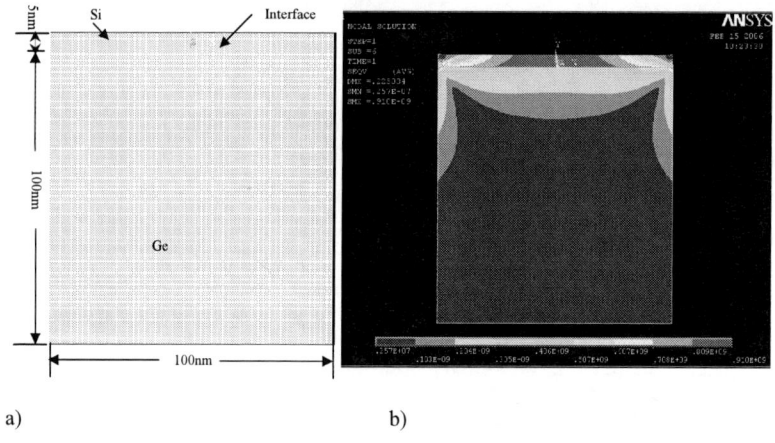

a)   b)

Figure 6. a) Illustration of sample geometry. b) Normal stress distribution for T=1000°C.

Conclusions

The procedure of local strain variation on microelectronic device structures with convergent beam electron diffraction (CBED) was illustrated. The procedure consists of 4 parts: high order Laue zone (HOLZ) line detection by the Hough transform, HOLZ line position simulation with the kinematic theory, lattice parameter refinement by examination of the goodness-of-fit function (minimization of χ^2), and theoretical strain state calculation with finite element modeling (FEM). The actual strain variation in a TEM foil of a CMOS device structure was shown. The preliminary FE simulation on a strained Si/SiGe heterostructure illustrates that FE modeling will enable us to quantify the the strain state within a device.

References

1. A. Toda, N. Ikarashi, and H. Ono, *J. Cryst. Growth*, **210**, 341 (2000)
2. A. Toda, N. Ikarashi, H. Ono, and K. Okonogi, *Appl. Phys. Lett.* **80 (13)**, 2278 (2002)
3. M. Kim, J. M. Zuo, and G. Park, *Appl. Phys. Lett.* **84 (12)**, 2181 (2004)
4. M. Kim, J. M. Zuo, and G. Park, *Appl. Phys. Lett.* **84 (12)**, 2181 (2004)
5. H. Lakner, B. Bollig, S. Ungerechts, and E. Kubalek, *J. Phys. D: Appl. Phys.* **29** 1767 (1996)
6. S. Kramer, J. Mayer, C. Witt, et al., Ultramicroscopy 81 245 (2000)
7. R. Wittmann, C. Rarzinger, and D. Gerthsen, *Ultramicroscopy* **70** 145 (1998)
8. J. M. Zuo, Ultramicroscopy **41**, 211 (1992)
9. S. J. R.zeveld, J. M. Howe, and S. Schmauder, *Acta Metall. Mater.* **40** S173 (1992)
10. F. Banhart, Ultramicroscopy **56**, 233 (1994)
11. S. L. Toh, K. P. Loh, C. B. Boothroyd, K. Li, H. Ang, and L. Chan, *J. Vac. Sci. Technol. B* **23 (3),** 940 (2005)
12. L. Clement, R. Pantel, L. F. Tz. Kwakman, and J. L. Rouviere, *Appl. Phys. Lett.* **85 (4)**, 651 (2004)
13. D. H. Ballard, *"Generalising the Hough transform to detect arbitrary shapes", Pattern Recognition*, **13**, pp. 111-122, (1981)
14. R. C Gonzalez, R. E. Woods, and S. L. Eddins, *Digital Image processing using MATLAB*, Upper Saddle River, N. J. Pearson Prentice Hall, (2004)
15. G. W. Awcock, and R. Thomas, *Applied Image Processing*, The Macmillan Press, NH (1995)
16. S. Kramer, and J. Mayer, *J. Microscopy*, **194 (1)**, 2 (1999)
17. P. M. Jones, G. M. Rackham, and J. W. Steeds, *Proc. Roy. Soc.* (London) A **354** 197 (1977)
18. Y. P. Lin, D M. Bird, and R. Vincent, *Ultramicroscopy*, **27** 233-240 (1989)
19. J. M. Zuo, *Ultramicroscopy*, **41** 221-223 (1991)
20. D. C. Hanselman, Mastering MATLAB 7, Pearson/Prentice Hall, Upper Saddle River, NJ, c2005
21. J. C. H. Spence and J. M Zuo, *Electron Microdiffraction*, Plenum Press, New York, (1992)
22. D. B. Williams, and C. B. Carter, Transmission Electron Microscopy: a Textbook for Materials Science III, Plenum Press, New York p 321 (1996)
23. S. Moaveni, Finite Element Analysis: Theory and Application with Ansys, Pearson/Prentice Hall, Upper Saddle River, NJ, c2003

24. W. M. Ashmawi, M. A. Zikry, K. Wang, and R. R. Reeber, J. Crystal Growth **266,** 415 (2004)
25. J. Li, D. Anjum, R. Hull, G. Xia, and J. L. Hoyt, *Appl. Phys. Lett.* **87**, 222111 (2005)
26. Y. C. Yeo, and J. Sun, *Appl. Phys. Lett.* **86**, 023103 (2005)

ECS Transactions, 2 (2) 559-568 (2006)
10.1149/1.2195691, copyright The Electrochemical Society

Analysis of Nano-scale Stress in Strained Silicon Materials and Microelectronics Devices by Energy-filtered Convergent Beam Electron Diffraction

Peng Zhang [a,b,*], Andrei A. Istratov [a,b], Haifeng He [c], Joel W. Ager [b],
Chris Nelson [c], Eric Stach [d], John Mardinly [e], Christian Kisielowski [c], Eicke R. Weber [a,b],
and John C.H. Spence [c]

[a] Department of Materials Science and Engineering, University of California,
Berkeley, California 94720, USA
[b] Materials Sciences Division, Lawrence Berkeley National Laboratory,
Berkeley, California 94720, USA
[c] National Center for Electron Microscopy, Lawrence Berkeley National Laboratory,
Berkeley, California 94720, USA
[d] School of Materials Engineering, Purdue University, West Lafayette,
Indiana 47907, USA
[e] Materials Technology Department, SC9-7, Intel Corporation, Santa Clara,
California 95054, USA
([*] Corresponding author. Email address: pzhang@lbl.gov)

The convergent beam electron diffraction (CBED) technique of
transmission electron microscopy (TEM) has excellent capabilities
for strain detection at high spatial resolution. Here we report strain
measurements in bulk ε-Si/SiGe/Si and in a strained 35nm PMOS
device in which SiGe acts as the source and drain. CBED
measurements of the composition of the relaxed SiGe buffer are in
quantitative agreement with Raman spectroscopy. For the PMOS
device, CBED measured a uniaxial compressive stress of 1.12GPa
in the channel. However, it was found that even in the cross-
sectional TEM samples with thicknesses greater than 300nm, the
intrinsic surface strain relaxation was often so severe that no
recognizable high-order Laue zone lines in the CBED patterns
could be collected. The amorphization of both free surfaces of
the TEM sample to a range of about 80nm is proposed to minimize
the impact of surface strain relaxation for future studies.

Introduction

It is known that elastic strain can be used to increase both hole and electron mobility
in silicon[1]. Thus, the use of strained silicon for the fabrication of complementary metal-
oxide-semiconductor (CMOS) devices makes possible gains in device performance at the
same scaling node. For this reason, strained silicon technologies have attracted intense
attention in recent years as a means to enhance performance of sub-100nm generations of
devices. Strain in the channel region of CMOS devices has been introduced either locally
by using selective epitaxial growth of SiGe alloy for PMOS[2-4] and high stress film
deposition for NMOS[2,3], or globally, e.g. by using bulk strained silicon wafers (ε-
Si/SiGe/Si)[5] and their derivatives such as strained silicon on insulator (SSOI) and strained
silicon on silicon germanium on insulator (SGOI). The evolution of strained silicon
technology has exceeded the predictions outlined by the 2001 International Technology
Roadmap for Semiconductors and has also posed new characterization requirements, one

of which is the measurement of local strain in the Si channel at tens of nanometers spatial resolution[6].

In the development of strained silicon technology, researchers have relied heavily on numerical simulations[2,4,7-9] to predict the local strain profile in individual devices. To the best of our knowledge, there are very few reports on the experimental measurement of strain profile near the channel of a strained silicon device. In two recent articles, high-resolution transmission electron microscopy (HRTEM)[10] and electron diffraction contrast imaging (EDCI)[11] were applied for local strain measurements in strained silicon devices. Nevertheless, it is generally believed that in HRTEM samples, which have a thickness of only several tens of nanometers[12], strain should be relaxed and thus the lattice images should be interpreted with precaution as they may not represent the strained lattice in a real device with sufficient accuracy. Very recently a group of researchers[11] reported high-resolution strain mapping by combining quantitative EDCI, three-dimensional finite element modeling (FEM), and electron diffraction contrast simulations. However, in this approach, the FEM requires the independent measurements of the intrinsic stress values of each material (e.g. by curvature measurement) in the device structure. Thus the application of this technique for a routine strain measurement would appear to be limited.

The convergent beam electron diffraction (CBED)[13,14] technique of transmission electron microscopy is probably the most suitable technique that can determine local strain at nanoscale spatial resolution and with high strain detection sensitivity. The shift of the high-order Laue zone (HOLZ) deficient lines in the transmission disk of the convergent electron beam is very sensitive to local lattice parameters. By using the χ^2 procedure[15] to match experimental and simulated HOLZ patterns, a strain of the order of 2×10^{-4} can be detected reliably. Recently, CBED has been successfully applied to measure highly localized strain near shallow trench isolation (STI) device structures[16-18].

However, we expect that several factors might limit the application of the CBED technique to the analysis of strain in highly strained nano-materials and sub-100nm devices. First, it should be kept in mind that in CBED analysis, a column of materials in the path of electron beam contribute to the HOLZ pattern. Moreover, a relatively thick (often > 200 nm) crystal with homogeneous properties (e.g. a zero or very small strain gradient) is required along the e-beam path for the formation of sharp HOLZ lines. Therefore, if we want to analyze strain along the source-drain direction, the *width* of the channel needs to be larger than 200nm, which exceeds the typical channel width of modern MOS devices. Thus a test structure rather than a real device is often necessary for CBED experiments. Secondly, most MOS devices are fabricated on (100) silicon and have channel direction along <011> orientation, which is the cut-line direction for cross-sectional TEM (XTEM) sample preparation. For conventional TEM analysis, the e-beam is directed along [110], which is an orientation not suitable for CBED strain analysis because the HOLZ lines are shielded by dynamical scattering patterns. It is thus necessary to tilt the sample to a low symmetry axis, which will degrade the spatial resolution due to projection effect[17].

Another obstacle for the application of CBED strain analysis is that in the electron-transparent thin foils required for TEM, elastic strain relaxation in the sample will change the strain state, making the strain measurement less reliable. This effect has been observed as split HOLZ lines in the CBED patterns[19,20]. By analyzing the splitting of HOLZ lines and modeling the lattice bending in TEM sample, it is possible to recover the strain profile of the unthinned sample[20]. However, it is often found that the HOLZ lines are blurred severely and even become invisible at the areas of interest because the local lattice planes are distorted too much. For this reason, strain analysis is limited to specific regions where the sample deformation is small and HOLZ lines are clean[19,21]. The strain relaxation in strained-layer TEM sample has been modeled analytically[12,22,23] and by using FEM[24,25]. It is found that the degree of strain relaxation depends on the amount of stress initially in the bulk sample, and the ratio between TEM sample thickness (t) and the thickness of the strained layer (Λ). A commonly accepted criterion is that when t/Λ value is sufficiently large (> 10), relaxation will mainly happen at the surfaces where electron beam enters and leaves the TEM sample[24-27]. This criterion implies that thick samples (>300 nm) are desirable for CBED analysis. On the other hand, the use of thick samples for CBED analyses increases the probability of electron inelastic scattering mainly due to plasmon losses (energy loss of the order of tens of eV) which will increase the diffused background of CBED pattern and lead to blurred HOLZ lines. The large-angle convergent beam electron diffraction (LACBED) technique can filter most of the inelastic electrons by using selected-area aperture as the angular filter[13,14] but typically the crystalline areas that are involved in the diffraction is relatively large (about 100 nm in diameter) thus the spatial resolution is not good enough for modern microelectronics devices and an energy filter is preferred. The combination of scanning transmission electron microscopy (STEM), energy filter, and CBED is an ideal choice and makes the spot-to-spot strain analysis possible[13].

In this work, CBED is applied to measure strain in bulk strained silicon and a locally strained 35nm PMOS device similar to that reported in Refs. 3 and 4. Complementary strain results on some samples were obtained by Raman spectroscopy. We will report our preliminary results obtained from the new Zeiss Libra® transmission electron microscope at the National Center for Electron Microscopy of the Lawrence Berkeley National Laboratory. The strain relaxation in XTEM is demonstrated by finite element modeling and CBED measurements. The possibility of minimizing the negative impact of surface strain relaxation in a TEM sample will be discussed.

Experimental

The Libra® Energy Filter 200 kV Transmission Electron Microscope combines several features including a Köhler Illumination system, a monochromator, a high-angle annular dark-field (HAADF) detector, and an in-column OMEGA energy filter. For CBED experiments, a dark field STEM image is collected first in scanning mode and then the electron beam is positioned to the area of interest under spot mode. A relatively large spot size between 5 nm to 8 nm is used for good electron intensity for thick TEM samples. The CBED pattern is collected by the CCD camera underneath the column. By using a 20eV energy selection slit, plasmon peaks are excluded from pattern formation. It is possible that cooling the double-tilt holder with liquid nitrogen might be used to further reduce the phonon scattering background and hydrogen-carbon contaminations. For our measurements, we used a regular double-tilt holder to minimize sample drift. The CBED pattern is collected near the silicon [340] axis. The determination of lattice parameters carried out is by using the Automatic Strain Analysis by CBED (ASAC) module of the analySIS® software developed by Soft Imaging System GmbH. The kinematical calculation of HOLZ lines for [340] axis has been validated by comparing the simulated results with dynamical simulation[17]. Use of this automated procedure for CBED analysis has been discussed elsewhere by Armigliato et al[16,17].

The bulk strained silicon sample with ε-Si/Si$_{0.75}$Ge$_{0.25}$/Si$_{1-y}$Ge$_y$/Si structure was characterized by Raman spectroscopy. A 457,9 nm laser was used as the light source and the probe depth in silicon was about 300 nm[28]. As discussed below, from the Si-Si phonon peaks from strained silicon and SiGe buffer layer, the Ge composition and the biaxial strain in ε-Si can be calculated[29-31].

Results and Discussion

Energy-filtered CBED Measurements on Strained Silicon Material

We have conducted CBED measurements on ε-Si/Si$_{0.75}$Ge$_{0.25}$/Si$_{1-y}$Ge$_y$/Si, precharacterized by Raman spectroscopy. Fig. 1 shows the low-magnification cross-sectional dark field STEM image of the strained silicon sample. The strained silicon cap is 30nm thick and the homogeneous 2.8µm Si$_{0.75}$Ge$_{0.25}$ buffer is grown on a 2.2µm dislocated SiGe graded buffer. The sample is prepared by focused ion beam and a wedge-shaped TEM sample with varying thickness is produced.

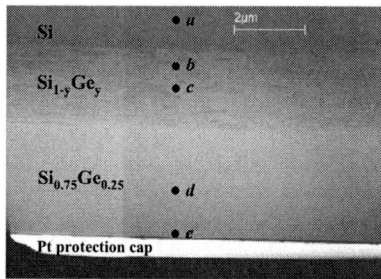

Figure 1. Cross-sectional dark field STEM image for a strained silicon sample. TEM sample is prepared by focused ion beam. The strained silicon layer is 30nm thick; the homogeneous Si$_{0.75}$Ge$_{0.25}$ buffer layer is 2.8µm thick and the graded SiGe buffer is 2.2µm thick. CBED patterns for points a-e are given in Fig.3(a)-(e) accordingly. Point e is in Si$_{0.75}$Ge$_{0.25}$ and is 30nm from the interface with ε-Si cap.

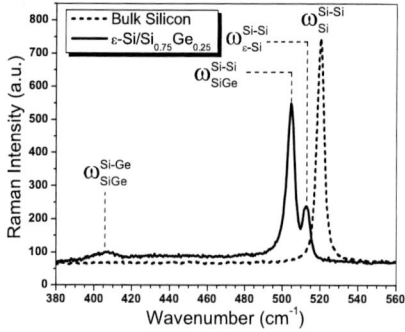

Figure 2. Raman spectra from a bulk silicon (short dashed line) wafer and a ε-Si/Si$_{0.75}$Ge$_{0.25}$/Si sample with 30nm ε-Si cap (solid line) as shown in Fig.1. The laser has a wavelength of 457.9nm.

The Raman spectra for the strained silicon sample and a bulk silicon reference sample are given in Fig.2. The Ge composition x and amount of mismatch strain ε in strained silicon can be calculated from the Raman spectra as follows:

$$\omega_{\varepsilon\text{-Si}}^{\text{Si-Si}} = \omega_{\text{Si}}^{\text{Si-Si}} - b \cdot \varepsilon \qquad [1]$$

$$\omega_{\text{SiGe}}^{\text{Si-Si}} = \omega_{\text{Si}}^{\text{Si-Si}} - c \cdot x \qquad [2]$$

where $\omega_{\text{Si}}^{\text{Si-Si}} = 520\text{cm}^{-1}$ is the phonon frequency of Si-Si vibration in bulk silicon; $\omega_{\varepsilon\text{-Si}}^{\text{Si-Si}}$ is that in strained silicon; and $\omega_{\text{Si-Ge}}^{\text{Si-Si}}$ is that in $\text{Si}_{1-x}\text{Ge}_x$. The phonon shift coefficients[29,30] are: $b=784\text{cm}^{-1}$ and $c=62\text{cm}^{-1}$. The calculated strain in ε-Si is $\varepsilon=0.956\%$ and x=25.1%. The lattice constant corresponding to measured strain is $a_{\varepsilon\text{-Si}} = 5.4829\text{Å}$. The Ge composition of the relaxed SiGe layer agrees well with the value given by the wafer vendor and ε-Si layer takes the lattice constants of the homogeneous $\text{Si}_{0.75}\text{Ge}_{0.25}$ buffer. The relaxed SiGe lattice constant ($a_{\text{SiGe}} = 0.5483$ nm) is calculated based on the equation[32]:

$$a(x) = 0.54310 + 0.01992 \cdot x + 0.002733 \cdot x^2 \text{ nm} \qquad [3]$$

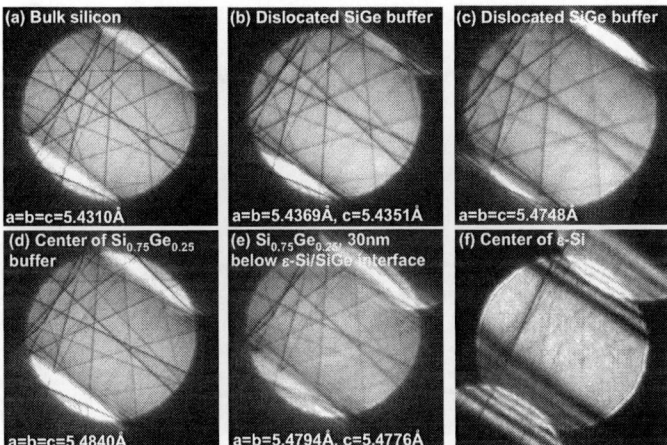

Figure 3. CBED patterns collected from corresponding points (a)-(e) as indicated in Fig.1. (f) is taken in the center of the strained silicon cap. The results of the fitted parameters are indicated in the patterns.

Fig.3(a)-(e) show the energy-filtered CBED patterns obtained at points a-e in Fig.1. All patterns are collected near the [340] projection. Based on CBED patterns taken from bulk silicon, the effective electron accelerating voltage is found to be 199.57kV by using the ASAC package. Fixing the effective voltage, the software package then fits the lattice parameters $(a,b,c,\alpha,\beta,\gamma)$ for other areas of interest by using a χ^2 procedure. Since the epitaxial growth was along the c-direction, constraints on the fit could be added based on the symmetry of the structure. Nevertheless, in our simulation we intentionally relaxed all constraints, thus six parameters are fitted. The simulations suggest that $\alpha = \beta = \gamma = 90°$ for all points being measured. Comparing Fig.3(b)-(d), one can clearly see that the lattice constants increase with increasing Ge composition. It is also observed (Fig.3(b)) that the SiGe alloy does not keep the cubic structure near the interface of bulk silicon and dislocated SiGe buffer. At the center of the $Si_{0.75}Ge_{0.25}$ buffer (Fig.3(d)), CBED measurements yield a lattice constant of 5.4840Å, which is in an excellent agreement with the value 5.4830Å determined by Raman spectroscopy. This suggests that the $Si_{0.75}Ge_{0.25}$ homogeneous buffer layer has cubic structure and there is little amount of, if any, residual strain after the epitaxial growth is done. However, in the $Si_{0.75}Ge_{0.25}$ region within about 80nm away from ε-Si/SiGe interface, our measurements systematically suggest that the lattice constants a, b are less than 5.4840Å. For example, about 30nm away from the interface (CBED pattern shown in Fig.3(e), local TEM sample thickness is about 300nm according to our dynamical simulations), the SiGe lattice constants are $a = b = 5.4794$Å$, c = 5.4776$Å. The decrease of SiGe lattice constants originates from the stress partition between the the ε-Si cap and SiGe buffer. The stress in the area near ε-Si/SiGe also relaxes upon thinning of the sample to electron transparency. Due to the relaxation of stress, the lattice planes will bend and generate a strain gradient along the electron beam direction. If the strain gradient is moderate, one can notice the splitting of HOLZ lines[20], which is also constantly observed in our experiments. In a highly strained area where the stress is of the order of GPa, the stress relaxation often makes the HOLZ lines unrecognizable as shown in Fig.3(f) even though the XTEM sample thickness (t) is about ten times the thickness of the strained silicon layer (W). The strain relaxation phenomenon is an impediment for the application of the CBED to the measurement of local strain.

Finite Element Modeling of Strain Relaxation in XTEM Sample

Two-dimensional FEM has demonstrated that the larger the t/Λ value, the less is the stress relaxation in a strained layer superlattice. In the past, FEM studies were mainly focused on multilayer periodic structures[12,22-25] and there are few reports on the strain relaxation in strained silicon XTEM samples. Here we use 2D FEM to demonstrate how stress is relaxed and distributed in an XTEM strained silicon sample. We assume that a ε-Si thin film is grown on a fully relaxed SiGe buffer. The thickness of this relaxed SiGe buffer is much larger than that of the ε-Si cap. This assumption is true for a typical strained silicon wafer which has about 2μm of relaxed SiGe buffer. As a simplification, a Young's modulus of 130GPa is selected for silicon and the Poisson's ratio is 0.28 for both silicon and SiGe. A lattice mismatch of ε=1% was introduced by artificially setting a differential thermal expansion coefficient of 10^{-4} for ε-Si and decreasing its temperature from 0 °C to -100 °C. The FEM calculation was carried out using the FEAP package[33]. In Fig. 4 we plot the simulated contour of σ_{yy} component superimposed on the elements for a XTEM sample with $t/W=10$. For a 30nm ε-Si cap and 300nm thick XTEM sample, more than 25% stress has relaxed in over half of the strained silicon layer. In the middle of the strained silicon cap, significant strain relaxation extends over 60nm from both free sides of this TEM sample. Over 1/3 of the crystal along the e-beam path has significant strain relaxation and has lattice parameters different than the other part of the strained material. It is also noticed that in the region near ε-Si/SiGe interface, the initially relaxed SiGe buffer is stressed compressively. The inhomogeneity of strain in the crystal along the electron beam is expected to yield a poor CBED pattern as we have shown in Fig.3(f).

e-beam

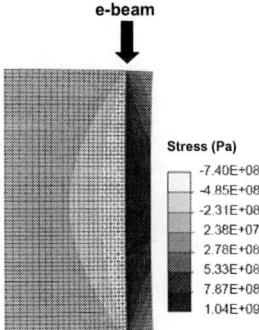

Stress (Pa)
-7.40E+08
-4.85E+08
-2.31E+08
2.38E+07
2.78E+08
5.33E+08
7.87E+08
1.04E+09

Figure 4. Simulated contour of stress component σ_{yy} along electron beam direction in a strained silicon XTEM sample. The initial biaxial tensile strain in the silicon cap is 10^{-2}. The distortion of sample is exaggerated for clarity.

Energy-filtered CBED Measurements on Locally Strained Device

Here we report our preliminary results on the CBED measurement of the stress in a locally stressed 35nm PMOS test device. The device structure and dimension are similar to that reported by other groups[3,4]. The channel width is 2μm. Energy-filtered CBED patterns were collected near the [340] axis, using the OMEGA energy filter to exclude most of the inelastic background from the diffraction pattern (all points in the diffraction pattern are elastically filtered simultaneously). It was found that when the XTEM sample thickness is about 250nm, no recognizable HOLZ patterns could be collected, which we attribute to strain relaxation in the thin foil. In this work, the sample thickness is about 700nm. When the sample is that thick, the inelastic scattering background will blur the CBED pattern and an energy filter is necessary. This effect is demonstrated in Fig.5. The two patterns are collected from the same place in bulk silicon. The pattern in Fig.5(a) is collected without applying the energy-selection slit below the OMEGA energy filter. For Fig.5(b), a slit-width of 20eV is applied to select the zero-loss electrons and the improvement of pattern quality is clear.

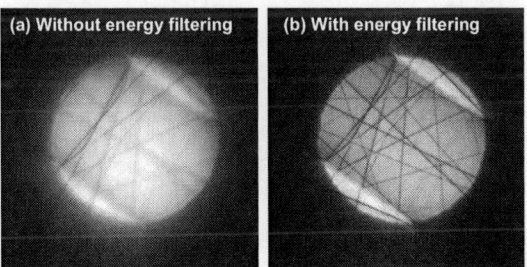

(a) Without energy filtering   (b) With energy filtering

Figure 5. (a) CBED pattern from the bulk silicon region. Energy-selection slit is not used so both elastic and in elastic electrons are collected for pattern formation. (b) Energy-filtered CBED pattern from the same point. Pattern is collected after (a). An energy selection slit width of 20eV is used.

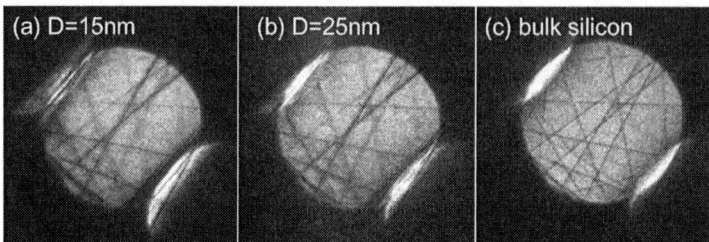

Figure 6. CBED patterns collected from the same locally strained silicon 35nm PMOS device at different positions: (a) D=15nm below the gate oxide; (b) D=25nm below the gate oxide ; and (c) bulk silicon at 1.3μm below the device area.

TABLE I. Fitted lattice parameters for patterns in Fig.6.σ_{xx} is the stress component along channel length direction and σ_{yy} is along channel width direction.

D (nm)	a (Å)	b (Å)	c (Å)	α (degree)	β (degree)	γ (degree)	σxx (GPa)	σyy (GPa)	χ2
15	5.4150	5.4150	5.4330	90.00	90.00	89.66	-1.12	-0.19	19.57
25	5.4240	5.4240	5.4270	89.96	90.04	89.86	-0.54	-0.15	2.34

CBED patterns (shown in Fig.6) were collected at two distances D (15nm and 25nm) below the gate oxide. The fitted parameters and χ^2 values are listed in Table I. At D=15nm below the oxide, the stress components are $\sigma_{xx} = -1.12$GPa along channel direction, and $\sigma_{yy} = -0.19$GPa along the channel width direction. At D=25nm, the stress components are $\sigma_{xx} = -0.54$GPa and $\sigma_{yy} = -0.15$GPa . The ratio of these two components agrees well with that predicted independently by finite element analysis[4]. The measured values of stress are quite plausible, but it should also be noted that the CBED patterns are relatively noisy (large χ^2 value) and significant errors might arise from the relatively poor quality of HOLZ lines. The repeatability of the measurement is now being evaluated by measuring stress at adjacent devices in the same die.

Conclusions

The CBED technique has been applied to measure lattice parameters in nano-scale strained silicon materials and devices. The in-column OMEGA filter efficiently removes the inelastic electrons and thus enables us to collect sharp HOLZ lines for subsequent χ^2 fitting. For a high-precision ($\sim 10^{-4}$) strain measurement, a computer program like the ASAC module of analySIS® is necessary for an objective analysis.

In the strain relaxed SiGe buffer of the strained silicon material, excellent agreement was obtained between the measurements given by CBED technique and Raman spectroscopy. A compressive stress of 1.12GPa along the channel direction in a locally stressed 35nm PMOS device was measured. Nevertheless, it has been found that in the proximity of the highly strained area where the stress intensity is of the order of GPa, strain relaxation is a bottleneck for the application of the CBED technique. The intrinsic strain relaxation in TEM thin foils gives an abrupt strain gradient along the path of the electron beam. This phenomenon results in blurry HOLZ lines or even yields unrecognizable HOLZ patterns. In the measurement of lattice parameters in strained silicon material, we find that strain relaxation may prevent a successful collection of HOLZ patterns in the strained silicon cap even when the TEM sample thickness

(t=300nm) and the thickness of the strained silicon layer (W=30nm) ratio is as high as $t/W = 10$. According to our finite element modeling, there is a significant strain gradient in the range of 60nm from both free surfaces of the TEM sample.

We might have a better chance to measure the stress in the strained area if we could find a way to shield the scattering of electrons from the partially relaxed crystalline columns. Here we propose using ion implantation to amorphorize both surfaces of the TEM sample to a range of about 80nm. It is expected that the amorphous layer would contribute a white noise to the CBED pattern rather than the formation of HOLZ lines which are due to the scattering of electrons by crystalline planes. The type of ions, implantation dosage, and implantation energy are under investigation.

Acknowledgements

The funding for this project was provided by SiWEDS, an industry-university collaborative research center (www.siweds.org) sponsored by the National Science Foundation. The bulk strained wafer was provided by Lawrence Semiconductor Research Laboratory, Inc. The PMOS device was provided by a SiWEDS industrial member. The authors acknowledge the use of the facilities of the National Center for Electron Microscopy, Lawrence Berkeley Lab, which is supported by the U.S. Department of Energy under Contract #DE-AC02-05CH11231. JCHS and HH acknowledge the support of Department of Energy under Contract #DOE. FG03-02ER45596. JWA was supported by the Director, Office of Science, Office of Basic Energy Science, Division of Materials Sciences and Engineering, of the US Department of Energy. We would like to thank Drs. A. Armigliato, M. Wibbelt, and J. Krieger for establishing the ASAC loaning license provided by Soft-Imaging Systems. PZ acknowledges useful suggestions from Prof. R. Gronsky. We acknowledge the assistance provided by the TEM group of Intel at Santa Clara, California.

References

[1] M. V. Fischetti, Z. Ren, P. M. Solomon, M. Yang and K. Rim, *J. Appl. Phys.*, **94**, 1079 (2003).

[2] S. E. Thompson, M. Armstrong, C. Auth, M. Alavi, M. Buehler, R. Chau, S. Cea, T. Ghani, G. Glass, T. Hoffman, C. H. Jan, C. Kenyon, J. Klaus, K. Kuhn, Z. Y. Ma, B. Mcintyre, K. Mistry, A. Murthy, B. Obradovic, R. Nagisetty, P. Nguyen, S. Sivakumar, R. Shaheed, L. Shiften, B. Tufts, S. Tyagi, M. Bohr, and Y. El-Mansy, *IEEE Trans. Electron Devices*, **51**, 1790 (2004).

[3] P. Bai, C. Auth, S. Balakrishnan, M. Bost, R. Brain, V. Chikarmane, R. Heussner, M. Hussein, J. Hwang, D. Ingerly, R. James, and J. Jeong, *IEDM Tech. Dig.*, 657 (2004).

[4] L. Smith, V. Moroz, G. Eneman, P. Verheyen, F. Nouri, L. Washington, M. Jurczak, O. Penzin, D. Pramanik and K. De Meyer, *IEEE Electron Device Lett.*, **26**, 652 (2005).

[5] M. L. Lee, E. A. Fitzgerald, M. T. Bulsara, M. T. Currie and A. Lochtefeld, *J. Appl. Phys.*, **97**, 011101 (2005).

[6] International Technology Roadmap for Semiconductors (2005), http://public.itrs.net/

[7] H. Kawasaki, K. Ohuchi, A. Oishi, O. Fujii, H. Tsujii, T. Ishida, K. Kasai, Y. Okayama, K. Kojima, K. Adachi, N. Aoki, T. Kanemura, D. Hagishima, M.

Fujiwara, S. Inaba, K. Ishimaru, N. Nagashima, and H. Ishiuchi, *IEDM Tech. Dig.*, 169 (2004).

[8] S.M. Cea, M. Armstrong, C. Auth, T. Ghani, M.D. Giles, T. Hoffmann, R. Kotlyar, P. Matagne, K. Mistry, R. Nagisetty, B. Obradovic, R. Shaheed, L. Shifren, M. Stettler, S. Tyagi, X. Wang, C. Weber, and K. Zawadzki, *IEDM Tech. Dig.*, 963 (2004).

[9] Y. C. Yeo and J. S. Sun, *Appl. Phys. Lett.*, **86** (2005).

[10] K. W. Ang, K. J. Chui, V. Bliznetsov, C. H. Tung, A. Du, N. Balasubramanian, G. Samudra, M. F. Li and Y. C. Yeo, *Appl. Phys. Lett.*, **86,** 093102 (2005).

[11] J. Li, D. Anjum, R. Hull, G. Xia and J. L. Hoyt, *Appl. Phys. Lett.*, **87,** 222111 (2005).

[12] M. M. J. Treacy and J. M. Gibson, *J. Vac. Sci. Technol. B*, **4,** 1458 (1986).

[13] J. C. H. Spence and J. M. Zuo, *Electron Microdiffraction*, Plenum Press, New York, (1992).

[14] J.-P. Morniroli, *Large-Angle Convergent-Beam Electron Diffraction Applications to Crystal Defects*, Société Française des Microscopies, Paris, (2002).

[15] J. M. Zuo, *Ultramicroscopy*, **41,** 211 (1992).

[16] A. Armigliato, R. Balboni, G. P. Carnevale, G. Pavia, D. Piccolo, S. Frabboni, A. Benedetti and A. G. Cullis, *Appl. Phys. Lett.*, **82,** 2172 (2003).

[17] A. Armigliato, R. Balboni and S. Frabboni, *Appl. Phys. Lett.*, **86,** 063508 (2005).

[18] M. Kim, J. M. Zuo and G. S. Park, *Appl. Phys. Lett.*, **84,** 2181 (2004).

[19] S. L. Toh, K. P. Loh, C. B. Boothroyd, K. Li, C. H. Ang and L. Chan, *J. Vac. Sci. Technol. B*, **23,** 940 (2005).

[20] L. Clément, R. Pantel, L. F. T. Kwakman and J. L. Rouviere, *Appl. Phys. Lett.*, **85,** 651 (2004).

[21] R. W. Carpenter and J. C. H. Spence, *Acta Crystallogr A*, **38,** 55 (1982).

[22] X. F. Duan, *Appl. Phys. Lett.*, **66,** 2247 (1995).

[23] H. Shimotahira and K. Nakamura, *Appl. Phys. Lett.*, **87,** 121907 (2005).

[24] C. R. Chen, Y. Liu and S. X. Li, *Mater. Sci. Eng. A*, **265,** 146 (1999).

[25] C. R. Chen and S. X. Li, *J. Mat. Sci.*, **35,** 1145 (2000).

[26] C. T. Chou, S. C. Anderson, D. J. H. Cockayne, A. Z. Sikorski and M. R. Vaughan, *Ultramicroscopy*, **55,** 334 (1994).

[27] A. Armigliato, R. Balboni and S. Frabboni, *Eur. Phys. J. Appl. Phys.*, **27,** 49 (2004).

[28] D. E. Aspnes, in Properties of Crystalline Silicon, R. Hull, Eds., p. 677, INSPEC, London, (1999).

[29] J. C. Tsang, P. M. Mooney, F. Dacol and J. O. Chu, *J. Appl. Phys.*, **75,** 8098 (1994).

[30] L. H. Wong, C. C. Wong, J. P. Liu, D. K. Sohn, L. Chan, L. C. Hsia, H. Zang, Z. H. Ni and Z. X. Shen, *Jpn. J. Appl. Phys.*, **44,** 7922 (2005).

[31] S. Nakashima, T. Yamamoto, A. Ogura, K. Uejima and T. Yamamoto, *Appl. Phys. Lett.*, **84,** 2533 (2004).

[32] R. Hull, in Germanium Silicon: Physics and Materials, R. Hull and J. C. Bean, Eds., *Semiconductor and Semimetals*, Vol. 56, p. 101, Academic Press, San Diego, CA, (1999).

[33] R. L. Taylor, FEAP - A Finite Element Analysis Program, 7.5 ed. (University of California at Berkeley, 2004).

ECS Transactions, 2 (2) 569-577 (2006)
10.1149/1.2195692, copyright The Electrochemical Society

Threading vs Misfit Dislocations in Strained-Si/SiGe Heterostructures: Preferential Etching and Minority Carrier Transient Spectroscopy

Jinggang Lu, Renhua Zhang, and George Rozgonyi
Materials Science and Engineering Department, North Carolina State University, Raleigh, North Carolina, USA 27695

Eugene Yakimov and Nikolai Yarykin
Institute of Microelectronics Technology RAS, Chernogolovka, Moscow region, 142432 Russia;

Mike Seacrist
MEMC Electronic Materials, St. Peters, MO 63376

Dislocations in sSi/SiGe heterostructures have been studied structurally by preferential defect etching and electrically by minority carrier transient spectroscopy (MCTS). It is shown that a two-step etching procedure will selectively delineate threading dislocations without interference from misfit dislocations at the buried sSi/SiGe interface. The effect of etching solution temperature is also evaluated, and found to dramatically improve the contrast of threading dislocation etch pits by employing an ice bath etching solution. Complementary electrical characterization of the sSi/SiGe heterostructures has been performed using optically excited MCTS to reveal minority carrier peaks associated with extended defects. The characteristics of corresponding deep levels and their dependence on excitation pulse duration allows us to assign MCTS peaks to dislocations in the strained Si and/or uniform composition SiGe layers.

Introduction

Due to the higher electron and hole mobilities in the strained-Si (sSi) channel layer than those of bulk Si, and excellent compatibility to modern complementary metal-oxide-semiconductor (CMOS) technology, strained Si is a promising candidate to replace traditional bulk Si in high-speed CMOS fabrication processes[1-2] The sSi can be realized by growing a thin epitaxial layer of Si on a strain-relaxed SiGe virtual substrate, creating a biaxial tensile strain in the sSi layer which can be readily tailored by adjusting the Ge content in the "substrate". However, since threading(TD) and interfacial misfit(MD) dislocations in sSi/SiGe layers will degrade device yield and performance, [3-4] determining the density and electrical activity of these distinct defects is critical for these devices.

Processing techniques, such as the use of a linear or terraced graded SiGe buffer layer [5-6], or chemical-mechanical polishing [7], have been utilized to reduce the threading dislocation density in the heterostructures to below 10^5 cm^{-2}. Since it is difficult to use transmission electron or atomic force microscopy to characterize the TD density due to their small field of view, preferential etching/Nomarski optical microscopy [8] has become the method of choice due to its large sampling area. In this article, TDs in a thin sSi layer have been characterized by optimized temperature dependent and two-step preferential etching, thereby enabling the selective delineation of TDs that exit the top sSi layer, while suppressing the interference produced by MDs at the buried sSi/SiGe interface. [9]

For practical CMOS device applications, dislocations located in the strained Si layer and near the interface of strained Si/SiGe are of most interest. Due to the nanoscale sSi layer thickness and low doping concentration, the depletion region of electrical junctions (either MOS capacitors or Schottky diodes) usually extends beyond the sSi layer, penetrating the SiGe layers. Due to the greatly reduced sensitivity to defects located within the depletion region, it is difficult to study the electrical properties of dislocations located inside strained Si layer by conventional Deep Level Transient Spectroscopy (DLTS) technique.[10] In addition, the charging of dislocations via capture of majority carriers impedes the effective filling of other majority carriers.[11] On the other hand, filling of charged extended defects with minority carriers is an effective alternative electrical diagnostic probe. For these reasons, in the present paper we have applied Minority Carrier Transient Spectroscopy (MCTS) using optical excitation to examine dislocation-related minority carrier traps in strained Si/SiGe structures.

Experimental Samples and Procedures

A strained-Si/$Si_{0.7}Ge_{0.3}$/Si heterostructural wafer, containing a 20nm sSi layer on top of a stepwise graded SiGe layer, has been used to illustrate single- and two-step etching[12-13] procedures for selective TD delineation. The sample was first etched in a diluted Schimmel solution $CrO_3(1M):HF(49\%):H_2O$ in a volume ratio of 1:2:6 for 20s. While this single-step etching is able to delineate individual TDs, it fails to isolate TDs adjacent to or overlapping etch groove traces of interfacial MDs. Thus, a two-step etching procedure was used in order to suppress the MD interference. The first step is a short duration etch in a diluted Secco [14] or Schimmel [15] solution to enable a pipe to form preferentially at TDs exiting the top sSi layer, without removing the whole sSi layer. This is followed by a selective SiGe etch with a 1:2:3 volume ratio solution of $HF(49\%):H_2O_2(30\%):HAc$, which selectively etches the underlying SiGe in the vicinity of TDs pipe, while essentially not attacking the Si cap[16].

The impact of adjusting solution temperature to adjust the relative etch rates of TD, sSi, and SiGe, thereby enhancing TD delineation, was also examined. The etch rates of a diluted Secco etchant, $K_2Cr_2O_7(0.03M):HF(37\%)$ in a 5:2 volume ratio, was characterized using a sample consisting of 40nm sSi on top of 600nm $Si_{0.78}Ge_{0.22}$. With part of the sample covered by wax, etch rates of sSi and SiGe at different temperatures were determined via step height measurements with a Tencor Alpha-500 profilometer. The solution temperature was measured with a Teflon coated thermocouple.

The heterostructures examined by MCTS consisted of four layers deposited on a highly doped p-Si substrate: a Si epilayer with boron concentration of about $10^{16}cm^{-3}$, then a thick graded $Si_{1-x}Ge_x$ layer, followed by a thick uniform $Si_{1-x}Ge_x$ (20% germanium content), and finally a strained-Si top layer on top of the relaxed uniform SiGe layer, see Fig. 1. Two sets of structures from different wafer suppliers with different strained-Si and SiGe layer thickness and carrier concentration were used in this MCTS study. In Set A, the thicknesses of graded SiGe, uniform SiGe and strained Si layers were 3000, 2000 and 115 nm, respectively, with a carrier concentration in SiGe layers lower than $10^{14}cm^{-3}$. In Set B, the thicknesses of graded SiGe, uniform SiGe and strained Si layers are 670, 1500 and 73.5 nm, respectively, and carrier concentration in SiGe layers of about $10^{15}cm^{-3}$. As-grown and thermally-annealed samples of both sets were studied. Note that partial strain relaxation of the top strained-Si layer could occur during thermal annealing. Set A consisted of as-grown sample A0 and annealed samples A1-A3 thermally annealed at 1100°C for 8, 30 and 120 seconds, respectively. Set B consisted of as-grown sample B0 and annealed sample B1 thermally oxidized at 800°C resulting in a 10 nm oxide.

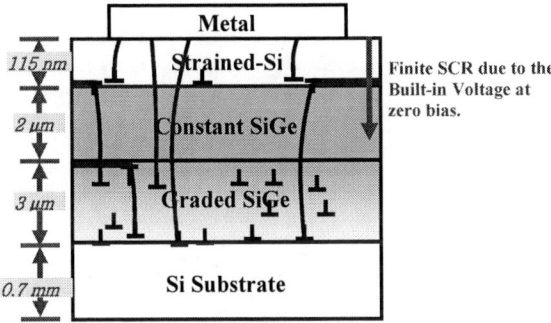

Figure 1. Schematic of the sample structure. Note that under small bias, the space charge region (SCR) covers the strained-Si layer and a portion of the constant SiGe layer.

For the electrical measurements, Schottky contacts were prepared by Al evaporation on the surface of strained-Si layer. Capacitance-Voltage (C-V) and MCTS measurements were carried out using a Bio-Rad DLTS 8000 system in the temperature range from 50 K to 300 K. In MCTS measurements, minority carriers captured by dislocations were excited by an 850 nm laser pulse from a GaAs/GaAlAs double heterostructure, while the laser pulse duration (t_P) was varied from 0.005 to 10 ms. All measurements were carried out at small reverse biases so that the depletion region covered the strained-Si layer as well as a portion of the uniform SiGe layer (Fig. 1), as determined from the C-V characteristics.

Results and Discussion

Preferential Defect Etching

Single- and Two-step Etching at Room Temperatures. Figure 2 is a micrograph of the 30% Ge sample etched in a diluted Schimmel solution (CrO_3(1M): HF(49%): H_2O in a volume ratio of 1:2:6) for 20s. Note that the strongly defined truncated linear structures are due to MDs generated at the sSi/SiGe interface since the 20nm sSi layer thickness is above the ~14nm critical thickness for strain relaxation.[17] When the sSi layer thickness is above a critical value for strain relaxation, MDs are generated at the sSi/SiGe interface through gliding of the pre-existing TD arms originated from the dense MDs in the SiGe graded layer. The truncated structure suggests that the MDs generated at the sSi/SiGe interface block the orthogonal gliding threading arms.[18] The observation of the interfacial MD traces indicates that the capping sSi layer has been removed. Note that TDs away from the MDs can be clearly identified. However, etch pits overlapping the MD segments cannot be well resolved, although there is a definite indication of the alignment of TDs along MDs evident from the truncated structure. Evidently, the single-step etching process failed to delineate TDs adjacent to the interfacial MDs due to their strong interference. In order to selectively delineate only the TDs exiting the sSi layer, it is important to avoid etching away the whole sSi layer, thereby suppressing interference from the MD traces.

Because of the 10 to 30 nm thin sSi layer, under-cutting of the SiGe buffer is needed in order to effectively enlarge the etch pits for optical microscopic examination. In other words, the ideal etching procedure must selectively attack TDs *and* SiGe while preserving the nanoscale defect-free sSi layer. Since both Schimmel and Secco solutions etch SiGe faster than pure Si (if the Ge content is not too high), the SiGe undercutting contributes to a better pit definition in the single-step etching process. To fully utilize the under-cutting mechanism, a distinct second-step etching which selectively attacks SiGe has been used, as shown in Fig. 3. The first step is a short duration etch in a diluted Secco or Schimmel solution to enable a pipe to preferentially form at TDs in the top sSi layer. This process allows a thin sSi layer to remain as a cap on the underlying SiGe layer. After the TDs have been "opened-up", the sample is further etched in a $HF(49\%):H_2O_2(30\%):HAc$ solution, which selectively etches the underlying SiGe in the vicinity of TD pipes, but essentially does not attack the Si cap. Thus, the TD etch pit contrast is enhanced and well resolved under an optical microscope.

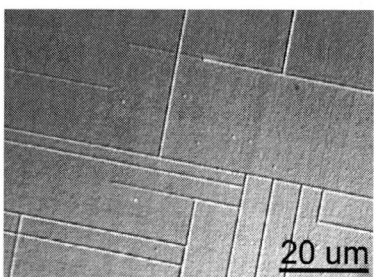

Figure 2. Optical image of a 30% Ge sample after 20s diluted Schimmel etch.

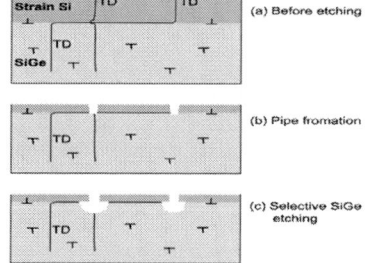

Figure 3. Illustration of the two-step etching procedure.

Figure 4(a) is a Nomarski optical image of the 30% Ge sample after a sequential 4s diluted Schimmel and 2 min selective SiGe etch. Note that TDs are well resolved without removal of the whole sSi layer (no truncated MDs). The same sample was subsequently Schimmel etched for an additional 5 sec, which removed the sSi layer and delineated the sSi/SiGe interfacial MDs, see Fig. 4(b). Note that most of the TDs are aligned along truncated linear MD traces. Note also that these MDs are imaged independently of the dense cross-grid of MDs located in the underlying buried graded SiGe layers.

The success of the two-step etching procedure relies on precise control of the first step etching time. It must be long enough to open-up a "pipe" at each TD exiting the sSi layer while being short enough to preserve a reasonable portion of the sSi layer which then functions as a mask during etching of the underlying SiGe layer. Although the two-step etching procedure is very useful in differentiating TDs from MDs at the sSi/SiGe interface; unfortunately, it is not always a robust/reproducible method, since the etch rate of the sSi layer is a strong function of solution temperature, sSi strain level/Ge concentration, and the fact that different types of TDs may exhibit different etching responses. Thus, it is not always possible to open-up all TDs during the first step etching. In addition, for samples with Ge contents below 20%, the second-step etching does not have a high etching selectivity for SiGe over sSi. In the following, we show that single-step etching using a chilled solution provides a useful alternate approach, since contrast enhancing SiGe under-cutting is improved.

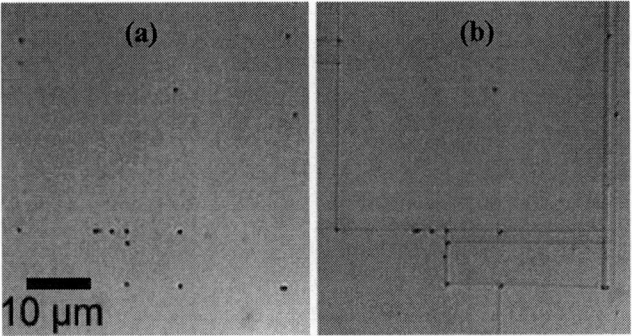

Fig. 4(a) Nomarski optical image after two-step etch; (b) the same area after an additional 5 sec diluted Schimmel etch to expose the sSi/SiGe interface.

Effect of Etching Solution Temperature. It is well known that a diffusion controlled etching results in chemical polishing, while a reaction-controlled etching is needed for preferential defect delineation[8]. If the Secco or Schimmel etchants are not highly diluted, the reaction is generally reaction-controlled and extended defects can therefore be resolved. The etching process associated with SiGe under-cutting discussed above is somewhat unique because it relies on diffusion of the reactants/products through the TD pipes. Since the TD pipes are narrow, then diffusion is likely to be the rate limitation factor. However, because a solution diffusion process is generally associated with a lower activation energy (weaker temperature dependence) than the corresponding reaction process, a normally diffusion-controlled etching process will shift to reaction-controlled one by decreasing the solution temperature. Therefore, SiGe undercut etching is likely to be improved at low temperatures, allowing the high etching selectivity of SiGe over Si to be fully realized by overcoming the rate limitation pipe diffusion factor.

To examine the impact of solution temperatures, we first characterized the etch rates of a diluted Secco etchant ($K_2Cr_2O_7$ (0.03M): HF(37%) in 5: 2 volume ratio) using a sample consisting of 40nm sSi on top of 600nm $Si_{0.78}Ge_{0.22}$. Figure 5 shows the step height vs etching time. The etch rates for sSi and $Si_{0.78}Ge_{0.22}$, as determined from the slopes, are 0.6 and 6.5 nm/sec at 22^0C, and 0.17 and 1.2 nm/sec at 2^0C, respectively. Note that the etch rates for Si and $Si_{0.78}Ge_{0.22}$ decreased by a factor of 3.5 and 5.4, respectively, and the etch selectivity of SiGe over Si decreases from 11 to 7 for a solution temperature decrease from 22 to 2^0C. The results suggest that, if the pipe diffusion process is in fact not a rate limitation factor, then the SiGe under-cutting (or the etch pit definition) will be worse with decreasing temperatures.

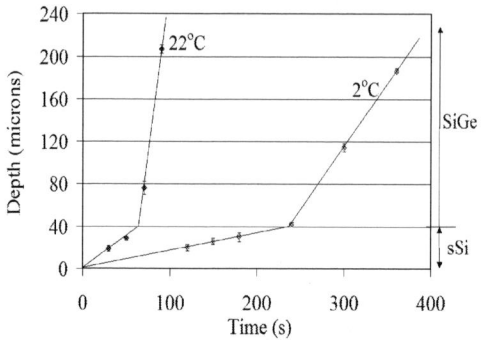

Figure 5. Etching depth vs time for a diluted Secco etchant at 2 and 22°C, respectively.

To clarify whether the pipe diffusion is the etch rate limitation factor and whether the under-cutting of SiGe will be improved at lower solution temperatures, a 25nm sSi/ $Si_{0.78}Ge_{0.22}$ sample was etched for 120 sec at 2°C and 35 sec at 2°C, respectively. The etching times were chosen such that essentially the same amount of sSi was removed at each temperature, which is an important prerequisite for comparison of etch pit contrasts. A comparison of Figs. 6(a) and (b) shows that the contrast of the 2°C etched sample is much better than the room temperature sample, demonstrating that a more efficient SiGe undercutting occurs at low temperatures and that diffusion of etchant/reactant through the TD pipes is indeed the most important rate limitation factor.

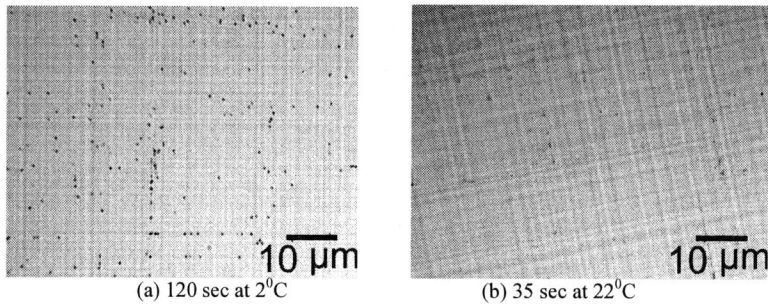

(a) 120 sec at 2°C   (b) 35 sec at 22°C

Figure 6. Nomarski images of a sSi/ $Si_{0.78}Ge_{0.22}$ sample etched at (a) 2°C for 120 sec, and (b) at 22°C for 35 sec.

Electrical Characterization

Figure 7 shows typical MCTS spectra (ΔC) measured on annealed strained-Si/SiGe heterostructure B1 using different optical pulse durations. Note that with an increase of pulse duration, the peak amplitude increases and the position of its maximum shifts to lower temperature. On the as-grown structure B0, the MCTS signal is lower than that for B1 and its peak position has a small (about 20°) shift to higher temperatures. The MCTS spectra measured on sample Set A exhibited essentially the same behavior, i.e., the position of MCTS maximum signal shifts with annealing from about 230 K for A0 to about 200 K for A3, and all spectra demonstrate a dependence on optical pulse duration similar to that shown in Fig. 7. For all strained-Si/SiGe heterostructures studied, this dependence is well described logarithmically, as shown in Fig. 8. In conventional DLTS, the logarithmic dependence of signal on excitation pulse duration, i.e., electrical pulse, and a shift of maximum position to lower temperatures, are characteristic features for dislocation-related levels, indicating a band-like nature of defect states[19]. In MCTS using optical pulse excitation, it is not so obvious that dislocation charging should follow a logarithmic dependence because minority carriers (electrons) created by the optical pulse drift in the depletion region. Despite the drift velocity, the dislocation charge could still affect both effective capture radius and distribution of electrons. And the observation of logarithmic dependence on optical pulse duration excludes the possibility that the observed spectra are associated with homogeneously distributed point defects.

The $\Delta C/C$ relation in all structures examined is found to decrease with bias increase, indicating that the associated defects are not distributed homogeneously in depth, but are either located in a thin near-surface layer or their density increases in this region. Thus, threading dislocations must also be excluded as a unique source for the signal observed, although if their properties are similar to those of MDs they may constitute a portion of the MCTS signal.

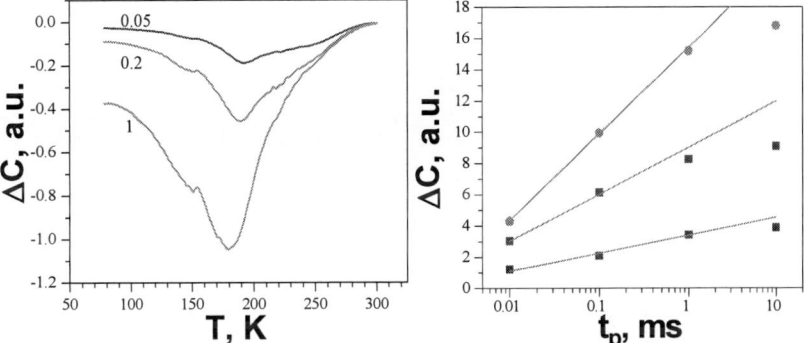

Figure 7. MCTS spectra of structure B1 for t_p = 0.05, 0.2 and 1 ms. e = 50 s⁻¹

Figure 8. Dependence of MCTS signal on t_p for A0, A1 and A3 structures.

Similar MCTS signals were observed by us in other strained Si/SiGe structures supporting the assumption that this signal is associated with the dislocations commonly observed in Si/SiGe heterostructures. An increase in defect concentration with annealing also supports this assumption because it is consistent with an increase of dislocation density via thermally induced strain relaxation. The MCTS peak observed consists of a few peaks that could be easily seen at larger rate windows. Moreover, in some structures complex spectrum is revealed also at larger excitation pulse duration, see Fig. 9. This prevents the correct determination of energy levels of corresponding defects. Nevertheless, their emission rates are very close to those of well known family of dislocation-related C-lines revealed in plastically deformed Si [20, 21] (Fig. 10) that also correlates with the assumption about the dislocation nature of electron traps observed. It should be mentioned that the concentration of centers observed achieves a value of 10^{13}cm^{-3} even if their distribution is assumed to be homogeneous. Really, it could be essentially higher. The EBIC contrast from dislocations in the structures studied could be revealed only at low temperatures. That allows us to estimate the deep level concentration density along dislocations as lower than 10^7cm^{-1}. From this estimation it follows that if the observed minority traps are associated with dislocations their density should exceed $10^6\text{-}10^7 \text{cm}^{-2}$.

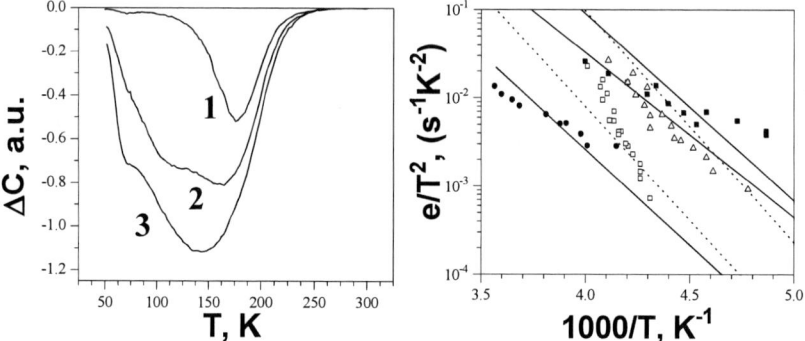

Figure 9. MCTS spectra obtained on B1 structure with pulse duration 0.01 (1), 0.1 (2) and 1 ms (3). e =50 s⁻¹

Figure 10. Thermal emission rate (T^2 corrected) as a function of inverse temperature for different structures studied. Corresponding dependences for C-family obtained in Ref [20-21] are shown by dotted and solid lines, respectively.

The shift of peak position with annealing is not understood yet but it also could be associated with band structure changes due to strain relaxation. Besides, if different lines observed in the MCTS spectra are associated with different dislocation types, strain relaxation could change their relative amount affecting the MCTS spectrum.

Conclusion

When the sSi layer thickness is above the critical thickness for strain relaxation and MDs are generated at the sSi/SiGe interface, single-step etching process fails to delineate TDs adjacent to the interfacial MDs due to their strong interference. A two-step etching procedure has been optimized to delineate TDs without the interference of these MDs. The first step is a short duration etch in a diluted Secco or Schimmel solution to enable a pipe to form preferentially at TDs. Then, the sample further went through a selective SiGe etch during with the pits are enlarged through SiGe under-cutting.

The impact of solution temperatures on TD delineation was also examined. It was found that the contrast of TD etch pits improves dramatically with decreasing solution temperatures, although the etch selectivity of SiGe over Si decreases. This result suggests that diffusion of etchant/reactant through the TD pipes is an important rate limitation factors controlling the SiGe under-cutting process.

Optically excited MCTS was used for electrical characterization of strained Si/SiGe heterostructures. The minority carrier peaks associated with extended defects are revealed. The characteristics of corresponding level and the dependence on excitation pulse duration allow associating the MCTS peak with dislocations in the strained Si and/or uniform SiGe.

References

1 K. Rim, J. L. Hoyt, and J. F. Gibbons, IEEE Trans. Electron Devices **47**, 1406 (2000).
2 K. Mistry, M. Armstrong, C. Auth, S. Cea, T. Coan, T. Ghani, T. Hoffmann, A. Murthy, J. Sandford, R. Shaheed, K. Zawadzki, K. Zhang, S. Thompson and M. Bohr., Symp. VLSI Tech. Dig., 50, 2004.
3 R. Hull, J. C. Bean, C. Buescher, J. Appl. Phys. **66**, 5837 (1989).
4 L. M. Giovane, H. –C. Luan, A. M. Agarwal, L. C. Kimerling, Appl. Phys. Lett. **78**, 541 (2001).
5 L. Di Gaspare, E. Palange, G. Capellini, and F. Evangelisti, J. Appl. Phys. **88**(1), 120-123 (2000).
6 A. D. Capewell, T. J. Grasby, T. E. Whall, and E. H. C. Parker, Appl. Phys. Lett., **81**(25), 4775-4777 (2002).
7 M. T. Currie, S. B. Samavedam, T. A. Langdo, C. W. Leitz, and E. A. Fitzgerald, Appl. Phys. Lett. **72**(14), 1718-1720 (1998).
8 G. A. Rozgonyi, in *Encyclopedia of Materials: Science and Technology*, edited by S. Mahajan, Amsterdam: Elsevier Science Ltd, 2001, pp. 8524-8533.
9 J. Lu, A. Czerwinski, L. Kordas, W. Zhao, and G. Rozgonyi, Proceeding of NIST Meeting on Characterization and Metrology for ULSI Technology, Richardson, Texas, March 15-18, 2005, p633-637.
10 N. Yarykin, R. Zhang, and G. A. Rozgonyi, ICMNE (2005).
11 P. N. Grillot, S. A. Ringel, E. A. Fitzgerald, G. P. Watson, and Y. H. Xie, J. Appl. Phys. **77**, 3248 (1995).
12 S. W. Bedell, D. K. Sadana, K. Fogel, H. Chen, and A. Domenicucci, Electrochemical and Solid-State Letters, 7 (5) G105-G107 (2004)
13 S. W. Bedell, H. Chen, D. K. Sadana1, K. Fogel, and A. Domenicucci, MRS PV **809**, (2004), B1.5.1
14 F. Secco d'Aragona, *J. Electrochem. Soc.*, **119** (7), 948 (1972).
15 D. G. Schimmel, J. Electrochem. Soc. **123**, 734 (1976).
16 T. K. Carns, M. O. Tanner, K. L. Wang, J. Electrochem. Soc. **142** (1995) 1260.
17 S. B. Samavedam, W. J. Taylor, J. M. Grant, J. A. Smith, P. J. Tobin, A. Dip, A. M. Phillips, and R. Liu, J. Vac. Sci. Technol. **B** 17(4), 1424 (1999).
18 E. A. Stach, R. Hull, R. M. Tromp, F. M. Ross, M. C. Reuter, and J. C. Bean, Phil. Mag. **A** 80 (9), 2159-2200 (2000).
19 W. Schröter, J. Kronewitz, U. Gnauert, F. Riedel, M. Seibt, Phys. Rev. B **52**, 13726, (1995).
20 P. Omling, E.R. Weber, L. Montelius, H. Alexander, J. Michel, Phys. Rev. **B**, 23, 6571 (1985).
21 D. Cavalcoli, A. Cavallini, E. Gombia, Phys. Rev. B **56**, 10208 (1997).

578

Author Index

Abe, T.	287	Hashiba, Y.	287
Ager, J.	559	Hattori, T.	413
Aid, S. R.	287	He, H.	559
Aihara, M.	523	Hendersen, T. M.	205
Akhmetov, V.	461, 471	Hirano, Y.	391
		Hong, Y.	515
Babu, S.	515	Hourai, M.	95, 109
Bartlett, R.	205	Hu, G.	485
Bersuker, G.	205	Huang, J.	541
Bethers, U.	363	Huber, W.	57
Borucki, L.	515	Hutcheson, G.	3
Chidambaram, P.	541	Ikeda, S.	313
Choi, J.	491	Imaoka, K.	391
Claeys, C.	349, 503	Inaba, S.	317
		Inoue, F.	391
Dabrowski, J.	247	Inoue, N.	453, 461, 471
Dai, X.	89	Irwin, R.	541
Danielson, D. T.	375	Ishimaru, K.	317
Dantz, D.	363	Istratov, A.	559
De Meyer, K.	349		
Diaz, C. H.	341	Jia, T.	89
Draina, J.	135	Jones, P.	541
Duscher, G.	549		
		Karen, A.	453
Economikos, L.	515	Kashima, K.	453, 471
Eifuku, K.	453	Kato, M.	471
Eneman, G.	349	Keswani, M. K.	515
		Kihara, T.	109
Fabry, L.	413	Kim, M.	541
Falster, R.	61	Kimerling, L.	375
Fandel, D.	135	Kirino, Y.	155
Ferrell, J.	135	Kisielowski, C.	559
Fujiwara, M.	317	Kissinger, G.	247
Fukuda, T.	155	Kitagawa, S.	471
Fukutani, S.	261	Kohno, M.	155
		Koizumi, M.	453
Gilles, D.	57	Koontz, E.	541
Goesele, U.	185	Korkin, A.	205
Goldstein, M.	515	Kramer, S.	135
Greer, J.	205	Krause, R.	167
Gupta, P.	123	Krivokapic, Z.	329
		Kulkarni, M.	123, 213, 229
Haag, M.	167	Kushner, V. A.	491
Habuka, H.	523		
Haeckl, W.	77	Lee, H.	515

Lee, J.	167	Sakuraba, M.	287
Lin, M.	329	Sattler, A.	247
Liu, A.	485	Schellenberger, M.	33
Loo, R.	349	Schmidt, M.	167
Lu, J.	569	Scholz, M.	349
		Schroder, D.	491
Ma, T.	485	Seacrist, M.	569
Mardinly, J.	559	Seto, S.	287
Massoud, H. Z.	189	Seuring, C.	247
Masumoto, K.	453	Shao, B.	485
Maszara, W. P.	329	Shiba, S.	261
Matsumoto, S.	287	Shimamune, Y.	287
Matsushita, D.	317	Shingu, K.	453
Michel, J.	375	Shinomiya, M.	453
Misra, D.	503	Simoen, E. R.	349, 503
Moriya, K.	471	Sinno, T.	77
Mueller, T.	247, 363	Spence, J.	559
Murota, J.	287	Srinivasan, P.	503
		Stach, E.	559
Nakabayashi, Y.	287	Sueoka, K.	95, 261
Nakamura, K.	275	Sugimura, W.	95, 109
Nakamura, K.	27	Suzuki, K.	523
Nakashima, K.	471		
Nakatsu, M.	461	Takahashi, T.	453
Nango, N.	471	Takayanagi, M.	317
Nara, Y.	27	Takeda, R.	471
Nelson, C.	559	Takenawa, T.	453
Nutsch, A.	433	Takeuchi, T.	523
		Tanahashi, K.	391
Oechsner, R.	33, 433	Tang, S.	541
Ogura, A.	155	Tao, M.	299, 401
Ohara, S.	261	Tawada, M.	523
Ono, T.	95, 109, 471	Thornton, T.	491
Ootsuka, F.	27	Tomioka, J.	275
		Toyonaga, K.	287
Packan, P.	185	Tsai, P.	167
Pantisano, L.	503	Tu, H.	89, 485
Park, D.	11		
Patri, U.	515	Umeno, S.	95
Pfeffer, M.	33	Urushido, H.	471
Pfeiffer, G.	167		
Pfitzner, L.	33	Verheyen, P.	349
Philipossian, A.	515	Virbulis, J.	363
		von Ammon, W.	77, 247, 363
Richter, H.	247	Voronkov, V. V.	61
Rodder, M.	313		
Roeder, G.	33	Wada, K.	287
Rozgonyi, G.	549, 569	Wang, Y.	541
Ryu, B.	11	Watanabe, M.	155

Watanabe, T.	317
Weber, E.	539, 559
Weijtmans, J.	541
Wise, R.	539, 541
Wu, Z.	89
Xiang, Q.	329
Yagi, H.	453
Yakimov, E.	569
Yamada-Kaneta, H.	391
Yamazaki, K.	391
Yamazaki, Y.	415
Yang, J.	491
Yang, X.	299, 401
Yarykin, N.	569
Zhang, G.	89
Zhang, P.	559
Zhang, R.	569
Zhao, W.	549
Zhou, Q.	89
Zhu, J.	401
Zhuang, Y.	515